Advances in Intelligent Systems and Computing

Volume 359

Series editor

Janusz Kacprzyk, Polish Academy of Sciences, Warsaw, Poland
e-mail: kacprzyk@ibspan.waw.pl

About this Series

The series "Advances in Intelligent Systems and Computing" contains publications on theory, applications, and design methods of Intelligent Systems and Intelligent Computing. Virtually all disciplines such as engineering, natural sciences, computer and information science, ICT, economics, business, e-commerce, environment, healthcare, life science are covered. The list of topics spans all the areas of modern intelligent systems and computing.

The publications within "Advances in Intelligent Systems and Computing" are primarily textbooks and proceedings of important conferences, symposia and congresses. They cover significant recent developments in the field, both of a foundational and applicable character. An important characteristic feature of the series is the short publication time and world-wide distribution. This permits a rapid and broad dissemination of research results.

Advisory Board

More information about this series at http://www.springer.com/series/11156

Hoai An Le Thi · Tao Pham Dinh
Ngoc Thanh Nguyen
Editors

Modelling, Computation and Optimization in Information Systems and Management Sciences

Proceedings of the 3rd International Conference
on Modelling, Computation and Optimization in
Information Systems and Management Sciences
- MCO 2015 - Part I

 Springer

Editors
Hoai An Le Thi
Laboratory of Theoretical and
    Applied Computer Science
University of Lorraine-Metz
France

Ngoc Thanh Nguyen
Division of Knowledge Management
    Systems
Wroclaw University of Technology
Poland

Tao Pham Dinh
Laboratory of Mathematics
National Institute for Applied
    Sciences-Rouen
France

ISSN 2194-5357   ISSN 2194-5365   (electronic)
Advances in Intelligent Systems and Computing
ISBN 978-3-319-18160-8   ISBN 978-3-319-18161-5   (eBook)
DOI 10.1007/978-3-319-18161-5

Library of Congress Control Number: 2015937024

Springer Cham Heidelberg New York Dordrecht London

Printed on acid-free paper

Springer International Publishing AG Switzerland is part of Springer Science+Business Media
(www.springer.com)

Preface

This volume contains 86 selected full papers (from 181 submitted ones) presented at the MCO 2015 conference, held on May 11–13, 2015 at University of Lorraine, France.

MCO 2015 is the third event in the series of conferences on Modelling, Computation and Optimization in Information Systems and Management Sciences organized by LITA, the Laboratory of Theoretical and Applied Computer Science, University of Lorraine.

The first conference, MCO 2004, brought together 100 scientists from 21 countries and was a great success. It included 8 invited plenary speakers, 70 papers presented and published in the proceedings, "Modelling, Computation and Optimization in Information Systems and Management Sciences", edited by Thi Hoai An and Pham Dinh Tao, Hermes Sciences Publishing, June 2004, 668 pages, and 22 papers published in the European Journal of Operational Research and in the Journal of Global Optimization. The second conference, MCO 2008 was jointly organized by LITA and the Computer Science and Communications Research Unit, University of Luxembourg. MCO 2008 gathered 66 invited plenary speakers and more than 120 scientists from 27 countries. The scientific program consisted of 6 plenary lectures and of the oral presentation of 68 selected full papers as well as 34 selected abstracts covering all main topic areas. Its proceedings were edited by Le Thi Hoai An, Pascal Bouvry and Pham Dinh Tao in Communications in Computer and Information Science 14, Springer. Two special issues were published in Journal of Computational, Optimization & Application (editors: Le Thi Hoai An, Joaquim Judice) and Advance on Data Analysis and Classification (editors: Le Thi Hoai An, Pham Dinh Tao and Ritter Guntter).

MCO 2015 covered, traditionally, several fields of Management Science and Information Systems: Computer Sciences, Information Technology, Mathematical Programming, Optimization and Operations Research and related areas. It will allow researchers and practitioners to clarify the recent developments in models and solutions for decision making in Engineering and Information Systems and to interact and discuss how to reinforce the role of these fields in potential applications of great impact. It would be a timely occasion to celebrate the 30th birthday of DC programming and DCA, an efficient approach in Nonconvex programming framework.

Continuing the success of the first two conferences, MCO 2004 and MCO 2008, MCO 2015 will be attended by more than 130 scientists from 35 countries. The International

Scientific Committee consists of more than 80 members from about 30 countries all the world over. The scientific program includes 5 plenary lectures and the oral presentation of 86 selected full papers as well as several selected abstracts covering all main topic areas. MCO 2015's proceedings are edited by Le Thi Hoai An, Pham Dinh Tao and Nguyen Ngoc Thanh in Advances in Intelligent Systems and Computing (AISC), Springer. All submissions have been peer-reviewed and we have selected only those with highest quality to include in this book.

We would like to thank all those who contributed to the success of the conference and to this book of proceedings. In particular we would like to express our gratitude to the authors as well as the members of the International Scientific Committee and the referees for their efforts and cooperation. Finally, the interest of the sponsors in the meeting and their assistance are gratefully acknowledged, and we cordially thank Prof. Janusz Kacprzyk and Dr. Thomas Ditzinger from Springer for their supports.

We hope that MCO 2015 significantly contributes to the fulfilment of the academic excellence and leads to greater success of MCO events in the future.

March 2015

Hoai An Le Thi
Tao Pham Dinh
Ngoc Thanh Nguyen

DC Programming and DCA:
Thirty Years of Developments

The year 2015 marks the 30th birthday of DC (Difference of Convex functions) programming and DCA (DC Algorithm) which were introduced by Pham Dinh Tao in 1985 as a natural and logical extension of his previous works on convex maximization since 1974. They have been widely developed since 1994 by extensive joint works of Le Thi Hoai An and Pham Dinh Tao to become now classic and increasingly popular.

DC programming and DCA can be viewed as an elegant extension of Convex analysis/Convex programming, sufficiently broad to cover most real-world nonconvex programs, but no too in order to be able to use the powerful arsenal of modern Convex analysis/Convex programming. This philosophy leads to the nice and elegant concept of approximating a nonconvex (DC) program by a sequence of convex ones for the construction of DCA: each iteration of DCA requires solution of a convex program. It turns out that, with appropriate DC decompositions and suitably equivalent DC reformulations, DCA permits to recover most of standard methods in convex and nonconvex programming. These theoretical and algorithmic tools, constituting the backbone of Nonconvex programming and Global optimization, have been enriched from both a theoretical and an algorithmic point of view, thanks to a lot of their applications, by researchers and practitioners in the world, to model and solve nonconvex programs from many fields of Applied Sciences, including Data Mining-Machine Learning, Communication Systems, Finance, Information Security, Transport Logistics & Production Management, Network Optimization, Computational Biology, Image Processing, Robotics, Computer Vision, Petrochemicals, Optimal Control and Automatic, Energy Optimization, Mechanics, etc. As a continuous approach, DC programming and DCA were successfully applied to Combinatorial Optimization as well as many classes of hard nonconvex programs such as Variational Inequalities Problems, Mathematical Programming with Equilibrium Constraints, Multilevel/Multiobjective Programming.

DC programming and DCA were extensively developed during the last two decades. They were the subject of several hundred articles in the high ranked scientific journals and the high-level international conferences, as well as various international research projects, and were the methodological basis of more than 50 PhD theses. More than

90 invited symposia/sessions dedicated to DC programming & DCA were presented in numerous international conferences. The ever-growing number of works using DC programming and DCA proves their power and their key role in Nonconvex programming/Global optimization and many areas of applications.

In celebrating the 30th birthday of DC programming and DCA, we would like to thank the founder, Professor Pham Dinh Tao, for creating these valuable theoretical and algorithmic tools, which have such a wonderful scientific impact on many fields of Applied Sciences.

Hoai An Le Thi
General Chair of MCO 2015

Organization

MCO 2015 is organized by the Laboratory of Theoretical and Applied Computer Science, University of Lorraine, France.

Organizing Committee

Conference Chair

Hoai An Le Thi   University of Lorraine, France

Members

Lydia Boudjeloud	University of Lorraine, France
Conan-Guez Brieu	University of Lorraine, France
Alain Gély	University of Lorraine, France
Annie Hetet	University of Lorraine, France
Vinh Thanh Ho	University of Lorraine, France
Hoai Minh Le	University of Lorraine, France
Duy Nhat Phan	University of Lorraine, France
Minh Thuy Ta	University of Lorraine, France
Thi Thuy Tran	University of Lorraine, France
Xuan Thanh Vo	University of Lorraine, France
Ahmed Zidna	University of Lorraine, France

Program Committee

Program Co-chairs

Hoai An Le Thi	University of Lorraine, France
Tao Pham Dinh	National Institute for Applied Sciences-Rouen, France

Members

El-Houssaine Aghezzaf	University of Gent, Belgium
Tiru Arthanari	University of Auckland, New Zealand
Adil Bagirov	University of Ballarat, Australia
Younès Bennani	University Paris 13-University Sorbonne Paris Cité, France
Lyes Benyoucef	University of Aix-Marseille, France
Lydia Boudjeloud	University of Lorraine, France
Raymond Bisdorff	University of Luxembourg, Luxembourg
Pascal Bouvry	University of Luxembourg, Luxembourg
Stéphane Canu	INSA–Rouen, France
Emilio Carrizosa	Universidad de Sevilla, Sevilla, Spain
Suphamit Chittayasothorn	King Mongkut's Institute of Technology Ladkraban, Thailand
John Clark	University of York, UK
Rafael Correa	Universidad de Chile, Santiago, Chile
Van Dat Cung	Grenoble INP, France
Frédéric Dambreville	DGA, France
Mats Danielson	Stockholm University, Sweden
Mohamed Djemai	University of Valenciennes, France
Sourour Elloumi	National School of Computer Science for Industry and Business, France
Alexandre Dolgui	Mines Saint-Etienne, France
Love Ekenberg	Stockholm University, Sweden
Ronan M.T. Fleming	University of Luxembourg, Luxembourg
Fabián Flores-Bazán	University of Concepción, Chile
Yann Guermeur	CNRS, France
Nicolas Hadjisavvas	University of the Aegean, Greece
Mounir Haddou	INSA-Rennes, France
Jin-Kao Hao	University of Angers, France
Duong-Tuan Hoang	University of Technology of Sydney, Australia
Van-Ngai Huynh	University of Quy Nhon, Vietnam
Fadili Jalal	University of Caen-IUF, France
J.Jung Jason	Chung-Ang University, South Korea
Joaquim Judice	University Coimbra, Portugal
Kang-Hyun Jo	University of Ulsan, South Korea
Pang Jong-Shi	University of Southern California, USA
Djamel Khadraoui	CRP H. Tudor, Luxembourg
Arnaud Lallouet	University of Caen, France
Aron Larson	Stockholm University, Sweden
Hoai Minh Le	University of Lorraine, France
Van Cuong Le	University Paris 1 Panthéon-Sorbonne, France
Duan Li	Chinese University of Hong Kong, Hong Kong
Yufeng Liu	University of North Carolina at Chapel Hill, USA

Gerhard-Wilhelm Weber	Middle East Technical University, Turkey
Adilson Elias Xavier	Federal University of Rio de Janeiro, Rio de Janeiro, Brazil
Jack Xin	University of California at Irvine, USA
Adnan Yassine	University of Le Havre, France
Daniela Zaharie	West University of Timisoara, Romania
Ahmed Zidna	University of Lorraine, France

Organizers of Special Sessions/Workshops

Combinatorial Optimization

| Viet Hung Nguyen | University Pierre and Marie Curie, France |

DC Programming and DCA: Thirty Years of Developments

| Hoai An Le Thi | University of Lorraine, France |

Dynamic Optimization

| Patrick Siarry | University of Paris-Est Créteil, France |

Global Optimization and Semi-Infinite Optimization

| Mohand Ouanes | University of Tizi-Ouzou, Algeria |

Maintenance and Production Control Problems

| Nidhal Rezg, Hajej Zied | University of Lorraine, France |
| Ali Gharbi | ETS Montreal, Canada |

Modeling and Optimization in Computational Biology

| Tobias Marschall | Saarland University, Germany |

Modeling and Optimization in Financial Engineering

| Duan Li | The Chinese University of Hong Kong |

Numerical Optimization

| Adnan Yassine | University of Le Havre, France |

Optimization Applied to Surveillance and Threat Detection

Frédéric Dambreville   ENSTA Bretagne & DGA, France

Post Crises Banking and Eco-finance Modelling

Duc Pham-Hi   ECE Paris, France

Spline Approximation & Optimization

Ahmed Zidna   University of Lorraine, France

Technologies and Methods for Multi-stakeholder Decision Analysis in Public Settings

Aron Larsson, Love Ekenberg,   Stockholm University, Sweden
    Mats Danielson

Variational Principles and Applications

Q. Bao Truong   Northern Michigan University, US
Christiane Tammer   Faculty of Natural Sciences II,
                                        Institute of Mathematics, Germany
Antoine Soubeyran   Aix-Marseille University, France

Sponsoring Institutions

University of Lorraine (UL), France
Laboratory of Theoretical and Applied Computer Science, UL
UFR Mathématique Informatique Mécanique Automatique, UL
Conseil Général de la Moselle, France
Conseil Régional de Lorraine, France
Springer
IEEE France section
Mairie de Metz, France
Metz Métropole, France

Contents

Part IV: Modelling and Optimization in Financial Engineering

Part V: Multiobjective Programming

Part VI: Numerical Optimization

Part VII: Spline Approximation and Optimization

Part VIII: Variational Principles and Applications

Part I
Combinatorial Optimization and Applications

A New Variant of the Minimum-Weight Maximum-Cardinality Clique Problem to Solve Conflicts between Aircraft

Thibault Lehouillier, Jérémy Omer, François Soumis, and Guy Desaulniers

Group on Research in Decision Analysis
3000, Côte-Sainte-Catherine Road
Montreal, QC H3T 2A7, Canada
{thibault.lehouillier,jeremy.omer,francois.soumis,
guy.desaulniers}@polymtl.ca

Abstract. In this article, we formulate a new variant of the problem of finding a maximum clique of minimum weight in a graph applied to the detection and resolution of conflicts between aircraft. The innovation of the model relies on the cost structure: the cost of the vertices cannot be determined a priori, since they depend on the vertices in the clique. We apply this formulation to the resolution of conflicts between aircraft by building a graph whose vertices correpond to a set of maneuvers and whose edges link conflict-free maneuvers. A maximum clique of minimal weight yields a conflict-free situation involving all aircraft and minimizing the costs induced. We solve the problem as a mixed integer linear program. Simulations on a benchmark of complex instances highlight computational times smaller than 20 seconds for situations involving up to 20 aircraft.

Keywords: Air Traffic Control, Conflict Resolution, Maximum Clique, Mixed Integer Linear Programming.

1 Introduction

Developing advanced decision algorithms for the air traffic control (ATC) is of great importance for the overall safety and capacity of the airspace. Resolution algorithms for the air conflict detection and resolution problem are relevant especially in a context of growing traffic, where capacity and safety become an issue. Indeed, a simulation-based study performed by Lehouillier et al. [1] shows that the controllers in charge of the traffic in 2035, which will have increased by 50%, would have to solve on average 27 conflicts per hour in a busy sector.

Maintaining separation between aircraft is usually referred to as the air conflict detection and resolution (CDR) problem. A conflict is a predicted loss of separation, i.e., when two aircraft are too close to each other regarding predefined horizontal and vertical separation distances of 5NM and 1000ft respectively. To solve a conflict, the controllers issue maneuvers that can consist of speed, heading or altitude changes. Given the current position, speed, acceleration and the

© Springer International Publishing Switzerland 2015
H.A. Le Thi et al. (eds.), *Model. Comput. & Optim. in Inf. Syst. & Manage. Sci.*,
Advances in Intelligent Systems and Computing 359, DOI: 10.1007/978-3-319-18161-5_1

predicted trajectory of a set of aircraft, the CDR problem corresponds to identifying the maneuvers required to avoid all conflicts while minimizing the costs induced.

The CDR problem is one of the most widely studied problems in air traffic management. For a comprehensive coverage of the existing literature, the reader may refer to the review in Martìn-Campo thesis [2]. Exact methods include optimal control, which can be associated with nonlinear programming. However, these methods suffer from the sensitivity to the starting point of the resolution and the high computational time. Mixed integer linear and nonlinear programming (MILPs and MINLPs) techniques are often considered. Omer and Farges [3] present a time-discretization of optimal control. Omer [4] also develops a space discretization using the points of interest for the conflict resolution. Pallottino et al. [5] develop MILPS solving the problem with speed changes and constant headings or with heading changes and constant speeds. Alonso-Ayuso et al. [6] develop a MILP that considers speed and altitude changes. However, MINLPs suffer from high computational times and do not give any optimality guarantee in finite time. Besides, the hypotheses made in MILPs to have linear constraints may not work in all situation. Several heuristics were developed to find a solution rapidly. Examples of techniques developed include ant colony algorithms like in Durand and Alliot [7], variable neighborhood searches (see Alonso-Ayuso et al. [8]). Other fast methods include particle swarm optimization, prescribed sets or neural networks. Heuristics find a solution rapidly, but the hypotheses can be restrictive and the convergence is not guaranteed. Graph theory is seldom used in ATC. Generally, conflicts between aircraft are modeled by a graph whose vertices represent the different aircraft and whose edges link pairs of conflicting aircraft, like in Vela [9]. Barnier and Brisset [10] assign flight levels to aircraft with intersecting routes by looking for maximum cliques in a graph where a proper coloring of the vertices defines an assignment of all aircraft to a set of flight levels.

The model presented in this article uses the concept of a clique in a graph, which is a subset of the vertices where each pair of elements is linked by an edge. Finding a maximum clique in an arbitrary graph is a well-known optimization problem that is $\mathcal{NP}$-hard. The problem has been thoroughly studied and several methods, both exact and heuristic, have been developed. For a comprehensive coverage on the subject, one can refer to Bomze et al. [11] and Hao et al. [12].

We formulate the air conflict detection and resolution problem as a new variant of the problem of finding a maximum clique of minimum weight in a graph. To this end, we build a graph whose vertices represent a set of possible maneuvers and where a clique yields a conflict-free solution involving all the aircraft. On the one hand, our model is innovative due to the cost structure for the vertices. With this model, we can maintain a reasonable size for the graph built, hence reducing the computational time. On the other hand, our model significance relies on its flexibility: a modification of the problem constraintes or objective function do not jeopardize the validity of the mathematical framework developed.

Being flexible is critical in ATC: in addition to being able to cover more ground, it will allow meaningful comparisons with existing models in the literature.

2   Problem Formulation

2.1   Modeling Aircraft Dynamics

To model the flight dynamics, we use the three-dimensional point-mass model presented in the BADA user manual [13]. Aircraft follow their planned 4D trajectory, which is a sequence of 4D points requiring time and space accuracy, leaving the remainder of the trajectory almost unconstrained. The non-compliance with this contract costs penalty fees to companies. As a consequence, an aicraft needs to recover its initial 4D trajectory after performing a maneuver. We assume that the planned speed for an aircraft corresponds to its nominal speed, i.e., the speed minimizing the fuel burn rate per distance unit traveled using the model described in [13].

Maneuvers are performed dynamically as described in [14], where the author states that the typical acceleration during a speed adjustment is in the order of 0.4kn/s. Heading changes are approximated by a steady turn of constant rate and radius. The changes of flight level are performed with a vertical speed, whose computation is detailed in [13], as a function of the thrust, drag, and true airspeed.

2.2   On Cliques and Stables

Let $\mathcal{G} = (\mathcal{V}, \mathcal{E})$ be an undirected, simple graph with a vertex set $\mathcal{V}$ and an edge set $\mathcal{E} \subseteq \mathcal{V} \times \mathcal{V}$.

A *clique* in graph $\mathcal{G}$ is a vertex set $\mathcal{C}$ with the property that each pair of vertices in $\mathcal{C}$ is linked by an edge:

$$\mathcal{C} \subseteq \mathcal{V} \text{ is a clique} \Leftrightarrow \forall (u, v) \in \mathcal{C} \times \mathcal{C}, (u, v) \in \mathcal{E} \tag{1}$$

A *maximum* clique in $\mathcal{G}$ is a clique that is not a subset of any other clique in $\mathcal{G}$. The cardinality of a maximum clique of $\mathcal{G}$ is called *clique number* and is denoted by $w(\mathcal{G})$. Let $c : \mathcal{V} \to \mathbb{R}$ be a vertex-weight function associated with $\mathcal{G}$. A *maximum* clique of *minimum-weight* in $\mathcal{G}$ is a maximum clique $\mathcal{C}$ that minimizes $\sum_{v \in \mathcal{C}} c(v)$.

A *stable set* $\mathcal{S} \subseteq \mathcal{V}$ is a subset of vertices no two of which are adjacent. A *bipartite* graph is a graph whose vertices can be partitionned into two distinct stable sets $\mathcal{V}_1$ and $\mathcal{V}_2$. Each edge of the graph connects one vertex of one stable to a vertex in the other stable. This concept is extended to $k-partite$ graphs, where the vertex set is partitionned into k distinct stable sets.

2.3   Graph Construction

In this subsection, we introduce the graph $\mathcal{G} = (\mathcal{V}, \mathcal{E})$ used to model the CDR problem.

Defining the Vertices. Let $\mathcal{F} = [\![1; n]\!]$ denote the set of the considered aircraft. We define $\mathcal{M} = \cup_{f=1}^{n} \mathcal{M}_f$ as the set of the possible maneuvers, $\mathcal{M}_f$ being the set of maneuvers for aircraft $f \in \mathcal{F}$. We consider both horizontal and vertical maneuvers of the following types:

- *NIL* refers to the *null* maneuver, i.e., when no maneuver is performed;
- H_θ corresponds to a heading change by an angle $\theta \in [-\pi; \pi]$;
- S_δ corresponds to a relative speed change of $\delta\%$;
- $V_{\delta h}$ denotes a change of δh flight levels.

A maneuver $m \in \mathcal{M}$ is described as a triplet $(\delta\chi_m, \delta V_m, \delta FL_m)$ corresponding to the heading, speed and flight level changes induced by m. The set of vertices is defined as $\mathcal{V} = [\![1; |\mathcal{M}|]\!]$, where $|\mathcal{M}|$ is the cardinality of set $\mathcal{M}$. We note $\mathcal{V}_f$ the set of vertices corresponding to aircraft f.

In emergency scenarios where the feasibility of the problem can be an issue, it is possible to introduce n vertices corresponding to costly emergency maneuvers to ensure the feasibility of the problem. However, since the feasibility was not an issue for the tested instances, those vertices were not considered in this article. The weight of the vertices correspond to the fuel consumption induced by the corresponding maneuvers. We give further detail in Subsection 2.3.

Defining the Edges. Let $(i, j) \in \mathcal{V} \times \mathcal{V}$ be a pair of vertices representing maneuvers $(m_i, m_j) \in \mathcal{M} \times \mathcal{M}$ of aircraft $(f_i, f_j) \in \mathcal{F} \times \mathcal{F}$. For $i \neq j$, we write $m_i \square m_j$ when no conflict occurs if aircraft f_i follows maneuver m_i while aircraft f_j performs maneuver m_j. The set of edges $\mathcal{E}$ corresponds to the pairs of maneuvers performed by two different aircraft without creating conflicts:

$$\mathcal{E} = \{(i, j) \in \mathcal{V} \times \mathcal{V}, i \neq j : m_i \square m_j\} \qquad (2)$$

It is important to note that there is no edge between two different maneuvers of a given aircraft, which yields Proposition 1.

Proposition 1. *For all $f \in \mathcal{F}, \mathcal{V}_f$ is a stable set, i.e there is no edge linking two distinct vertices of $\mathcal{V}_f$. Hence, the graph $\mathcal{G}$ is $|\mathcal{F}|$-partite.*

Let $(i, j) \in \mathcal{V} \times \mathcal{V}$ be a pair of vertices representing maneuvers $(m_i, m_j) \in \mathcal{M} \times \mathcal{M}$ of aircraft $(f_i, f_j) \in \mathcal{F} \times \mathcal{F}$. The methodology used to compute if the edge (i, j) is added to $\mathcal{G}$ is described with the following notations:

- $\mathcal{T}$: time horizon for the conflict resolution;
- $\boldsymbol{p}_{f_i}(t) \in \mathbb{R}^3$: position vector of aircraft f_i at time t. $p_{f_i,x}(t)$ $p_{f_i,y}(t)$ and $p_{f_i,z}(t)$ denote respectively the abscissa, ordinate and altitude components of the position vector;
- $\boldsymbol{s}_{f_i}(t) \in \mathbb{R}^3$: speed vector of aircraft f_i at time t. $s_{f_i,x}(t)$ $s_{f_i,y}(t)$ and $s_{f_i,z}(t)$ denote respectively the abscissa, ordinate and altitude components of the speed vector;
- $\boldsymbol{a}_{f_i}(t) \in \mathbb{R}^3$: acceleration vector of aircraft f_i at time t. $a_{f_i,x}(t)$ $a_{f_i,y}(t)$ and $a_{f_i,z}(t)$ denote respectively the abscissa, ordinate and altitude components of the acceleration vector;
- $\boldsymbol{p}_{f_j}(t), \boldsymbol{s}_{f_j}(t)$ and $\boldsymbol{a}_{f_j}(t)$ are also defined following the same notations.

The definition of the maneuvers m_i and m_j applied to f_i and f_j is used to project the aircraft trajectory over time. Aircraft f_i and f_j are said to be separated at time t if and only if at least one of constraints (3) and (4) holds. In this paper we choose $D_{h,min} = 5NM$ and $D_{v,min} = 1000ft$.

$$d^h_{f_if_j}(t)^2 = (p_{f_i,x}(t) - p_{f_j,x}(t))^2 + (p_{f_i,y}(t) - p_{f_j,y}(t))^2 \geq D^2_{h,min} \tag{3}$$

$$d^v_{f_if_j}(t)^2 = (p_{f_i,z}(t) - p_{f_j,z}(t))^2 \geq D^2_{v,min} \tag{4}$$

At any time $t \in \mathcal{T}$, either none, one or both aircraft are maneuvering. $\mathcal{T}$ can thus be divided into intervals where both f_i and f_j have a constant acceleration. For each interval, we compute the time at which the aircraft are the closest to verify if the separation constraints hold. Let $\mathcal{T}_k$ be one of these intervals. Consider f_i and $t_0 \in \mathcal{T}$ be the starting time of maneuver m_i. If we assume that maneuver m_i is applied with a constant acceleration, we obtain the position and the speed vector of f_i at time $t_0 + t$ with t such that $t - t_0 \leq |\mathcal{T}_k|$:

$$\boldsymbol{p}_{f_i}(t_0 + t) = \boldsymbol{p}_{f_i}(t_0) + (t - t_0)\boldsymbol{s}_{f_i}(t_0) + \frac{(t - t_0)^2}{2}\boldsymbol{a}_{f_i}(t_0) \tag{5}$$

$$\boldsymbol{s}_{f_i}(t_0 + t) = \boldsymbol{s}_{f_i}(t_0) + (t - t_0)\boldsymbol{a}_{f_i}(t_0) \tag{6}$$

Let $\boldsymbol{p}^h_{f_if_j}$ (respectively $\boldsymbol{s}^h_{f_if_j}$, $\boldsymbol{a}^h_{f_if_j}$) denote respectively the horizontal position, the speed and the acceleration of aircraft f_j relatively to aircraft f_i. We define

$$d^h_{f_if_j}(t + \tau) = ||\boldsymbol{p}^h_{f_if_j}(t + \tau)||$$

$$= ||\boldsymbol{p}^h_{f_if_j}(t) + \tau\boldsymbol{s}^h_{f_if_j}(t) + \frac{\tau^2}{2}\boldsymbol{a}^h_{f_if_j}(t)||$$

where $\tau \geq 0$.

Let $\tau_{f_if_j} \in \underset{\tau \geq 0}{\operatorname{argmin}}\, d^h_{f_if_j}(t + \tau)^2$, and $t^h_{f_if_j} \in \underset{t \in \mathcal{T}}{\operatorname{argmin}}\, d^h_{f_if_j}(t)^2$.

We have: $t^h_{f_if_j} = \begin{cases} 0 & \text{if } \tau_{f_if_j} = 0 \\ |\mathcal{T}| & \text{if } \tau_{f_if_j} \geq |\mathcal{T}_k| \\ \tau_{f_if_j} & \text{otherwise} \end{cases}$

Aircraft f_i and f_j are horizontally separated during interval $\mathcal{T}$ if and only if (7) holds:

$$d^h_{f_if_j}(t^h_{f_if_j})^2 \geq D^2_{h,min} \tag{7}$$

By a similar reasoning, aircraft f_i and f_j are vertically separated during interval $\mathcal{T}$ if and only if (8) holds:

$$d^v_{f_if_j}(t^v_{f_if_j})^2 \geq D^2_{v,min} \tag{8}$$

If either (7) or (8) holds when aircraft f_i and f_j apply maneuvers m_i and m_j, then an edge is created between i and j. As explained in 2.1, it is important that every aircraft initiates a safe return towards its initial trajectory once the conflict is avoided. For each edge, we compute the minimum time necessary before one or both aircraft can recover their initial trajectories. The cost of the recovery of a trajectory is detailed in Subsection 2.3.

Application to the CDR Problem. As mentioned in Section 1, given the current position, speed, acceleration and the planned trajectories of a set of aircraft, solving the CDR problem consists in finding a conflict-free set of maneuvers that minimizes the costs. Proposition 2 links the cliques in $\mathcal{G}$ to the CDR problem:

Proposition 2. *Let $\mathcal{C}$ be a clique in graph $\mathcal{G}$. Then $\mathcal{C}$ represents a set of conflict-free maneuvers for a subset of $\mathcal{F}$ of cardinality $|\mathcal{C}|$.*

Proposition 2 shows that finding a set of conflict-free maneuvers for $\mathcal{F}$ is equivalent to finding a clique of $\mathcal{G}$ of cardinality $|\mathcal{F}|$. We derive the following theorem:

Theorem 1. *If a conflict-free solution exists, then $\omega(\mathcal{G}) = |\mathcal{F}|$. Otherwise, $\omega(\mathcal{G})$ is the maximum number of flights involved in a conflict-free situation.*

We define the problem $\mathrm{CDR}_{\mathcal{M}}$ as the restriction of the CDR problem to the set of maneuvers $\mathcal{M}$. Using both Proposition 2 and Theorem 1, we can state anew the $\mathrm{CDR}_{\mathcal{M}}$ problem as follows: solving the $\mathrm{CDR}_{\mathcal{M}}$ problem consists in finding a clique of maximum cardinality and minimal cost in graph $\mathcal{G}$. In fact, we consider a new variant of a clique problem where the weight associated with a vertex is not known a priori and rather depends on the edges induced by the clique. Indeed, the cost associated with a maneuver depends on the duration that this maneuver will be performed before returning towards the planned trajectory. Because this duration depends on the maneuvers selected for the other aircraft, it cannot be determined a priori and must be computed as the maximum duration needed to avoid a loss of separation with all other aircraft given their chosen maneuvers. To handle such vertex costs, we first define edge costs.

Computing the Cost of the Edges. The cost measure chosen for this article corresponds to the extra fuel consumption induced by the maneuvers, i.e., the additional fuel required to return to the 4D trajectory after the maneuver is performed. We use the model given in [13]. For a jet commercial aircraft f, the fuel consumption by time and distance unit is given by (9) and (10):

$$C_{t,f}(t) = c_{1,f}\left(1 + \frac{V_f(t)}{c_{2,f}}\right) F_{T,f}(t) \tag{9}$$

$$C_{d,f}(t) = \frac{C_{t,f}(t)}{V_f(t)} \tag{10}$$

where $c_{1,f}$ and $c_{2,f}$ are numerical constants depending on the type of aircraft f.

We compute the cost of an edge $e = (i,j)$ linking two vertices representing two maneuvers of aircraft f_i and f_j, denoted m_i and m_j, as a pair constituted of the extra fuel costs for both f_i and f_j, denoted $C_i^{(i,j)}$ and $C_j^{(i,j)}$. The additional consumed fuel corresponds to the performed maneuver along with the fuel required to recover the inital 4D trajectory. After a change of speed of $\delta\%$ during a period δt, the aircraft recovers its 4D trajectory by making the opposite change of speed during δt. After a change of direction $\delta\chi$ during a period δt, the aircraft performs a turn with an angle θ_r in order to recover its physical trajectory along with a change of speed to retrieve the 4D trajectory. The cost induced is the extra fuel burnt when the aircraft flies at the recovery speed and the fuel burnt on the extra distance induced by the maneuver. For a flight level change, we compute the extra cost as the difference of consumption between the different flight levels, along with the cost of changing twice of flight level. The distance flown is also longer, and this extra distance is also accounted for.

Computing the Cost of the Vertices. Several techniques can be followed in order to determine the vertices cost. The basic one would be to discretize the duration of the maneuver, and to create the vertices accordingly. In this situation, computing the costs would be straight-forward. However, the drawback of this method is that the graph built is huge, which could result in a difficult resolution. We choose to follow another structure of cost because it is more compact in terms of graph size.

Let us consider a vertex i which corresponds to a maneuver m_i for an aicraft f_i. The cost of each edge linking i to one of its neighbors j, associated to a maneuver m_j for aircraft f_j, corresponds to f_i applying m_i during a time t_i^j, which depends on m_j. Time t_i^j is the minimum time during which f_i must apply m_i in order to avoid any conflict if one or both aircraft return to their initial trajectory. Following maneuver m_i for a duration t_i^j induces a cost $C_i^{(i,j)}$. If i is part of the maximum clique $\mathcal{C}$ to be determined, we need to establish the time t_i during which maneuver m_i is actually applied in order to determine its cost c_i. t_i is obtained by:

$$t_i = \max_{j \in \mathcal{V} \cap \mathcal{C}} t_i^j \tag{11}$$

As a consequence, we have that c_i is the cost of aircraft f_i applying m_i during t_i. If i is not part of the maximum clique $\mathcal{C}$, then no constraint is imposed on the cost c_i. As detailed in Section 3, the optimization model will automatically force the value of c_i to 0. To conclude, we have that for any $i \in \mathcal{V}$:

$$c_i = \begin{cases} \max_{j \in \mathcal{V} \cap \mathcal{C}} C_i^{(i,j)} & \text{if } i \in \mathcal{C} \\ 0 & \text{otherwise} \end{cases}$$

2.4   Illustrative Example

For the sake of clarity, an illustrative example with three aircraft is given in Figure 1. If each aircraft follows its planned trajectory, conflicts will happen between the blue aircraft and the two others. For this example, we assume that, in addition to the null maneuver, only two heading changes ($\pm 30°$) are allowed. We build the CDR graph shown in Figure 1(b). The graph is 3-partite, as the vertex set is partitionned into 3 stable sets of 3 vertices each. Solving the CDR is then equivalent to searching for a minimum-weight clique of 3 vertices, i.e., a triangle.

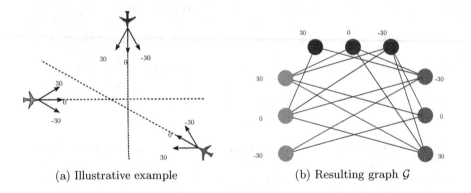

(a) Illustrative example   (b) Resulting graph $\mathcal{G}$

Fig. 1. Illustrative example with three aircraft

3 Methodology

Determining the cost of a vertex i is very specific, since it is correlated to whether or not i belongs to a maximum clique $\mathcal{C}$. As a consequence, the algorithms usually used in existing librairies dedicated to graph theory cannot be used for our model. We formulate the problem as a mixed-integer linear program using the following variables:

$$- x_i = \begin{cases} 1 & \text{if vertex } i \text{ is part of the maximum clique} \\ 0 & \text{otherwise} \end{cases}$$

$- c_i \in \mathbb{R}_+$ is the cost of vertex i.

We describe the clique search by the following linear integer program:

$$\text{minimize} \sum_{i \in \mathcal{V}} c_i \tag{12}$$

$$\text{subject to } x_i + x_j \leq 1, \forall (i,j) \notin \mathcal{E} \tag{13}$$

$$\sum_{i \in \mathcal{V}} x_i = |\mathcal{F}| \tag{14}$$

$$c_i \geq C_i^{(i,j)}(x_i + x_j - 1), \forall (i,j) \in \mathcal{E} \tag{15}$$

$$x_i \in \{0;1\}, \forall i \in \mathcal{V} \tag{16}$$

$$c_i \in \mathbb{R}_+, \forall i \in \mathcal{V} \tag{17}$$

The objective function (12) minimizes the cost of the maneuvers. (13) are clique constraints, and constraint (14) exploits Theorem 1 defining the cardinality of the maximum clique. Constraints (15) are used to compute the cost of the vertices: if a vertex is in the maximum clique, then its cost must be greater than its cost on all edges connecting it to other vertices in the clique.

4 Results

All tests were performed on a computer equipped with an Intel Core i7-3770 processor, 3.4 GHz, 8-GB RAM. The algorithms were implemented in C++ and using CPLEX 12.5.1.0[1].
The headings of the tables presented in these section are given as follows:

- *case*: case configuration;
- $|\mathcal{F}|$: number of aircraft;
- $|\mathcal{V}|$: number of vertices;
- $|\mathcal{E}|$: number of edges;
- $d = \frac{2|\mathcal{E}|}{|\mathcal{V}|(|\mathcal{V}|-1)}$: graph density;
- n: number of variables;
- m: number of constraints;
- z_{ip}: optimal value for the problem;
- *nodes*: number of branch-and-bound nodes;
- t_{lp}: time (in seconds) to continuous relaxation of the MILP;
- t_{ip}: time (in seconds) to obtain the z_{ip} value;

4.1 Benchmark Description

The benchmark used for this study gathers three types of instances. The first set is roundabout instances $\mathcal{R}_n$, where n aircraft are distributed on the circumference of a 100NM radius and fly towards the center at the same speed and altitude. The second set is crossing flow instances $\mathcal{F}_{n,\theta,d}$, where two trails of n aircraft separated by d nautical miles intersect each other with an angle θ. The last type of instance is a grid $\mathcal{G}_{n,d}$ constituted of two crossing flow instances $\mathcal{F}_{n,\frac{\pi}{2},d}$ with a 90° angle, one instance being translated 15NM North-East from the other. An example of these instances is given on Figure 2.

4.2 Computational Results

The first set of simulations considers only horizontal maneuvers, with relative speed changes of ±3% and ±6% and heading changes of ±5°, ±10°, ±15°. The graph remains small when one considers this set of maneuvers, and their small magnitude makes them less costly. Nevertheless, if these values were to be inefficient to solve all the conflicts, we could introduce maneuvers of larger magnitude.

Table 1 gathers information about the graph $\mathcal{G}$, the MILP and the main computational results. The solution time for the continuous relaxation is very small, but the quality of the relaxation is mediocre. Indeed, the fractional solution of the linear relaxation chooses two maneuvers for each aircraft with a value of 0.5. Constraints (15) force the cost of each vertex to be 0, yielding an optimal value of 0 and a gap of 100%. Results also display short solution times: problems known to be complex with 20 aircraft are solved to optimality in less than 15 seconds. This result is very satisfying since the density of the graph is high.

---

[1] See the IBM-ILOG CPLEX v12.5. User's manual for CPLEX.

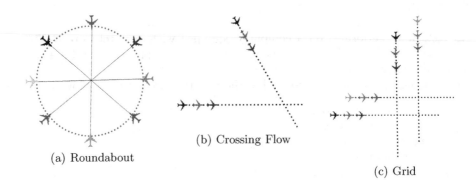

(a) Roundabout

(b) Crossing Flow

(c) Grid

Fig. 2. Examples

Table 1. Dimensions of the instances and computational results

Instance type	Case	Graph $\mathcal{G}$				MILP		Resolution			
		$|\mathcal{F}|$	$|\mathcal{V}|$	$|\mathcal{E}|$	d	m	n	z_{ip}	nodes	t_{lp}	t_{ip}
Roundabout	$\mathcal{R}_2$	2	22	90	0.39	44	225	3.71	6	0	0.02
	$\mathcal{R}_4$	4	44	492	0.52	88	1073	14.98	73	0	0.02
	$\mathcal{R}_6$	6	66	1194	0.56	132	2521	22.7	0	0.01	0.15
	$\mathcal{R}_8$	8	88	2184	0.57	176	4545	31.05	47	0.01	0.53
	$\mathcal{R}_{10}$	10	110	3430	0.57	220	7081	112.7	208	0.05	1.56
	$\mathcal{R}_{12}$	12	132	4944	0.57	264	10153	189.27	581	0.09	3.41
	$\mathcal{R}_{14}$	14	154	6720	0.57	308	13749	224.75	183	0.1	6.98
	$\mathcal{R}_{16}$	16	176	8896	0.57	352	18145	261.44	162	0.15	9.5
	$\mathcal{R}_{18}$	18	198	11358	0.57	396	23113	636.7	257	0.21	12.1
	$\mathcal{R}_{20}$	20	220	14027	0.58	440	28461	740.6	210	0.27	3.2
Flows	$\mathcal{F}_{5,30,10}$	10	110	4522	0.75	220	9265	49.08	405	0.02	1.5
	$\mathcal{F}_{5,45,10}$	10	110	4518	0.75	220	9257	41.29	535	0.02	1.52
	$\mathcal{F}_{5,60,10}$	10	110	4478	0.75	220	9177	34.49	238	0.02	1.39
	$\mathcal{F}_{5,75,10}$	10	110	4492	0.75	220	9205	30.66	496	0.02	1.34
	$\mathcal{F}_{5,90,10}$	10	110	4528	0.76	220	9277	28.28	269	0.02	1.41
Grids	$\mathcal{G}_{2,3,10}$	12	132	6645	0.78	264	13555	57.65	564	0.01	3.64
	$\mathcal{G}_{2,5,10}$	20	220	19724	0.82	440	39889	121.92	2740	0.2	12.7

In the second simulation set, we introduce altitude maneuvers: aircraft are allowed to move to an adjacent flight level. Table 2 reports the main results. The values of the optimal solutions for the roundabout instances remain the same, highlighting that it is optimal to make simple turns instead of changing flight levels. For the crossing flows and the grid instances, it is more efficient for some aircraft to change their flight level instead of turning or changing their speed. As a consequence, the solutions are less expensive. Solution times tend to slightly increase, but the solution can still be computed in a short time. These results are promising since the instances tested are denser than real-life instances.

Table 2. Dimensions of the instances and computational results

		Graph $\mathcal{G}$				MILP		Resolution			
Instance type	Case	$|\mathcal{F}|$	$|\mathcal{V}|$	$|\mathcal{E}|$	d	m	n	z_{ip}	nodes	t_{lp}	t_{ip}
	$\mathcal{R}_2$	2	26	116	0.36	52	285	3.71	7	0	0.05
	$\mathcal{R}_4$	4	52	832	0.63	104	1769	14.98	272	0	0.41
	$\mathcal{R}_6$	6	78	2076	0.69	156	4309	22.7	498	0.01	0.25
	$\mathcal{R}_8$	8	104	3840	0.72	208	7889	31.05	328	0.01	1.12
Roundabout	$\mathcal{R}_{10}$	10	130	6080	0.73	260	12421	112.7	499	0.05	1.41
	$\mathcal{R}_{12}$	12	156	9096	0.75	312	18505	189.27	798	0.09	3.42
	$\mathcal{R}_{14}$	14	182	12208	0.74	364	24781	224.75	532	0.1	6.88
	$\mathcal{R}_{16}$	16	208	16416	0.76	416	33249	261.44	467	0.15	11.45
	$\mathcal{R}_{18}$	18	234	20772	0.76	468	42013	636.7	682	0.21	14.12
	$\mathcal{R}_{20}$	20	260	25760	0.77	520	52041	740.6	845	0.27	8.12
	$\mathcal{F}_{5,30,10}$	10	130	5102	0.75	260	9956	43.37	784	0.02	1.58
	$\mathcal{F}_{5,45,10}$	10	130	5098	0.75	260	9874	38.45	889	0.02	2.13
Flows	$\mathcal{F}_{5,60,10}$	10	130	5059	0.75	260	9845	29.78	645	0.02	1.96
	$\mathcal{F}_{5,75,10}$	10	130	5134	0.75	260	9899	30.11	897	0.02	1.78
	$\mathcal{F}_{5,90,10}$	10	130	5199	0.76	260	10078	23.01	540	0.02	1.45
Grids	$\mathcal{G}_{2,3,10}$	12	156	7320	0.79	312	15087	45.18	945	0.01	4.12
	$\mathcal{G}_{2,5,10}$	20	260	20945	0.83	520	45878	99.73	3847	0.2	16.7

5   Conclusions

A new variant of the problem of finding a maximum clique of minimum weight in a graph and its application to aircraft conflict resolution have been presented. The innovation of the model comes from the cost structure: the costs of the vertices cannot be determined a priori since they depend on the vertices in the clique. As a consequence, we model the problem as a MILP. The model performs well, since complex instances involving up to 20 aircraft are solved to optimality in near real-time. The design of the model is flexible, meaning that tuning some parameters of the model will allow meaningful comparisons with existing models.

These conclusions validate the model as a basis for further research. For instance, techniques reducing the size of the graph are of interest. Adding uncertainties is also a meaningful extension of the model. Additional benchmarks including real-life instance and random scenarios will be necessary in order to challenge the model.

References

1. Lehouillier, T., Omer, J., Soumis, F., Allignol, C.: Interactions between operations and planning in air traffic control. In: Proceedings of the 2nd International Conference of Research in Air Transportation, Istanbul (2014)
2. Campo, M., Javier, F.: The collision avoidance problem: methods and algorithms. Ph.D. dissertation (2010)
3. Omer, J., Farges, J.-L.: Hybridization of nonlinear and mixed-integer linear programming for aircraft separation with trajectory recovery. IEEE Transactions on Intelligent Transportation Systems 14(3), 1218–1230 (2013)

4. Omer, J.: A space-discretized mixed-integer linear model for air-conflict resolution with speed and heading maneuvers (2014)
5. Pallottino, L., Feron, E.M., Bicchi, A.: Conflict resolution problems for air traffic management systems solved with mixed integer programming. IEEE Transactions on Intelligent Transportation Systems 3(1), 3–11 (2002)
6. Alonso-Ayuso, A., Escudero, L.F., Martin-Campo, F.J., Javier, F.: Collision avoidance in air traffic management: a mixed-integer linear optimization approach. IEEE Transactions on Intelligent Transportation Systems 12(1), 47–57 (2011)
7. Durand, N., Alliot, J.-M., Noailles, J.: Automatic aircraft conflict resolution using genetic algorithms. In: Proceedings of the Symposium Applied Computing, Philadelphia (1996)
8. Alonso-Ayuso, A., Escudero, L.F., Martín-Campo, F.J., Mladenović, N.: A vns metaheuristic for solving the aircraft conflict detection and resolution problem by performing turn changes. Journal of Global Optimization, 1–14 (2014)
9. Vela, A.E.: Understanding conflict-resolution taskload: implementing advisory conflict-detection and resolution algorithms in an airspace (2011)
10. Barnier, N., Brisset, P.: Graph coloring for air traffic flow management. Annals of Operations Research 130(1-4), 163–178 (2004)
11. Bomze, I.M., Budinich, M., Pardalos, P.M., Pelillo, M.: The maximum clique problem. In: Handbook of combinatorial optimization, pp. 1–74. Springer (1999)
12. Wu, Q., Hao, J.-K.: A review on algorithms for maximum clique problems. European Journal of Operational Research (2014)
13. User manual for the Base of Aircraft Data (BADA), Eurocontrol, Tech. Rep. 11/03/08-08 (2011)
14. Paielli, R.A.: Modeling maneuver dynamics in air traffic conflict resolution. Journal of Guidance, Control, and Dynamics 26(3), 407–415 (2003)

An Adaptive Neighborhood Search for k-Clustering Minimum Bi-clique Completion Problems

Mhand Hifi, Ibrahim Moussa, Toufik Saadi, and Sagvan Saleh

EPROAD EA 4669, Université de Picardie Jules Verne,
7 rue du Moulin Neuf, 80000 Amiens, France
{firstname.name}@u-picardie.fr

Abstract. In this paper, we propose to solve the k-clustering minimum bi-clique completion problem by using an adaptive neighborhood search. An instance of the problem is defined by a bipartite graph $G(V = (S, T), E)$, where V (resp. E) denotes the set of vertices (resp. edges) and the goal of the problem is to determine the partition of the set S of V into k clusters (disjoint subsets) such that the number of the edges that complete each cluster into a bi-clique, according to the vertices of T, should be minimized. The adaptive search is based upon three complementary steps: (i) a starting step that provides an initial solution by applying an adaptation of Johnson's principle, (ii) an intensification step in which both exchanging and k-opt strategies are introduced and, (iii) a diversification step that tries to explore unvisited solutions' space. The method is evaluated on benchmark instances taken from the literature, where the provided results are compared to those reached by recent methods available in the literature. The proposed method remains competitive and it yields new results.

Keywords: Bi-clique, combinatorial optimization, heuristic, local search.

1 Introduction

In this paper, we investigate the use of the neighborhood search for solving the so-called k-*Clustering minimum Bi-clique Completion Problem* (noted k-CmBCP). An instance of k-CmBCP is defined by a bipartite graph $G(V, E)$, where V is the set of n vertices such that $V = S \cup T$ and $S \cap T = \emptyset$ and, E denotes the set of edges. We recall that if (S, T) is a bi-clique graph, then each vertex of S (resp. T) is related to all the vertices of T (resp. S). Moreover, if the graph is formed with k bi-partite graphs (clusters), i.e., $\Big((S_1, T_1), (S_2, T_2), \ldots, (S_k, T_k)\Big)$, then all vertices of each couple (S_j, T_j), $j = 1, \ldots, k$, are interconnected. Because $G(V, E)$ is a general directed graph, looking for a k-clustering is equivalent to searching for a k bi-partite subgraphs of G with a minimum additional edges that do not belong to E. Similarly, the goal of the problem is to find the best partition of the set S into k clusters with a minimum additional edges.

© Springer International Publishing Switzerland 2015
H.A. Le Thi et al. (eds.), *Model. Comput. & Optim. in Inf. Syst. & Manage. Sci.*,
Advances in Intelligent Systems and Computing 359, DOI: 10.1007/978-3-319-18161-5_2

As described in Gualandi *et al.* [4], such a problem has several real applications such as bundling channels for multicast transmissions. In such application, given a set of demands of services from customers, the aim consists of determining the k multicast sessions that is able to partition the set of the demands. Moreover, each of the considered service has to belong to a multicast session while each costumer can appear in several sessions.

The k-Clustering minimum Bi-clique Completion Problem (noted k-CmBCP) was first introduced by Faure *et al.* [2] in which the authors proved its NP-hardness. To our knowledge, very few paper dressing the KCmBCP are available in the literature. Among these papers, we cite the paper of Faure *et al.* [2] in which the authors proposed an integer linear programming model for solving small sized instances to optimality. In the same paper, a column generation-based heuristic has been also presented, where some large-scale instances of k-CmBCP have been tackled. Gualandi [3] tackled the problem by using a hybrid method; that is, an approach that combines constraint programming and semidefinite programming. Finally, Gualandi *et al.* [4] designed a special branch-and-price based method in order to accelerating the search process and improving the quality of the provided upper bounds when using the Cplex solver.

In this paper, we propose to solve the k-CmBCP by using an adaptive neighborhood search. Such an approach can be viewed as a variant of both methods proposed in Hifi and Michrafy [5] and a simplest version of Shaw's [7] large neighborhood search. We recall that an instance of k-CmBCP is characterized by a bipartite graph and the objective is to divide the first set S of vertices into k disjoint clusters, where the number of edges that must be added to form the k bi-cliques (representing the cost of the problem), should be minimum.

The rest of the paper is organized as follows. In Section 2, the k-CmBCP is first illustrated on an example and later its mathematical model is given. Section 3 discusses the tailored adaptive neighborhood search for approximately solving the k-CmBCP. The performance of the proposed algorithm is evaluated in Section 4, where its obtained results are compared to those reached by recent algorithms available in the literature. Finally, Section 5 summarizes the contribution of the paper.

2   The k-CmBCP

In this section, k-CmBCP is first illustrated throughout an example with $k = 2$. Second and last, the mathematical formulation of the k-CmBCP is given.

Fig. 1 illustrates a small example representing a k-CmBCP: the graph $G = (S, T)$ is defined by $S = \{1, ..., 4\}$ and $T = \{5, ..., 8\}$ and, the aim is to search the 2-clustering for G. First, a feasible solution (cf. Fig. 1.(a)) can be built by partitioning the first set S as follows: $S_1 = \{1, 2\}$ and $S_2 = \{3, 4\}$, respectively. Second, because each partition forms a bipartite graph with its corresponding links belonging to the second set T, then it is necessary to add some connections between the cluster at hand and its corresponding links. In this case, Fig. 1.(b) shows, for both clusters, the added links (represented by the dashed edges for

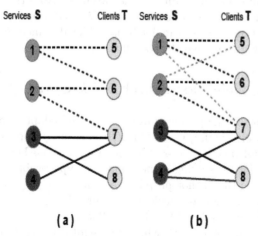

Fig. 1. Illustration of an instance of 2-CmBCP and its constructed feasible solution

the first cluster S_1 and the bold edges for S_2) we need to make the solution feasible for k-CmBCP. Third and last, one can observe that the resulting cost is equal to 3, where the first cluster S_1 needs 2 additional edges and 1 for the second cluster S_2.

Formally the k-CmBCP can be stated as an Integer Linear Programming Model (ILPM):

$$\min \sum_{p \in K} \sum_{(i,j) \in \overline{E}} z_{ijp} \tag{1}$$

$$\text{s.t.} \tag{2}$$

$$x_{ip} + y_{jp} \leq 1 + z_{ijp}, \ \forall (i,j) \in \overline{E}, \ \forall p \in K \tag{3}$$

$$\sum_{p \in K} x_{ip} = 1; \ \forall i \in S \tag{4}$$

$$x_{ip} \leq y_{jp}, \ \forall (i,j) \in E, \ \forall p \in K \tag{5}$$

$$x_{ip}, y_{jp} \in \{0,1\}, \ \forall i \in S, \ \forall j \in T, \ \forall p \in K \tag{6}$$

$$z_{ijp} \in \{0,1\}, \forall (i,j) \in \overline{E}, \forall p \in K \tag{7}$$

where the objective function Equation (1) minimizes the number of edges that completes each induced bipartite subgraph into a biclique, in others words, it minimizes the total cost of all clusters (i.e., the number of edges added in this case). Equation (3) ensures the link between both variables x_{ip} and y_{jp} by setting z_{ijp} to 1 whenever the other two variables are fixed to 1. Equation (4) ensures that each vertex belonging to the set S should be assigned to a single cluster. Equation (5) forces each vertex $j \in T$, which is adjacent to another vertex $i \in S$ and assigned to the pth cluster, to be affected to the same cluster (Of course, we recall that some vertices can be assigned to several clusters). Finally, Equations (6) and (7) ensure the integrality of the variables. Note that, the limit of the formulation described above (equations (1) to (7)) is that any permutation of the indices p provide s the same optimal solution. The aforementioned issue was tackled by Fahle [1] by introducing the so-called symmetry-breaking constraints.

3   An Adaptive Neighbor Search for k-CmBCP

In this section, we discuss the principle of the proposed adaptive neighborhood search for approximately solving the k-CmBCP. The used approach is mainly based upon three complementary steps. Such an approach has been already used in Hifi and Michrafy [5] for solving some knapsack type problems. It has also used by Shaw [7], where the author proposed a general purpose.

Herein, an adaptive method is considered, where the first step is applied in order to build a quick starting feasible solution. The second step tries to improve the quality of the starting solution by using the so-called *intensification strategy*. The aforementioned strategy alternatively combines both exchanging and 2-opt procedures. The third step introduces a *diversification strategy* in order to handle some sub-spaces with an efficient manner: both degrading and re-optimizing strategies are introduced for searching next feasible spaces.

3.1   A Starting Solution for k-CmBCP

There exists several ways for defining a starting solution when using heuristics. Herein, a simple greedy procedure is considered: it is based on the principle of Johnson's algorithm (cf., Johnson [6]). Indeed, in a bipartite graph, which can be considered as a special case of the set covering problem, an adaptation of Johnson's algorithm may be described as follows:

(i) Set $S' = S$,
(ii) Order all vertices of the first set S' in decreasing order of their degrees,
(iii) Select a vertex of S' having the greatest degree and put it into the current cluster (or into a new cluster to open),
(iv) Define a reduced graph G' by removing the last vertex from S',
(v) Stop if $S' = \emptyset$, repeat steps (ii)-(iv) otherwise.

Of course, the above procedure is used in order to partition the set S of the bipartite graph G. Then, in order to provide the cost of the starting solution reached by the greedy procedure, an adding step is used in order to complete each created cluster with the additional edges that forms the k-bicliques for G.

3.2   Improving the Quality of the Solutions: Intensification Step

Generally, improving a solution at hand requires to introduce a local search, where its aim is to perform an interesting search on a series of neighborhoods. Herein, the used intensification strategy is is based on combining both *exchanging*

Algorithm 1. Improving the quality of the current solution

Require: A starting feasible solution S with its added edges (noted e_{nb}).
Ensure: A (new) feasible solution S'.

1: $ok \leftarrow$ `false`;
2: **while** $\big(\text{not}(ok)$ and (a local stopping condition is not met)$\big)$ **do**
3:    **for all** $k_j \in K$ **do**
4:       **for all** $s_i \in k_j$ **do**
5:          **for all** $k_\ell \in K$, $\ell \neq j$, **do**
6:             Put s_i (from k_j) in k_ℓ;
7:             Let S' be a new solution with its corresponding objective function e^ℓ_{nb};
8:             **if** $(e^\ell_{nb} < e_{nb})$ **then**
9:                $S \leftarrow S'$ and $ok =$`true`;
            else
10:                Exit with the best solution of the neighborhood S^*, if a local stopping condition is met.
11:             **end if**
12:          **end for**
13:       **end for**
14:    **end for**
15: **end while**

---

strategy and a *special k-opt procedure*. The exchanging is employed between some services (vertices) of the current clusters (forming the set S), whereas the k-opt procedure considers several moves between clusters.

Algorithm 1 describes the main steps of the intensification procedure which is used for improving the quality of the solutions at hand. Indeed, the input of the algorithm is the solution provided by the greedy procedure (with its objective value e_{nb} that measures the number of edges added to complete the solution) described in Section 3.1. Second, the main loop `while` (line 2) is used for stopping the search whenever the local search is able to provide a new feasible solution with a better value (in this case, a boolean, namely ok, is introduced in order to stop the loop). Of course, the stopping criteria is also consolidated with a runtime limit; that is, a runtime that the loop can take for exploring a series of neighborhoods (this parameter is experimentally defined). Third, an exchanging step is considered in lines from 3 to 6: in this case, two cases can be considered:

1. A vertex belonging to a cluster can be exchanged with another vertex that belongs to another one; the solution is updated when the new solution improves the quality of the best solution of the neighborhood.
2. When the previous step fails to provide a better solution, then a 2-opt procedure is introduced as follows: (a) exchange two vertices from a selected cluster and move them to other clusters (when it is possible) and, (b) recombine the clusters for providing a k-CmBCP's feasible solution.

Fourth and last, k-CmBCP's feasible solution is updated (line 9) if the current objective function improves the quality of the solution in the visited neighborhood.

3.3   Applying a Diversification Step

Now we are going to show how to simulate the diversification principle for searching a series of new solutions. Indeed, the intensification search can improve the solutions when it explores a series of neighborhoods, or when these neighborhoods are very close. Moreover, changing the space search induces, in some cases, the exploration of new neighborhoods and so, some new and better solutions may emerge. Herein, an adaptive neighborhood search is introduced, which is based upon *degrading strategy* and *re-optimization strategy* (as used in Hifi and Michrafy [5]). After applying both strategies, the intensification step (cf. Algorithm 1) is recalled in order to improve the quality of each solution at hand.

Now we are going to show how to combine both strategies for searching a series of new solutions. Let F be the current feasible solution obtained at the first step (or at internal of the iterative search). Let α be a constant, such that $\alpha \in [0, 100]$, denoting the percentage of free vertices of the set S, i.e., some vertices are free according to the current solution F . Then, the diversification procedure can be considered as an alternate approach which is performed by calling the following two procedures:

- *Degrading procedure*: Consider $F^{(\alpha)}$ as the free vertices' set associated to the current solution F, where $\alpha\%$ of vertices belonging to S are unassigned. Let $\overline{S}^{(\alpha)}$ be the complementary set of $S^{(\alpha)}$ according to the set of services S.
- *Re-optimization procedure*: Complete the current partial solution by using Johnson's algorithm on the subset $\overline{S}^{(\alpha)}$.

Note that, on the one hand, both degrading and re-optimizing procedures are combined in order to produce a series of partial feasible solutions of k-CmBCP and their improved ones. On the other hand, at each time, one can re-call Algorithm 1 in order to improve the quality of the solutions.

The main steps of such a procedure is detailed in Algorithm 2; that is, a process trying to simulate the diversification principle when varying both degrading and re-optimizing procedures. First, Algorithm 2 starts by setting the initial solution equals to the solution obtained by Algorithm 1 (of course, the provided solution can also reached at each internal search). The main loop (line 2: it is used to stop the search when the stopping criteria is performed; herein, a maximum number of iterations is considered) describes the principe of the diversification procedure when combining both degrading and re-optimizing strategies. In this case, it always starts with the best solution obtained up to now (initially, the solution is that reached by Algorithm 1). Next, the degrading procedure is called (line 6) in a local loop for randomly removing $\alpha\%$ of the vertices from the current solution and trying to improve the quality of the loc al solution by the re-optimization procedure (line 7). The local loop serves to search the best solution on a series of neighborhoods an so, a small number of degrading / re-optimizing are introduced (herein, ten degrading / removing calls are considered). At each step of the local loop, the new provided solution is improved by calling

---

Algorithm 2. An adaptive neighborhood search

Require: A starting feasible solution S_{start} of value $V(S_{start})$
Ensure: An approximate solution S_{best} of value $V(S_{best})$

1: $S_{best} = S_{start}$;
2: **while** (a stopping condition is not met) **do**
3:     $S' = S_{best}$;
4:     **while** (a local stopping condition is not met) **do**
5:         $S_{local} = S'$;
6:         Apply the **Removing strategy** on the current solution S_{local};
7:         Call the **Re-optimizing strategy** for completing the current solution;
8:         Let S' be the new obtained solution;
9:         Call Algorithm 1 for improving the current solution S';
10:         **If** $(V(S' < V(S_{best})))$, **then** $S_{best} = S'$;
11:     **end while**
12: **end while**

Algorithm 1 (lines 8 and 9). Finally, the process is repeated with the best solution found to far, where the algorithm stops with the best k-CmBCP's approximate feasible solution when the maximum number of iterations is performed.

4   Computational Results

This section investigates the effectiveness of the proposed Adaptive Neighborhood Search (noted ANS) on some instances extracted from Gualandi *et al.* [4]. Preliminary results obtained by the proposed ANS are compared to the best results available in the literature and to those reached by the ILPM (cf. Section 2) when solved with the Cplex solver [8] version 12.5 (limited to one hour and with the default settings). The proposed ANS was coded in C++ and tested on Pentium Core Duo 2.9 Ghz

Table 1 describes the main characteristics of the set of instances tested. Column 1 of the table indicates the instance label. Columns from 2 to 4 display cardinalities of both sets S (services) and T (costumers) and the number of clustering k needed. Finally, column 5 tallies the density (d) of the graph associated to each instance. Note also that these instances represent six group of instances, where each group is composed of three available instances and it depends on the variation of the number of services ($|S|$), costumers ($|T|$), the clusters (k) and the density d of its corresponding graph.

4.1   Parameter Settings

In the preliminary results, two main parameters should be taken into account by ANS (cf., Algorithm 2): the stopping criteria and the percentage of the removed vertices belonging to the current solution. Of course, in order to maintain the diversity of the solutions, we set the local stopping condition to twenty, which

Table 1. Description of the instances' characteristics

Inst.	S	T	k	d
I1.1	50	50	5	0.3
I1.2	50	50	5	0.5
I1.3	50	50	5	0.7
I2.1	50	50	10	0.3
I2.2	50	50	10	0.5
I2.3	50	50	10	0.7
I3.1	80	80	5	0.3
I3.2	80	80	5	0.5
I3.3	80	80	5	0.7
I4.1	80	80	10	0.3
I4.2	80	80	10	0.5
I4.3	80	80	10	0.7
I5.1	100	100	5	0.3
I5.2	100	100	5	0.5
I5.3	100	100	5	0.7
I6.1	100	100	10	0.3
I6.2	100	100	10	0.5
I6.3	100	100	10	0.7

means that Algorithm 2 exists either by an improved solution reached before attaining the maximum number of iterations fixed or by the best solution reached in the neighborhood.

Because the improved procedure (cf., Algorithm 1) works when each solution is submitted to degradation and re-optimization, then the maximum number of global iterations was fixed to fifty. Finally, three values for the parameters α were considered: $\alpha \in \{2\%; 5\%; 10\%\}$. On the one hand, limited computational results showed that the algorithm with the value $\alpha = 2\%$ requires more runtime for improving some instances of the literature. On the other hand, when the algorithm is run with $\alpha = 10\%$, the obtained results becomes, in some cases, worse than those published in the literature. Only for $\alpha = 5$ (or closest to the aforementioned value) the algorithm seems to get more interesting results. Hence, the choice with $\alpha = 5$ is adopted for the rest of this paper.

Table 2 displays the variation of Av_Sol representing the average value of all solutions realized by ANS over all treated instances by fixing the same average runtime (1000 seconds, on average). It shows the variation of Av_Sol denoting the average value of solutions reached by ANS over all treated instances: the first column tallies the average results with $\alpha = 2\%$ and the third column displays the average results for $\alpha = 5\%$. One can observe, globally for the same runtime, the average quality of the solutions realized is better when fixing α to 5%.

Table 2. Effect of α on the quality of the final solutions

#Inst	Variation of α	
	2%	5%
I1.1	1319	1319
I1.2	1078	1072
I1.3	672	671
I2.1	938	938
I2.2	876	875,4
I2.3	577	575
I3.1	3820.2	3815
I3.2	2862	2862
I3.3	1769.2	1768.4
I4.1	3202	3202
I4.2	2571	2571
I4.3	1618.4	1618
I5.1	6248	6248
I5.2	4655.6	4650
I5.3	2842	2842
I6.1	4298	4298
I6.2	4298.4	5427.2
I6.3	2644	2644
Av_Sol	2644.4	2633.11

4.2 Effect of the Degrading and Re-optimizing Procedures

This section evaluates the effect of the proposed method based upon degrading and re-optimizing strategies. Recall that the method works as follows. First, it deteriorates the starting / current solution by removing $\alpha\%$ of the variables fixed to one. Second, it re-optimizes the partial solution reached by using a greedy procedure and improving it using an intensification search. Third and last, it re-iterates the same resolution on the last solution until either no better solution is obtained or a maximum number of iterations is performed.

Table 3 shows the results realized by ANS and those extracted from Gualandi *et al.* [4] (noted GMM) and the solutions given by the Cplex solver 12.5 (of course, because the Cplex is tailored for solving the problems to optimality, then the runtime limit was extended to one hour). Column 1 represents the instance label. Column 2 displays the solution value reached by the Cplex solver and column 3 tallies Gualandi *et al.*s's solution values (representing the best solutions of the literature). Columns from 4 to 8 display the five solutions provided by ANS representing the five trials. Finally, columns 9 and 10 show the average solution values over the five trials and the best solutions reached by ANS over these five trials.

Table 3. Performance of ANS vs Cplex and GMM algorithm

#Inst	Cplex V_{Cplex}	GMM V_{GMM}	ANS's solution values					Av_Sol	Best
			1	2	3	4	5		
I1.1	1423	1321	1319	1319	1319	1319	1319	1319	1319
I1.2	1091	1072	1072	1072	1072	1072	1072	1072	1072
I1.3	676	672	671	671	671	671	671	671	671
I2.1	1135	938	938	938	938	938	938	938	938
I2.2	930	876	875	876	876	875	875	875.4	875
I2.3	587	577	575	575	575	575	575	575	575
I3.1	4166	3819	3815	3815	3815	3815	3815	3815	3815
I3.2	2931	2862	2862	2862	2862	2862	2862	2862	2862
I3.3	1775	1769	1768	1769	1768	1768	1769	1768.4	1768
I4.1	3735	3202	3202	3202	3202	3202	3202	3202	3202
I4.2	2675	2571	2571	2571	2571	2571	2571	2571	2571
I4.3	1634	1618	1618	1618	1618	1618	1618	1618	1618
I5.1	6645	6248	6248	6248	6248	6248	6248	6248	6248
I5.2	4716	4658	4650	4650	4650	4650	4650	4650	4650
I5.3	2854	2842	2842	2842	2842	2842	2842	2842	2842
I6.1	6119	4298	4298	4298	4298	4298	4298	4298	4298
I6.2	9111	5428	5427	5428	5427	5427	5427	5427.2	5427
I6.3	2704	2644	2644	2644	2644	2644	2644	2644	2644
Average	3050.39	2634.17	2633.06	2633.22	2633.11	2633.06	2633.11	2633.11	2633.06

The analysis of Table 3 follows.

1. First, one can observe the inferiority of Cplex solver since it is not able to match the best solutions of the literature.
2. Second, the best method of the literature (GMM) realizes 10 best solutions out of 18, representing a percentage of 55.56% of the best solutions. In fact, GMM matches all the ten instances, but there is no value better than the new solutions produced by ANS (cf., the last column of Table 3).
3. Third, among the five trials realized by ANS, one can observe that the average values (see the last line of Table 3) are better than GMM's average value. Indeed, ANS is able to realize an average value varying from 2633.22 (trial 2) and 2633.06 (trials 1 and 4) whereas GMM realizes an average value of 2634.17.
4. Fourth and last, regarding the best solutions realized by ANS (cf., column 10), one can observe that ANS is able to reach eight new solution values, which represents a percentage of more than 44% of the tested instances.

5   Conclusion

In this paper, an adaptive neighborhood search was proposed for approximately solving the k-clustering minimum bi-clique completion problem. The method is based upon degrading and re-optimizing strategies. The first strategy is used in order to diversify the search process and the second one is employed for repairing the configuration at hand. Both strategies were completed by an intensification procedure; that is, a neighboring search which tries to improve the quality of a series of provided solutions. Computational results showed that the proposed algorithm performed better than both Cplex solver and more recent method available in the literature by yielding high-quality solutions: it improved most than 44% of the best solutions of the literature.

References

1. Fahle, T.: Simple and Fast: Improving a Branch-And-Bound Algorithm for Maximum Clique. In: Möhring, R.H., Raman, R. (eds.) ESA 2002. LNCS, vol. 2461, pp. 485–498. Springer, Heidelberg (2002)
2. Faure, N., Chrétienne, P., Gourdin, E., Sourd, F.: Biclique Completion Problems for Multicast Network Design. Discrete Optimization 4, 360–377 (2007)
3. Gualandi, S.: *k*-Clustering Minimum Biclique Completion via a Hybrid CP and SDP Approach. In: van Hoeve, W.-J., Hooker, J.N. (eds.) CPAIOR 2009. LNCS, vol. 5547, pp. 87–101. Springer, Heidelberg (2009)
4. Gualandi, S., Maffioli, F., Magni, C.: A Branch-and-Price Approach to k-Clustering Minimum Biclique Completion Problem. International Transactions in Operational Research 20, 101–117 (2013)
5. Hifi, M., Michrafy, M.: A Reactive Local Search-Based Algorithm for the Disjunctively Knapsack Problem. Journal of the Operational Research Society 57, 718–726 (2006)
6. Johnson, D.S.: Approximation Algorithms for Combinatorial Problems. Journal of Computer and System Sciences 9, 256–278 (1974)
7. Shaw, P.: Using Constraint Programming and Local Search Methods to Solve Vehicle Routing Problems. In: Maher, M.J., Puget, J.-F. (eds.) CP 1998. LNCS, vol. 1520, pp. 417–431. Springer, Heidelberg (1998)
8. Cplex Solver: IBM, ILOG, http://www.ilog.com/products/cplex/

An Integer Programming Model for Branching Cable Layouts in Offshore Wind Farms

Arne Klein[1*], Dag Haugland[1], Joanna Bauer[1], and Mario Mommer[2]

[1] Department of Informatics, University of Bergen, Bergen, Norway
{arne.klein,dag.haugland,joanna.bauer}@ii.uib.no
[2] Modellierung und Systemoptimierung Mommer GmbH,
Heidelberg, Germany
msm@msmommer.de

Abstract. An integer programming model for minimizing the cabling costs of offshore wind farms which allows for branching of the cables is developed. Model features include upper bounds on the number of cable branches made at any wind turbine, upper bounds on the cable loads, and exclusion of crossing cable segments. The resulting model resembles the capacitated minimum spanning tree problem, with the addition of degree and planarity constraints. Numerical experiments with realistic wind farm data indicate that the benefit from branching is small when using only one cable type, but is up to 13% if allowing for two different cable types.

Keywords: offshore wind farms, cable routes, integer programming, constrained minimum spanning tree.

1 Introduction

Offshore wind energy is becoming an increasingly more important energy source. Up to now, the main development is taking place in Northern Europe, with 6562 MW out of the global installed capacity of 7045 MW installed in Europe at the end of 2013 [6]. The by far most important countries in this respect are the United Kingdom and Denmark, with PR China, Belgium, Germany, Netherlands and Sweden following. The installed capacity in other countries is negligible.

Starting with a yearly annual installed capacity of only 4 MW in 2000, the industry has been steadily growing to 1567 MW being newly installed in 2013 [4]. With an estimated total installed offshore wind capacity of 23500 MW in 2020, which is almost four times the capacity installed at the end of 2013, the industry is expected to continue its quick growth [5].

During the planning and construction phase of an offshore wind farm, the decision on how to choose the cabling routes has a significant influence on the total cost of the cabling, as the cable as well as the trenching in the seabed cost per meter are considerably higher offshore than onshore. There are usually

---

[*] Corresponding author.

© Springer International Publishing Switzerland 2015
H.A. Le Thi et al. (eds.), *Model. Comput. & Optim. in Inf. Syst. & Manage. Sci.*,
Advances in Intelligent Systems and Computing 359, DOI: 10.1007/978-3-319-18161-5_3

one or more substations, and each turbine has to be connected to at least one of these. However, multiple turbines can be connected in series. The maximum number of turbines on one series circuit is determined by the cable type, and the maximum power which can be transported by it. Usual numbers of turbines per cable are between 4 and 8.

This opens an optimization problem of finding cable routes between turbines and substations with minimum total cable length. If taking into account different cable types with respectively different power capacities, the objective changes to minimizing cable cost.

Optimization of cable routes in offshore wind farms was recently addressed by Bauer and Lysgaard [1], who suggested a model with hop-indexed variables, resembling a planar open vehicle routing problem. Reflecting the fact that cable lines are not allowed to cross each other, planarity constraints apply to the model. For a given cable capacity, as well as fixed turbine and substation locations, the objective of the model in [1] is to find the cable routes of minimum total length.

An important assumption of [1] is that the turbines are connected to substations along paths. That is, with the exceptions of the turbine closest to and most remotely from the substation, all turbines have a direct link with exactly two other turbines. In practice, however, it is in some cases possible to branch the power cables at the turbine locations without significant additional effort or cost, which opens the possibility of a further reduction of the total required cable length. This has been done for example in the Walney 1 offshore wind farm [2], which is located on the Northwestern English coast. The branching option is not captured in [1].

We present a new optimization model incorporating all features of [1]. In addition, our model allows for branching of the power cables at the wind turbine locations. Reflecting practical limitations, our model accepts an upper bound on the number of branches that can be made at any turbine location, which is dependent on the cable setup and connection possibilities at a turbine. We will refer to this bound as the branching capacity. We assume that no additional cost is connected to branching within the branching capacity. Following [1], there is an upper bound, referred to as the cable capacity, on the number of turbines to be connected by one cable (to one substation), and no two cable lines may cross because of the applied cable trenching methods.

The problem, which is presented in detail in section 2, can be defined in terms of a graph where the node set represents turbines and substations, in addition to an imaginary node referred to as the root, representing the electrical grid to which the turbines will supply power. We use the root node for setting the problem into context with existing literature, but not in our model formulation, as we do not optimize the grid connection. The edges represent possible connections between turbines, between turbines and substations, and between substations and the grid. Each node but the root is associated with a point in the plane, and each edge but those incident to the root have a weight equal to the Euclidean distance between its end nodes (the weights of all edges incident to the root are negative with sufficiently large absolute value). In this graph, we are asking for a spanning tree of minimum weight, satisfying the following constraints:

- The number of nodes in each subtree below the root does not exceed the cable capacity,
- The degree of each turbine node is no more than the branching capacity plus one,
- When embedding the edges as straight lines between its end nodes, intersections occur only between edges incident to the same node.

The problem under study combines the features of several well-studied graph optimization problems. In particular, this applies to the *capacitated* minimum spanning tree problem [10], where bounds on the subtree sizes are introduced. In the special case where the bounds are equal for all nodes, which we assume, the problem version is referred to as the *unitary demand* version. The *degree-constrained* minimum spanning tree problem [9] addresses the issue of branching capacities. Adding the degree constraint to the minimum spanning tree problem renders the problem NP-hard [11], at least for branching capacity no larger than 3, proving that also our problem is NP-hard.

The remainder of this text is organized as follows: In the next section, we develop an integer programming model based on a set of variables suggested by Gouveia [8]. After introducing model and notation, we continue in section 3 with a presentation of the numerical results obtained from an implementation of our model. We give the optimal solutions for different cable capacities, and by comparing to results from the model allowing only linear cabling [1], we determine cost savings obtainable by allowing branches.

2   An Integer Programming Model for Minimizing Cable Lengths

In this section, we develop an integer programming model for the cabling problem outlined in Section 1. We start with a description of the model parameters and variables and their meaning in the offshore wind farm context of the problem. We also relate them to their respective equivalents in the capacitated minimum spanning tree problem. After the introduction of the variables we continue using the wind energy context in further discussions.

Consider a graph with nodes set $V = V_c \cup V_d$, where V_c represents a given set of wind turbines, and V_d represents power substations. The edge set $E \subseteq V^2$ of the graph represents the possible connections between a turbine and a substation or another turbine. The corresponding arc set is denoted $A_E = \{(i,j) : \{i,j\} \in E\}$. Each edge and arc has an associated cost $c_{ij} \forall (i,j) \in A_E$. We assume that the edge cost is proportional to the Euclidean distance between the locations of the end nodes of the edge, which implies that the costs are symmetric, i.e., $c_{ij} = c_{ji}$. However, validity of the model below does not depend on this assumption.

The maximum cable capacity $C \in \mathbb{N}^+$ is the maximum number of turbines which can be connected by the chosen cable type. It is dependent on the type of

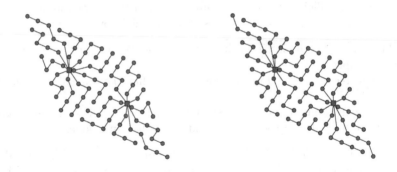

Fig. 1. Sheringham Shoal layout for C=5; left WM1, right WMB1; circles are turbine locations, squares are substation locations

cable and determined by a combination of the power output of a single turbine, the current carrying capacity of the cable, as well as different kinds of power losses which occur in cables.

The maximum number of cable branches at a turbine location, including all incoming and outgoing cables, is defined by the branching capacity $m \in \mathbb{N}^+$.

The set of crossing routes $\chi \subset E^2$ is defined such that $\{\{i,j\},\{u,v\}\} \in \chi$ if edges $\{i,j\}$ and $\{u,v\}$ cross each other.

Gavish [7] suggested a single-commodity flow model for the capacitated minimum spanning tree problem, which later was proved to be equally strong as the model by Gouveia [8]. It can be argued that the latter model is more appealing since the number of constraints is smaller, and that it exclusively has binary variables. It is proved [8] that by use of binary variables indexed by arcs $(i,j) \in A_E$ and feasible subtree sizes $h = 1, \ldots, C$, no continuous variables are needed in the formulation. Following the idea of [8], the integer programming model below contains only binary decision variables.

As our main focus is to analyze what cost reductions can be obtained when branched cable routes are allowed, rather than finding the strongest possible model formulation, we do not incorporate valid inequalities such as those proposed in [8]. For the same reason, we neither make any attempt to integrate in our model any of the more recent contributions to strong formulations for the capacitated minimum spanning tree problem, which have emerged since [8]. Interested readers are referred to e.g. the thesis by Ruiz y Ruiz [12].

Define the decision variable $x_{ij}^h \in \{0,1\}$ $\forall (i,j) \in A_E, h = 1, \ldots, C$ such that it takes the value 1 if the solution contains the edge $\{i,j\}$, with j closer than i to some substation in the tree, and if h turbines (including i) are connected to the substation via i. If not all conditions are met, $x_{ij}^h = 0$. That is, $x_{ij}^h = 1$ indicates that (i,j) is an arc in the spanning tree pointing towards the root, and the subtree rooted at i contains h nodes. Note that in the original capacitated minimum spanning tree formulation [8], the decision variables are defined such that arcs point from the root towards the leaves. Our definition is however adopted for the purpose of consistency with [1].

$$\min \quad \sum_{(i,j)\in A_E} \sum_{h=1}^{C} c_{ij} x_{ij}^{h} \tag{1}$$

$$\text{s.t.} \quad \sum_{(i,j)\in A_E} \sum_{h=1}^{C} x_{ij}^{h} = 1 \qquad \forall\, i \in V_c \tag{2}$$

$$\sum_{(i,j)\in A_E} \sum_{h=1}^{C-1} h x_{ij}^{h} - \sum_{(j,k)\in A_E} \sum_{h=1}^{C} h x_{jk}^{h} = -1 \qquad \forall\, j \in V_c \tag{3}$$

$$\sum_{(i,j)\in A_E} \sum_{h=1}^{C} x_{ij}^{h} \le m \qquad \forall\, j \in V_c \tag{4}$$

$$\sum_{h=1}^{C} \left(x_{ij}^{h} + x_{ji}^{h} + x_{uv}^{h} + x_{vu}^{h} \right) \le 1 \qquad \forall\, \{\{i,j\},\{u,v\}\} \in \chi \tag{5}$$

$$x_{ij}^{h} \in \{0,1\} \quad \forall\, (i,j) \in A_E,\, h = 1,\dots,C \tag{6}$$

$$x_{ij}^{C} = 0 \qquad \forall\, (i,j) \in A_E \cap \{V \times V_c\} \tag{7}$$

We minimize the total cost or distance over all used routes in the objective function (1). Constraint (2) assures that each turbine has exactly one outgoing cable directed towards some substation. By equation (3), the load of the cable outgoing from turbine j equals the sum of the cables entering j, plus the load 1 of turbine j. The cable load is defined as the number of turbines that connect to some substation via the cable.

A maximum number of branches per turbine location is defined in (4). The planarity constraints are defined in (5) and assure that no cables cross each other.

As it is also of interest to investigate optimal cable routes with two different cable types available, each of these with different cost and capacity, we introduce a second model. Assume that if the load of any cable $(i,j) \in A_E$ is no more than $Q \in \mathbb{N}^+$, where $Q < C$, the connection cost is $q_{ij}\forall(i,j) \in A_E$. A reasonable assumption is that $q_{ij} < c_{ij}$, but validity of our model does not require this to hold.

The resulting formulation of the alternative objective function is given by:

$$\min \quad \sum_{(i,j)\in A_E} \left[\sum_{h=1}^{Q} q_{ij} x_{ij}^{h} + \sum_{h=Q+1}^{C} c_{ij} x_{ij}^{h} \right] \tag{8}$$

s.t. (2) – (7). In the computational experiments reported in the next section, we also apply this objective function to the model from [1] for comparing our results.

3 Numerical Experiments

The models suggested in Section 2, as well as the model from [1], are implemented in Python using the Python CPLEX library with CPLEX 12.6.1.0 as the integer programming solver. Default options with multithreading disabled are used.

The non-crossing constraints of the models are implemented via a lazy constraint callback, which only adds the corresponding non-crossing constraint if the solution contains the respectively crossing routes. This is a necessity resulting from the large number of constraints, increasing with $O(|V|^4)$.

All computational experiments are performed on an i7-4600U CPU with 8GB of RAM.

We choose four different real wind farm layouts as the data base for our numerical experiments. In addition to Barrow, Sheringham Shoal and Walney 1, which have also been used in [1], we also use the data from Walney 2 [3]. With the turbine and substation locations of the respective farm layouts, the distance between the turbines are computed and subsequently taken as the edge cost c_{ij}. We allow all possible connections in all wind farms, i.e. $E = V^2$. The branching capacity is set to $m = 3$ in all tests.

Table 1. General information on test case wind farms

Wind farm	Number of turbines	Number of substations
Barrow	30	1
Sheringham Shoal	88	2
Walney 1	51	1
Walney 2	51	1

In the following, we refer to the original model formulation without branching from [1] as WM1, and to our model formulation from equation (1) – (7), which allows for branching, as WMB1. Both of these models use the distances between turbines c_{ij} as the cost in the objective function.

The computational results in Table 2 are computed within 15 minutes each. Values for Sheringham Shoal for capacities $C \geq 6$ are not reported, as no optimal solution is found after 1.5 hours of calculation for model WMB1. Computations for $C \geq 8$ are also not possible for a part of the other wind farms within one hour and less than 8 GB of memory consumption and thus not included.

The optimality gap in the root node g_{WM1} and g_{WMB1} does not follow a systematic pattern, while the number of processed nodes n_{WM1} and n_{WM1} is generally higher for the WMB1 model which includes branching.

Table 2. Optimal values for models with one cable type

C	WM1	g_{WM1}	n_{WM1}	Walney 1 WMB1	g_{WMB1}	n_{WMB1}	ρ
4	47802	1.67	1248	47654	59.84	4757	0.31
5	43539	0.04	1	43421	8.24	2676	0.27
6	41587	26.39	483	41420	0.85	554	0.40
7	40789	19.27	1221	40620	62.76	963566	0.42

C	WM1	g_{WM1}	n_{WM1}	Walney 2 WMB1	g_{WMB1}	n_{WMB1}	ρ
4	62233	0.86	75	62061	62.04	1418	0.28
5	56572	4.74	8620	56258	3.32	69440	0.56
6	52228	2.15	327	51943	1.70	1887	0.55
7	49788	4.84	666	49568	69.89	422671	0.44

C	WM1	g_{WM1}	n_{WM1}	Barrow WMB1	g_{WMB1}	n_{WMB1}	ρ
4	23568	1.00	62	23568	5.90	406	0.00
5	20739	6.26	114	20738	3.85	326	0.00
6	18375	0.00	1	18374	0.00	1	0.01
7	17781	4.67	564	17781	6.76	7116	0.00

C	WM1	g_{WM1}	n_{WM1}	Sheringham Shoal WMB1	g_{WMB1}	n_{WMB1}	ρ
4	69222	0.00	1	68937	32.19	1087	0.41
5	64828	60.50	6530	64365	60.93	211005	0.72
6	62031	62.18	5585				
7	60667	17.42	8152				

C is the cable capacity. The columns WM1 and WMB1 are the optimal values given in m. The relative improvement ρ of the objective value is calculated by $\rho = \frac{WMB1}{WM1} - 1$ with the WM1 and WMB1 values of the respective row and given in %. g_{WM1} and g_{WMB1} are the optimality gaps at the root nodes of the respective model, given in %, and n_{WM1} and n_{WMB1} the corresponding number of nodes processed for the solution.

For the investigated wind farms and cable capacities the relative savings in cable length are below 1% in all cases. It is thus only marginally useful to apply branching in a wind farm if $E = V^2$, and there is only one cable type available.

The Barrow offshore wind farm turns out to be particularly unsuitable for branching. The reason for this is that the turbines are located in several rows, with a significant larger spacing tangential to this row. This favors connecting the turbines sequentially without branching into another row.

For the investigation of the models allowing for two different cable types, we set the cost of the cable with lesser capacity q_{ij} to the distance between the turbines, and increase the cost for the cable type with larger capacity by a factor f, such that $c_{ij} = fq_{ij}$.

Table 3. Optimal values for models with two cable type

f	C	Q	Walney 1			Walney 2			Sheringham Shoal		
			WM2	WMB2	ρ	WM2	WMB2	ρ	WM2	WMB2	ρ
1.5	5	2	57055	54958	3.82	74547	72863	2.31	83598	80264	4.15
1.5	5	3	52416	52108	0.59	68770	68165	0.89	75882	75164	0.96
1.5	5	4	47040	46874	0.35	61481	61309	0.28	69182	68897	0.41
1.5	6	2	55460	52027	6.60	70421	67578	4.21			
1.5	6	3	51434	49887	3.10	65874	64436	2.23			
1.5	6	4	46846	46641	0.44	60483	60312	0.28			
1.5	6	5	43400	43282	0.27	55875	55663	0.38			
1.7	5	2	62372	59028	5.67	81520	78184	4.27	89857	84685	6.11
1.7	5	3	54178	54095	0.15	72018	71354	0.93	77896	77591	0.39
1.7	5	4	47410	47243	0.35	62129	61957	0.28	69222	68937	0.41
1.7	6	2	60754	55589	9.29	77197	72839	5.98			
1.7	6	3	53831	52660	2.22	70668	68644	2.95			
1.7	6	4	47410	47198	0.45	61600	61417	0.30			
1.7	6	5	43491	43374	0.27	56341	56080	0.47			
1.7	7	2	60205	52873	13.87	74572	[1]68828	8.35			
1.7	7	3	53647	49354	8.70	53646	49354	8.70	77391	[3]72486	6.51
1.7	7	4	47410	47066	0.73	61409	61001	0.67			
1.7	7	5	43491	43374	0.27	56276	[2]56075	0.36			
1.7	7	6	41587	41420	0.40	52170	51943	0.44			

f is the cost multiplier and C and Q are the cable capacities. The columns WM2 and WMB2 are the optimal values given in m. The relative improvement ρ is calculated by $\rho = \frac{\text{WMB2}}{\text{WM2}} - 1$ with the WM2 and WMB2 values of the respective row and wind farm and given in %. The values with footnotes were not solved to optimality, but with the following optimality gaps: 1) 1.85%, 2) 1.40%, 3) 0.33%

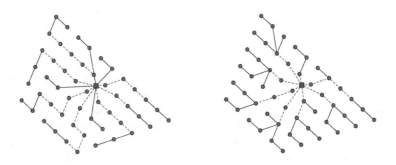

Fig. 2. Walney 1 layout for $C = 7$, $Q = 2$, $f = 1.7$; left WM2, right WMB2; circles are turbine locations, squares are substation locations, dashed lines are cables with higher capacity

In practice, the actual choice of f depends on the wind farm in question, as well as the cable capacities C and Q. It is indicated in [1] that a factor of about $f = 1.7$ is suitable for $Q = 8$ and $C = 5$. We choose this value as well as a slightly lower value of $f = 1.5$ for our numerical experiments, to also investigate the influence of the relative cost factor f.

All numerical results in Table 3 are computed within one hour of computation time each. The computations for WMB2 are more expensive than for WM2, and are thus the limiting factor. Most calculations with WM2 were completed within of less than 10 seconds. The computations for Sheringham Shoal are limited to $C =$ because of the computation time limit, except for the sample with $C = 7, Q = 3$. The optimality gap at the root node and the number of processed nodes are not included for the sake of brevity, as they are similar to the results in Table 2.

The numerical results show that relative savings increase in average with an increasing difference $d = C - Q$ in cable capacities. Only for $d \geq 2$ and $Q \leq 3$ relative savings of more than 1% are achieved by applying branching. The highest computed saving of 13.87% by using branching in the two cable type formulation makes a significant difference in the total cabling cost (see fig. 2).

A selection of Q, C values is computed for $f = 1.5$ to investigate the influence of this parameter on the relative improvement. There is no systematic difference observable between $f = 1.5$ and $f = 1.7$. The optimal values are in the same order of magnitude, but fluctuating in both directions.

4   Conclusion and Further Work

The results in this article show that it is of advantage to consider branching cable layouts for offshore wind farms utilizing two or possibly more different kinds of cable types. As the relative cost improvements are below 1% for farms with only one cable type, branching is not a necessity in these cases.

Possible further work on this model includes a more detailed cost modeling of the cost parameters c_{ij} and q_{ij}. This can take into account the physical location of cables, as well as length, and non-straight routes, for example because of seabed topography. In addition, the investigation of layouts with forbidden cable routes, i.e. $E \subset V^2$, is of interest for the modeling of real offshore wind farms.

References

1. Bauer, J., Lysgaard, J.: The Offshore Wind Farm Array Cable Layout Problem – A Planar Open Vehicle Routing Problem. Journal of the Operational Research Society (to appear)
2. Dong Energy: Walney 1 Offshore Wind Farm,
   http://www.seafish.org/media/527826/walney_kingfisher_flyer_lres.pdf
3. Dong Energy: Walney 2 Offshore Wind Farm,
   http://www.seafish.org/media/527868/walney_2_lres.pdf
4. EWEA: The European offshore wind industry - key trends and statistics 2013,
   http://www.ewea.org/stats/eu-offshore-2013

5. EWEA: Wind energy scenarios for 2020, `http://www.ewea.org/publications/reports/wind-energy-scenarios-for-2020/`
6. Global Wind Energy Council (GWEC): Global Wind Report 2013, `http://www.gwec.org`
7. Gavish, B.: Topological Design of Centralized Computer Networks: Formulations and Algorithms. Networks 12, 355–377 (1982)
8. Gouveia, L.: A $2n$ Constraint Formulation for the Capacitated Minimal Spanning Tree Problem. Operations Research 43(1), 130–141 (1995)
9. Narula, S.C., Ho, C.A.: Degree-constrained Minimum Spanning Tree. Computers & Operational Research 7, 239–249 (1980)
10. Papadimitriou, C.H.: The Complexity of the Capacitated Tree Problem. Networks 8, 217–230 (1978)
11. Papadimitriou, C.H., Vazirani, U.V.: On two geometric problems related to the travelling salesman problem. Journal of Algorithms 5, 231–246 (1984)
12. Ruiz y Ruiz, H.E.: The Capacitated Minimum Spanning Tree Problem. PhD Thesis, Department of Statistics and Operations Research, Universitat Politiécnica de Catalunya, Barcelona, Spain (2012)

Belief Propagation for MiniMax Weight Matching

Mindi Yuan, Shen Li, Wei Shen, and Yannis Pavlidis

Walmart Labs and University of Illinois at Urbana-Champaign, Champaign, IL, USA
{myuan,wshen,yannis}@walmartlabs.com,shenli3@illinois.edu

Abstract. In this paper, we investigate the MiniMax Weight Matching (MMWM) problem in a bipartite graph. We develop a belief propagation algorithm of message complexity $O(n^3)$, which can find such a matching in a graph of size n upon uniqueness of the optimum. The algorithm is one of a very few fully polynomial time solutions where belief propagation algorithms are proved correct. Since the algorithm can be distributed, the convergence time then drops to n unit time if each edge can simultaneously compute and pass the messages within one unit time at each iteration.

1 Introduction

Belief propagation (BP) algorithms (or message passing algorithms) on various graphical models (GM) have been used in areas like modern coding theory, artificial intelligence, statistics, and neural networks. The two basic versions of BP, sum-product algorithms and max-product algorithms [8] are developed corresponding to the two main problems in probabilistic inference: evaluating the marginal distribution and maximum *a posteriori* (MAP). They are known to converge to the correct solutions if the GM's are cycle free. For single-cycle graphs, the correctness and convergence of BP are investigated in [1] and [10], while they are still open problems for arbitrary GM's.

However, even for GM's with cycles, the belief propagation algorithms are observed to perform surprisingly well in many cases, some of which are with rigorous proof of optimality and convergence. For example, for the maximum weight matching (MWM) problem in a bipartite graph, Bayati *et al.* [2] formulated a max-product algorithm by calculating the MAP probability on a well defined GM, which encodes the objective and constraints of the optimization problem. Shortly after, Bayati and Y. Cheng [3] independently simplified the max-product algorithm to obtain two essentially same algorithms, which reduced the message complexity by order of two. Moreover, in [12] and [11], Yuan et. al. proposed message passing algorithms for a constrained assignment problem and the generalized assignment problem respectively.

Although some belief propagation algorithms can converge to the optimum in finite iterations, the running time of them is actually pseudo-polynomial even if the problem itself has other fully polynomial time solutions, like the MWM

© Springer International Publishing Switzerland 2015

H.A. Le Thi et al. (eds.), *Model. Comput. & Optim. in Inf. Syst. & Manage. Sci.*,
Advances in Intelligent Systems and Computing 359, DOI: 10.1007/978-3-319-18161-5_4

mentioned above. The convergence time of BP usually depends on the difference of the optimum and the second best solution. This difference could be exponential on the input size, resulting in pseudo-polynomial running time. In [5], the authors developed a BP algorithm for the minimum cost flow problem. (In fact, maximum weight matching is a special case of the minimum cost flow problem.) They also presented a fully polynomial time randomized approximation scheme for the problem. However, as they said themselves in the paper, the "near optimal" solution is "rather fuzzy".

Our algorithm is therefore one of a very few fully polynomial time solutions where belief propagation algorithms are proved correct. We will derive the iterative belief propagation algorithm and show both its correctness and convergence. There are works on maximum (weight) matching, which is a well studied problem. There are also works on minimax grid matching [9][6]. However, these problems are slightly different from the problem in this paper. According to the best of our knowledge, there are few works directly on the minimax weight matching problem, hence almost no distributed approaches for MMWM, either.

In practice, a story for MMWM can be: One job consists of n tasks to be processed on n computers. In order to minimize the job processing time, one can minimize the maximum processing time of the tasks. For example, for MapReduce [4] jobs, the Reduce jobs usually can not start until all the Map jobs are finished. To minimize the total processing time for Map jobs, we could optimize the scheduling using MMWM. In addition, our algorithm can be implemented on Pregel [7], a system for large-scale graph processing.

The rest of the paper is organized as follows. Section 2 defines and describes the problem. Section 3 derives the belief propagation algorithm. The proof of correctness and convergence for the algorithm is given in section 4. We generalize the algorithm in section 5 and conclude in section 6.

2   Problem Description

Let $G = (T, S, E)$ be a symmetric complete bipartite graph. T and S are the sets of n nodes in the two partitions respectively, i.e., $T = \{T_1, T_2, ..., T_n\}$ and $S = \{S_1, S_2, ..., S_n\}$. E is the set of edges between T and S. That is, $(T_i, S_j) \in E, 1 \le i, j \le n$. Assign a weight w_{ij} to each edge (T_i, S_j).

Definition 1. In a symmetric complete bipartite graph, a minimax weight matching is a perfect matching whose maximum weight is minimized.

Mathematically, if $p = \{p(1), p(2), ..., p(n)\}$ is a permutation of $\{1, 2, ..., n\}$, then the collection of n edges $\{(T_1, S_{p(1)}), (T_2, S_{p(2)}), ..., (T_n, S_{p(n)})\}$ is a perfect matching. Denote both the permutation and the corresponding matching by p. The maximum weight of the matching p, represented by W_p, is

$$W_p = \max_{1 \le i \le n} w_{ip(i)}. \tag{1}$$

The MMWM p^* is the optimal matching defined as

$$p^* = \arg\min_p W_p. \tag{2}$$

Here is an example of MMWM. If the weight matrix is $\mathbf{w} = \begin{pmatrix} 2\ 1\ 4 \\ 4\ 2\ 1 \\ 3\ 4\ 2 \end{pmatrix}$, where $\mathbf{w}(i,j)(1 \leq i,j \leq 3)$ is the weight for edge (T_i, S_j). It is not hard to find that only one MMWM exists in this case, lying at the diagonal, i.e., $p^* = \{(T_1, S_1), (T_2, S_2), (T_3, S_3)\}$ with $W_{p^*} = 2$.

3 Algorithm

In this section, we define the messages, derive the updating rules, and develop the MMWM algorithm.

Before that, we first introduce the concept of a computation tree. Figure 1 shows a 3 by 3 original graph. The associated computation tree is drawn in Figure 2. Note that the tree can be extended arbitrarily in depth, so long as all the relations (who connects to whom) are kept among the nodes.

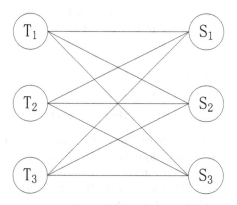

Fig. 1. Original graph

For each edge e on the computation tree, define a function $m(e)$ on the subtree below e *with* e included. Let this function return the maximum weight of the MMWM on that subtree. Note a perfect tree matching is just one case of the general graph matchings. That is, each vertex except some leaves, must attach to one and only one associated edge. Similarly, define $m'(e)$ on the subtree below e but *without* e included. Let this function return the maximum weight of the MMWM on the corresponding subtree. The main idea of the algorithm is to 1) first let every edge separately make its own decision about itself to be chosen or not in the optimal tree matching, and 2) at the end gather all these individual

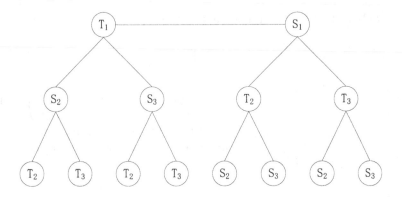

Fig. 2. Computation tree for Figure 1

decisions to form a global optimal matching in the original graph. In order to realize this idea, one intuitive way is to check the difference $m(e) - m'(e)$ for each root edge, like (T_1, S_1) in Figure 2. If it is greater than 0, the maximum weight of MMWM on the tree including e will surpass that without e. We should hence discard e in order to minimize the maximum weight. Otherwise, choose it.

To enable these final decisions for each edge, we should keep passing that difference $(m(e) - m'(e))$ along the computation tree. Then at the root edge e_r, combine the messages passed up from each of its endpoints left (m_l from T_1) and right (m_r from S_1):

$$
\begin{aligned}
& m(e_r) - m'(e_r) \\
&= \max\{m_l(e_r), m_r(e_r)\} \\
& \quad - \max\{m'_l(e_r), m'_r(e_r)\}
\end{aligned}
\tag{3}
$$

which is also known as the *belief* of the root edge. However, according to Eq. (3), if only passing the difference along the tree, like $m_l(e_r) - m'_l(e_r)$ or $m_r(e_r) - m'_r(e_r)$, we are unable to compute $\max\{m_l(e_r), m_r(e_r)\} - \max\{m'_l(e_r), m'_r(e_r)\}$ at the final step. So we should pass at least both the $m(e)$ and $m'(e)$, not just their difference. In the following, these two messages are showed to be enough.

Now consider the updating rules at each node. For an $n \times n$ graph, each node, except the leaves, will have $n - 1$ children in the corresponding computation tree. Refer to Figure 3 for visualization. Mark the edges between the node and all its children as $i_1, i_2, ..., i_p$ and the edge between this node and its father as o.

Let $w(e)$ return the weight of edge e. Then

$$
m(o) = \max\{w(o), \max_{1 \leq m \leq p} m'(i_m)\}
\tag{4}
$$

$$
m'(o) = \min_{1 \leq m \leq p} \{\max\{m(i_m), \max_{n \neq m} m'(i_n)\}\}
\tag{5}
$$

Eq. (4) computes the message for the maximum weight of the minimax tree matching including edge o. In order to form a valid matching, if o is already

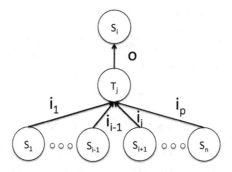

Fig. 3. Messages passing through T_j

included, all the $n-1$ children must be excluded. That is why the second maximization in Eq. (4) is taken on all the $m'(i_m)$'s, not the $m(i_m)$'s. Eq. (5) updates the message for the maximum weight of the minimax tree matching without edge o. Every node, except the leaves, must belong to one edge in the perfect matching. Since o is not chosen in this case, we thus have to pick up one edge from the $n-1$ children, while abandoning all that child's $n-2$ brothers. In order to find the minimax matching at the end, we take a minimization of all the maximum weights in Eq. (5).

Last but not least, let us consider the initialization. Since leaves have no children, the output messages from the leaves should be $m(o) = w(o)$ and $m'(o) = \min(\mathbf{w})$. Recall that $\mathbf{w}$ is the weight matrix. Actually, we can also set $m'(o)$ to be any value smaller than $\min(\mathbf{w})$.

Here is the entire algorithm:

---

(1)Initialization:

$$m^{(0)}_{T_i \to S_j} = w_{ij} \tag{6}$$

$$m'^{(0)}_{T_i \to S_j} = \min(\mathbf{w}) \tag{7}$$

$$m^{(0)}_{S_j \to T_i} = w_{ij} \tag{8}$$

$$m'^{(0)}_{S_j \to T_i} = \min(\mathbf{w}) \tag{9}$$

(2)Messages at the kth iteration:

$$m^{(k)}_{T_i \to S_j} = \max\{w_{ij}, \max_{m \neq j} m'^{(k-1)}_{S_m \to T_i}\} \tag{10}$$

$$m'^{(k)}_{T_i \to S_j} = \min_{m \neq j}\{\max\{m^{(k-1)}_{S_m \to T_i}, \max_{n \neq m,j} m'^{(k-1)}_{S_n \to T_i}\}\} \tag{11}$$

$$m^{(k)}_{S_j \to T_i} = \max\{w_{ij}, \max_{m \neq i} m'^{(k-1)}_{T_m \to S_j}\} \tag{12}$$

$$m'^{(k)}_{S_j \to T_i} = \min_{m \neq i}\{\max\{m^{(k-1)}_{T_m \to S_j}, \max_{n \neq m,i} m'^{(k-1)}_{T_n \to S_j}\}\} \tag{13}$$

(3)Decisions at the end of kth iteration:

$$D_{ij}^{(k)} = \max\{m_{T_i \to S_j}^{(k)}, m_{S_j \to T_i}^{(k)}\} - \max\{m_{T_i \to S_j}'^{(k)}, m_{S_j \to T_i}'^{(k)}\} \tag{14}$$

If $D_{ij}^{(k)} < 0$, choose the edge (T_i, S_j); Otherwise, not.

---

4   Proof of Correctness and Convergence

Theorem 1. For an n by n complete bipartite graph, if the minimax weight matching is unique, the algorithm converges to the optimal solution after $k \geq n$ iterations.

To prove this, we can equivalently investigate the following two statements:

1.1 The root edge $e_r = (T_i, S_j)$ is not a part of the MMWM in the original bipartite graph but $m(e_r) < m'(e_r)$, meaning the root edge is chosen. We will show this is impossible for $k \geq n$;

1.2 The root edge $e_r = (T_i, S_j)$ is a part of the MMWM in the original bipartite graph but $m(e_r) > m'(e_r)$, meaning the root edge is not chosen. We will show this is impossible for $k \geq n$.

Proof. For Statement 1.1, Let Ω be the set of all the edges constructing the perfect tree matching with maximum weight $m(e_r)$. Ω is therefore a tree matching (All the nodes on the tree, except some leaves, connect to exactly one edge) and $e_r \in \Omega$ by definition. Let X^* be the MMWM in the original graph. Map X^* to the computation tree to get a tree matching Ω^*. By the assumption in Statement 1.1, $e_r \notin \Omega^*$. Construct an alternating path P with respect to Ω and Ω^* on the tree starting from e_r. Then continuously augment P by selecting each time two edges alternately from Ω^* and Ω, until it reaches the leaves. For example, consider the toy example in Figure 1 and 2. Suppose the matching constructing Ω is $\{(T_1, S_1), (T_2, S_2), (T_3, S_3)\}$, while the MMWM in the original graph constructing Ω^* is $\{(T_1, S_2), (T_2, S_1), (T_3, S_3)\}$. The alternating path can then be $P = [...T_2 \leftarrow S_2 \leftarrow T_1 \leftrightarrow S_1 \to T_2 \to S_2...]$. This path P starts from the root edge $T_1 \leftrightarrow S_1$ as shown in the middle of the square brackets. For the first *alternative augmenting*, it has to pick up two edges from Ω^* since the root edge belongs to Ω. $(S_2 \leftarrow T_1)$ and $(S_1 \to T_2)$ are hence selected. After that, the path goes through $(T_2 \leftarrow S_2)$ and $(T_2 \to S_2)$, which belong to Ω again. This process proceeds until P reaches the leaves. According to the definitions of Ω^* and Ω, such a path is guaranteed to exist.

Modify Ω to Ω' by adding edges belonging to $P \cap \Omega^*$ and removing those from $P \cap \Omega$. Again by definition, $e_r \notin \Omega'$. We know $\max(\Omega) = m(e_r)$ ($\max(X)$ returns the maximum weight of edge set X by definition and $m'(e_r) \leq \max(\Omega')$ because Ω' is just one instance of the tree matchings without e_r. (Recall that $m(e_r)$ is the maximum weight of the minimax tree matching with e_r, and $m'(e_r)$ is that without e_r). At this point, if we can show $\max(\Omega') \leq \max(\Omega)$, then:

$$m'(e_r) \leq \max(\Omega') \leq \max(\Omega) = m(e_r) \tag{15}$$

and a contradiction is immediately obtained, because at the very beginning the assumption is $m(e_r) < m'(e_r)$. Since the only difference between Ω' and Ω is the edges in P, we then only need to show:

$$\max(P \cap \Omega^*) < \max(P \cap \Omega). \tag{16}$$

It is certainly not true for some starting iterations. We are interested in when it must be true. To investigate that, first map P back to the original graph. Note that, if we walk through P step by step in the original graph, we will get a series of cycles at the beginning and at most one trailing path at the end. For instance, consider mapping the path constructed above back to its original graph Figure 1. Suppose this path is only of length five (alternatively augmented twice). Walking through this path from one endpoint to the other $T_2 \to S_2 \to T_1 \to S_1 \to T_2 \to S_2$, we get a cycle at the beginning ($T_2 \to S_2 \to T_1 \to S_1 \to T_2$) and a path ($T_2 \to S_2$) at the end. Next we will show that once mapping P back forms a cycle in the original graph, the algorithm must converge to the optimum.

Assume one cycle has already appeared after mapping P back to the original graph. Denote the edges in the cycle by C. Now prove by contradiction. If $\max(P \cap \Omega^*) \geq \max(P \cap \Omega)$, then $\max(\mathbf{X}^*) \geq \max(P \cap \Omega^*) \geq \max(P \cap \Omega) \geq \max(C \cap P \cap \Omega)$. By switching the edges in the cycle, we can therefore use the edges $C \cap P \cap \Omega$ and $\mathbf{X}^*/(C \cap P \cap \Omega^*)$ to construct another matching with maximum weight at most $\max(\mathbf{X}^*)$. It is impossible, because we assume the MMWM is unique. Consequently, $\max(P \cap \Omega^*) < \max(P \cap \Omega)$.

For Statement 1.2, the reasoning is very similar, hence omitted.

The entire proof will end after answering this question: When does a cycle have to appear in the original graph after mapping path P back to it? The answer is simply, "n". The length of the path P after k iterations is $2k + 1$ and the total number of nodes is $2n$. $\lfloor \frac{2k+1}{2n} \rfloor \geq 1$ yields $k \geq n$. Consequently, the algorithm will converge in n iterations. □

Additionally, the number of iterations needed for convergence of the algorithm K_a actually does not equal to that for the messages K_m. What can be derived directly is:

$$K_a \leq K_m. \tag{17}$$

Intuitively, when the messages converge, the decision matrix D will stay the same, which leads to the convergence of the algorithm. However, the convergence of the algorithm, i.e., all the signs of the cells in the decision matrix will never change anymore, does not necessarily ensure the convergence of the messages. Another question thus arises: Will the messages converge? An intuitive answer is yes, because both $m(e_r)$ and $m'(e_r)$ are nondecreasing and bounded by $\max(\mathbf{w})$, when the number of iterations k increases.

As a result, the message complexity of the algorithm is $O(n^3)$ for a graph of size $O(n)$. To see this, consider a bipartite graph with n nodes in each partition, hence n^2 edges in total. Each edge needs n iterations to make a correct decision. n^3 messages are therefore exchanged before finding the minimax weight matching. Additionally, each of the messages can be computed in at most $O(n^2)$ time.

As a result, it is a fully polynomial time belief propagation algorithm! Upon distributed implementation, if every edge can compute and pass the messages simultaneously within one unit time in each iteration, the MMWM can then be found in n unit time.

5   Extension to General MMWM

In the previous sections, we require the graph to be symmetric and complete. In this section, we show the algorithm still works in an asymmetric and/or incomplete bipartite graph as long as the MMWM is unique.

In order to apply the algorithm, repair the graph to a complete and symmetric one by adding dummy nodes and edges:

- Asymmetric to symmetric:
  Add dummy nodes to ensure same number of nodes in each side/partition of the original bipartite graph.

- Incomplete to complete:
  Add edges from the side with initially less nodes to the other side:
  - For the dummy nodes and the nodes initially without any edge, add edges with arbitrary weights as long as they are less than the minimum of the edge weights in the original graph;

  - For the nodes initially with at least one edge connecting to the other side, add edges with arbitrary weights as long as they are greater than the maximum of the edge weights in the original graph.

Now the original algorithm is once again able to find the unique MMWM. For the result, among the selected edges, we can just keep the ones which are initially in the graph, while discarding all the "artificial" ones. The proof of correctness is straightforward and hence omitted here.

6   Conclusion

In this paper, we proposed a belief propagation algorithm to solve the MMWM problem. The algorithm can find the unique optimum with message complexity $O(n^3)$. This is one of a very few fully polynomial time solutions where belief propagation algorithms are proved correct.

References

1. Aji, S.M., Horn, G.B., McEliece, R.J.: On the convergence of iterative decoding on graphs with a single cycle. In: Proc. IEEE Int. Symp. Information Theory, Cambridge, MA, USA, p. 276 (1998)

2. Bayati, M., Shah, D., Sharma, M.: Maximum weight matching via max-product belief propagation. In: Proc. IEEE Int. Symp. Information Theory, Cambridge, pp. 1763–1767 (2005)
3. Cheng, Y., Neely, M., Chugg, K.M.: Iterative message passing algorithm for bipartite maximum weighted matching. In: Proc. IEEE Int. Symp. Information Theory, Cambridge, pp. 1934–1938 (2006)
4. Dean, J., Ghemawat, S.: Mapreduce: simplified data processing on large clusters. Communications of the ACM 51(1), 107–113 (2008)
5. Gamarnik, D., Shah, D., Wei, Y.: Belief propagation for min-cost network flow: Convergence and correctness. Operations Research 60, 410–428 (2012)
6. Leighton, F.T., Shor, P.: Tight bounds for minimax grid matching, with applications to the average case analysis of algorithms. In: Proc. of ACM Symposium on Theory of Computing, pp. 91–103 (1986)
7. Malewicz, G., Austern, M.H., Bik, A.J., Dehnert, J.C., Horn, I., Leiser, N., Czajkowski, G.: Pregel: a system for large-scale graph processing. In: ACM SIGMOD International Conference on Management of Data, pp. 135–146 (2010)
8. Pearl, J.: Probabilistic Reasoning in Intelligent Systems: Networks of Plausible Inference. Morgan Kaufmann, San Francisco (1988)
9. Shor, P.W., Yukich, J.E.: Minimax grid matching and empirical measures. Ann. Probab. 19, 1338–1348 (1991)
10. Weiss, Y.: Correctness of local probability propagation in graphical models with loops. Neural Comput. 12, 1–42 (2000)
11. Yuan, M., Jiang, C., Li, S., Shen, W., Pavlidis, Y., Li, J.: Message passing algorithm for the generalized assignment problem. In: 11th IFIP International Conf. on Network and Parallel Computing, pp. 423–434 (2014)
12. Yuan, M., Shen, W., Li, J., Pavlidis, Y., Li, S.: Auction/belief propagation algorithms for constrained assignment problem. In: Conf. on Algorithms and Discrete Applied Mathematics, pp. 238–249 (2015)

Decision Support System for the Multi-depot Vehicle Routing Problem

Takwa Tlili* and Saoussen Krichen

LARODEC, Institut Supérieur de Gestion Tunis, Université de Tunis, Tunisia

Abstract. This paper is concerned with designing an integrated transportation solver for multi-depot vehicle routing problem with distance constraints (MD-DVRP). The MD-DVRP is one of the most tackled transportation problems in real-world situations. It is about seeking the vehicle routes that minimize the overall travelled distance. To cope with the MD-DVRP, a Decision Support System (DSS) is designed based on the integration of Geographical Information System (GIS) and the Local Search (LS). The DSS architecture as well as its performance are checked using a real world case.

Keywords: Multi-depot vehicle routing problem, Metaheuristic, Local search, Decision Support System.

1 Introduction

In recent decades, many studies have contributed a considerable progress in order to study Vehicle Routing Problems (VRPs), due to its efficiency to solve logistic distribution problem. However, the VRP should be adjusted when it addresses more than one depot, that is the real life situation. The Multi Depot VRP (MDVRP) considers cases where there is more than one depot. Each vehicle departs from a depot to serve a set of customers. After the delivery, each vehicle returns to the depot where they started. Every customer is to be served by one vehicle on one occasion, and no vehicle can be loaded exceeding its maximum capacity. Since the MDVRP is NP-hard (Garey & Johnson, 1979), the literature on exact approaches is sparse (Baldacci & Mingozzi, 2009 and Contardo & Martinelli, 2014). Most authors have focused on the development of approximate methods to find near-optimal solutions quickly. Example approximate approaches include Ant colony optimization (Narasimha et al., 2013), memetic algorithm (Luo & Chen, 2014), Granular tabu search (Escobar et al., 2014), and Variable neighborhood search (Kuo & Wang, 2012). A recent survey paper (Karakatič & Podgorelec, 2015) summarizes the Genetic algorithms designed for solving the MDVRP. Consider an industrial firm with multiple depots with predetermined locations. Each depot has an enough capacity to store all the customers' demands. A fleet of vehicles with limited capacity is used to deliver the goods from depots to customers. The MD-DVRP consists in designing a set

---

* Corresponding author.

© Springer International Publishing Switzerland 2015   47
H.A. Le Thi et al. (eds.), *Model. Comput. & Optim. in Inf. Syst. & Manage. Sci.*,
Advances in Intelligent Systems and Computing 359, DOI: 10.1007/978-3-319-18161-5_5

of least cost vehicle routes such that, (1) Each customer is served by only one vehicle, (2) Each route starts and ends at the same depot, (3) The total length of each route must not exceed the distance constraint assigned to each vehicle, and (4) Each depot must not surpass assigned vehicles. Due to the complexity of VRPs (Lenstra & Kan, 1981), there has been a growing interest in the use of decision support systems (DSS). The integration of a metaheuristic into a DSS increases the capability to analyze and handle transportation problems.

In the last decades, the world has witnessed a great scientific revolution, thanks to the growth of information technologies and particularly the digital information known by its accuracy, up to date and efficiency. Various softwares have utilized it. Among the most popular ones, we can refer to Geographical Information Systems (GIS) as a main tool that provides the necessary functions to collect, manage, analyze and generate spatial data. It can generate sophisticated geographical output easy to be understood in order to guide appropriately the decision makers in their works (Li et al., 2003).

The main contributions of this paper are (i) to solve a novel variant of the MD-DVRP which is the MDVRP with distance constraints using the ILS metaheuristic. (ii) to model mathematically the MD-DCVRP. (iii) to propose a DSS based on GIS to aid decision makers on solving the MD-DCVRP. (iii) to validate the designed DSS using a Tunisia real case study.

The remaining of the paper is organized as follows: Section 2 describes mathematically the problem. Section 3 provides a description of the resolution methodology. Section 4 presents a detailed explanation of the proposed DSS architecture. Section 5 describes the computational results. Section 6 details the case study.

2   Mathematical Model

The MD-DVRP is defined as an undirected graph $G = (A, E)$ where a node $j \in A$ corresponds to either a customer or a depot and an edge $e \in E$ expresses a path between a pair of nodes. Let $E = \{1, ..., n\}$ be the customers set and $F = \{1, ..., m\}$ the set of depots. Each customer has a demand q_j $(j \in \{1, 2, 3, ..., n\})$ to be delivered by a vehicle k. Every vehicle is characterized by a maximum capacity C_k and can travel a maximum distance P_k $(k \in \{1, 2, 3, ..., p\})$. The travelled distance between node i and node j is d_{ij}. We state in what follows the decision variable used for the development of the MD-DVRP model.

$$x_{ijk} = \begin{cases} 1 \text{ if the arc } (i, j) \text{ is traversed by the vehicle } k \\ 0 \text{ elsewhere} \end{cases}$$

The MD-DVRP can be expressed mathematically as follows.

$$Min \quad Z(x) = \sum_{i \in A} \sum_{j \in A \setminus \{i\}} \sum_{k=1}^{p} d_{ij} \times x_{ijk} \qquad (1)$$

Subject to

$$\sum_{j \in A \setminus \{i\}} \sum_{k=1}^{p} x_{jik} = 1, \forall i \in E \tag{2}$$

$$\sum_{j \in A \setminus \{i\}} \sum_{k=1}^{p} x_{ijk} = 1, \forall i \in E \tag{3}$$

$$\sum_{i \in F} \sum_{j \in E} x_{ijk} \leq 1, \forall 1 \leq k \leq p \tag{4}$$

$$\sum_{j \in E} x_{ijk} = \sum_{j \in E} x_{jik}, \forall 1 \leq k \leq p, \forall i \in F \tag{5}$$

$$\sum_{i \in S} \sum_{j \in S} x_{ijk} \leq |S| - 1, \forall 1 \leq k \leq p, \forall S \subset E \tag{6}$$

$$\sum_{i \in A} \sum_{j \in E \setminus \{i\}} x_{ijk} \times q_j \leq C_k, \forall 1 \leq k \leq p \tag{7}$$

$$\sum_{i \in A} \sum_{j \in E \setminus \{i\}} x_{ijk} \times d_{ij} \leq P_k, \forall 1 \leq k \leq p \tag{8}$$

$$\sum_{k=1}^{p} \sum_{j \in E} x_{ijk} \leq v_i, \forall i \in F \tag{9}$$

- The objective function (1) expresses the total distance traveled by all vehicles that must be minimized.
- Constraints (2) and (3) state that each customer is supplied only once.
- Constraints (4) ensure that each vehicle is assigned to only one depot.
- Constraints (5) guarantee that each vehicle originates and terminates at the same depot.
- Constraints (6) eliminate the sub-tours.
- Constraints (7) explain that the vehicle capacity should be respected, e.g, the total customers demands assigned to a vehicle should not exceed its maximum capacity.
- Constraints (8) describe that each vehicle should not exceed a distance threshold.
- Constraints (9) express that the maximum number of vehicles for each depot should be respected.

3   Resolution Methodology

Since MD-DVRP is at least as difficult as MDVRP which is known to be NP-hard problem, therefore, we propose a hybrid optimization heuristic method called HH, based on Local Search (LS) approach as we are motivated by its success for a wide range of hard optimization problems such as the Travelling Salesman Problem (TSP) and the MDVRP. The main idea of LS is to improve the current solution by searching for a better one in its neighborhood. Our proposed algorithm includes two iterated phases: construction and improvement.

3.1   Construction Phase

Once the HH parameters, such as the depots number and customers coordinates have been set, the assignment sub-phase is performed. In this sub-phase, we attribute each customer to the nearest depot after calculating the distance slacks of all depots and choosing the shortest one. Distances are obtained using Euclidean method. Then, the routing sub-phase is achieved. It consists of assigning customers belonging to the same depot to different routes. For this purpose, we use the Clark and Wright saving method which is considered as a promising one. This method consists of creating a saving matrix for each depot. For example, for a depot d, the saving value for customers i and j is calculated by equation 10.

$$S(i,j) = Distance(d,i) + Distance(d,j) - Distance(i,j) \qquad (10)$$

Then, customers with greatest saving value are assigned to the same route without exceeding the vehicle capacity and the maximum distance to travel. Once the routing sub-phase is accomplished, we proceed to the scheduling sub-phase. It yields a better customers sequencing of each route. For this, we use the Iterative Improvement Insertion. This algorithm attempts to enhance the customers scheduling of each route. In fact, it consists of selecting a customer randomly without repetition and trying to insert him in the best position (which has the lowest cost). This is repeated for all customers.

3.2   Improvement Phase

Once the initial solution is generated by the construction phase, we try to yield better results by combining a set of optimization heuristics following these steps:
① **Balancing Step:** The purpose of this algorithm is to balance the customers number distributed over the routes belonging to the same depot. In other words, it spreads the customers over the routes in a way that their number becomes roughly equal for each route. This step allows to deteriorate the objective function while respecting the capacity constraint and the maximum distance to travel. It allows to browse a larger search space and so to have a better chance to avoid local minima.
② **Exchange Step:** This procedure involves two types of moves that are intra-depot and inter-depot.

- The first move consists of selecting two customers randomly from the same depot and belonging to different routes. If the permutation between these two clients yields better solution, the exchange is established.
- The second move has the same principle of the first one except that the exchange involves two customers that belong to different depots. The permutation between the clients is done if the move improves the current solution.

③ **Iterative Improvement Insertion:** This is the same scheduling as mentioned in the initial phase.

Algorithm 1. Iterative Improvement Insertion

Require: π_0: customer table
Ensure: Best solution found
1: improve=true;
2: **while** improve=true **do**
3:    improve=false;
4:    **for** k=1 ...p **do**
5:       Remove a customer k at random from π_0 without repetition;
6:       π_1 = Best permutation by inserting k in any possible position of π_0;
7:       **if** $C_{max}(\pi_1) \leq C_{max}(\pi_0)$ **then**
8:          $\pi_0 = \pi_1$;
9:          improve=true;
10:       **end if**
11:    **end for**
12: **end while**

---

4   Decision Support System Architecture

The DSS is a flexible and interactive tool that aims to facilitate the human decision-making process in complex environments. The conceptual architecture of the proposed DSS for solving the MD-DVRP is shown in Fig. 1. The DSS integrates the ILS solver to get a near-optimal solution and a GIS that acts as a result viewer. The methodology can be summarized in five interconnected steps. Such steps are stated as follows.

① Retrieve or create geographical data related to highways, depots, customers and vehicles from GIS layers. Customers and depots layers are presented as point and highway layers as lines.

② Generate a minimal distance matrix using Dijkstra shortest path algorithm (Dijkstra, 1959). It contains minimal distances between customers and depots. These distance are gathered using GIS.

③ Assign each customer to the nearest depot. For this, we compare distance between each customer and existing depots, in order to select the nearest one.

④ Establish routing and programming vehicles plan from each depot using LS metaheuristic. Thereby, we obtain the routing planning, customers sequencings and vehicles number.

⑤ Generate the geographical report (itinerary for each vehicle) that guides appropriately the decision makers. It describes each vehicle itinerary, starting from a particular depot to serve a set of customers optimally. The whole scenario is summarized in figure 1.

5   Computational Results

HH algorithm is implemented in C++ language, based on 2 GHz Intel Core Duo processor with 3 GO RAM under Windows 7. To evaluate the performance of our work, we test it on a set of 22 MD-DVRP benchmarks from P01 to P22, known as Cordeau's instances. These instances are listed in table 1 where:

− I defines the instance name.

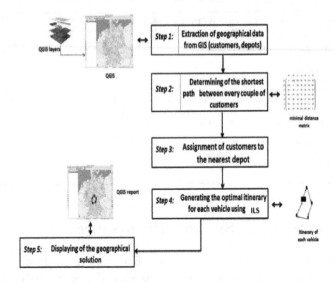

Fig. 1. DSS architecture based on GIS

- n is the number of customers.
- m presents the number of depots.
- v mentions the number of vehicles assigned for each depot.
- Q indicates the capacity of each vehicle.
- D defines the maximum distance to be traveled by each vehicle.

To measure the performance of our algorithm, we apply the following metrics.

- $C(s)$: Best found total travelled distance
- $C^*(s)$: Best known solution (Cordeau benchmark)
- GAP: percentage between the best known solution $C^*(s)$ and the best found solution C(s).

Table 1. Instances parameters of Cordeau's instances

I	n	m	v	Q	D	I	n	m	v	Q	D
P01	50	4	4	80	∞	P12	80	2	5	60	∞
P02	50	4	2	160	∞	P13	80	2	5	60	200
P03	75	3	3	140	∞	P14	80	2	5	60	180
P04	100	2	8	100	∞	P15	160	4	5	60	∞
P05	100	2	5	200	∞	P16	160	4	5	60	200
P06	100	3	6	100	∞	P17	160	4	5	60	180
P07	100	4	4	100	∞	P18	240	6	5	60	∞
P08	249	2	14	500	310	P19	240	6	5	60	200
P09	249	3	12	500	310	P20	240	6	5	60	180
P10	249	4	8	500	310	P21	360	9	5	60	∞
P11	249	5	6	500	310	P22	360	9	5	60	200

Table 2. Experimental results of the HH on Cordeau's instances

I	$C^*(s)$	$C(s)$	GAP	I	$C^*(s)$	$C(s)$	GAP
P01	576.87	606.11	5.06	P12	1318.95	1390.04	5.44
P02	473.53	496.45	4.84	P13	1318.95	1400.2	6.16
P03	641.19	675.32	5.32	P14	1360.12	1554.25	13.5
P04	1001.59	1062.6	6.09	P15	2505.42	2766.3	10.4
P05	750.03	782.34	4.32	P16	2572.23	2885.45	12.18
P06	876.50	910.13	3.8	P17	2709.09	2969.45	9.61
P07	885.8	904.4	2.09	P18	3702.85	4202.3	13.48
P08	4437.68	4784.2	7.8	P19	3827.06	4137.07	8.09
P09	3900.22	4102.22	5.17	P20	4058.07	4466.56	9.65
P10	3663.02	3960.01	8.13	P21	5474.84	5770.05	5.39
P11	3554.18	4036.55	13.57	P22	5702.16	6523.85	13.65

Table 2 illustrates the computational results obtained through our approach HH on 23 instances used by Cordeau et al, (1997) in terms of the best known solution $C^*(s)$, best optimal solution $C(s)$ and gap GAP. From table 2, we can point out that, for a considerable number of instances, the results of our approach are near to those of the best known solution. We can also notice that the gap varies between 2.09 and 13.65.

6   Real Case Study

In this section, we present a real case study for our approach. We take for our work a data base example of Ezzahra city in Tunisia that will be applied on QGIS software to generate its related map. In order to obtain this output, QGIS combines different layers where each one describes a specific theme such sea, building, road and islet.

The available database of the studied area contains a layer of roads which do not allow the movement from one customer to another. So, we create a new vector layer that is the Transport network consisting on a set of arcs connecting customers. This layer represents the roads by which a vehicle can deliver items to a set of customers.

Let us consider an example of 10 customers and 3 depots dispersed around the Ezzahra city as shown in figure 2.

Different setting parameters of the problem are presented in table 3 and customers' demands are in table 4. In order to pick out the shortest path between

Table 3. Description of example parameters

Parameters	
Vehicle capacity (kg)	250
Number of vehicles for each depot	2
Maximum distance traveled by each vehicle (m)	2350

Table 4. Customer's demands

Customer	C_1	C_2	C_3	C_4	C_5	C_6	C_7	C_8	C_9	C_{10}
Demand	80	60	70	95	150	80	60	70	60	120

each couple of customers, we opt for an additional plugin integrable on QGIS called ROAD GRAPH that calculates the minimum distances between two points on any polyline layer and illustrates this path over the road network. By means of ROAD GRAPH, we obtain the following matrix that summarizes all shortest paths between each pair of customers and depots. It consists of round distances in meter.

$$
\begin{array}{c}
\begin{array}{cccccccccccccc}
& D_1 & D_2 & D_3 & C_1 & C_2 & C_3 & C_4 & C_5 & C_6 & C_7 & C_8 & C_9 & C_{10}
\end{array}\\
\begin{array}{c}
D_1\\ D_2\\ D_3\\ C_1\\ C_2\\ C_3\\ C_4\\ C_5\\ C_6\\ C_7\\ C_8\\ C_9\\ C_{10}
\end{array}
\left(\begin{array}{ccccccccccccc}
0 & 1132 & 1556 & 1359 & 807 & 639 & 1261 & 381 & 374 & 1376 & 579 & 275 & 1880\\
 & 0 & 2455 & 760 & 252 & 1455 & 907 & 784 & 425 & 180 & 500 & 1407 & 2779\\
 & & 0 & 2175 & 2014 & 1000 & 731 & 1750 & 1374 & 2640 & 2063 & 1594 & 324\\
 & & & 0 & 657 & 1440 & 1444 & 1101 & 1135 & 968 & 1411 & 1645 & 2285\\
 & & & & 0 & 1014 & 1248 & 444 & 712 & 626 & 754 & 988 & 2089\\
 & & & & & 0 & 714 & 750 & 374 & 1620 & 1063 & 594 & 1196\\
 & & & & & & 0 & 1464 & 1088 & 1874 & 1777 & 1499 & 841\\
 & & & & & & & 0 & 521 & 1069 & 313 & 630 & 1946\\
 & & & & & & & & 0 & 1475 & 834 & 411 & 1570\\
 & & & & & & & & & 0 & 519 & 1699 & 2816\\
 & & & & & & & & & & 0 & 948 & 2259\\
 & & & & & & & & & & & 0 & 1790\\
 & & & & & & & & & & & & 0
\end{array}\right)
\end{array}
$$

Fig. 2. Solution of the example

Figure 2 presents a geographical view of the obtained results after solving the example. It illustrates the best traveling path for each vehicle, while taking into account the capacity restriction, maximum distance to be traveled by each one and the limited number of vehicles assigned to each depot. This map is used to guide vehicles drivers to serve customers through the shortest itinerary presented as yellow arrows. In the example of depot 3, two vehicles are used. The first one leaves the depot to serve in the order C_9, C_3 then C_6 with a cost of 1617. The second vehicle serves respectively C_5 and C_8 with a cost of 1273. Both of these vehicles come back to the depot D_1 after serving the corresponding customers.

7 Conclusion

In this paper the Multi Depot Vehicle Routing Problem (MD-DCVRP) is evoked and solved using the Iterated Local Search approach (ILS). In order to better visualize the obtained results and make it more intuitive, we proposed to combine the ILS with a GIS to design a Decision Support System (DSS). The proposed DSS provides assistance to operating managers in transportation logistics. To assess the efficiency of our framework, we proposed to solve an application of a Tunisian case.

References

1. Baldacci, R., Mingozzi, A.: A unified exact method for solving different classes of vehicle routing problems. Mathematical Programming 120, 347–380 (2009)
2. Chen, P., Huang, H.-k., Dong, X.-Y.: Iterated variable neighborhood descent algorithm for the capacitated vehicle routing problem. Expert Systems with Applications 27, 1620–1627 (2010)
3. Contardo, C., Martinelli, R.: A new exact algorithm for the multi-depot vehicle routing problem under capacity and route length constraints. Discrete Optimization 12, 129–146 (2014)
4. Cordeau, J.-F., Gendreau, M., Laporte, G.: A tabu search heuristic for periodic and multi-depot vehicle routing problems. Networks 2, 105–119 (1997)
5. Escobar, J.W., Linfati, R., Toth, P., Baldoquin, M.G.: A hybrid granular tabu search algorithm for the multi-depot vehicle routing problem. Journal of Heuristics 20, 483–509 (2014)
6. Garey, M., Johnson, D.: Computers and intractability: A guide to the theory of np-completeness, 1st edn. W. H. Freeman (1979)
7. Glover, F., Laguna, M.: Tabu search. Kluwer Academic Publishers, Boston (1997)
8. Karakatič, S., Podgorelec, V.: A survey of genetic algorithms for solving multi depot vehicle routing problem. Applied Soft Computing 27, 519–532 (2015)
9. Li, H., Kong, C.W., Pang, Y.C., Shi, W.Z., Yu, L.: Internet-based geographical information systems system for e-commerce application in construction material procurement. Journal of Construction Engineering and Management 129, 689–697 (2003)
10. Luo, J., Chen, M.-R.: Improved shuffled frog leaping algorithm and its multiphase model for multi-depot vehicle routing problem. Expert Systems with Applications 41, 2535–2545 (2014)
11. Narasimha, K.V., Kivelevitch, E., Sharma, B., Kumar, M.: An ant colony optimization technique for solving min–max multi-depot vehicle routing problem. Swarm and Evolutionary Computation 13, 63–73 (2013)

Iterated Tabu Search for the Mix Fleet Vehicle Routing Problem with Heterogenous Electric Vehicles

Ons Sassi[1], Wahiba Ramdane Cherif-Khettaf[1], and Ammar Oulamara[2]

[1] University of Lorraine - LORIA, Nancy, France
{ons.sassi,wahiba.ramdane}@loria.fr
[2] University of Lorraine, Ile de Saulcy, Metz, France
ammar.oulamara@loria.fr

Abstract. In this paper, we address the vehicle routing problem with mixed fleet of conventional and heterogenous electric vehicles, denoted VRP-MFHEV. This problem is motivated by a real-life industrial application and it is defined by a mixed fleet of heterogenous Electric Vehicles (EVs) having distinct battery capacities and operating costs, and identical Conventional Vehicles (CVs) that could be used to serve a set of geographically scattered customers. The EVs could be charged during their trips at the depot and in the available charging stations, which offer charging with a given technology of chargers and propose different charging costs. EVs are subject to the compatibility constraints with the available charging technologies and they could be partially charged. The objective is to minimize the number of employed vehicles and to minimize the total travel and charging costs. To solve the VRP-MFHEV, we propose a Multi-Start Iterated Tabu Search (ITS) based on Large Neighborhood Search (LNS). The LNS is used in the tabu search of the intensification phase and the diversification phase of the ITS. Different implementation schemes of the proposed method including best-improvement and first-improvement strategies, are tested on generalized benchmark instances. The computational results show that ITS produces competitive results, with respect to results obtained in previous studies, while the computational time remains reasonable for each instance. Moreover, using LNS in the intensification phase of ITS seems improving the generated solutions compared to using other neighborhood search procedures such as 2opt.

Keywords: Electric vehicle routing problem, Electric vehicle charging, Meta-heuristics, Iterated Tabu Search, Large Neighborhood Search, Optimization.

1 Introduction

Nowadays, many cities aim at keeping their streets safe for everyone and saving the environment while encouraging sustainable driving options such as ride sharing [1] and Electric Vehicles (EVs) use [2]. In fact, EVs may decrease

© Springer International Publishing Switzerland 2015
H.A. Le Thi et al. (eds.), *Model. Comput. & Optim. in Inf. Syst. & Manage. Sci.*,
Advances in Intelligent Systems and Computing 359, DOI: 10.1007/978-3-319-18161-5_6

transportation-related emissions and provide for less dependence on foreign oil. However, electric cars are still facing many weaknesses related to the high purchase prices, battery management and charging infrastructure.

Several research groups are currently working on EV related problems and the progress in the field is important. The problem of energy-optimal routing is addressed in [3]. In [4], the authors formulate the Green Vehicle Routing Problem (GVRP) as a Mixed Integer Linear Program (MIP) and propose two constructive heuristics to solve this problem. An overview of the GVRP is given in [5]. Schneider et al. [6] combine a Vehicle Routing Problem with the possibility of charging a vehicle at a station along the route. They introduce the Electric Vehicle Routing Problem with Time Windows and Recharging Stations (E-VRPTW). E-VRPTW aims at minimizing the number of employed vehicles and total traveled distance. In [7], the Electric Vehicle Routing Problem with Time Windows and Mixed Fleet to optimize the routing of a mixed fleet of EVs and Conventional Vehicles (CVs) is addressed. To solve this problem, an Adaptive Large Neighborhood Search algorithm that is enhanced by a local search for intensification is proposed. Almost the same problem is addressed in [8]. The only difference here is the fact of considering a heterogenous fleet of vehicles that differ in their transport capacity, battery size and acquisition cost. An Adaptive Large Neighbourhood Search with an embedded local search and labelling procedure for intensification is developed. In [9], the authors present a variation of the electric vehicle routing problem in which different charging technologies are considered and partial EV charging is allowed.

In this paper, we extend earlier work in the literature with several important new life aspects regarding the constraints and objective function. In fact, we consider the electric vehicle routing problem with mixed fleet of conventional and heterogenous electric vehicles that involves the design of a set of minimum cost routes, starting and terminating at the depot, which services a set of customers. We consider different charging technologies and partial EV charging. In addition to the routes construction, EVs charging plans should be determined. We are also concerned with new constraints. Firstly, EVs are not necessarily compatible with all charging technologies. Secondly, charging stations could propose different charging costs even if they propose the same charging technology and they are subject to operating time windows constraints. Our objective function is also different. In fact, we aim at minimizing total operating and charging costs involved with the use of a mixed fleet. Our overall objective is to provide enhanced optimization methods for EV charging and routing that are relevant to the described constraints.

The new constraints described above are inspired by a real-life problem that has been first addressed in the framework of a French national R&D project led by many companies and research laboratories. Furthermore, this study follows on from the work presented in [10] where exact and heuristic methods were presented to solve the joint EV scheduling and charging problem. In [11] and [12], a mixed integer programming model, heuristic approaches and an Iterated Local Search metaheuristic were proposed to solve this problem.

Within this study, we propose an Iterated Tabu Search based on a Large Neighborhood Search (ITS-LNS). ITS has been used in the literature to solve NP-hard problems such as the Traveling Salesman Problem [13]. To the best of our knowledge, the ITS hasn't never been used to solve the electric vehicle routing problem.

The remainder of the paper is organized as follows. In Section 2, we introduce the notation in detail. In Section 3, our solving approach is presented. Section 4 summarizes the computational results. Concluding remarks are given in Section 5.

2   Problem Description and Notation

We define the VRP-MFHEV on a complete, directed graph $G = (V', A)$. V' denotes the set of vertices composed of the set V of n customers, the set F of external charging stations and other chargers located at the depot $F = \{1, \ldots, f\}$. The set of arcs is denoted by $A = \{(i, j) \mid i, j \in V', i \neq j\}$.

Our optimization time horizon $[0, T]$, which represents typically a day, is divided into T equidistant time periods, $t = 1, \ldots, T$, each of length δ, where t represents the time interval $[t - 1, t]$. All customers have to be served during $[0, T]$. A nonnegative demand q_i is associated with each customer $i \in V$, this represents the quantity of goods that will be delivered to this customer. With each customer we also associate a service time s_i. Each arc $(i, j) \in A$ is defined by a distance $d_{i,j}$ and a nonnegative travel time $t_{i,j}$ required to travel $d_{i,j}$. When an arc (i, j) is traveled by an EV, it consumes an amount of energy $e_{i,j}$ equal to $r \times d_{ij}$, where r denotes a constant energy consumption rate.

The chargers in charging station f are available during the time window $[a_f, b_f]$. Accordingly, the EV must wait if it arrives at charging station f before time a_f.

Each charging station $f \in F$ can deliver a maximum charging power p_f (kW) and proposes a charging cost c_f expressed in (euros/kWh). Note that, within this study, we consider that the charging stations could propose three different charging technologies: (i) Level 1 charger which is the slowest charging level that provides charging with a power of 3.7 kW; (ii) Level 2 charger offers charging with a power of 22 kW and (iii) Level 3 charger which is the fastest charging level that delivers a power of 53 kW.

We consider a set $M_{\mathrm{EV}} = \{1, \ldots, m_{\mathrm{EV}}\}$ of EVs and a set $M_{\mathrm{CV}} = \{m_{\mathrm{EV}} + 1, \ldots, m_{\mathrm{EV}} + m_{\mathrm{CV}}\}$ of Combustion Engine Vehicles (CVs), needed to serve all customers. Each EV k operates with a battery characterized by its nominal capacity of embedded energy CE_k(kWh). Each EV (CV) is characterized by a maximum capacity Q^{EV} (Q^{CV}) which represents the maximum quantity of goods that could be transported by the vehicle. Denote by FC^{EV} (FC^{CV}) (euros/ day) the fixed costs related to EVs (CVs). Denote by OC_k^{EV} (OC^{CV}) the operating costs (euros/km) related to the maintenance of EV k (CV), accidents, etc. Thus, if an arc (i, j) is traveled by an EV k (CV), this has an operating cost denoted by $cost_{i,j,k}^{\mathrm{EV}}$ ($cost_{i,j}^{\mathrm{CV}}$) and is computed as: $cost_{i,j,k}^{\mathrm{EV}} = d_{i,j} \times OC_k^{\mathrm{EV}}$ ($cost_{i,j}^{\mathrm{CV}} = d_{i,j} \times OC^{\mathrm{CV}}$).

Each customer $i \in V$ should be visited, by either an electric or conventional vehicle, exactly once during $[0, T]$. Each charging station could be visited as many times as required or not at all. When charging is undertaken in a charging station f, it is assumed that only the required quantity of energy is injected into the EV battery. Thus, EVs could be partially charged.

Since we consider many charging technologies (slow and fast charging), we should also consider the fact that not all EVs technologies are compatible with fast charging. Thus, when we plan the charging of an EV, only the charging stations proposing compatible charging technologies should be considered. A feasible solution to our problem is composed of a set of feasible routes assigned to adequate vehicles and a feasible EVs charging planning. A feasible route is a sequence of nodes that satisfies the following constraints:

- Each route must start and end at the depot;
- the overall amount of goods delivered along the route, given by the sum of the demands q_i for each visited customer, must not exceed the vehicle capacity (Q^{EV} or Q^{CV});
- the total duration of each route, calculated as the sum of all travel durations required to visit a set of customers, the time required to charge the vehicle during the interval $[0, T]$, the service time of each customer and, eventually, the waiting time of the EV if it arrives at a charging station before its opening time, could not exceed T;
- no more than m_{EV} EVs and m_{CV} CVs are used;
- each customer should be visited once between 0 and T;
- the following charging constraints are satisfied:
    - when charging is undertaken, each EV should be charged with a compatible charging technology;
    - at each charging station f, charging could only be undertaken during its operating time window $[a_f, b_f]$;
    - the battery capacity constraints should be satisfied

We seek to construct a minimum number of routes such that all customers are served, all EVs are optimally charged and the total cost of routing and charging is minimized. The objective function, measured in monetary units, consists in minimizing five costs: (i) the routing cost that depends on the number of kilometers traveled by each vehicle and the vehicle operating cost, (ii) the charging cost engendered by charging EVs in the charging stations during $[0, T]$, (iii) the vehicles total fixed cost and (vi) the total cost engendered by the waiting time of the EVs if they arrive at a charging station before its opening time.

3   Iterated Tabu Search Based on Large Neighborhood Search

Our ITS-LNS algorithm uses a Tabu search based on a Large Neighborhood Search (LNS) in the intensification phase and a LNS in the diversification phase. The LNS was first proposed by Shaw ([14]), and later adapted by Pisinger and

Ropke ([15]). The ITS algorithm was first proposed by Alfonsas Misevicius to solve the Traveling Salesman Problem [13]. It combines intensification and diversification mechanisms to avoid getting stuck in local optima. The intensification phase searches for good solutions in the neighbourhood of a given solution, whereas diversification is responsible for escaping from local optima and moving towards new regions of the search space. The ITS algorithm iterates five procedures: (i) Initial solution generation; (ii) a Tabu Search which improves a given solution; (iii) a Diversification Mechanism that generates a new starting point through a perturbation of the solution returned by the Tabu Search; (iv) an Acceptance Criterion that specifies if the solution should be accepted or not and (v) a Stopping Criterion that specifies when the ITS procedure should stops. In the following, Algorithm 1 describes our multi-start ITS-LNS algorithm in detail.

The next subsections describe the different ITS-LNS procedures in detail.

Algorithm 1. ITS-LNS Algorithm

1: **Input:** A graph $G = (V', A)$ and a set of $m_{EV} + m_{CV}$ vehicles
2: **Output:** A set of routes assigned to at most $m_{EV} + m_{CV}$ vehicles
3: Let $Max\_Restart$ be the maximum number of iterations to be executed starting from a new initial solution
4: Let max_{impr} be the maximum number of consecutive diversification phases allowed without improvement of the current best solution
5: Let $Record$ be the value of the best solution obtained
6: Initially, $Record =: +\infty$, $impr := 0$, $restart := 0$, $n_{Iter} := 0$
7: **while** $restart < Max\_Restart$ **do**
8:     Generate an initial solution and let s_0 be this solution
9:     $s_1 := TS - LNS(s_0)$
10:    $record := cost_{s_1}$
11:    **while** $n_{Iter} < max_{Iter}$ AND $impr < max_{impr}$ **do**
12:        $s' := Perturbation\_based\_LNS(s_1)$
13:        $s'_1 := TS - LNS(s_1)$
14:        **if** $cost_{s'_1} < (1 + Dev) \times record$ **then**
15:            $s_1 := s'_1$
16:        **end if**
17:        **if** $cost(s'_1) < Record$ **then**
18:            $Record := cost(s'_1)$, $impr := 0$
19:        **else**
20:            $impr := impr + 1$
21:        **end if**
22:        $n_{Iter} := n_{Iter} + 1$
23:    **end while**
24:    $restart := restart + 1$
25: **end while**

3.1   Initial Solution Generation

An initial feasible solution is generated with a Charging Routing Heuristic (CRH). For each vehicle, the CRH assigns a route composed of a set of customers and eventually a set of charging stations. The CRH starts from an empty solution and extends it iteratively until a complete solution is constructed. While at least one vehicle is still available, the heuristic selects an EV k having a maximum battery capacity (or a CV if no EV is available). Then, it inserts iteratively the customers into an active route at the position causing minimal increase in tour cost until a violation of capacity or battery capacity of the selected EV occurs. The heuristic anticipates, when possible, any violation due to the battery capacity constraint by inserting charging stations during the tour construction. The best charging station is selected among the compatible and available charging stations belonging to the neighborhood $V(i)$ of the current node i, where $V(i)$ is the set of all nodes within the circle defined by the center i and the radius α; where α is the maximum distance that could be traveled by the EV using its current state of charge. If a violation of one of the constraints occurs, the current route is assigned to the selected vehicle, another EV with a maximum battery capacity is selected and a new route is activated.

When a customer could not be reached using any of the available EVs, it is assigned to the CV engendering the minimal cost increase in the solution cost while satisfying the capacity and the total route duration constraints, until at most the predefined number of routes is constructed.

Algorithm 2 gives more details about the CRH heuristic.

3.2   Intensification Phase

The intensification phase uses a Tabu Search procedure that is based on a Large Neighborhood Search (TS-LNS). In this paper, we use a fixed length of tabu list and we make tabu any moves that have been involved in the solutions that have been visited in the recent past. The following parameters are useful in the TS-LNS:

- h: the size of the tabu list Tab
- $Iter$: parameter that controls the size of the main loop of the TS-LNS
- $IterLNS$: parameter that specifies the number of times the LNS should be repeated
- $trial$: parameter that specifies the number of times the *Random Insertion Method (RIM)* (described below) should be repeated in order to find the best improvement insertion
- Num: parameter that controls the size of the neighborhood list that will be used in the LNS

The TS-LNS procedure (see Algorithm 3) restarts $Iter$ times and for each new solution, it seeks for the best solution by performing $IterLNS$ iterations of the LNS procedure which is based on the following destroy and repair strategies. A node j and a set of $Num - 1$ additional nodes located the nearest possible

---

Algorithm 2. Charging Routing Algorithm

1: **Input:** A graph $G = (V', A)$ and a set of $m_{EV} + m_{CV}$ empty routes
2: **Output:** A set of routes assigned to at most $m_{EV} + m_{CV}$ vehicles
3: **while** the maximum number of routes is not yet reached AND there exists at least one customer that is not yet served **do**
4:    Select the EV k with the highest battery capacity among all available EVs not yet assigned
5:    **while** the total route duration is less than T and the total amount of goods delivered along the route is less than Q^{EV} **do**
6:       Sort the list of nodes randomly and let $V(i)$ be the set of all neighbors of node i not yet visited and that could be visited using the remaining battery energy of the current vehicle
7:       **if** $V(i)$ contains at least one customer and either the depot or a charging station $f \in V(j) \cap F(j)$ **then**
8:          select a node j from $V(i)$ such that $cost^{EV}_{i,j,k}$ is minimal
9:       **else if** ($V(j)$ is empty or it contains only customers or incompatible charging stations) AND (charging is possible) **then**
10:          the vehicle should get charged before visiting j, in that case insert the compatible charging station with the lowest cost while ensuring that this charging station will be available when the EV arrives at this station
11:       **else**
12:          Assign i to the CV having a sufficient capacity and engendering a minimum insertion cost
13:       **end if**
14:    **end while**
15: **end while**

---

to j (in terms of costs), are randomly selected (the selected neighbors may be in different routes). This neighborhood of Num nodes is then ejected from the solution. The ejected nodes are then re-inserted back into the partial solution using the Random Insertion Method . Here, we distinguish the First-Improvement (FI) and the Best-Improvement (BI) strategies. In best-improvement, a large neighborhood is explored and the best solution is returned. That means that, for each list of ejected nodes, random permutations of nodes are tested and the permutation that leads to the best solution is saved. In first-improvement, the first permutation improving the initial solution is saved.

For each list of ejected nodes, the RIM procedure is repeated *trial* times and, at the end, the ejected nodes are re-inserted in the route positions engendering either the best or the first improvement in the solution cost. If the solution becomes infeasible, we insert a new charger, having the lowest cost, in the route while ensuring that the constraints related to the compatibility of the charging stations with the EV are satisfied. If it is not possible to insert the ejected node in an already constructed route, a new route that contains this node and the depot may be created. In that case, the vehicle ownership cost is added to the total route cost.

When all customers have been re-inserted back into the solution, the new solution is compared with the original solution. If the resulting solution is better than the original solution, then the next iteration continues with the new solution. Otherwise, the next iteration continues with the original solution. After $IterLNS$ runs, the best solution found by LNS during the search is reported. This solution is accepted if it is not tabu. TS-LNS restarts.

Algorithm 3. Tabu Search based on LNS procedure

1: **Input:** a solution S
2: **Output:** new solution produced S'
3: Let Tab be the tabu list and set $S' := S$
4: Initially, $best\_cost := cost(S)$
5: **for** $i = 0$ to $Iter$ **do**
6:    **for** $j = 0$ to $IterLNS$ **do**
7:        Eject a list of Num nodes from the solution S'
8:        **for** $k = 0$ to $trial$ **do**
9:            Insert the ejected nodes in the cheapest route positions that are not in the Tabu list, following the RIM and let S'' be the obtained solution
10:           **if** $cost(S'') < best\_cost$ **then**
11:               $S''^* := S''$
12:               **if** $(First\_improvement)$ **then**
13:                   goto 17
14:               **end if**
15:           **end if**
16:           Eject again the list of Num nodes
17:        **end for**
18:        **if** $cost(S''^*) < cost(S')$ **then**
19:            $best\_cost := cost(S''^*)$
20:        **end if**
21:    **end for**
22:    $S' := S''^*$
23:    update the Tabu list Tab
24: **end for**

In the following, we detail the Random Insertion Method.

RIM. This method selects randomly a node among the list of ejected nodes and inserts it in the position that generates the minimal cost increase in the total solution cost. If the insertion of a customer in a given route position leads to a violation of the vehicle capacity or total time constraints, this route position will not be accepted. However, if the insertion of a customer in a given route position still satisfies the vehicle capacity and total time constraints but leads to a violation of the energy constraints (in the case where the EV needs more energy to serve this customer or the time planned for charging decreases since it depends on the opening time windows of the charging stations), this method tries to repair the solution by inserting chargers in the route while ensuring the compatibility between the EV and the chargers and satisfying the charging stations' operating time windows constraints.

3.3 Diversification Phase

The solution generated by the TS-LNS procedure is perturbed to avoid stopping at a local optimum. The diversification mechanism uses the LNS but it explores a larger neighborhood space than the one explored by the TS-LNS. In fact, during the diversification phase, the current solution is destroyed by ejecting a larger number of nodes than the number of nodes ejected in the intensification phase. The diversification phase consists in the following steps:

- Eject a random list of $Num_{perturb}$ nodes such that $Num_{perturb} > Num$.
- Inject randomly the ejected nodes.

3.4 Acceptance Criterion

To escape from a current locally optimal solution, non improving-solutions could be accepted. Our acceptance criterion is based on the mechanism of accepting non-improving solutions used by the Record-to-Record algorithm [16]. During the run of the ITS-LNS procedure, any solution is accepted if its objective value is lower than $(1 + Dev) \times Record$, where the $Record$ is the value of best solution obtained and Dev is a parameter. Initially, $Record$ is equal to the initial objective function. During the search process, $Record$ is updated with the objective value of the best solution so far.

4 Computational Experiments and Discussion

Our methods were implemented using C++. All experiments were carried out on an Intel Xeon E5620 2.4GHz processor, with 8GB RAM memory. We conducted numerical experiments on generalized E-VRPTW Benchmark Instances proposed in [6], denoted VRP-MFHEV. Each VRP-MFHEV instance is composed of 100 customers and 21 charging stations proposing different charging technologies and different charging costs. The fleet of vehicles is composed of 50 CVs, 25 EVs having 22 (kWh) battery packs and 25 EVs having 16 (kWh) battery packs.

4.1 Computational Results

The computational results concerning the different implementation schemes are summarized in Table 1. Table 2 presents the results generated by the Tabu Serach procedure and the ILS algorithm presented in [11]. For each implementation scheme, the entries show the average gap (Gap(%)) and the average computational time (CPU (s)) for each instance category. The Gap of a generated solution

(S) is calculated in relation to the initial solution $(Gap = \frac{S_{CRH}-S}{S})$. In Table 1, we evaluate the influence of the different components of the ITS-LNS on the quality of the generated solutions. In Table 2, we compare the results obtained by the ITS-LNS algorithm with those obtained by the TS-LNS, ITS-2opt and the Iterated Local Search algorithm.

Preliminary experiments carried out allowed us to fix the values of many parameters of the algorithm. For all the experiments, the parameter max_{Iter} was fixed at 10000 iterations, max_{impr} at 100, $trial$ at 50 and h at 7. For each instance, we tested our methods following different implementation schemes obtained by varying the parameters of the ITS-LNS algorithm. The different implementation schemes are represented by the quadruplets $(Num, Num_{perturb}, Max\_Restart, imp)$ where $imp = 0$ if the FI strategy is chosen and $imp = 1$ if the BI strategy is chosen.

Table 1. Influence of the different ITS-LNS parameters on the quality of generated solutions

Instance	(2,5,1,0)		(3,5,1,0)		(4,5,1,0)		(2,6,1,0)		(3,6,1,0)		(4,6,1,0)	
	Gap	CPU(s)	Gap	CPU(s)	Gap	CPU(s)	Gap	CPU(s)	Gap	CPU(s)	Gap	CPU(s)
C1	21.77	12.57	25.94	14.39	30.38	15.04	22.26	12.96	25.38	13.12	30.35	9.71
C2	26.21	13.07	29.72	14.34	36.75	11.67	26.21	9.86	31.51	11.11	36.67	16.13
R1	33.81	14.58	39.41	12.32	45.87	13.60	35.38	15.44	40.88	16.30	45.45	14.55
average	27.26	13.40	31.69	13.68	37.65	13.43	27.95	12.75	32.59	13.51	37.49	13.46
Instance	(2,7,1,0)		(5,7,1,0)		(7,8,1,0)		(2,5,1,1)		(2,5,2,0)		(2,5,3,0)	
	Gap	CPU(s)	Gap	CPU(s)	Gap	CPU(s)	Gap	CPU(s)	Gap	CPU(s)	Gap	CPU(s)
C1	21.48	15.23	38.89	16.13	43.21	16.03	21.57	118.88	98.15	22.59	98.60	33.24
C2	25.65	12.15	42.73	13.11	48.77	18.26	25.18	105.13	103.83	24.32	106.71	37.53
R1	34.52	10.87	52.59	11.54	60.17	16.14	34.15	136.43	128.89	23.90	133.62	36.93
average	27.21	12.75	44.73	13.59	50.71	16.81	26.96	120.14	110.29	23.60	112.97	35.90

Table 2. Performance of ITS-LNS compared to TS, ILS and ITS-2opt

Instance	TS		ILS		ITS-LNS		ITS-2opt	
	Gap	CPU(s)	Gap	CPU(s)	Gap	CPU(s)	Gap	CPU(s)
C1	21.85	0.018	21.35	14.08	22.26	12.96	21.97	13.16
C2	24.83	0.0175	25.42	9.58	26.21	9.86	25.73	9.87
R1	32.50	0.017	32.41	14.34	35.38	15.44	33.67	16.22
average	26.39	0.0175	26.39	13.11	27.95	12.75	27.12	13.08

4.2   Discussion

The computational results presented in Table 1 show that the value $diff = Num_{Perturb} - Num$ and the size of the list of ejected nodes Num impact the quality of the solutions generated by the ITS-LNS algorithm. In fact, better solutions are obtained when $diff$ is small $(diff = 1, diff = 2)$ and Num is high. In the case where $Num = 7$ and $Num_{perturb} = 8$ and for the instances $R1$, the ITS-LNS improves by around 60% the initial solutions. However, in the case where $Num = 2$ and $Num_{perturb} = 5$, the ITS-LNS improves by only 33% the initial solutions. Among the 9 first implementation schemes, the configuration (7,8,1,0) seems to be the best in terms of quality of the generated solution with an average gap of 50.71%, and the computational time remains reasonable (16.81 s). Moreover, ITS-LNS with BI seems to generate better solutions for the instance class $R1$ but it consumes much more computational time than ITS with FI. For the instance classes $C1$ and $C2$, ITS-LNS with FI strategy generates better solutions than ITS-LNS with BI strategy. Furthermore, it is obvious that the ITS-LNS with restart improves significantly the quality of the generated solutions. In fact, the ITS-LNS improves the generated solutions up to 128% in the case where $Max\_Restart = 2$, and up to 133% in the case where $Max\_Restart = 3$. The computational time remains acceptable for both cases.

The computational results presented in Table 2 show that the TS-LNS generates acceptable solutions in a very short computational time (less than 0.02 s). Moreover, the ITS-LNS seems producing competitive results with respect to the results generated by the Iterated Local Search. In fact, the ITS-LNS improves almost all solutions generated by the ILS algorithm, while the computational time remains reasonable for each instance. Furthermore, we can notice that using the LNS procedure in the intensification phase of ITS-LNS improves the obtained results compared to using more classical neighborhood search procedure such as 2opt search.

5   Conclusion

In this paper, we considered a new real life vehicle routing problem with mixed fleet of conventional and heterogenous electric vehicles, denoted VRP-MFHEV. This problem extends the electric vehicle routing problem studied in the literature by considering simultaneously a mixed fleet of CVs and heterogenous EVs, compatibility constraints between EVs and chargers, several charging technologies and different charging costs. To solve this problem, we developed a Multi-Start Iterated Tabu Search which uses a LNS in the intensification and diversification phases. Our method was tested on generalized benchmark instances. The computational results show that ITS-LNS produces competitive results, with respect to results obtained in previous studies, while the computational time remains reasonable for each instance. Moreover, we concluded that using LNS in the intensification phase improves the generated solutions compared to other neighborhood search procedures such as 2opt search. As further work, we will study lower bounds and we will relax our problem and compare our results with those of the literature.

References

1. Aissat, K., Oulamara, A.: A posteriori approach of real-time ridesharing problem with intermediate locations. In: Proceedings of ICORES 2015 (2015)

2. Sassi, O., Oulamara, A.: Simultaneous electric vehicles scheduling and optimal charging in the business context: Case study. In: IET, IET (2014)
3. Artmeier, A., Haselmayr, J., Leucker, M., Sachenbacher, M.: The optimal routing problem in the context of battery-powered electric vehicles. In: Workshop CROCS at CPAIOR-10, 2nd International Workshop on Constraint Reasoning and Optimization for Computational Sustainability (2010)
4. Erdogan, S., Miller-Hooks, E.: A green vehicle routing problem. Transport. Res. Part E 48, 100–114 (2012)
5. Lin, C., Choy, K., Ho, G., Chung, S., Lam, H.: Survey of green vehicle routing problem: Past and future trends. Expert. Syst. Appl. 41, 1118–1138 (2014)
6. Schneider, M., Stenger, A., Goeke, D.: The electric vehicle routing problem with time windows and recharging stations. Technical report, University of Kaiserslautern, Germany (2012)
7. Goeke, D., Schneider, M., Professorship, D.S.E.A.: Routing a mixed fleet of electric and conventional vehicles. Technical report, Darmstadt Technical University, Department of Business Administration, Economics and Law, Institute for Business Studies, BWL (2014)
8. Hiermann, G., Puchinger, J., Hartl, R.F.: The electric fleet size and mix vehicle routing problem with time windows and recharging stations. Technical report, Working Paper (accessed July 17, 2014), http://prolog.univie.ac.at/research/publications/downloads/Hie_2014_638.pdf (2014)
9. Felipe, Á., Ortuño, M.T., Righini, G., Tirado, G.: A heuristic approach for the green vehicle routing problem with multiple technologies and partial recharges. Transportation Research Part E: Logistics and Transportation Review 71, 111–128 (2014)
10. Sassi, O., Oulamara, A.: Joint scheduling and optimal charging of electric vehicles problem. In: Murgante, B., et al. (eds.) ICCSA 2014, Part II. LNCS, vol. 8580, pp. 76–91. Springer, Heidelberg (2014)
11. Sassi, O., Cherif-Khettaf, W.R., Oulamara, A.: Multi-start iterated local search for the mixed fleet vehicle routing problem with heterogenous electric vehicles. In: Ochoa, G., Chicano, F. (eds.) EvoCOP 2015. LNCS, vol. 9026, pp. 138–149. Springer, Heidelberg (2015)
12. Sassi, O., Ramdane Cherif-Khettaf, W., Oulamara, A.: Vehicle routing problem with mixed fleet of conventional and heterogenous electric vehicles and time dependent charging costs. Technical report, hal-01083966 (2014)
13. Misevičius, A.: Using iterated tabu search for the traveling salesman problem. Information Technology and Control 32, 29–40 (2004)
14. Shaw, P.: Using constraint programming and local search methods to solve vehicle routing problems. In: Maher, M.J., Puget, J.-F. (eds.) CP 1998. LNCS, vol. 1520, pp. 417–431. Springer, Heidelberg (1998)
15. Pisinger, D., Ropke, S.: Large neighborhood search. In: Handbook of metaheuristics, pp. 399–419. Springer (2010)
16. Li, F., Golden, B., Wasil, E.: A record-to-record travel algorithm for solving the heterogeneous fleet vehicle routing problem. Computers & Operations Research 34, 2734–2742 (2007)

Optimal Migration Planning of Telecommunication Equipment

Alain Billionnet[1], Sourour Elloumi[1], and Aurélie Le Maître[2]

[1] CEDRIC-ENSIIE 1, Square de la Résistance, 91025 Evry, France
[2] GDF SUEZ 361, avenue du Président Wilson, 93210 Saint-Denis La Plaine, France

Abstract. Frequent upgrades of equipment in the telecommunications industry occur due to the emergence of new services or technological breakthroughs. In this work, we consider a network where each client is linked to a site and handled by a card located on that site. A technological migration has to be undertaken within a short horizon of a few years and it consists of replacing all the existing cards by cards of a new generation within a fixed number of years. For practical considerations, all the cards of a site must be replaced in the same year. Furthermore, we can assume that, because of new offers, the number of clients per site is decreasing during the planning horizon. This enables us to reuse the new cards that are not used any more once some clients have left. The optimization problem consists of deciding, for each year, which sites are migrated and how many cards are bought or reused, in order to minimize the total cost. We present an exact solution for this problem, based on an integer linear programming formulation.

Keywords: Telecommunications, migration, Integer programming, Practice of OR.

1 An Industrial Issue

A telecommunication operator such as France Telecom is often confronted with the emergence of new technologies that implies adaptation of the existing networks. This problem occurred recently with the growth of Voice Over Internet Protocol (or VoIP), which is now one of the most important ways of conveying information. In the initial state of the network for fixed telecommunications, France Telecom is using PSTN technology (Public Switched Telephone Network) and, in order to reduce operating costs, the operator wishes to replace this technology by VoIP within a predetermined horizon of a few years. This means that in each site of the network, the PSTN equipment connected to the client has to be replaced by a VoIP compatible card. The overall operation has to be realized at minimal cost. So, we have to determine the optimal replacement year for each site in order to reduce the total cost. We will call migration the equipment replacement process in a site and we will assume that we need one VoIP card per client.

Furthermore, a particular feature has to be considered. Indeed, with competition between providers and emergence of new offers inside France Telecom as

H.A. Le Thi et al. (eds.), *Model. Comput. & Optim. in Inf. Syst. & Manage. Sci.*,
Advances in Intelligent Systems and Computing 359, DOI: 10.1007/978-3-319-18161-5_7

well, clients are more and more subject to cancel their standard contract for offers that combine TV, Internet and phone services. These new offers are directly using VoIP technology. As a consequence, the number of clients concerned by the technological change is decreasing. This raises an issue: when a new expensive VoIP card has been set up for a client who leaves afterwards, the investment is lost. To secure a return on this investment, cards that become unused before the end of the planning horizon can be reused. Such cards are disconnected and installed again for other clients on a site not yet migrated to VoIP. Taking into account this possibility, we have to determine for each year of the planning horizon, how many VoIP cards are bought and how many of them are reused. As reusing a card is less expensive than buying a new one, we will try to maximize the number of reused cards. This specificity is responsible for the problem's difficulty. In fact, without the possibility of reusing cards, we could determine the optimal year of migration for each site independently and would be closer to classical equipment replacement problems (see [1], [4] and [5]). However, an important difference between our problem and equipment replacement is that we change the equipment only once.

A Basic Example. In order to illustrate this problem, we propose a simple example of a network composed of only two sites. Without dealing with cost issues, we present a feasible solution. In Table 1, we give the number of cards, i.e. of clients, on each site per year during the five-year planning horizon.

Table 1. A basic 2-site example with decreasing number of cards on each site

Year	1	2	3	4	5
Number of clients of site 1	4	4	3	2	1
Number of clients of site 2	5	4	3	3	2

Figure 1 shows a possible solution for the basic example. In this solution, the first site migrates at year 2 and the second site at year 4. A site uses PSTN cards before its migration and VoIP cards after its migration. When VoIP cards become useless for a site because its number of clients (needed cards) decreases, they are disconnected and placed in a global stock available for further migrations of other sites. On the first year, no site is migrated so no VoIP card is needed. On the second year, we have to migrate site 1 and the stock of VoIP cards is empty, so we buy 4 new VoIP cards, corresponding to the current number of clients on site 1. From the second to the third year, one client on site 1 has left, so the corresponding VoIP card is disconnected and put into the stock. At the beginning of the fourth year, another client has left from site 1, so another VoIP card is placed into the stock. At this same year, site 2 has to be migrated and has three clients. As reused cards are cheaper than new ones, we take two cards from the stock, we buy a new VoIP card and we connect the three cards to site 2.

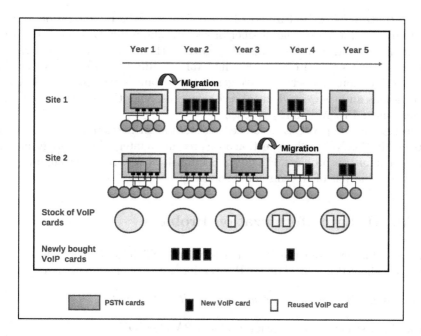

Fig. 1. A possible solution for the basic example of Table 1. Site 1 is migrated year 2 and site 2 is migrated year 4

Eventually, at the beginning of the fifth year, two more VoIP cards are put into the stock, even if they are no more needed, as all the sites have already been migrated. Such a stock could be used for example to replace some faltering equipment.   □

Another specificity is that a tax has to be paid each year for each kind of equipment set up in the network: on old PSTN equipment as well as on new VoIP cards. This tax represents a percentage of the equipment purchasing price. Therefore, when a site is migrated to VoIP, tax on the PSTN equipment is no more paid. In contrast, a tax on new equipments (VoIP cards) is to be paid. Regarding the old equipments (PSTN cards), the amount of tax payable for each site is divided into two parts: a fixed part and a part associated with each card. Note that these two parts are site-dependent. As tax is proportional to the equipments purchase price, and old PSTN equipements were very expensive at the time they were bought, paying the tax on new VoIP equipments is actually cheaper than on the old PSTN equipments.

To conclude, we have two types of decisions to make: the year at which each site is migrated, and the number of cards bought or reused every year. Our objective is to minimize the total operating costs. We call this optimization problem MigR (for Migration with Reuse of cards).

Target Instances to be Solved. France Telecom fixed the migration horizon to 5 years and estimated the number of sites to about 10000. Even if they did not provide us with real instances, they gave us their specifications in terms of number of clients per site and year, intervals of different costs, ... We precise these specifications in Section 4. We generated instances corresponding to these specifications and, as we well show, succeeded in solving these instances with an Integer Linear Programming approach.

The rest of the paper is organized as follows: In Section 2, we show that problem MigR is NP-hard. In Section 3 we present an integer linear programming formulation for this problem. Finally, in Section 4, we report some computational results and analysis of the computed solutions.

2   An NP-hard Optimization Problem

Using a reduction from the Partition problem, which is NP-complete [3], we prove now the NP-hardness of MigR, even for a 2-year planning horizon.

Theorem 1. *The optimization problem MigR is NP-hard.*

Proof. We first define the decision problem PARTITION.

PARTITION: Let $A = \{a_1, ..., a_p\}$ be a set of p elements and denote by $s(a_i) \in \mathbb{N}$ the weight of a_i. Let $B = \sum_{a_i \in A} s(a_i)$. Is there a subset A' of A such that $\sum_{a_i \in A'} s(a_i) = \dfrac{B}{2}$?

Let *Part* be an instance of PARTITION. From this instance, let us now construct an instance $DMigR2$ of the decision problem associated to MigR with a 2-year planning horizon.

$DMigR2$: Let $S = \{S_1, ..., S_p\}$ be the set of p sites of a network such that the number of clients of any site S_i is equal to $2s(a_i)$, the first year, and equal to $s(a_i)$, the second year. Let $P > 1$ be the price of a new VoIP card. The reuse cost of a card is 0. The tax gain corresponding to the installation of a new card is of 1 per card the first year and of 0 per card the second year. Is there a migration solution of total cost less than or equal to $PB - B$?

- Let us first prove that, if there exist A' such that $\sum_{a_i \in A'} s(a_i) = \dfrac{B}{2}$, we can build a solution of $DMigR2$ with a cost less than or equal to $PB - B$. On the first year, we migrate all sites S_i such that $a_i \in A'$. The remaining sites are migrated on the second year. In this solution, we need to buy exactly B cards on the first year. The associated price is PB and the associated tax gain is B. On the second year, since half of the migrated clients have left, $\dfrac{B}{2}$ cards become available. They can be used for the non-migrated sites. These remaining sites require precisely $\dfrac{B}{2}$ cards. So, no additional cost is needed on the second year, and the total cost of this solution is $PB - B$.

- Now, let us prove that from any solution Sol of $DMigR2$ with a cost less than or equal to $PB - B$, we can build a set A' such that $\sum_{a_i \in A'} s(a_i) = \dfrac{B}{2}$.

  In solution Sol, all the sites are migrated on the first or second year. Let A' be the set composed of the elements a_i such that, in Sol, site S_i is migrated on the first year. Let X be the number of cards bought and installed on the first year. This means that $\sum_{a_i \in A'} s(a_i) = \dfrac{X}{2}$.

- Now, we prove that $X = B$. At the end of the first year, $\frac{X}{2}$ cards are disconnected, since half of the clients have left. At year 2, there remains only $\frac{2B-X}{2} = B - \frac{X}{2}$ clients to migrate.

  - Suppose that $X > B$. The cost associated to the first year is $PX - X$. The cost associated to the second year is non negative since there is no tax gain. Hence, the total cost is at least $PX - X$ which is strictly greater than $PB - B$ because $P > 1$. This contradicts the fact that the cost of Sol is less than or equal to $PB - B$.

  - Suppose that $X < B$. On the second year, $\frac{X}{2}$ cards become available and the remaining sites require $B - \frac{X}{2}$ cards. So, $B - X$ new cards have to be bought at a price equal to $P(B - X)$. Hence, the total cost is equal to $PX - X + PB - PX = PB - X$. As $PB - X > PB - B$, this contradicts the fact that the cost of Sol is less than or equal to $PB - B$.

  So, $X = B$ and we proved that A' satisfies $\sum_{a_i \in A'} s(a_i) = \dfrac{B}{2}$.

3   Integer Linear Programming Formulation

We formulate problem MigR by an integer linear program.

We are given the following data:

m	: number of years of the planning horizon dedicated to the migration
p	: number of sites to be migrated
$n_{i,t}$	: number of clients (PSTN or VoIP cards needed) on site i at year t
P_t	: price of a VoIP card at year t
IA_t	: cost of reusing a VoIP card at year t
I_t	: unit installation cost of VoIP cards at year t
CC_i	: fixed cost due to the use of PSTN cards for site i
CI_i	: unit cost of a PSTN card for site i
Tax_t	: tax at year t, to be paid on the total costs of VoIP or PSTN cards

We denote by M the set $\{1, ..., m\}$ and by P the set $\{1, ..., p\}$.

The aim is to plan the migration of p sites over a horizon of m years. The migration of a site consists of the replacement of all its PSTN card by VoIP cards. At year t, if site i is migrated, it requires $n_{i,t}$ VoIP cards. Each card is either bought at price P_t or reused for a cost IA_t. The actual cost associated with a new card is P_t plus the taxes that have to be paid starting from year t. Let $PT_t = P_t(1 + \sum_{k=t}^{m} Tax_k)$ be that actual cost. An additional cost comes from the installation of the cards at the unit cost I_t. However, a gain comes from the fact that the company will stop paying taxes on the replaced PSTN cards. Starting from year t and for any following year k until the end of the planning horizon, the telecom company will save, for each site i, $Tax_k(CC_i + CI_i n_{i,k})$. For convenience, we will denote by $GT_{i,t}$ the total saving of the company if site i is migrated at year t, i.e. $GT_{i,t} = \sum_{k=t}^{m} Tax_k(CC_i + CI_i n_{i,k})$. As the number of cards per site is decreasing over the years, a balance has to be found between a late migration policy that allows to buy less cards and an early migration policy that will allow to save more taxes on PSTN cards.

Coming back to the cost of buying a VoIP card, we can observe that there is no storage cost. It may be more interesting to buy a card a few years before the year it is needed. Hence, in an optimal migration policy, we can consider that the cost of a VoIP card at year t is not PT_t but the smallest PT_k where k varies from 1 to t. In the following, we will consider $PR_t = min_{k=1...t} PT_k$ as the cost of a first use of a VoIP card at year t, regardless to its acquisition year. It can be easily checked that PR_t decreases over the years.

We now introduce three sets of variables:
$x_{i,t} \in \{0,1\}$: binary variable equal to 1 if and only if site i is migrated at year t
$a_t \in \mathbb{N}$: number of new VoIP cards needed at year t
$r_t \in \mathbb{N}$: number of reused VoIP cards at year t

We formulate MigR by the following integer linear program:

$$(PL0) \begin{cases} \min \sum_{t \in M} PR_t a_t + \sum_{t \in M} IA_t r_t + \sum_{t \in M} I_t \sum_{i \in P} n_{i,t} x_{i,t} \quad - \sum_{t \in M} \sum_{i \in P} GT_{i,t} x_{i,t} & (1) \\[2mm] \text{s.t.} \\[1mm] \sum_{t \in M} x_{i,t} = 1 & \forall i \in P & (2) \\[2mm] r_1 = 0 & & (3) \\[2mm] a_t + r_t = \sum_{i \in P} n_{i,t} x_{i,t} & \forall t \in M & (4) \\[2mm] \sum_{k \leq t} r_k \leq \sum_{k < t} \sum_{i \in P} (n_{i,k} - n_{i,t}) x_{i,k} & \forall t \in M - \{1\} & (5) \\[2mm] x_{i,t} \in \{0,1\} & \forall i \in P, \forall t \in M \\[1mm] a_t, r_t \in \mathbb{N} & \forall t \in M \end{cases}$$

The objective function (1) is composed of the purchase and tax cost, the reuse cost, the installation cost, and the gain related to the savings of taxes on PSTN cards.

The main constraints (2) concern the year of migration of the sites and impose that each site is migrated during the horizon. Constraint (3) says that at the beginning of the horizon, the stock of available cards is empty. At each year, the number of VoIP cards needed is equal to the sum of new and reused cards installed. This is guaranteed by Constraints (4). Eventually, we cannot reuse more cards than those available in the stock. Hence, each year, the number of cards reused since the beginning of the horizon must be lower than the number of cards that are no longer used. This is imposed by Constraints (5).

In order to lighten our formulation, we suppress in Constraints (4) variables a_t that can be viewed as slack variables and we substitute $\left(\sum_{i \in P} n_{i,t} x_{i,t} - r_t\right)$ to a_t in the objective function. We obtain the following integer linear program:

$$
(PL1) \begin{cases}
\min \sum_{t \in M}(IA_t - PR_t)r_t + \sum_{t \in M}\sum_{i \in P}(PR_t + I_t)n_{i,t}x_{i,t} \quad - \sum_{t \in M}\sum_{i \in P} GT_{i,t}x_{i,t} \\
\text{s.t.} \\
\sum_{t \in M} x_{i,t} = 1 & \forall i \in P \\
r_1 = 0 \\
\sum_{k \leq t} r_k \leq \sum_{k < t}\sum_{i \in P}(n_{i,k} - n_{i,t})x_{i,k} & \forall t \in M - \{1\} \\
r_t \leq \sum_{i \in P} n_{i,t}x_{i,t} & \forall t \in M \\
x_{i,t} \in \{0,1\} & \forall i \in P, \forall t \in M \\
r_t \in \mathbb{N} & \forall t \in M
\end{cases}
$$
(6)

We get a compact linear integer programming formulation for problem MigR.

By inspecting the row matrix of problem $PL1$, one can observe that the sub-matrix associated to variables r_t is a 0-1 matrix and it has the consecutive-ones property. This sub-matrix is hence totally unimodular (see for example [6]). Further, as the $n_{i,t}$ coefficients are integers, Constraints (6) can be replaced by simple non negativity conditions on the r_t variables. The r_t variables will have integer values in an optimal solution of the obtained problem. Another consequence is that, if the migration years are known (i.e. the $x_{i,t}$ variables are fixed), the problem of finding an optimal solution for buying or reusing VoIP cards can be solved by linear programming and therefore in polynomial time. Recall that the decision problem with the x and r variables is NP-hard. It follows that the difficult part of the decision is only in determining the x variables values, i.e. the migration year of each site.

4 Computational Results

We now focus more precisely on the problem solution. We implemented the mathematical problem $(PL1)$ using the modeling language AMPL [2] and solved it by the mixed-integer linear solver of CPLEX 12.5 with all parameters set to their default value. The experiments have been carried out on a PC with an Intel Core i5-2540M processor having 8 Go of RAM and running Windows 7.

Specifications of the Target Instances. We generate 5 instances following the specifications given by France Telecom and inspired from real-life instances.

The number of periods (years) in the planning horizon, m, is equal to 5. The number of sites, p, is equal to 10000.

For each site i, we first randomly generate $n_{i,1}$, the number of clients -or PSTN cards- in the first year in the interval [50, 2500]. Then, for any following year t, we generate a random coefficient ρ in the interval [0.5, 1] and set $n_{i,t} = \rho n_{i,t-1}$. The obtained $n_{i,t}$ are decreasing over the years as required in our assumptions on the problem.

For the other data, we consider that the variation from a year to the following one are small. We first generate P_1, IA_1, I_1, CC_1, CI_1 and Tax_1 in the intervals which bounds are given in Table 2. Then, iteratively for each of them and for each year, we generate a small coefficient ν in the interval [0.95, 1.05] and multiply it by the data of the previous year.

Table 2. Intervals used for random generation of the initial values of data

	P_1	IA_1	I_1	CC_1	CI_1	Tax_1
lower bound of the interval	40	2	2	10000	50	0.03
upper bound of the interval	70	5	10	50000	100	0.05

Solution Results and Analysis for the Target Instances. We randomly generate 5 instances following the specifications of the target instances. In Figure 2, each curve represents an instance and shows the evolution of the cumulated number of migrated sites over the years of the horizon, in the obtained optimal solution. This cumulated number is always equal to 10000 at the end of the horizon meaning that all the sites are migrated. We can observe that the curves are very different. For some instances, like Instance 3, the optimal solutions is to migrate very few sites in the beginning of the horizon and to perform more migrations at the end of the horizon. But for others, like Instance 5, it is the opposite. None of these optimal solutions consists to simply migrate all the sites at the end of the horizon. The cost of these simple solutions is 62% higher in average than the cost of the optimal solutions.

Other aspects of the optimal solutions of the 5 instances are illustrated in Figure 3. For each year, 3 columns are represented. The first one is the number of clients of all the sites, as given in the data. Following our assumption, this number is decreasing over the years. The second column gives the cumulative number of migrated clients since the beginning of the horizon, in the optimal solution. The third column gives the cumulative number of purchased cards since the beginning of the horizon, in the optimal solution. By definition, the two last numbers are increasing over the years. We can make the following observations:

- Except for the first year, the number of purchased cards is always lower than the number of migrated clients. This is due to the fact that, starting from year 2, some cards bought on the previous years may be reused.
- On the last year, the cumulative number of purchased cards is equal or very close to the number of clients. As this number is the smallest over

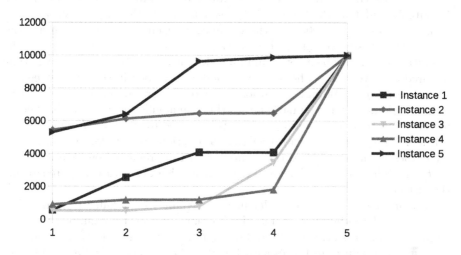

Fig. 2. Optimal solutions: Cumulative number of migrated sites from the beginning of the horizon

the years, it represents a lower bound on the VoIP cards to be purchased. This means that, in the optimal solution, we almost never buy useless cards. Nevertheless, the reuse possibility allows us to migrate more clients.

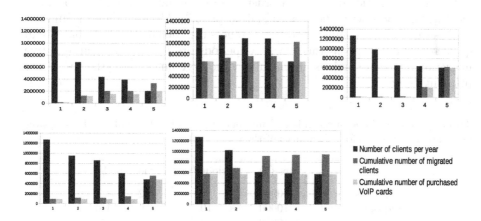

Fig. 3. For each of the 5 instances, number of clients per year, cumulative number of migrated clients, and cumulative number of purchased VoIP cards

Let us finally report that CPLEX does not need to perform branching for any of the instances. The average solution time over the 5 instances is 2 seconds. Theses averages are also mentioned in Table 3.

Influence of the Number of Sites and the Number of Periods on the Resolution of the Problem. As we saw above, in the target instances, the number of sites p is 10000 and the planning horizon m is 5 years. In a series of additional tests, we varied these parameters to show their influence on the resolution of the problem by examining, in each case, the computation time required to solve $(PL1)$, the number of nodes developed in the search tree and the structure of the obtained solutions. Our results are summed up in Table 3.

In additional test 1, we first assumed that the number of sites remained equal to 10000 but that the planning horizon is increased to 10 years. We find that to consider a 10-period horizon rather than a 5-period horizon does not introduce difficulties in the resolution: the computation time remains low. We also note that, for the five considered instances, a large number of sites are migrated during the last period.

In additional test 2, we consider instances comprising again 10000 sites, but with a very large number of periods. We can assume, for example, that the planning horizon is always 5 years but that the period is not the year but the month. This produces instances with 10000 sites and 60 periods. We can observe that this large number of periods does not make the resolution impossible. In fact, the computation times remain relatively low. We can also note that few periods are concerned by an effective migration and that these periods are mainly at the beginning and at the end of the planning horizon.

In additional test 3, we study the influence of the number of sites on the difficulty of the problem. For this we consider 5 instances with 100000 sites and a management horizon of 5 periods. The results suggest that the significant increase in the number of sites does not make the problem more difficult to solve.

Finally, in additional test 4, in order to test the limits of our approach, we consider instances with 20000 sites and 60 periods. The computation time increases significantly compared to the instances considered previously but is still quite reasonable. Also in this case, few periods are concerned by an effective migration and these periods are mainly at the beginning and at the end of the planning horizon.

Table 3. Solution time in seconds and number of nodes for different numbers of sites and periods (averages for 5 instances of the same size)

	Sites (p)	Periods (m)	time	nodes
target instances	10000	5	2	0
additional test 1	10000	10	9	161
additional test 2	10000	60	105	156
additional test 3	100000	5	17	0
additional test 4	20000	60	574	0

Influence of the Costs. We examined the impact of increase the cost of reusing a VoIP card on optimal solutions. For this, we have drawn the values IA_1 in the same interval as that of P_1 ([40,70]) keeping all other intervals. We observed that the increase in IA_1 led to optimal solutions where the migrations focus on one or two periods: the first one, the last one or another period. In addition, the average decrease in the cost, compared to the alternative of regularly migrate the sites, is about 30% over 5 instances with 10000 sites and 5 periods.

We also examined the impact of increase the values of Tax_t on optimal solutions. For this, we have strongly increased the values of Tax_1 without changing the values of other parameters. We observed that this increase in Tax_1 did not significantly alter the structure of the optimal solutions.

Possible Extensions of Our Approach. The fact of having been able to formulate the problem with an integer linear program makes it easy to consider several extensions. We give two examples below.

First, suppose that, besides the costs already considered, it is interesting for the company to migrate as soon as possible a client from the old technology to the new one which offers much more possibilities. To take account of this, we can consider for example that at each period t of the planning horizon a unit revenue, g_t, is associated with each client still present and already migrated. In this case the optimal solution of the problem is obtained by solving the program $(PL1)$ in which the quantity $\sum_{t \in M} g_t \sum_{i \in P} n_{i,t} \sum_{k=1}^{t} x_{i,k}$ is added to the objective function.

A second extension of the model may be to impose the migration of a certain number of sites, m_t, at each period t. For example, it may be interesting for many reasons to balance the workload related to the migration over time. This new aspect of the problem can easily be taken into account by adding to $(PL1)$ the constraints $\sum_{i \in P} x_{i,t} = m_t \quad \forall t \in M$. Some experiments we conducted with the modified model showed that the introduction of these new constraints does not significantly slow the resolution of the problem.

5   Conclusion

To sum up, we studied a very specific industrial problem arising in telecommunication networks and proved its NP-hardness. Several integer linear programming formulations are conceivable for this problem. We proposed a relatively simple one which is very efficient, since it provides an exact solution for real-life instances within an average time of 2 seconds. We provide an analysis of the solution, we show the impact of changing several parts of the data, and we explore possible extensions of the considered problem. The different tests we performed showed that the retained formulation was robust. Indeed, the model can be extensively modified (objective function, additional constraints, different sizes) without compromising the achievement of optimal solutions in a reasonable time. An important asset of the approach is that it uses exclusively standard, commercially available software.

Acknowledgments. The authors would like to thank the referees for their comments and suggestions that helped us to improve the first version of the article. They are also grateful to Francis Sourd for helpful discussions.

This work was carried out during the PhD of Aurélie Le Maître at Orange Labs, Issy-Les-Moulineaux, France.

References

1. Brown, D.C.: Facing the challenges of equipment replacement. Public Works 136(11), 21 (2005)
2. Fourer, R., Gay, D.M., Kernighan, B.W.: AMPL: A Modeling Language for Mathematical Programming. Duxbury Press (2003) ISBN 978-0-534-38809-6
3. Garey, M.R., Johnson, D.S.: Computers and Intractability: a Guide to the Theory of NP-completeness. W.H. Freeman and Co., San Francisco (1979)
4. Hartman, J.C., Murphy, A.: Finite-horizon equipment replacement analysis. IIE Transactions 38(5), 409–419 (2006)
5. Hopp, W.J., Nair, S.K.: A model for equipment replacement due to technological obsolescence. European Journal of Operational Research 63(2), 207–221 (1992)
6. Nemhauser, G.L., Wolsey, L.A.: Integer and Combinatorial Optimization. Wiley and Sons (1999)
7. Wolsey, L.A.: Integer Programming. Wiley and Sons (1998)

Optimal Ordering of Tests
with Extreme Dependencies

Daniel Berend[1], Shimon Cohen[3], Solomon E. Shimony[2], and Shira Zucker[4]

[1] Departments of Mathematics and Computer Science,
Ben-Gurion University, Beer Sheva, Israel
[2] Department of Computer Science,
Ben-Gurion University, Beer Sheva, Israel,
[3] Orbotech, Yavne, Israel
[4] Department of Computer Science,
Sapir academic College, Shaar Hanegev, Israel

Abstract. Optimization of testing strategies has numerous facets. Here we examine the case where tests are one-sided perfect - thus an optimal strategy consists of a sequence of tests, and the remaining problem is to find an optimal (w.r.t. time, or any other resource) ordering of a given set of tests. In prior work, we examined conditions under which statistically independent test sequences can be optimized under precedence constraints. This paper examines conditions under which one can efficiently find an optimal ordering of tests with statistical dependencies. We provide low-order polynomial time algorithms for special cases with non-trivial dependency structures.

1 Introduction

Classifications of objects in real time is one of the most important activities to be applied in many types of applications. The variety of such applications ranges from factory inspection lines where fault product manufacturing needs to be detected, through robotic applications which need to determine whether an object is an obstacle or not, to security systems that need to detect real time security threats and act upon them. An important method that addresses this issue for image classification is the feature cascade architecture for rapid object detection [5].

The *basic sequence optimization problem* is defined as follows.

Problem 1. Let $X = \{x_1, x_2, \ldots, x_n\}$ be a set of tests for a certain property, such as a defect. Each test may return either "reject" or "don't know". Each object is tested by a sequence of tests, until it is rejected by some test, or there are no more relevant tests. The tests are assumed to be one-sided perfect, i.e., a "reject" means that the tested object does not have the desired property, with no errors possible. Also, we are given the execution time t_i for each test, and the "reject" probability r_i of each x_i. The problem is to find the ordering of the tests which minimizes the expected runtime.

In the basic sequence optimization problem, we assume that there is no structure of any sort, i.e. that all sequences are allowed, that all tests are statistically independent, and that exactly one test must reject in order to detect the desired property. Adding precedence constraints was thoroughly examined in [1], where we use some properties taken from [3] and show that Problem 1 with precedence constraints on the tests is NP-complete. This paper examines the case without precedence constraints, but where tests are statistically dependent, in which case the reject probability is conditional on the outcome of previous tests.

Dependency may lead to some of the tests being irrelevant, in which case they will be dropped from the sequence. For example, if one test x is dominated by another test, or by a union of tests, which were already executed and returned "don't know", then redundant test x is dropped. (Here, x is dominated by $y_1 \cup y_2 \cup \ldots \cup y_k$ if each time x rejects a test, at least one of the tests $y_1, \ldots, y_k$ rejects it. See Definition 2.) Nevertheless, we assume that testing continues as long as no previous test in the sequence has rejected, and there still exist unperformed tests with a positive probability for rejection (given the results of earlier tests in the sequence).

The problem of dependent test sequence optimization is NP-hard, even under the simplifying assumption of no precedence constraints [2]. We therefore focus on some special cases with extreme dependencies, as follows. The first type of extreme dependency is one test *dominating* another, simpler test. Namely, it rejects in every case in which the simpler test rejects, as well as in some additional cases. (The point in having the simpler test is that it usually takes less time, despite being less precise, and thus it may make sense to employ this test.) The second type of extreme dependency is where two tests may be useful only in *disjoint* circumstances. For example, one test rejects only if some measurable feature f takes on some value f_1, while another test rejects only if f takes some other value f_2.

In this paper we present some algorithms for cases of extreme dependencies ("disjoint" tests, and "dominating" tests), and combinations thereof that can be optimized efficiently. Section 2 presents the main results. Later sections are devoted for the proofs.

2   Main Results

When referring to a test x_i, we indicate its execution time by t_i. The notation r_i is used to indicate its prior rejection probability, i.e. the rejection probability given no previous tests. Conditional rejection probabilities are implied by context, e.g. if a dominating test x_j was previously run and reported "don't know" then the rejection probability of x_i in this context is zero.

Let $Q(x) = \frac{r_x}{t_x}$ be the *quality* of a test x.

We begin with the simplest case – where all tests are disjoint.

Definition 1. *Tests x_1 and x_2 are disjoint if the events "x_1 rejects" and "x_2 rejects" cannot both occur (for the same object).*

Theorem 1. *Let $x_1, \ldots, x_n$ be pairwise disjoint tests. The minimal expected execution time is obtained by sorting the tests in non-increasing order of quality.*

Our next result deals with the situation where all tests are comparable in terms of the cases they are able to resolve. More formally, consider

Definition 2. *A test x_2 dominates a test x_1, which we denoted by $x_1 \subset x_2$, if it always rejects (at least) every item rejected by x_1.*

Clearly, this property is (weakly) anti-symmetric. Obviously, in this case $r_1 \leq r_2$. Note that, if $t_2 \leq t_1$, then it will never make sense to perform x_1 at all. Thus, we may assume that $t_1 < t_2$.

Definition 3. $X = (x_1, x_2, \ldots, x_n)$ *is a domination chain (of tests) if x_j dominates x_i for $1 \leq i < j \leq n$.*

Theorem 2. *Given a domination chain of n tests, Algorithm 1 (see Section 4 below) finds the ordering of tests for which the expected execution time is minimal. The runtime of the algorithm is $O(n \lg^2 n)$.*

We proceed to the case where there are two disjoint chains of dominating tests.

Let $X = (x_1, \ldots, x_n)$ and $Y = (y_1, \ldots, y_m)$, $n \geq m$, be domination chains, where each test in X is disjoint from each test in Y.

Remark 1. The optimal solution is not necessarily a merge of the optimal solutions of each chain separately. For example, let $X = (x_1, x_2)$ and $Y = (y_1)$, where $r_{x_1} = 0.5, t_{x_1} = 20$, $r_{x_2} = 0.6, t_{x_2} = 35$, and $r_{y_1} = 0.25, t_{y_1} = 2$. Then the solution for the chain X by itself is (x_2), but the solution for the two disjoint chains is (y_1, x_1, x_2).

Theorem 3. *Given two domination chains of tests, $X = (x_1, \ldots, x_n)$ and $Y = (y_1, \ldots, y_m)$, where each test in X is disjoint from each test in Y, the algorithm described in Section 5 below, finds the ordering of tests for which the expected execution time is minimal. The runtime of the algorithm is $O(n^2 \lg^2 n)$.*

Naturally, the dynamic programming algorithm for one and two chains can be generalized to solve K domination chains of tests. We believe that the improved results presented in the last theorems, can also be generalized to any number of K chains, but have not implemented it.

We now proceed to the case where we have some combination of independence and disjointness as follows. Let d be an integer. Assume we have d sets of tests, $S_1, S_2, \ldots, S_d$, where each two tests belonging to the same set are disjoint, and each two tests from different sets are independent. We present an algorithm for arranging the tests so that the expected execution time will be minimal.

Theorem 4. *Given sets $S_i = \{x_{i1}, x_{i2}, \ldots, x_{in_i}\}, 1 \leq i \leq d$, as described above, Algorithm 5 finds an optimal ordering in time $O(n \lg n)$, where $n = \sum_{r=1}^{d} n_r$.*

3   Proof of Theorem 1

Performing the tests according to the original ordering, the expected duration will be

$$t_1 + (1 - r_1)t_2 + (1 - r_1 - r_2)t_3 + \ldots + (1 - \sum_{i=1}^{n-1} r_i)t_n.$$

Assume in contradiction that there exists an optimal solution which is not sorted according to the quality of tests. Then there exists two tests x_i, x_j such that x_j follows immediately after x_i in this solution, and $Q_i < Q_j$. Denote by E_{ij} the execution time of this ordering, and by E_{ji} the execution time of the same ordering, replacing only the order of x_j and x_i. Let A (B, respectively) be the ordering before (after, respectively) performing x_i (x_j, respectively). Let T_A (t_B, respectively) be, as usual, the execution time of A (B, respectively). Then,

$$E_{ij} = t_A + (1 - r_A)t_i + (1 - r_A - r_i)t_j + (1 - r_A - r_i - r_j)t_B$$

$$E_{ji} = t_A + (1 - r_A)t_j + (1 - r_A - r_j)t_i + (1 - r_A - r_j - r_i)t_B,$$

and

$$E_{ij} - E_{ji} = r_j t_i - r_i t_j,$$

which is greater than 0 if and only if $\frac{r_j}{t_j} > \frac{r_i}{t_i}$, which is true if and only if $Q_j > Q_i$, in contradiction to our first assumption.

4   Proof of Theorem 2

Assume there exists a sequence of tests $(x_1, x_2, \ldots, x_j)$ with optimal ordering ending with x_j. Let L_j be such an ordering and let T_j be its expected runtime.

It will be convenient for us to denote by x_0 a virtual test with $t_0 = r_0 = 0$.

Before presenting our best result for a single domination chain of tests, we mention briefly a slower algorithm, which will help us in presenting some other results in the paper. This algorithm, whose runtime is $O(n^2)$, is a dynamic programming algorithm. For each $1 \le i \le n$, it records the optimal list L_i and its expected duration T_i. We start with $T_0 = 0$, and then conclude each T_i in turn by $T_i = \min_{0 \le j \le i-1} \{T_j + (1 - r_j)t_i\}$. Recording for each i the j for which the minimum in the last formula is attained, we easily get the optimal ordering.

The list obtained from L by appending x_i to it is denoted by $L \cdot x_i$. Let $\text{pred}(k) := j$ if $L_k = L_j \cdot x_k$. If $L_k = (x_k)$, then $\text{pred}(k) = 0$.

Lemma 1. *The function* $\text{pred}(k)$ *is non-decreasing.*

Proof. Let $\text{pred}(k) = i$. Thus, $T_k = T_i + (1 - r_i)t_k$. Similarly, the duration of a solution $L_j \cdot x_k$, for any $j < k$, is $T_j + (1 - r_j)t_k$. Since $\text{pred}(k) = i$,

$$T_i + (1 - r_i)t_k \le T_j + (1 - r_j)t_k. \tag{1}$$

Let $\mathrm{pred}(\ell) = j$, where $\ell > k$. Therefore,

$$T_j + (1 - r_j)t_\ell \leq T_i + (1 - r_i)t_\ell. \tag{2}$$

Adding (1) and (2) by sides, we get that

$$(1 - r_i)(t_k - t_\ell) \leq (1 - r_j)(t_k - t_\ell),$$

and since $\ell > k$, this is true if and only if $r_j \geq r_i$, which means that $j \geq i$.

The improved algorithm we present below is based on the following three facts:

1. The function $\mathrm{pred}(k)$ is non-decreasing.
2. We keep only places in which $\mathrm{pred}(k)$ is changed.
3. We use binary search in order to find $\mathrm{pred}(k)$ for some specific k.

Thus, finding $\mathrm{pred}(k)$ is done in an efficient way, which finds the optimal solution in a faster way.

Let us now present the details of our algorithm.

Algorithm 1 builds a table $(A[i,j])_{1 \leq i \leq 3, 1 \leq j \leq \mathsf{val}}$. In the first row of A we list those values of k from 2 and above for which $\mathrm{pred}(k) > \mathrm{pred}(k-1)$. Here, val is the number of changes in the sequence $(\mathrm{pred}(k))_{k=1}^{n}$. The other two rows are defined by $A[2,j] = \mathrm{pred}(A[1,j])$ and $A[3,j] = T_{A[2,j]}$.

Example 1. Suppose $n = 5$, and the execution time and rejection probability of the tests are given by Table 1. One can check that A is given by Table 2.

Table 1. Tests for Example 1

j	0	1	2	3	4	5
t_j	0	0.01	1.1	2	109	10,000
r_j	0	0.01	0.02	0.03	0.04	0.05

Table 2. Table A for the tests of Example 1

k	2	4
$\mathrm{pred}(k)$	1	3
$T_{\mathrm{pred}(k)}$	0.01	1.99

Algorithm 1 invokes Algorithm 2, which actually fills the table. After the table is filled, Algorithm 1 uses it to create the required list L_n. It invokes the procedure $\mathrm{findPrev}(k)$, which finds by binary search the largest index p such that $A[1,p] \leq k$. This procedure returns the value $A[2,p]$, which is equal to $\mathrm{pred}(k)$. Clearly, its runtime is $O(\lg \mathsf{val}) = O(\lg n)$.

Algorithm 2, which builds Table A, works as follows. At each step $l \leq n - 1$, it finds the index k such that:

1. In the problem $(x_1, x_2, \ldots, x_l, x_k)$ we have $\mathrm{pred}(k) = l$.
2. For $k' \in [l+1, k-1]$, in the problem $(x_1, x_2, \ldots, x_l, x_{k'})$ we have $\mathrm{pred}(k') < l$.

If no such k exists, then $k = \infty$. Note that an identical proof to that of Lemma 1 yields that, if $k'' > k$, then in $(x_1, x_2, \ldots, x_l, x_{k''})$ we have $\mathrm{pred}(k'') = l$. To this end, the algorithm first checks whether in the problem $(x_1, x_2, \ldots, x_l, x_{A[1,\mathrm{val}]})$ it is better to perform x_l instead of $x_{A[2,\mathrm{val}]}$ before $x_{A[1,\mathrm{val}]}$, namely

$$T_l + (1 - r_l)t_{A[1,\mathrm{val}]} < A[3, \mathrm{val}] + (1 - r_{A[2,\mathrm{val}]})t_{A[1,\mathrm{val}]}, \tag{3}$$

where the calculation of T_l is performed by Algorithm 4. If not, then according to Lemma 1 we have $k > A[1, \mathrm{val}]$, and we can use binary search to find k in the range $[A[1, \mathrm{val}] + 1, n]$. If (3) holds then $k \leq A[1, \mathrm{val}]$, and we find k in the range $[A[1, v-1]+1, A[1, v]]$ by binary search. Here, $v \leq \mathrm{val}$ is the minimal index such that $T_l + (1 - r_l)t_{A[1,v]} < A[3, v] + (1 - r_{A[2,v]})t_{A[1,v]}$, and is again found by binary search. After the required k was found, the algorithm updates the table with $A[1, \mathrm{val}] = k$, $A[2, \mathrm{val}] = l$ and $A[3, \mathrm{val}] = T_l$.

At each step l of the algorithm, for each test x_k we know the value of $\mathrm{pred}(k)$ among all tests $0, \ldots, l$, namely in the problem $(x_1, x_2, \ldots, x_l, x_k)$. In particular, at the last step $l = n - 1$, for each test x_k we know $\mathrm{pred}(k)$ among all tests $0, \ldots, n-1$, namely in the problem $(x_1, x_2, \ldots, x_{n-1}, x_n)$, as required.

```
Solve1Chain(x₁,...xₙ)
Input: A single domination chain of tests
Output: Lₙ

A ← BuildTableA(x₁,...xₙ)
S ← new stack
k ← n
while k > 0
    push(S, k)
    k ← findPrev(k)
pop all elements from S into the queue Lₙ
return Lₙ
```

Algorithm 1. Computation of the optimum for one chain of tests

Example 2. Table 3 defines a domination chain of tests. The tables in Figure 1 exemplify the execution of Algorithm 2 on this chain. For $6 \leq l \leq 8$, we get the same table as in the previous step, except for the value in $A[2, 3]$, which is enlarged by 1 each time. Hence, these steps are omitted. Note that the algorithm gives $L_{10} = (3, 8, 10)$.

```
BuildTableA(x₁, ... xₙ)
Input: A single domination chain of tests
Output: A

// initialize:
  A[1, 1] ← n
  A[2, 1] ← 0
  A[3, 1] ← 0
  val ← 1 // index of last valid column in A
  for l ← 1 to n − 1 // find k s.t. pred(k) = l and pred(k′) < l
    if OptimalTime(A, 1, val, l) + (1 − rₗ)t_{A[1,val]} >
                              A[3, val] + (1 − r_{A[2,val]})t_{A[1,val]}
      k ← min{k′ ∈ [A[1, val] + 1, n] :
                  Time(A, l, k′) < Time(A, findPrev(k′), k′)}
      if k = ∞ then break // no k satisfying the inequality
      val + +
      A[1, val] ← k
      A[2, val] ← l
      A[3, val] ← OptimalTime(A, 1, val, l)
    else
      if val ≠ 1
        v ← min{v′ ∈ [1, val − 1] :
                  Time(A, l, A[1, v′]) < Time(A, A[2, v′], A[1, v′])}
        k ← min{k′ ∈ [A[1, v − 1] + 1, A[1, v]] :
                  Time(A, l, k′) < Time(A, findPrev(k′), k′)
        if k = ∞ then break
        val ← v
        A[1, val] ← k
        A[2, val] ← l
        A[3, val] ← OptimalTime(A, 1, val, l)
      if val = 1
        k ← min{k′ ∈ [0, A[1, val]] :
                  Time(A, l, k′) < Time(A, findPrev(k′), k′)
        if k < ∞
          A[1, val] ← k
          A[2, val] ← l
          A[3, val] ← tₗ
  return A
```

Algorithm 2. Build Table A

Time(A, l, j)
Input: The table A and two indices $l < j$ of tests
Output: The optimal time for performing $L_l \cdot x_j$

return OptimalTime$(A, 1, \text{val}, l) + (1 - r(x_l))t_j$

Algorithm 3. Find the time of performing $L_l \cdot x_j$

OptimalTime(A, k_1, k_2, l)
Input: The part of Table A from column k_1 up to column k_2, and an index l
satisfying $A[1, k_1] \le l < A[1, k_2]$
Output: T_l

$k \leftarrow$ maximal index in $[k_1, k_2 - 1]$ with $A[1, k] \le l$ //binary search
return $A[3, k] + (1 - r_{A[2,k]})t_l$

Algorithm 4. Find optimal time, according to the current Table A

Table 3. Chain of tests for Example 2

j	1	2	3	4	5	6	7	8	9	10
t_j	4	12	13	86	191	357	360	572	1330	10000
r_j	0.01	0.02	0.05	0.09	0.11	0.12	0.15	0.18	0.19	1

k	8
pred(k)	1
$T_{\text{pred}(k)}$	4

a)

k	8	9
pred(k)	1	2
$T_{\text{pred}(k)}$	4	12

b)

k	6
pred(k)	3
$T_{\text{pred}(k)}$	13

c)

k	6	10
pred(k)	3	4
$T_{\text{pred}(k)}$	13	86

d)

k	6	10
pred(k)	3	5
$T_{\text{pred}(k)}$	13	191

e)

k	6	10
pred(k)	3	8
$T_{\text{pred}(k)}$	13	556.4

f)

Fig. 1. Table A after running the algorithm on the tests of Example 2. (a) after step $l = 1$, (b) after step $l = 2$, (c) after step $l = 3$, (d) after step $l = 4$, (e) after step $l = 5$, (f) after step $l = 9$

4.1 Conclusion of the Proof

Algorithm 4 is actually a binary search and therefore its runtime is $O(\lg \text{val}) = O(\lg n)$. Hence, this is also the runtime of Algorithm 3. Hence, the runtime of the binary searches performed when finding the minima of k' or v' over the corresponding intervals of Algorithm 2, satisfy the recursion $T(n) = T(\frac{n}{2}) + O(\lg n)$, whose solution is $T(n) = O(\lg^2 n)$. The running time of Algorithm 2 is $O(n \lg^2 n)$. Algorithm 1 mainly invokes Algorithm 2 and therefore its runtime is $O(n \lg^2 n)$.

5 Proof of Theorem 3

The following propositions provide interesting properties of the optimal solution.

Proposition 5. *Let the optimal solution include two tests s' and s'', belonging to the same chain, such that s'' is performed right after s'. Let A be the sequence of tests performed before s', and B the set of tests performed after s'', and $\overline{x}$ ($\overline{y}$, respectively) be the last element of X (Y, respectively) in A. Then, $Q_{s'} \geq Q_{s''}$.*

Proposition 6. *Let the optimal solution include two tests $s' \in X$ and $s'' \in Y$, such that s'' is performed right after s'. Let A be the sequence of tests performed before s', and B the set of tests performed after s'', and $\overline{x}$ ($\overline{y}$, respectively) be the last element of X (Y, respectively) in A. Then, $Q_{s'} - \frac{r_{\overline{x}}}{t_{s'}} \geq Q_{s''} - \frac{r_{\overline{y}}}{t_{s''}}$.*

As in Section 4, we begin by presenting a dynamic programming algorithm, which runs in $O(n^3)$ time. The algorithm constructs two tables, opt_x and opt_y, where $\text{opt}_x(i,j)$ ($\text{opt}_y(i,j)$, respectively) records, for each $0 \leq i \leq n$ and $0 \leq j \leq m$, the optimal ordering of $x_1, \ldots, x_i, y_1, \ldots, y_j$ ending with x_i (y_j, respectively). (Note that the last element of Y in $\text{opt}_x(i,j)$ is y_j and the last element of X in $\text{opt}_y(i,j)$ is x_i.) Entries $\text{opt}_x(i,0), 0 \leq i \leq n$ (entries $\text{opt}_y(0,i), 0 \leq i \leq m$, respectively) are equal to the values computed by the algorithm described in Section 4, on Chain X (Y, respectively). Each entry $\text{opt}_x(i,j)$ is computed by $\text{opt}_x(i,j) = \min_{i'<i}\{\text{opt}_x(i',j) + (1 - r_{x_{i'}} - r_{y_j})t_{x_i}, \text{opt}_y(i',j) + (1 - r_{x_{i'}} - r_{y_j})t_{x_i}\}$. Entries $\text{opt}_y(i,j)$ are computed similarly. The optimal solution is the one corresponding to the lesser of $\text{opt}_x(n,m)$ and $\text{opt}_y(n,m)$.

Theorem 3 presents a better result. Note that the difference between the improved algorithm for two domination chains of Theorem 3 below and the improved algorithm for one domination chain (from Theorem 2) is similar to the difference between the dynamic programming algorithms for these two problems.

5.1 Conclusion of the Proof of the Theorem

Let $T_X(i,j)$ be the minimal expected duration of performing some of the tests $x_1, \ldots, x_i, y_1, \ldots, y_j$, where y_j must be performed and x_i must be performed last. Denote by $L_X(i,j)$ the list of tests performed in the corresponding optimal solution. $T_Y(i,j)$ and $L_Y(i,j)$ are defined analogically, but with y_j performed last.

Let $\mathrm{pred}_X(i,j) := k$ if k is the index of the latest test (in X or in Y) performed before x_i in $L_X(i,j)$. Let $\mathrm{pred}_Y(i,j) := k$ if k is the index of the latest test (in X or in Y) performed before y_j in $L_Y(i,j)$.

Lemma 2. *The function* $\mathrm{pred}_X(i,j)$, *considered as a function of i and restricted to those values of i for which the test right before x_i belongs to X, is non-decreasing. A similar result holds for the analogous restriction of the function* $\mathrm{pred}_Y(i,j)$, *of j.*

Proof. Let $\mathrm{pred}_X(i,j) = a$, $\mathrm{pred}_X(i',j) = a'$ where $i' > i$ and a, a' are indices of tests from X. We need to show that $a' \geq a$. We may assume that $a' < i$.

Since $\mathrm{pred}_X(i,j) = a$, the expected duration of the tests, ending with x_i, is

$$T_x(i,j) = T_x(a,j) + (1 - r_{x_a} - r_{y_j})t_{x_i}.$$

Similarly, the expected duration of the tests, ending with $x_{i'}$, is

$$T_x(i',j) = T_x(a',j) + (1 - r_{x_{a'}} - r_{y_j})t_{x_{i'}}.$$

Since $\mathrm{pred}_X(i,j) = a$ and $\mathrm{pred}_X(i',j) = a'$, we have

$$T_x(a,j) + (1 - r_{x_a} - r_{y_j})t_{x_i} \leq T_x(a',j) + (1 - r_{x_{a'}} - r_{y_j})t_{x_i}, \tag{4}$$

and

$$T_x(a',j) + (1 - r_{x_{a'}} - r_{y_j})t_{x_{i'}} \leq T_x(a,j) + (1 - r_{x_a} - r_{y_j})t_{x_{i'}}. \tag{5}$$

Adding (4) and (5) by sides, we get that

$$t_{x_i}(r_{x_{a'}} - r_{x_a}) \leq t_{x_{i'}}(r_{x_{a'}} - r_{x_a}). \tag{6}$$

Since $i' > i$, we have $t_{x_{i'}} > t_{x_i}$, so that (6) holds if and only if $r_{x_{a'}} \geq r_{x_a}$, which means that $a' \geq a$.

The proof for $\mathrm{pred}_Y(i,j)$ is equivalent.

The improved algorithm for two disjoint domination chains of tests is very similar to Algorithm 1, which deals with a single domination chain, and we will point out only at the differences.

In general, instead of building one table, as in Algorithm 1, here we have two chains and therefore build two tables. Each table here has one extra dimension, which integrates the tests of the other chain (as detailed in the next paragraph). Moreover, to find the optimal solution we have to choose the better out of the two tables.

Specifically, the algorithm builds two $3 \times n \times$ val tables A_X, A_Y, where val is as in Algorithm 1, i.e., val is the number of changes in the sequence $(\mathrm{pred}_X(i,j))_{i=1}^{n}$ or $(\mathrm{pred}_Y(i,j))_{j=1}^{m}$. For each $1 \leq j \leq m$, A_X keeps the minimal i such that $\mathrm{pred}_X(i,j) \geq \ell$, where ℓ is the index in the outer loop, as in Algorithm 1. Similarly to Algorithm 1, here we also keep $T_{\mathrm{pred}_X(i,j)}$ and $T_{\mathrm{pred}_Y(i,j)}$.

In order to find the expected optimal time, we have to check which one of x_n, y_m comes last. This is done by finding the values of $\operatorname{pred}_X(n, m)$ in A_X, $\operatorname{pred}_Y(n, m)$ in A_Y and their expected durations, and choose the better of these two options.

Each part of the algorithm is similar to that of the algorithm for one chain, with minor changes. For example, in the binary search, we have to search twice – in the two new tables, and choose the best option among the two.

Since the construction of Table A_X (respectively, A_Y) may be considered roughly as performing Algorithm 1 for each value of j (respectively, of i), the runtime here is $O(n)$ times as large as there, namely $O(n^2 \lg^2 n)$.

6   Sketch of the Proof of Theorem 4

It will be convenient to add to each set S_i a virtual test x_{i0}, with $t_{x_{i0}} = r_{x_{i0}} = 0$. For convenience, here we write $t(x_{ik})$ and $r(x_{ik})$ instead of $t_{x_{ik}}$ and $r_{x_{ik}}$. Given a set $X \subseteq S_i$ for some i, say $X = \{x_{ik_1}, x_{ik_2}, \ldots, x_{ik_m}\}$, where $k_1 < k_2 < \ldots < k_m$, then the rejection probability of X, namely the probability of rejecting if we only employ the tests in X, is $r(X) = \sum_{p=0}^{m} r(x_{ik_p})$, where $k_0 = 0$. The expected duration is $t(X) = \sum_{p=0}^{m} (1 - \sum_{q=0}^{p-1} r(x_{ik_q})) t(x_{ik_p})$. (Note that here we assume that the tests are performed not in the optimal order but rather in the "natural" order.) Let $r(A_{ij}) = \sum_{k=1}^{j-1} r(x_{ik})$, for $x_{ik} \in S_i \cap A$ and A an ordered list of the tests performed before x_{ij}.

For any test $x_{ik} \in S_i$ and set $A \subseteq \bigcup_{i=1}^{d} S_i$ (with $x_{ik} \notin A$), denote by $Q'(x_{ik}|A)$ the conditional quality of x_{ik}, given that all tests in $S_i \cap A$ failed:

$$Q'(x_{ik}|A) = \frac{Q(x_{ik})}{1 - r(S_i \cap A)}.$$

Let $t(D|A)$ be the expected duration of any list D of tests, given that all tests in A failed. If $A = \emptyset$, we can simply write $t(D)$.

Algorithm 5 sorts each set $S_i, 1 \leq i \leq d$, according to non-increasing order of quality. In the full paper we prove that the optimal ordering must agree with this ordering of each S_i by itself. At each step of the algorithm, it compares the last elements of all sets and finds the one with minimal conditional quality Q'. In our proof we show that this element must be the last among all those just tested.

To implement the choice of the element to be adjoined to the tail of the list at each stage efficiently, we maintain a minimum heap, consisting of the last element of each sorted set. The heap consists of those elements with least values of Q out of each set S_i that have not yet joined the list we construct. At each step, we extract the minimal element of the heap, according to Q', in $O(\lg d)$ time. This element is inserted into the end of the list L. For each element in the heap, we have to record the set it was obtained from, so that we can insert into the heap the next element from that set. Recall that at the beginning of each S_i we have a virtual element; we consider the conditional quality of this element as

---

FindOrder($S_1, S_2, \ldots, S_d$)
Input: d sets of tests, satisfying the requirements of Theorem 4
Output: an optimal ordering L

for $i = 1$ **to** d
    sort S_i according to non-increasing order of quality
    $n_i \leftarrow$ index of last element of S_i
    compute conditional quality Q' of each $S_i[n_i]$
$n \leftarrow \sum_{i=1}^{d} n_i$
$n' \leftarrow n$
$H \leftarrow$ BuildMinimumHeap($\{S_i[n_i']\}_{1 \le i \le d}$) by Q'
while $n' > 0$
    $L[n'] \leftarrow$ ExtractMin(H)
    let S_k be the set from which the minimum was obtained
    $n' \leftarrow n' - 1$
    $n_k \leftarrow n_k - 1$
    compute conditional quality Q' of $S_k[n_k]$
    Insert($S_k[n_k], H$)

Algorithm 5. Find an optimal ordering

infinite, so that the heap always consists of one element from each S_i, and thus is of size d. The algorithm stops after n elements are extracted from the heap and inserted into L.

Runtime of the Algorithm. Sorting each set costs $O(n_i \lg n_i)$. Thus, all the sorting is done in $O(n \lg n)$ time. The construction of a minimum heap takes $O(d)$ time. Now, each pop and push costs $O(\lg d)$, and is performed exactly n times. Therefore, the runtime of the algorithm is $O(n \lg n) + O(n \lg d) = O(n \lg n)$.

References

1. Berend, D., Brafman, R., Cohen, S., Shimony, S.E., Zucker, S.: Optimal Ordering of Independent Tests with Precedence Constraints. Discrete Applied Mathematics 162, 115–127 (2014)
2. Berend, D., Brafman, R., Cohen, S., Shimony, S.E., Zucker, S.: Optimal Ordering of Statistically Dependent Tests, preprint
3. Lawler, E.L.: Sequencing Jobs to Minimize total weighted Completion Time. Annals of Discrete Mathematics 2, 75–90 (1978)
4. Monma, C.L., Sidney, J.B.: Sequencing with series-parallel precedence constraints. Mathematics of Operations Research 4(3), 215–224 (1979)
5. Viola, P.A., Jones, M.J.: Rapid object Detection using a Boosted Cascade of Simple Features. CVPR 1, 511–518 (2001)

Optimization of Pumping Energy and Maintenance Costs in Water Supply Systems

Pham Duc Dai and Pu Li

Simulation and Optimal Processes Group, Institute of Automation and Systems
Engineering, Technische Universität Ilmenau
P.O. Box 100565, 98693 Ilmenau, Germany
{duc-dai.pham,pu.li}@tu-ilmenau.de

Abstract. We develop a general mixed-integer nonlinear programming
(MINLP) approach for optimizing the on/off operations of pumps in water supply systems with multiple reservoirs. The objective is to minimize
the pumping energy cost and, at the same time, the pump maintenance
cost should be kept at certain levels, which is achieved by constrain-
ing the number of pump switches. Due to the fact that pump switching
is represented by a non-smooth function it is impossible to solve the
resulting optimization problem by gradient based optimization methods. In this work, we propose to replace the switching function with
linear inequality constraints in the formulation of MINLP. The reformulated constraints not only restrict pump switching, but also tighten the
formulation by eliminating inefficient MINLP solutions. Two case studies with many different scenarios on the user-specified number of pump
switches are taken to evaluate the performance of the proposed approach.
It is shown that the optimized pump scheduling leads to the specified
number of pump switches with reduced pumping energy costs.

Keywords: optimal pump scheduling, water supply system, minlp,
pump switching.

1 Introduction

Due to its dramatic price increases in the recent years, the electricity cost of
pumping takes the most part of the total operating costs of water supply systems.
In the United States, the energy consumption by pumping is 5% of all generated
electricity and similarly high amount of energy consumption in the European
countries [8]. Many measures have been proposed to reduce the pumping energy
cost, among which the optimization of pump scheduling represents one of the
most effective approaches [3]. The basic idea of optimal pump scheduling for
water utilities is to utilize the advantages of low priced tariff periods and shift
the energy load in high priced tariff periods ([9], [3]). It can be shown that
the application of an appropriate optimal pump scheduling can save 10% of the
annual expenditure on energy and related costs [9].

Although optimal pump scheduling to minimize the operating cost is highly
desirable, it leads to a very difficult combinatorial optimization problem, since

© Springer International Publishing Switzerland 2015   93
H.A. Le Thi et al. (eds.), *Model. Comput. & Optim. in Inf. Syst. & Manage. Sci.*,
Advances in Intelligent Systems and Computing 359, DOI: 10.1007/978-3-319-18161-5_9

binary variables have to be introduced to represent the on/off operations of the pumps [9]. Optimal pump scheduling problems can be solved by mixed-integer linear programming (MINLP) [7], mixed- integer linear programming (MILP) [1], and simulation based optimization approaches [9].

Instead of using a MINLP solver, Mouatasim in [7] proposed to solve the optimization problem by a random perturbation of a reduced gradient method. Better results were shown than those obtained from solving the same problem by using a global MINLP solver. Although this solution approach is promising, it is only applied to solve small-scale pump scheduling problems up to 10 pumps with 10 binary variables [7]. In addition, the MINLP problem is only formulated for several time intervals without considering the dynamic changes of water levels in reservoirs and moreover, pump switching was not considered in the study.

To utilize the advantage of low priced energy tariff, a time horizon for 24 hours should be considered for optimal pump scheduling [3]. McCormick and Powell in [6] used a two-stage optimization to minimize the pumping energy and maintenance costs. In the first stage, the optimal pump scheduling is found by solving a MILP problem, and it is further improved towards reduction of the energy cost and number of pump switches in the second stage by using Simulated Annealing (SA). The Dynamic programming (DP), Scatter search, and Tabu search were also applied to optimize pump scheduling problems ([1],[9]).

Operating with excessive pump switches will cause wear and tear of the pumps. This will increase the maintenance and repair costs ([10], [6]). Thus an optimal pumping schedule should consider the pumping energy cost and the number of pump switches [10]. For this reason, a constraint to restrict pump switching is necessary. However, such a constraint is described by a non-smooth function [10], it cannot be used in the formulation of MINLP or MILP [6]. In ([6],[5]) a penalty function on pump switching is added to the objective function to address this issue.

The purpose of this paper is twofold. First, we develop a general MINLP model for optimization of pump scheduling problems in water supply systems with multiple reservoirs. Second, we propose to use linear inequality constraints instead of the non-smooth pump switching constraint in formulation of MINLP. The idea of this kind of formulation comes from solving mixed-integer optimal control problems [11]. To the best of our knowledge, this method has not been applied to the restriction on pump switching in formulating optimal pump scheduling problems. The resulting MINLP has a nonlinear objective function and linear inequality constraints and hence it can be efficiently solved by available MINLP solvers. The difference between our proposed approach on handling pump switches and the ones in ([5],[6]) lies in the fact that the maximum number of pump switches is clearly defined, while it is not the case when a penalty function is used as in ([5],[6]). Two case studies with different scenarios on specified number of pump switches with multiple reservoirs are taken to evaluate the proposed approach. Based on the optimal results, the operators can select the pump scheduling with both the pumping energy cost and the desirable number of pump switches.

The remainder of the paper is organized as follows. Section 2 presents the MINLP model for optimization of pump scheduling for water supply systems. Section 3 presents two case studies for determining optimal pump scheduling. Conclusions of the paper are provided in section 4.

2 Problem Definition and Solution Approach

We consider a water supply system with n pumps and n_r reservoirs depicted in Fig.1. It is supposed that the water demand pattern and electrical tariff are known. The MINLP problem for optimal pump scheduling is formulated in a time horizon $T=24$ (hours).

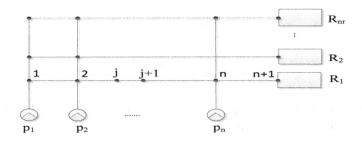

Fig. 1. A water supply system with multiple reservoirs

2.1 Pumping Energy Cost

For simplification, we at first formulate the MINLP for the system with one reservoir (i.e., R_1). The formulation of MINLP for the system with multiple reservoirs is then extended.

The electrical power consumption of pump i is calculated by [7]

$$P_i = \frac{\rho g z_i Q_i H_i}{\eta_i} \tag{1}$$

where Q_i is the flow of pump i (m^3/s); H_i is the total dynamic pump head (m); g is the acceleration to gravity (m/s^2); η_i is the efficiency of pump i; $z_i \in \{0,1\}$ is a binary variable representing on/off operation of the pump.

For a system with n pumps, the pumping energy cost in the time horizon T will be

$$E = \sum_{k=1}^{T} \sum_{i=1}^{n} P_{i,k} \gamma_k \Delta t_k \tag{2}$$

where γ_k is electrical tariff at time interval k; $\Delta t_k = 1$ (hour) is length of time interval k ($k = 1,...,T$).

In equation (1), the total dynamic head of pump i is calculated by [9]

$$H_i = H_{st} + H_r(Q) + \Delta H_f(Q) + \Delta H_m(Q) \tag{3}$$

where H_{st} is static head and it is equal to the difference between elevation of reservoir and pump discharge; $H_r(Q)$ is the water level in reservoir. It depends on the amount of pumping water, water demand, and initial water level in the reservoir; $\Delta H_m(Q)$ is local head loss of pump i; $\Delta H_f(Q)$ is the head loss across pipe sections from pumps to reservoirs (i.e., pipe sections from pump p_i to reservoir R_1 at index $n+1$ as shown in Fig.1). The sum of local and pipe section head losses can be approximated by [7]

$$\Delta H_f + \Delta H_m \simeq 1.1 \Delta H_{li} \tag{4}$$

with

$$\Delta H_{li} = \Delta H_{p_i,i} + \sum_{j=i}^{n} \Delta H_j \tag{5}$$

where $\Delta H_{p_i,i}$ is the head loss in the pipe section (p_i, i); ΔH_j is the head loss in the pipe section $(j; j+1)$, with $j = 1,...,n$, $i = 1,...,n$.

$$\Delta H_{p_i,i} = \frac{8\lambda_i L_i (z_i Q_i)^2}{g\pi D_i^5} \tag{6}$$

where L_i is the length of pipe section (p_i, i); D_i is the diameter of the pipe; Here we use the approximated value of friction factor $\lambda_i \approx 0.109$ [7]. In addition, head loss on pipe section $(j; j+1)$ is calculated by

$$\Delta H_j = \frac{8\lambda_j' L_j \left(\sum\limits_{m=1}^{j} z_m Q_m \right)^2}{g\pi D_j^5} \tag{7}$$

with $\lambda_j' \approx 0.093$ [7] and $\sum\limits_{m=1}^{j} z_m Q_m$ is the flow in the pipe section $(j; j+1)$. From the above equations, we obtain the equation of total head loss on pipe sections from pump p_i to reservoir R_1 is

$$\Delta H_{li} = \left(\frac{8\lambda_i L_i Q_i^2}{g\pi^2 D_i^5} \right) z_{i2} + \sum_{j=i}^{n} \frac{8\lambda_j' L_j \left(\sum\limits_{m=1}^{j} z_m Q_m \right)^2}{g\pi^2 D_j^5} \tag{8}$$

Replace (3), (8), and (1) into (2), the energy cost by pump i during time period Δt_k is

$$
E_{i,k} = \gamma_k \frac{\rho g(z_{i,k}Q_i)(H_{st}+H_{r,k}(z_{i,k},Q)+1.1\Delta H_{li})}{\eta_i}\Delta t_k
$$

$$
= \gamma_k \Delta t_k \frac{\rho g(z_{i,k}Q_i)}{\eta_i} \left\{ \begin{array}{l} H_{st} + H_{r,k}\left(z_{i,k},Q\right) \\ +1.1 \left[\left(\frac{8\lambda_i L_i Q_i^2}{g\pi^2 D_i^5}\right) z_{i,k}^2 + \sum\limits_{j=i}^{n} \frac{8\lambda_j^{'} L_j \left(\sum\limits_{m=1}^{j} z_{m,k}Q_m\right)^2}{g\pi^2 D_j^5} \right] \end{array} \right\} \quad (9)
$$

2.2 Linear Inequality Constraints

In this work, the hydraulic mass balance model is used to represent the equilibrium principle between the amount of water coming to and out of the reservoirs (i.e., water demand) ([1],[6], [7], [10]). The water levels $(H_{r,k})$ of reservoir r with cross-sectional area S_r is calculated by the following equation:

$$
H_{r,k} = H_{r,1} + \sum_{j=1}^{k-1} \frac{\Delta t_j}{S_r} \left(\sum_{i=1}^{n} (z_{i,j}Q_i) - Q_{r,j} \right) \quad (10)
$$

and it is bounded by the minimum and maximum allowable water levels (H_{min} and H_{max})

$$
H_{\min} \leqslant H_{r,k} \leqslant H_{\max} \quad (11)
$$

Moreover, the final water levels in reservoirs should be at least the initial ones. So that,

$$
H_{r,1} \leqslant H_{r,T} \quad (12)
$$

In order to reduce maintenance cost for pumps, a constraint on pump switching is to be introduced. In this study, we use the following constraint [11]

$$
\sum_{i=1}^{n} \sum_{k=1}^{T-1} |z_{i,k} - z_{i,k+1}| \leqslant N_{\max} \quad (13)
$$

This constraint is non-smooth since it contains the absolute term. However, it can be handled by a set of linear inequalities defining facets of feasible MINLP solution [11]

$$
\begin{array}{l} sw_{i,k} \geqslant z_{i,k} - z_{i,k+1} \\ sw_{i,k} \geqslant -z_{i,k} + z_{i,k+1} \\ \sum\limits_{i=1}^{n} \sum\limits_{k=1}^{T-1} sw_{i,k} \leqslant N_{max} \end{array} \quad (14)
$$

where $sw_{i,k} = |z_{i,k} - z_{i,k+1}|$; N_{max} is maximum number of pump switches and it is predefined. From the equation of energy cost (9) and constraints (11),(12), and (14), we have the following MINLP problem for optimal pump scheduling:

$$\min E = \sum_{k=1}^{T} \Delta t_k \gamma_k \left\{ \sum_{i=1}^{n} \frac{\rho g Q_i}{\eta_i} z_{i,k} \left[\begin{array}{l} H_{st} + H_{r,k}(z_i, Q) \\ +1.1 \left(\begin{array}{l} \left(\frac{8\lambda_i Q_i^2}{g\pi^2 D_i^5}\right) z_{i,k}^2 \\ + \sum_{j=i}^{n} \frac{8\lambda_j' L_k \left(\sum_{m=1}^{j} z_{m,k} Q_m\right)^2}{g\pi^2 D_k^5} \end{array} \right) \end{array} \right] \right\}$$

s.t.

$$H_{r,k} = H_{r,1} + \sum_{j=1}^{k-1} \frac{\Delta t_j}{S_r} \left(\sum_{i=1}^{n} (z_{i,j} Q_i) - Q_{r,j} \right) \qquad (15)$$

$$H_{r,1} \leqslant H_{r,T}$$
$$sw_{i,k} \geqslant z_{i,k} - z_{i,k+1}$$
$$sw_{i,k} \geqslant -z_{i,k} + z_{i,k+1}$$
$$\sum_{i=1}^{n} \sum_{k=1}^{T-1} sw_{i,k} \leqslant N_{max}$$
$$i = 1, ..., n; \quad k = 1, ..., T-1; \quad g = 1, .., k-1; z_{i,k} \in \{0, 1\}$$

The inequality constraints on the variables $z_{i,k}$ in (15) restrict number of pump switches to N_{max}. To simplify the expression, we further represent the objective function in the following form

$$E = \sum_{k=1}^{T} \Delta t_k \gamma_k \left\{ \sum_{i=1}^{n} \left[a_{i,k}(z) z_{i,k} + c_i z_{i,k}^3 + b_i z_{i,k} \left(\sum_{j=i}^{n} \frac{L_j \left(\sum_{m=1}^{j} z_{m,k} Q_m\right)}{D_j^5} \right) \right] \right\} \qquad (16)$$

where $a_{i,k}, b_i$, and c_i are defined as:

$a_{i,k}(z) = a_i (H_{st} + H_{r,k}(z_i, Q))$, $a_i = \frac{\rho g Q_i}{\eta_i}$, $c_i = \frac{8.8\rho\lambda_i L_i Q_i^3}{\pi^2 \eta_i D_i^5}$, $b_i = \frac{8.8\rho\lambda_i' Q_i}{\pi^2 \eta_i}$. The term $a_{i,k}$ can be further expressed as

$$\sum_{i=1}^{n} z_{i,k} a_{i,k} = \frac{\rho g Q_i}{\eta_i} \sum_{i=1}^{n} \left(z_{i,k} (H_{r,1} + H_{st}) + \sum_{g=1}^{k-1} \frac{\Delta t_g}{S_r} \left(\sum_{u=1}^{n} z_{iu,kg} Q_u - z_{i,k} Q_{r,g} \right) \right) \qquad (17)$$

where $z_{iu,kg} = z_{i,k} z_{u,g}$, $i = 1, ..., n; u = 1, ..., n; g = 1, ..., k-1$. Because $z_{i,k}$ is binary variable, we have $z_{i,k} = z_{i,k}^2 = z_{i,k}^3$ [12]. In this way, the objective function in (16) is simplified and generalized to the expression bellows:

$$E = \sum_{k=1}^{T} \Delta t_k \gamma_k \left\{ \begin{array}{l} \dfrac{\rho g Q_i}{\eta_i} \sum_{i=1}^{n} \left(\begin{array}{l} z_{i,k} \left(H_{r,1} + H_{st} \right) \\ + \sum_{g=1}^{k-1} \dfrac{\Delta t_g}{S_r} \left(\sum_{u=1}^{n} z_{iu,kg} Q_u - z_{i,k} Q_{r,g} \right) \end{array} \right) \\[2em] + \sum_{i=1}^{n} \left(c_i + b_i Q_i^2 \sum_{j=i}^{n} LD_j \right) z_{i,k} \\[2em] + \sum_{i=1,i<j<n}^{n-1} \left\{ \begin{array}{l} b_i \left(\sum_{l=j}^{n} LD_l \right) \left(Q_j^2 + 2 Q_i Q_j \right) + \\[1em] b_j \left(\sum_{l=j}^{n} LD_l \right) \left(Q_i^2 + 2 Q_i Q_j \right) \end{array} \right\} z_{ij,k} \\[2em] + 2 \left(\sum_{i=1,i<j<h<n}^{n} \left(\sum_{l=h}^{n} (LD_l) \left(\begin{array}{l} b_h Q_i Q_j \\ + b_i Q_h Q_j \\ + b_j Q_i Q_h \end{array} \right) z_{ijh,k} \right) \right) \end{array} \right\} \quad (18)$$

In the expression, we define $z_{ij,k} = z_{i,k} z_{j,k}$, $z_{ijh,k} = z_{i,k} z_{j,k} z_{h,k}$, and $LD_j = L_j / D_j^5$

2.3 The Formulation of MINLP for Water Supply System with Multiple Reservoirs

Now the objective function E in (18) is extended for a system with n_r reservoirs as follows:

$$E =$$

$$\sum_{k=1}^{T} \Delta t_k \gamma_k \sum_{r=1}^{n_r} \left\{ \begin{array}{l} \dfrac{\rho g Q_i}{\eta_i} \sum_{i=1}^{n} \left(\begin{array}{l} z_{i,r,k} \left(H_{r,1} + H_{st,r} \right) \\ + \sum_{g=1}^{k-1} \dfrac{\Delta t_g}{S_r} \left(\sum_{u=1}^{n} z_{iu,r,k,g} Q_u - z_{i,r,k} Q_{r,g} \right) \end{array} \right) \\[2em] + \sum_{i=1}^{n} \left(c_i + b_i Q_i^2 \sum_{j=i}^{n} LD_j \right) z_{i,r,k} + \\[2em] + \sum_{i=1,i<j<n}^{n-1} \left\{ \begin{array}{l} b_i \left(\sum_{l=j}^{n} LD_l \right) \left(Q_j^2 + 2 Q_i Q_j \right) + \\[1em] b_j \left(\sum_{l=j}^{n} LD_l \right) \left(Q_i^2 + 2 Q_i Q_j \right) \end{array} \right\} z_{ij,r,k} + \\[2em] 2 \left(\sum_{i=1,i<j<h<n}^{n} \left(\sum_{l=h}^{n} (LD_l) \left(\begin{array}{l} b_h Q_i Q_j \\ + b_i Q_h Q_j \\ + b_j Q_i Q_h \end{array} \right) z_{ijh,r,k} \right) \right) \end{array} \right\} \quad (19)$$

where $z_{i,r,k}$, a binary variable which is used to indicate whether pump i supplies water to reservoir r or not. In addition, following constraints are used to ensure that at a particular time interval (k) a switched on pump will only supply water to one of the reservoirs.

$$\sum_{r=1}^{n_r} z_{i,r,k} \leqslant 1, \quad i = 1, ..., n, k = 1, .., T \quad (20)$$

The linear inequality constraints on number of pump switches in (14) are extended as:

$$sw_{i,r,k} \geqslant z_{i,r,k} - z_{i,r,k+1}$$
$$sw_{i,r,k} \geqslant -z_{i,r,k} + z_{i,r,k+1}$$
$$\sum_{r=1}^{n_r} \sum_{i=1}^{n} \sum_{k=1}^{T-1} sw_{i,r,k} \leqslant N_{max} \tag{21}$$

To solve the optimization problem formulated above, we employ the MINLP solver BONMIN [2] in GAMS [4]. All the computation experiments in the following case studies are conducted on an Intel (R) Core (TM) 3.40GHz 2.99GB RAM desktop.

Table 1. Data for case study 1

a_i	b_i	c_i	$Q_i(m^3/s)$	L_k (m)	D_k (m)
86.33	0.005	4.8	0.04	938.6	0.25
59.2	0.005	1.89	0.035	1,936.3	0.35
65.18	0.005	4.331	0.04	1,352.6	0.4
55.42	0.005	4.069	0.035	1,191	0.45
70.89	0.005	1.414	0.05	3,684	0.5
64.94	0.005	2.456	0.05	864	0.6
68.09	0.005	0.074	0.05	2,381	0.6
81.61	0.005	5.276	0.07	331.1	0.7
32.07	0.0032	11.693	0.025	625	0.8
28.03	0.005	3.21	0.025	11,3	0.9

3   Case Studies

3.1   Case Study 1

We consider at first a water supply system comprising of ten pumps and one reservoir. The data for formulating the optimization problem modified from [7] is given in Table.1. The MINLP problem formulated has 240 binary variables and 507 linear constraints. The base water demand (Q_r) for the reservoir is assumed to be 0.35(m^3/s). The demand patterns are assumed to be 0.8 for periods from 1.00 a.m. to 6.00 a.m., 1.0 for periods from 7.00 a.m. to 20.00 , and 0.8 for periods from 21.00 to 24.00. The energy priced tariff is assumed to be 0.024($/kW) for periods from 1.00a.m. to 6.00a.m., and 0.1194 ($/kW) for periods from 7.00 a.m. to 24.00. The initial water level in the reservoir $(H_{r,0})$ is 15(m). The lower and upper bounds for water levels in the reservoir are 7.0 (m) and 28.0(m), respectively.Static heads (H_{st}) for all pumps are assumed to be 35(m).

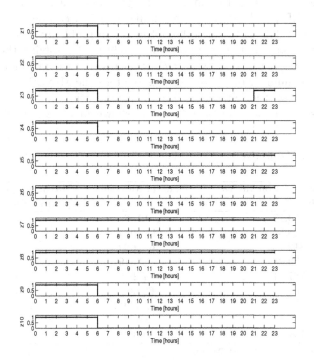

Fig. 2. Optimal pumping schedules with $N_{max}=7$

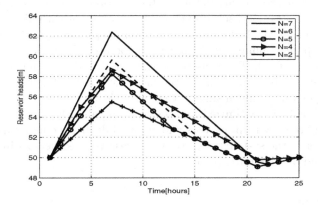

Fig. 3. Optimal head trajectories for different values of N_{max}

The energy costs with respect to the scenarios on the maximum number of pump switches are given in Table.2. It can be seen that the pump scheduling with higher allowable number of pump switches will result in a lower pumping energy cost, and the same is true reversely. Interestingly, as the allowable number of pump switches is larger than 6, the optimized pumping schedules produce the same pumping energy costs (see in Table.2). The computation time for solving each of the MINLP problems is also shown in Table 2. It can be seen that while BONMIN takes 14.352 (s) to solve the optimization problem without using the pump switching constraint, it requires only 1.529 (s) for solving the one with the pump switching constraint ($N_{max}=7$). It means that the introduction of the constraint on the number of pump switches tightens the MINLP by eliminating inefficient MINLP solutions and therefore the computation time can be reduced. The optimal solution (for the case $N_{max}=7$) shown in Fig.2 indicates that all

Table 2. Objective function values with number of pump switches

Objective function values ($/day)	Computation time (s)	Maximum number of pump switches (N_{max})
869.77	0.196	1
825.85	115.924	2
812.25	109.699	3
799.28	172.319	4
787.91	97.298	5
768.73	62.666	6
757.88	1.529	7
757.88	2.901	8
757.88	14.352	–

-: without pump switching constraint

pumps are scheduled to operate in the low tariff periods (e.g., 1 to 6). In the high tariff periods, the optimized scheduling will use the pumps with higher efficiency (e.g., low value of a_i) which are near the reservoir to operate. In particular, during the high tariff periods pumps 5,6,7, and 8 are operated, while pump 1 and 2 are switched off (see Fig.2). The reason for the priority of selecting pumps near reservoir to be switched on is due to the fact that the total head losses on the sections of pipes will be much smaller than those located far from the reservoir. As shown in Fig.3, the optimized pump scheduling also allows the reservoir to be filled during the low tariff periods and emptied during the high tariff periods to supply water to the systems. Moreover, the reservoir recovers its initial water level by the end of scheduling period.

3.2   Case Study 2

Now we extend the same system considered above with two reservoirs. The data of pumps are the same as used as in case study 1. The diameters of reservoir 1 and 2 are 15.0 and 10.0(m), respectively. The base water demands (Q_r) for reservoir 1 and reservoir 2 are assumed to be 0.2(m^3/s) and 0.15(m^3/s), respectively. The formulated MINLP has 480 binary variables, 1007 linear constraints. For solving the problem using GAMS, the time limitation for MINLP is set to 50000.0 (s). The results of energy costs corresponding to different N_{max} are given in Table.

Table 3. Pumping energy costs and maximum number of pump switches

Objective function values ($/day)	Maximum number of pump switches (N_{max})
672.626	5
654.321	10
652.87	13
639.637	15
637.665	20
635.522	25
633.646	−

−: without pump switching constraint

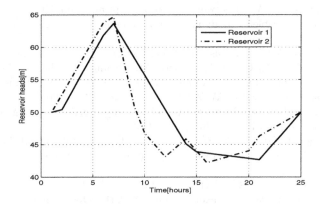

Fig. 4. Optimal head trajectories of both reservoirs with $N_{max}=20$

3. Again it can be seen that as the allowable number of pump switches increases, more pumping energy cost is saved. However, the energy cost deceases will be slow (about 2.0($/day)) as the number of pump switches is larger than 15. The optimal pump scheduling uses the pumps located near the reservoirs to be turned on instead of the ones far from the reservoirs. Similar to the results from the case study 1, due to the optimized pump scheduling, the pumps are operated intensively to pump water to both reservoirs during low priced tariff periods as shown in Fig. 4. And the stored water in the reservoirs is supplied to the system by the gravity of reservoirs during the high priced tariff periods; hence it significantly relieves operations of the pumps in these periods.

4   Conclusions

In this study, we developed a general MINLP model for optimizing the operations of pumps in water supply systems with multiple reservoirs. The optimized pump scheduling will result in a reduction of the pumping energy cost with a user-specified number of pump switches. We proposed to use linear inequalities defining the facets of the feasible MINLP solution for restricting the number of pump switches. It tightens the formulation of MINLP and helps to reduce the computational burden in solving the formulated MINLP problem. The efficiency of the proposed approach was demonstrated by determining optimal pump scheduling strategies in two case studies with different scenarios on allowable numbers of pump switches with different numbers of reservoirs. Our future work will concentrate on handling the number of pump switches allowable for each pump using the minimum up/down time constraints in MINLP problem.

References

1. Bene, J.G., Selek, I.: Water network operational optimization: Utilizing symmetries in combinatorial problems by dynamic programming. Civil Engineering 56(1), 51–61 (2012)
2. Bonami, P., Lee, J.: Bonmin user's manual (2009)
3. Boulos, P.F., Wu, Z., Orr, C.H., Moore, M., Hsiung, P., Thomas, D.: Optimal pump operation of water distribution systems using genetic algorithms. In: Proc., Distribution System Symp., AWWA, San Diego, pp. 23–25 (2001)
4. Brooke, A., Kendrick, D., Meeraus, A.: GAMS release 2.25: a user's guide. GAMS Development Corporation, Washington (1996)
5. Burgschweiger, J., Gnädig, B., Steinbach, M.C.: Optimization models for operative planning in drinking water networks. Optimization and Engineering 10(1), 43–73 (2009)
6. McCormick, G., Powell, R.S.: Derivation of near-optimal pump schedules for water distribution by simulated annealing. Journal of the Operational Research Society 55(7), 728–736 (2004)
7. El Mouatasim, A.: Boolean integer nonlinear programming for water multireservoir operation. Journal of Water Resources Planning and Management 138(2), 176–181 (2011)
8. Feldman, M.: Aspects of Energy Efficiency in Water Supply Systems. In: Proceedings of the 5th IWA Water Loss Reduction Specialist Conference, South Africa, pp. 85–89 (2009)
9. Hajji, M., Fares, A., Glover, F., Driss, O.: Water pump scheduling system using scatter search, Tabu search and neural networks the case of Bouregreg water system in Morocco. In: World Environmental and Water Resources Congress 2010@ sChallenges of Change, pp. 822–832 (2010)
10. Lansey, K.E., Awumah, K.: Optimal pump operations considering pump switches. Journal of Water Resources Planning and Management 120(1), 17–35 (1994)
11. Sager, S., Jung, M., Kirches, C.: Combinatorial integral approximation. Mathematical Methods of Operations Research 73(3), 363–380 (2011)
12. Watters, L.J.: Letter to the Editor-Reduction of Integer Polynomial Programming Problems to Zero-One Linear Programming Problems. Operations Research 15(6), 1171–1174 (1967)

Part II
DC Programming and DCA:
Thirty Years of Developments

A Based-DC Programming Approach
for Planning a Multisensor Multizone Search
for a Moving Target

Hoai An Le Thi[1], Duc Manh Nguyen[2,*], and Tao Pham Dinh[3]

[1] Laboratory of Theoretical and Applied Computer Science, UFR MIM,
University of Lorraine, Ile du Saulcy, 57045 Metz, France
[2] Department of Mathematics and Informatics,
Hanoi National University of Education, Cau Giay, Hanoi, Viet Nam
[3] Laboratory of Mathematics, National Institute for Applied Sciences - Rouen, 76801
Saint-Etienne-du-Rouvray, France
`hoai-an.le-thi@univ-lorraine.fr`, `nguyendm@hnue.edu.vn`, `pham@insa-rouen.fr`

Abstract. In this paper, we consider a well-known problem in the general area of search theory: planning a multisensor in multizone search so as to minimize the probability of non-detection of a moving target under a given resource effort to be shared. The solution method is based on a combination of the forward-backward split technique and DC programming. Numerical experiments demonstrate the efficiency of the proposed algorithm in comparison with the existing method.

Keywords: Search theory, Hierarchical optimization, Combinatorial optimization, DC programming and DCA, Nonlinear mixed 0-1 programming, Exact penalty.

1 Introduction

Search theory is defined by Cadre and Soiris [3] as a discipline that treats the problem of how a missing object can be searched optimally, when the amount of searching time is limited and only probabilities of the possible position of the missing object are given. The theory of how to search for missing objects has been a subject of serious scientific research for more than 50 years. It is a branch of the broader applied science known as operations research [5].

In fact, search theory was first established during World War II by the work of B. O. Koopman and his colleagues [10] in the Antisubmarine Warfare Operations Research Group (ASWORG). The applications of search theory were firstly made on military operations [19]. Koopman [9] stated that the principles of search theory could be applied effectively to any situation where the objective is to find a person or object contained in some restricted geographic area. After military applications, it was also applied to different problems such as surveillance, explorations, medicine, industry and search and rescue operations [8].

---

* Corresponding author.

© Springer International Publishing Switzerland 2015   107
H.A. Le Thi et al. (eds.), *Model. Comput. & Optim. in Inf. Syst. & Manage. Sci.*,
Advances in Intelligent Systems and Computing 359, DOI: 10.1007/978-3-319-18161-5_10

The aim of searching in the context of Aeronautical Search and Rescue (ASAR), for instance, is to find the missing aircraft effectively and as quickly as possible with the available resources [18].

In this paper we consider an extension of the problem investigated in [12] when the target is moving. The problem can be described as follows: suppose that a space of search is partitioned into zones of reasonable size. A unique sensor must be able to explore efficiently a whole zone. Each zone is itself partitioned into cells. A cell is an area in which every points have the same properties, according to the difficulty of detection (altitude, vegetation, etc.). Each sensor has its own coefficient of visibility over a cell. The visibility coefficients depend also on the kind of target that is searched. Here, there is a unique target to detect. This study considers a multi period search of a moving target. This means that information about the sensors and the target will be now indexed by time (the period index). The target prior is now trajectorial and we shall consider here a Markovian (target) prior. Furthermore, assuming that sensors act independently at the cell level. The target is said undetected for this multiperiod search if it has not been detected at any period of the search. The objective is allotting sensors to search zones and finding the search resources sharing of multisensor in multizone at each time period so as to minimize the probability of non-detection of a target.

This problem is very complicated because of the huge number of possible target trajectories. For a unique sensor, the problem has been theoretically solved in [17,20]; while extensions to double layered constraints have been considered in [7]. In practice, all feasible algorithms are based on a forward-backward split introduced by Brown [2]. In this case, although the forward-backward split technique (see [21]) allows us to simplify the main problem, the obtained subproblem is still very hard since it is hierarchical:

- At upper level: finding the best allotment of sensors to search zones (a sensor is allotted to a unique zone);
- At lower level: determining the best resource sharing for every sensor, in order to have an optimal surveillance over the allotted zone.

At the upper level, the objective function can be non-convex or implicitly defined via an algorithm applied to the lower level. This makes the problem very hard. In [16], Simonin et al. have proposed a hierarchical approach for solving this type of subproblem where a cross-entropy (CE) algorithm [1,15] has been developed for the upper level while an optimization method based on the algorithm of de Guenin [6] for detecting a stationary target has been used in the lower level. In [12], we introduced a new method for solving this type of subproblem. This is a *deterministic continuous optimization approach* based on DC (Difference of Convex functions) programming and DCA (DC optimization Algorithms). Specifically, we proposed a new optimization model that is a nonlinear mixed 0-1 programming problem. This problem was then reformulated as a DC (Difference of Convex functions) program via a penalty technique. DC programming and DCA (DC algorithm) ([11,13,14]) have been investigated for solving the resulting DC program. This motivates us to investigate the combination of the forward-backward split technique and our proposed method for solving this problem.

The paper is organized as follows. In Section 2, the problem statement and the classic "forward-backward split" are presented. In Section 3, we present our approach developed in [12] for solving the subproblem, then introduce a schema combining the forward-backward split and DCA. The numerical results are reported in Section 4 while some conclusions and perspectives are discussed in Section 5.

2   Problem Statement

First, let us introduce the notations employed in the remainder of the paper.
The search periods are indexed by $t \in \{1, 2, ..., T\}$
E : space of search, Z : number of zones, S : number of sensors
z : zone index
i : cell index
s : sensor index
α : prior on the initial location of the target
$\varphi_s^t(c_{i,z})$: quantity of resource of sensor s allotted to cell i of the zone z at the time t
Φ_s^t : quantity of resource available for sensor s to search the space at the time t
$w_{i,z,s}$: coefficient that characterizes the acuity of sensor s over cell i of the zone z (visibility coefficient)

We report below the problem statement described in [16] (see [16] for more details).

The space of search: the search space, named E, is a large space with spatially variable search characteristics. The search space E is divided into Z search zones, denoted $E_z, z = 1, 2, ..., Z$, each of them is partitioned into C_z cells, denoted $\{c_{i,z}\}_{i=1}^{C_z}$ so that:

$$E = \bigcup_{z=1}^{Z} E_z, E_z \cap Z_{z'} = \emptyset, \forall z \neq z',$$

$$E_z = \bigcup_{i=1}^{C_z} c_{i,z}, c_{i,z} \cap c_{j,z}, \forall i \neq j.$$

A cell $c_{i,z}$ represents the smallest search area in which the search parameters are constant. For example, it can be a part of land with constant characteristics (latitude, landscape). Each zone must have a reasonable size in order to be explored by a sensor within a fixed time interval.

The target: the target is hidden in one unit of the search space. Its location is characterized by a prior $\alpha_{i,z}$. Thus, we have

$$\sum_{z=1}^{Z} \sum_{i=1}^{C_z} \alpha_{i,z} = 1.$$

The means of search: means of search can be passive (e.g. IRST, ESM) or active sensors (radars). We will consider that searching the target will be carried out by S sensors. Due to operational constraints, each sensor $s \in S$ must be allotted to a unique search zone. For example, it could be the exploration time to share between units of a zone. At the lower level the amount of search resource allocated to the cell $c_{i,z}$ for the sensor s at the time t -if sensor s is allotted to zone E_z -is denoted $\varphi_s^t(c_{i,z})$. It can represent the time spent on searching the cell $c_{i,z}$ (passive sensor), the intensity of emissions or the number of pulses (active sensors), etc. Furthermore, each sensor s has at the time t a search amount $\varPhi_s^t$, it means that if sensor s is allotted to the zone E_z, we have the constraint:

$$\sum_{i=1}^{C_z} \varphi_s^t(c_{i,z}) \leq \varPhi_s^t.$$

To characterize the effectiveness of the search at the cell level, we consider the conditional non-detection probability $\bar{P}_s(\varphi_s^t(c_{i,z}))$ which represents the probability of not detecting the target given that the target is hidden in $c_{i,z}$ and that we apply an elementary search effort $\varphi_s^t(c_{i,z})$ on $c_{i,z}$. Some hypotheses are made to model $\bar{P}_s(\varphi_s^t(c_{i,z}))$. For all sensors, $\varphi_s^t(c_{i,z}) \mapsto \bar{P}_s(\varphi_s^t(c_{i,z}))$ is convex and non-increasing (law of diminishing return). Assuming independence of elementary detections, a usual model is $\bar{P}_s(\varphi_s^t(c_{i,z})) = \exp(-w_{i,z,s}\varphi_s^t(c_{i,z}))$, where $w_{i,z,s}$ is a (visibility) coefficient which characterizes the reward for the search effort put in $c_{i,z}$ by sensor s.

An additional assumption is that sensors act independently at the cell level which means that at the time period t if S sensors are allotted to $c_{i,z}$ the probability of not detecting a target hidden in $c_{i,z}$ is simply $\prod_{s=1}^{S} \bar{P}_s(\varphi_s^t(c_{i,z}))$.

At each time period t, let $m_t : \{1, 2, ..., S\} \rightarrow \{1, 2, ..., Z\}$ be a mapping allotting sensors to search zones. Our aim is to find both the optimal mappings m_t and the optimal local distributions φ_s^t in order to minimize the non-detection probability, i.e.,

$$F((m_t, \varphi_s^t)_{t=1}^T) = \sum_{\vec{\omega} \in \Omega} \alpha(\vec{\omega}) \prod_{t=1}^T \prod_{s \in m_t^{-1}(z)} \bar{P}_s\left(\varphi_s^t(\vec{\omega}(t))\right),$$

where Ω denotes the set of target trajectories, $\vec{\omega}$ a target trajectory in Ω, and $\vec{\omega}(t)$ is the cell of the target trajectory $\vec{\omega}$ at the time t. That leads to solve the following constrained problem [16]:

$$\begin{cases} \min_{(m_t, \varphi_s^t)_{t=1}^T} F((m_t, \varphi_s^t)_{t=1}^T) \\ \text{s.t.} \quad \forall t, \forall z, \forall s \in m^{-1}(z), \sum_{i=1}^{C_z} \varphi_s^t(c_{i,z}) \leq \varPhi_s, \\ \forall i \in z, \varphi_s^t(c_{i,z}) \geq 0, \\ \forall t, m_t \text{ mapping} : \{1, 2, ..., S\} \rightarrow \{1, 2, ..., Z\}. \end{cases} \quad (1)$$

In the next part, we give a brief presentation of the forward-backward split technique which allows us to decompose this problem into smaller problems. For more detail see [16].

Forward-backward Split. We can rewrite the objective function F as follows:

$$F((m_t, \varphi_s^t)_{t=1}^T) = \sum_{z=1}^Z \sum_{i=1}^{C_z} \beta_{i,z}^\tau \prod_{s \in m_\tau^{-1}(z)} \bar{P}_s(\varphi_s^\tau(c_{i,z})),$$

where

$$\beta_{i,z}^\tau = \sum_{\vec{\omega} \in \vec{\omega}_{i,z,\tau}} \alpha(\vec{\omega}) \prod_{\substack{1 \le t \le T \\ t \neq \tau}} \prod_{s \in m_t^{-1}(z)} \bar{P}_s\left(\varphi_s^t(c_{i_t, z_t})\right),$$

$$\vec{\omega}_{i,z,\tau} = \{\vec{\omega} \in \Omega : \vec{\omega}(\tau) = c_{i,z}\},$$

$$\vec{\omega} = (c_{i_1, z_1}, ..., c_{i_\tau, z_\tau}, ..., c_{i_T, z_T}),$$

$$\alpha(\vec{\omega}) = \alpha_{i_1, z_1} \prod_t^{T-1} \alpha_{t,t+1}(c_{i_t, z_t}, c_{i_{t+1}, z_{t+1}}).$$

Here, $\alpha_{t,t+1}(c_{i_t,z_t}, c_{i_{t+1},z_{t+1}})$ is probability the target move from cell c_{i_t,z_t} to the cell $c_{i_{t+1},z_{t+1}}$.

It remains to have a mean to calculate efficiently the $\beta_{i,z}^\tau$. To that aim, the trajectory Markov hypothesis is instrumental and we consider the following splitting of the $\beta_{i,z}^\tau$:

$$\beta_{i,z}^\tau = U_{i,z}^\tau . D_{i,z}^\tau,$$

where U and D are recursively defined by

$$U^\tau(i, z) = \sum_{j \in \tilde{z}} \alpha_{\tau-1,\tau}(j, i) \prod_{s \in m_{\tau-1}^{-1}(\tilde{z})} \bar{P}_s\left(\varphi_s^{\tau-1}(c_{j,\tilde{z}})\right) U^{\tau-1}(j, \tilde{z}),$$

$$D^\tau(i, z) = \sum_{j \in \tilde{z}} \alpha_{\tau,\tau+1}(j, i) \prod_{s \in m_{\tau+1}^{-1}(\tilde{z})} \bar{P}_s\left(\varphi_s^{\tau+1}(c_{j,\tilde{z}})\right) D^{\tau+1}(j, \tilde{z}).$$

In the above equations, we denote by $\tilde{z}$, the zones which can be attained conditionally to the hypothesis that the target is in the cell i of the zone z at the period τ and that it has a Markovian prior α. Such a forward-backward split was introduced by Brown [2].

Now, for a given τ, and considering that the $\beta_{i,z}^\tau$ are known, the multiperiod search problem is put in the situation: the target is "static" with prior $\beta_{i,z}^\tau$ (called subproblem). The formulation of subproblem can be described as follows:

$$\begin{cases} \min_{\{m, \varphi_s(c_{i,z})\}} \sum_{z=1}^Z \sum_{i=1}^{C_z} \beta_{i,z} \prod_{s \in m^{-1}(z)} \bar{P}_s(\varphi_s(c_{i,z})) \\ \text{s.t.} \quad \forall z, \forall s \in m^{-1}(z), \sum_{i=1}^{C_z} \varphi_s(c_{i,z}) \le \Phi_s, \\ \forall i \in z, \varphi_s(c_{i,z}) \ge 0, \\ m \text{ mapping} : \{1, 2, ..., S\} \to \{1, 2, ..., Z\}. \end{cases} \tag{2}$$

3   Solution Method for the Subproblem

In this section, we present briefly the DC programming approach developed in [12] for solving the subproblem. For more details, see [12].

3.1 DC Formulation of the Subproblem (2)

Let us introduce the allocation variable $u_{z,s}$ defined by

$$u_{z,s} = \begin{cases} 1 & \text{if the sensor } s \text{ is allotted to the zone } z, \\ 0 & \text{otherwise.} \end{cases}$$

Let variable $x_{i,z,s} = \varphi_s(c_{i,z})$ be the quantity of resource of the sensor s allotted to the cell $c_{i,z}$ in the zone z. We can rewrite (2) in the following form:

$$\begin{cases} \min_{x,u} f(x,u) = \sum_{z=1}^{Z} \sum_{i=1}^{C_z} \beta_{i,z} \exp(- \sum_{s=1}^{S} w_{i,z,s} x_{i,z,s} u_{z,s}) \\ \text{s.t. } x \in D, u \in M, \end{cases}$$

where $D = \{x = (x_{i,z,s}) \in \mathbb{R}_+^d : \sum_{i=1}^{C_z} x_{i,z,s} \leq \varPhi_s, z = 1, ..., Z, s = 1, ..., S, \}$,

$$d = S.(C_1 + C_2 + ... + C_z),$$

$$M = \{u = (u_{z,s}) \in \{0,1\}^{Z.S} : \sum_{z=1}^{Z} u_{z,s} = 1, s = 1, ..., S\},$$

which is a nonlinear mixed 0-1 programming problem. It is easy to see that the objective function of (P), say f, is convex in x for each fixed u, and similarly, it is convex in u for each fixed x. Moreover, f is infinitely differentiable.

Consider the function p and the bounded polyhedral convex set K defined, respectively, by:

$$p(u) = \sum_{z=1}^{Z} \sum_{s=1}^{S} u_{z,s}(1 - u_{z,s}),$$

and

$$K = \{u = (u_{z,s}) \in [0,1]^{Z.S} : \sum_{z=1}^{Z} u_{z,s} = 1, s = 1, ..., S\}.$$

We notice that p is finite and concave on $\mathbb{R}^{Z.S}$, non-negative on K and
$$M = \{u \in K : p(u) \leq 0\}.$$
Hence Problem (P) can be rewritten as

$$\alpha = \min\{f(x,u) : x \in D, \ u \in K, \ p(u) \leq 0\}.$$

The exact penalty result is given in the following theorem.

Theorem 1. *(see [12])*
(i) Let
$$t^0 = \max\{\|\nabla_u^2(f(x,u))\| : \ u \in [0,1]^n, x \in D\},$$
then $\forall t > t^0$ the problem (P)

$$\min\{f(x,u) : x \in D, u \in M\} \quad (P)$$

is equivalent to the next problem
$$\min\{f(x,u) + tp(u) : x \in D, u \in K\}. \quad (P_t)$$
in the following sense: they have the same optimal value and the same optimal solution set.
(ii) If (x^*, u^*) is a local solution to problem (P_t) then (x^*, u^*) is a feasible solution to the problem (P).

Note that the part (ii) in this theorem is very useful for DCA applied to (P_t) because DCA usually produces a local minimizer. In the sequel we will investigate DCA for solving problem (P_t) with $t^0 = \Phi\beta w^2 \sum_{s=1}^{S} \Phi_s$ ([12]).

3.2 DC Algorithm (DCA)

Outline of DC Programming and DCA: DC programming and DCA, which constitute the backbone of smooth/nonsmooth nonconvex programming and global optimization, have been introduced by Pham Dinh Tao in 1985 and extensively developed by Le Thi Hoai An and Pham Dinh Tao since 1994 to become now classic and increasingly popular ([11,13,14] and references therein). They address the problem of minimizing a function f which is a difference of convex functions on the whole space $\mathbb{R}^p$ or on a convex set $C \subset \mathbb{R}^p$. Generally speaking, a DC program takes the form

$$\alpha = \inf\{f(x) := g(x) - h(x) : x \in \mathbb{R}^p\} \quad (P_{dc}) \tag{3}$$

where g, h are lower semicontinuous proper convex functions on $\mathbb{R}^p$. Such a function f is called DC function, and $g - h$, DC decomposition of f while g and h are DC components of f.

The idea of DCA is simple: each iteration of DCA approximates the concave part $-h$ by its affine majorization (that corresponds to taking $y^k \in \partial h(x^k)$) and minimizes the resulting convex function (that is equivalent to determining $x^{k+1} \in \partial g^*(y^k)$).

Generic DCA scheme

Initialization: Let $x^0 \in \mathbb{R}^p$ be a best guess, $0 \leftarrow k$.

Repeat

 Calculate $y^k \in \partial h(x^k)$

 Calculate $x^{k+1} \in \arg\min\{g(x) - h(x^k) - \langle x - x^k, y^k \rangle : x \in \mathbb{R}^p\}$ (P_k)

 $k + 1 \leftarrow k$

Until convergence of x^k.

It is important to mention the following main convergence properties of DCA:

- DCA is a descent method (the sequences $\{g(x^k) - h(x^k)\}$ and $\{h^*(y^k) - g^*(y^k)\}$ are decreasing) *without linesearch*;
- If the optimal value α of the problem (P_{dc}) is finite and the infinite sequences $\{x^k\}$ and $\{y^k\}$ are bounded then every limit point x^* (resp. y^*) of the sequence $\{x^k\}$ (resp. $\{y^k\}$) is a critical point of $g - h$ (resp. $h^* - g^*$).
- DCA has a *linear convergence* for general DC programs.
- DCA has a finite convergence for polyhedral DC programs.

It is worth noting that the general DCA scheme for solving general DC programs is rather a philosophy than an algorithm. In fact, there is not only one DCA but infinitely many DCAs for a considered DC program. DCA's distinctive feature relies upon the fact that DCA deals with the convex DC components g and h but not with the DC function f itself. This fact is crucial for nonconvex nonsmooth programs for which DCA is one of the rare effective algorithms. The solution of a practical nonconvex program by DCA must have two stages: the search of an appropriate DC decomposition and the search of a good initial point. An appropriate DC decomposition, in our sense, is the one that corresponds to a DCA which is not expensive and has interesting convergence properties.

Description of the DCA Applied to (P_t)

From the computations, we have

$$||H(f)||_\infty \leq \max\{S\beta w^2 + S\beta w^2 \Phi + \beta w, 2\Phi\beta w^2 \sum_{s=1}^{S} \Phi_s + Nw\beta\}.$$

where

$$N = \max\{C_z : z = 1, ..., Z\}.$$

Hence, let $\rho := \max\{S\beta w^2 + S\beta w^2 \Phi + \beta w, 2\Phi\beta w^2 \sum_{s=1}^{S} \Phi_s + Nw\beta\}$, a DC formulation of (P_t) can be

$$\min \{g(x, u) - h(x, u) : (x, u) \in D \times K\}, \tag{4}$$

where

$$g(x, u) := \frac{\rho}{2}||(x, u)||^2 \text{ and } h(x, u) := \frac{\rho}{2}||(x, u)||^2 - f(x, u) - tp(u).$$

DCA applied to DC program (4) consists of computing, at each iteration k, the two sequences $\{(y^k, v^k)\}$ and $\{(x^k, u^k)\}$ such that $(y^k, v^k) \in \partial h(x^k, u^k)$ and (x^{k+1}, y^{k+1}) is an optimal solution of the next convex quadratic program :

$$\min \left\{ \frac{\rho}{2}||(x, u)||^2 - \langle (x, u), (y^k, v^k) \rangle : (x, u) \in D \times K \right\}$$

which can be decomposed into two smaller problems

$$\min \left\{ \frac{\rho}{2}||x||^2 - \langle x, y^k \rangle : x \in D \right\} \tag{5}$$

and

$$\min \left\{ \frac{\rho}{2}||u||^2 - \langle u, v^k \rangle : u \in K \right\}. \tag{6}$$

We are now in a position to summarize the DCA for solving Problem (P_t)

Step 1. Initialization: let (x^0, u^0) satisfy the constraints of the problem. Choose $\epsilon_1 > 0, \epsilon_2 > 0$ and $k = 0$.

Step 2. Compute $(y^k, v^k) = \nabla h(x^k, u^k)$, with

$$y^k = \rho x^k - \nabla_x f(x^k, u^k), \quad v^k = \rho u^k - \nabla_u f(x^k, u^k) + t(2u^k - e).$$

Step 3. Compute (x^{k+1}, u^{k+1}) by solving the two convex quadratic problems (5) and (6).

Step 4. Iterate Step 2 and 3 until

$$|(g - h)(x^{k+1}, u^{k+1}) - (g - h)(x^k, u^k)| \leq \epsilon_1(1 + |(g - h)(x^{k+1}, u^{k+1})|)$$

$$\text{or } ||(x^{k+1}, u^{k+1}) - (x^k, u^k)||_\infty \leq \epsilon_2(1 + ||(x^{k+1}, u^{k+1})||_\infty).$$

Theorem 2. *(Convergence properties of Algorithm DCA, for simplicity's sake, we omit here the dual part of these properties (see [12]))*

i) DCA generates the sequence $\{(x^k, u^k)\}$ such that the sequence $\{(g - h)(x^k, u^k)\}$ is decreasing convergent.

ii) The sequence $\{(x^k, u^k)\}$ converges to a KKT point for the problem (4).

3.3   The Combination of Forward-Backward Split Technique and DCA (FAB&DCA)

The multisensor multizone moving target algorithm takes the following form:

1. **Initialization:**
   $$\forall \tau, \forall z, \forall i, D_1^\tau(i, z) = 1,$$
   $$\forall k, \forall z, \forall i, U_k^1(i, z) = \alpha_{i,z}.$$

2. **Iteration** (k index):
   - **Iteration**(τ index)
     - $\forall z, \forall i$, compute the optimal allotment and resource sharing by DCA with prior
       $$\beta_{i,z}^\tau = U_k^\tau(i, z).D_k^\tau(i, z);$$
     - $\forall z, \forall i$, compute $U_k^{\tau+1}(i, z)$;

   - $\forall \tau, \forall z, \forall i$, compute $D_{k+1}^\tau(i, z)$;
   - **Stop:** when the search plan is no more improved.

4   Numerical result

Suppose that the search space is the lake of Laouzas in France [16]. The search space is divided into $Z = 4$ and there are $n = 30$ cells in each zone (see Figure 1). We assume that the target is Markovian and moves south east direction. The transition matrix describing the target motion is given in Figure 2 and is assumed to be constant over time. The search is carried out over four time periods by means of the six sensors. All sensors has the same amount of resource Φ for all time period. The coefficients are given in Table 1. We take five search plans.

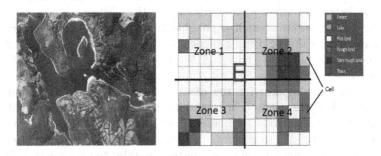

Fig. 1. An aerial photograph of the lake of Laouzas and its partition

The program is written by language C on Microsoft Visual C++ 2008 and implemented on a notebook with chipset Intel(R) Core(TM) Duo CPU 2.0 GHz, RAM 3GB.

The starting point (x^0, u^0) of DCA, where $x^0 = (x_{i,z,s}^0)$, $u^0 = (u_{z,s}^0)$ are chosen as follows: $x_{i,z,s}^0 = \frac{\Phi}{n}$, $u_{z,s}^0 = \frac{1}{Z}, s = 1, ..., S, z = 1, ..., Z, i = 1, ..., C_z$. The parameters ρ and t are dynamically adjusted during DCA's iterations. From an

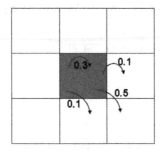

Fig. 2. The target transition probability

Table 1. Parameters for dataset

Type of Cell	Prior of target	Visibility of sensors					
		Sensor 1	Sensor 2	Sensor 3	Sensor 4	Sensor 5	Sensor 6
Forest	0.0085	0.4	0.5	0.6	0.8	0.5	0.1
Water	0.001	0.9	0.1	0.1	0.1	0.3	0.5
Plat Plan	0.0115	0.3	0.1	0.4	0.6	0.5	0.2
Rough plan	0.013	0.2	0.7	0.8	0.2	0.4	0.6
Very rough plan	0.014	0.1	0.6	0.7	0.1	0.3	0.5
Town	0	0.8	0.9	0.1	0.7	0.6	0.2

algorithmic point of view, the smaller the ρ and t are, the more efficient the DCA are (because the smaller the concave part of the objective function of Problem (P_t) is, the better its affine approximation is). Hence it would be efficient to choose the values t and ρ as small as possible. In our experiments, we start DCA with a quite small value of ρ and t (depending on the parameters Φ), and then increase both parameters ρ and t by 1% at each iteration of DCA. We take $\epsilon_1 = \epsilon_2 = 10^{-7}$.

In Table 2, we compare, for the case of $\Phi = 5$, the result obtained by FAB&DCA and by the method of combining the forward-backward split technique and the Cross-Entropy (CE) method (FAB&CE)(see [12,16]). The main idea of this CE method is to generate particular allotments of sensors to search zones that will be evaluated and then selected, in order to obtain a drawing law which will converge toward the optimal allotment. The parameters for CE are as follows: for each iteration, $\theta = 0.3$, the number of samples is $N = 50$, and the number of iterations is limited to 15. We observe that the FAB&DCA produced the better solutions than the FAB&CE while the CPU is shorter. Moreover, although we just pay attention to the results at the time 4 of the last search plan (iteration $k = 5$), FAB&DCA works better than FAB&CE does in all time of this search plan.

Table 2. Non-detection probability with $\Phi = 5$

Time	Sensors	FAB&CE method				Non-detection probability	CPU time	FAB&DCA method				Non-detection probability	CPU time
		Allotment S = 6						Allotment S = 6					
Time 1	Sensor 1	0	1	0	0	0.468681	18.57s	0	1	0	0	0.440446	9.36s
	Sensor 2	1	0	0	0			0	0	1	0		
	Sensor 3	0	0	1	0			0	0	1	0		
	Sensor 4	1	0	0	0			1	0	0	0		
	Sensor 5	1	0	0	0			1	0	0	0		
	Sensor 6	0	0	0	1			0	0	0	1		
Time 2	Sensor 1	1	0	0	0	0.386112		1	0	0	0	0.367013	
	Sensor 2	0	1	0	0			0	1	0	0		
	Sensor 3	0	0	1	0			0	1	0	0		
	Sensor 4	0	1	0	0			0	1	0	0		
	Sensor 5	0	1	0	0			0	0	0	1		
	Sensor 6	0	1	0	0			0	0	0	1		
Time 3	Sensor 1	0	0	0	1	0.326479		1	0	0	0	0.310935	
	Sensor 2	0	1	0	0			0	1	0	0		
	Sensor 3	0	0	1	0			0	1	0	0		
	Sensor 4	0	1	0	0			0	0	1	0		
	Sensor 5	0	0	0	1			0	0	0	1		
	Sensor 6	0	0	0	1			0	0	0	1		
Time 4	Sensor 1	0	0	1	0	0.272548		0	0	1	0	0.270107	
	Sensor 2	0	0	0	1			0	0	0	1		
	Sensor 3	0	0	0	1			0	0	0	1		
	Sensor 4	0	0	1	0			0	0	1	0		
	Sensor 5	0	0	0	1			0	0	0	1		
	Sensor 6	0	0	0	1			0	0	0	1		

Table 3 presents additional results with $\Phi = 10$ and $\Phi = 15$. FAB&DCA produced once again the better solutions than the FAB&CE while CPU time is shorter. Here, there is a negative relationship between Φ and the non-detection probability. When $\Phi = 15$, the probability of non-detection obtained by FAB&DCA is 0.078225.

Table 3. Relation of non-detection probability and amount of resource

Amount of resource Φ	Time	FAB&CE method		FAB&DCA method	
		Non-detection probability	CPU time	Non-detection probability	CPU time
5	Time 1	0.468681	18.57s	0.440446	9.36s
	Time 2	0.386112		0.367013	
	Time 3	0.326479		0.310935	
	Time 4	0.272548		0.270107	
10	Time 1	0.254927	18.43s	0.228051	9.00s
	Time 2	0.205480		0.191276	
	Time 3	0.173844		0.160008	
	Time 4	0.149615		0.137433	
15	Time 1	0.139648	18.41s	0.134255	14.17s
	Time 2	0.113025		0.111196	
	Time 3	0.098799		0.091861	
	Time 4	0.090995		0.078225	

5   Conclusion

We have presented a new approach based-DC programming for solving the problem planning a multisensor multizone search for a moving target. This method combines the forward-backward split technique with DCA which is applied to the DC formulation of the subproblem. The table 2 and 3 show that our approach can solve this problem more effectively than the FAB&CE method, an efficient algorithm for the considered problem. In a future work, we will investigate this approach for larger scale problems.

References

1. de Boer, P.T., Kroese, D.P., Mannor, S., Rubinstein, R.Y.: A Tutorial on The Cross-Entropy Method. Annals of Operations Research 134, 19–67 (2005)
2. Brown, S.S.: Optimal search for a moving target in discrete time and space. Operations Research 32(5), 1107–1115 (1979)
3. Cadre, J.P., Souris, G.: Searching Tracks. IEEE Transactions on Aerospace and Electronic systems 36(4), 1149–1166 (2000)
4. Dobbie, J.M.: Transfer of detection contacts to tracking contacts in surveillance. Operations Research 14, 791–800 (1966)
5. Frost, J.R.: Principles of search theory, part III: Probability density distributions. Response 17(3), 1–10 (1999c)
6. de Guenin, J.: Optimum distribution of effort: an extension of the Koopman theory. Operations Research 9(1), 1–7 (1961)
7. Hohzaki, R., Iida, K.: A concave minimization problem with double layers of constraints on the total amount of resources. Journal of the Operations Research Society of Japan 43(1), 109–127 (2000)
8. Haley, K.B., Stone, L.D. (eds.): Search Theory and Applications. Plenum Press, New York (1980)
9. Koopman, B.O.: Search and Screening: General Principles with Historical Applications. Pergamon Press, New York (1980)
10. Koopman, B.O.: Search and Screening: General Principle with Historical Applications. MORS Heritage Series, Alexandria (1999)
11. Le Thi, H.A., Pham Dinh, T.: The DC (difference of convex functions) Programming and DCA revisited with DC models of real world nonconvex optimization problems. Annals of Operations Research 133, 23–46 (2005)
12. Le Thi, H.A., Nguyen, D.M., Pham Dinh, T.: A DC programming approach for planning a multisensor multizone search for a target. Submitted in Computers & Operations Research 41, 231–239 (2014)
13. Pham Dinh, T., Le Thi, H.A.: Convex analysis approach to DC programming: Theory, Algorithms and Applications (dedicated to Professor Hoang Tuy on the occasion of his 70th birthday). Acta Mathematica Vietnamica 22, 289–355 (1997)
14. Pham Dinh, T., Le Thi, H.A.: DC optimization algorihms for solving the trust region subproblem. SIAM J. Optimization 8, 476–505 (1998)
15. Rubinstein, R.Y., Kroese, D.: The cross-entropy method: a unified approach to combinatorial optimization. Monté Carlo simulation, and machine learning. Springer, Berlin (2004)
16. Simonin, S., Cadre, J.P., Dambreville, F.: A Hierarchical Approach for Planning a Multisensor Multizone Search for a Moving Target. Computers & Operations Research 36(7), 2179–2192 (2009)
17. Stone, L.D.: Necessary and suffcient conditions for optimal search plans for moving targets. Mathematics of Operations Research (4), 431–440 (1979); Math. Sci. Net.
18. Stone, L.D.: What's happened in search theory since the 1975 Lanchester prize? Operations Research 37(3), 501–506 (1989)
19. Stone, L.D.: Theory of Optimal Search, 2nd edn. Operations Research Society of America, ORSA Books, Arlington (1989)
20. Stromquist, W.R., Stone, L.D.: Constrained optimization of functionals with search theory applications. Mathematics of Operations Research 6(4), 518–527 (1981)
21. Washburn, A.R.: Search for a moving target, The FAB algorithm. Operations Research 31(4), 739–751 (1983)

A DC Algorithm for Solving Quadratic-linear Bilevel Optimization Problems

Aicha Anzi and Mohammed Said Radjef

Laboratory of Modelisation and Optimization of Systems (LAMOS)
University of Bejaia 06000, Algeria
{anzi_aicha,msradjef}@yahoo.fr

Abstract. In this paper we propose an algorithm for solving bilevel programming problems, in which the upper level objective function is quadratic and the second level is a linear problem. First, we transform the problem into a corresponding single level optimization problem using the Karush-Kuhn-Tucker optimality conditions associated with the second level problem. Then, we solve the resulting problem by an algorithm which is a combination of the DC algorithm in DC programming and the exact penalty method.

Keywords: Bilevel programming, exact penalty, DCA, DC programming, KKT optimality conditions.

1 Introduction

Bilevel programming is a tool for modeling two level hierarchical decision making. This class of programs constitutes a branch of mathematical programming in which the constraints are, partially, determined by another optimization problem. The decision maker at the upper level is termed as the leader, and at the lower level the follower. The control of variables is partitioned between the decision makers who attempt to optimize their individual objectives. The leader goes first in order to optimize his/her objective function. The follower observes the leader's decision and constructs his/her decision. This kind of problems has a wide field of applications [1], [13] and [20] and has been increasingly addressed in the literature [10], [12], [15], [16] and [17].

Bilevel programs are nonconvex problems and have been proved to be NP-Hard problems [15] even if the objective and constraint functions are all linear. Numerous algorithms have been developed to solve bilevel problems. A classical approach proceeds by replacing the lower level's problem with its associated Karush-Kuhn-Tucker optimality conditions. The resulting single optimization problem is solved by a variety of techniques (see [9], [11], [14] and [18]).

In this paper, based on this approach, we propose an algorithm which is a combination between a nonconvex technique (the DC or Difference of Convex functions) and the exact penalty method, to solve the quadratic-linear bilevel programming problem. The DC programming and DCA have been developed by P.D. Tao and Le T.H. An since 1986 to solve nonconvex and nonsmooth

H.A. Le Thi et al. (eds.), *Model. Comput. & Optim. in Inf. Syst. & Manage. Sci.*,
Advances in Intelligent Systems and Computing 359, DOI: 10.1007/978-3-319-18161-5_11

programming problems (see [3], [5], [6] and [8]). This method has proved its efficiency from both theoretical and numerical viewpoints and has been successfully applied to a large number of nonconvex and nondifferentiable problems in various domains [2], [7] and [23].

The rest of the paper is organized as follows. Section 2 is devoted to the problem formulation. In section 3, we describe how to reformulate the problem via an exact penalty technique. Section 4 contains the application of DC programming and DCA for solving the resulting penalized problem. Computational results are presented in section 5, while some conclusion is presented in the last section.

2   Problem Formulation

We consider the bilevel quadratic linear problem (BQLP) formulated as follows:

$$\min_{x} \ F(x,y) = \frac{1}{2}(x,y)^T Q \begin{pmatrix} x \\ y \end{pmatrix} + c^T x + d^T y, \tag{1a}$$

$$s.t. \ \ A_1 x + B_1 y \le b_1, \tag{1b}$$

$$x \ge 0; \tag{1c}$$

$$\min_{y} \ f(x,y) = c_2^T x + d_2^T y, \tag{1d}$$

$$s.t. \ \ A_2 x + B_2 y \le b_2, \tag{1e}$$

$$y \ge 0, \tag{1f}$$

where $Q \in \mathbb{R}^{(n_1+n_2) \times (n_1+n_2)}$ is a symmetric indefinite matrix.

$x, c_2 \in \mathbb{R}^{n_1}$; $y, d_2 \in \mathbb{R}^{n_2}$; $b_1 \in \mathbb{R}^{m_1}$; $b_2 \in \mathbb{R}^{m_2}$; $A_1 \in \mathbb{R}^{m_1 \times n_1}$; $A_2 \in \mathbb{R}^{m_2 \times n_1}$; $B_1 \in \mathbb{R}^{m_1 \times n_2}$ and $B_2 \in \mathbb{R}^{m_2 \times n_2}$.

Now, we give some definitions of the (BQLP):

1. The constraints region

$$S = \{(x,y) \in \mathbb{R}^{n_1} \times \mathbb{R}^{n_2} : A_1 x + B_1 y \le b_1, A_2 x + B_2 y \le b_2, x \ge 0, y \ge 0\}.$$

2. The feasible region of the follower for each fixed $x \ge 0$

$$S(x) = \{y \in \mathbb{R}^{n_2} : B_2 y \le b_2 - A_2 x, y \ge 0\}.$$

3. The rational reactions set of the follower for each fixed $x \ge 0$

$$R(x) = \{y/y = \arg\min[f(x,\hat{y}) : \hat{y} \in S(x)]\}.$$

4. The inducible region

$$RI = \{(x,y) \in S, \ y \in R(x)\}.$$

The inducible region represents the feasible set over which the leader may optimize his/her objective function.

To ensure that problem (1) is well defined, it is assumed that S is nonempty and bounded.

There are mainly two ways to formulate a bilevel programming problem: the pessimistic formulation and the optimistic one. In this paper, we consider the optimistic formulation. In this case, an optimal solution of (BQLP) is defined as follows:

Definition 1. *A feasible solution* $(x^*, y^*) \in RI$ *is optimal for problem (1) if*

$$F(x^*, y^*) \le F(x, y), \ \forall (x, y) \in RI.$$

3 Reformulation and Notations

In this section, we transform the (BQLP) problem into a single optimization problem and give some useful notations.

Using the Karush-Kuhn-Tucker optimality conditions associated to the lower level's problem (1d)-(1f), we obtain the following equivalent problem:

$$\min_x \ F(x, y) = \frac{1}{2}(x, y)^T Q \begin{pmatrix} x \\ y \end{pmatrix} + c^T x + d^T y, \tag{2a}$$

$$A_1 x + B_1 y + e = b_1, \tag{2b}$$

$$A_2 x + B_2 y + w = b_2, \tag{2c}$$

$$B_2^T u - v = d_2, \tag{2d}$$

$$v^T y + u^T w = 0, \tag{2e}$$

$$x \ge 0, \ y \ge 0, \ u \ge 0, \ v \ge 0, \ w \ge 0, \ e \ge 0, \tag{2f}$$

where $e \in \mathbb{R}^{m_1}$, $u \in \mathbb{R}^{m_2}$, $v \in \mathbb{R}^{n_2}$ and $w \in \mathbb{R}^{m_2}$.

We now introduce some useful notations. Let

$z = (x^T, y^T, e^T, w^T, v^T, u^T)^T \in \mathbb{R}^n$, $p = (c^T, d^T, 0, 0, 0, 0)^T \in \mathbb{R}^n$,

$P = \begin{pmatrix} Q & 0 \\ 0 & 0 \end{pmatrix} \in R^{n \times n}$,

$E_u = (0, 0, 0, 0, 0, I_{m_2})$, $E_v = (0, 0, 0, 0, I_{n_2}, 0)$,
$E_w = (0, 0, 0, I_{m_2}, 0, 0)$, $E_y = (0, I_{n_2}, 0, 0, 0, 0)$,

$$A = \begin{pmatrix} A_1 & B_1 & I_m & 0 & 0 & 0 \\ A_2 & B_2 & 0 & I_{m_2} & 0 & 0 \\ 0 & 0 & 0 & 0 & -I_{n_2} & B_2^T \end{pmatrix} \in R^{m \times n}, \quad b = \begin{pmatrix} b_1 \\ b_2 \\ d_2 \end{pmatrix} \in R^m,$$

where I_k is $k \times k$ identity matrix ; 0 is zero matrix with appropriate dimension for each case, with $n = n_1 + 2n_2 + m_1 + 2m_2$; $m = m_1 + m_2 + n_2$.

Using these notations, we obtain:
$u^T w = (E_u z)^T (E_w z) = z^T (E_u^T E_w) z = z^T D^1 z$ and $v^T y = (E_v z)^T (E_y z) = z^T (E_v^T E_y) z = z^T D^2 z$.

It follows that: $u^T w + v^T y = z^T D^1 z + z^T D^2 z = z^T D z$ with $D^1 + D^2 = D$.

Note that the elements $d_{ij} (i = \overline{1,n}, \ j = \overline{1,n})$ of matrix D are all nonnegative. Setting $Dz = q(z)$, problem (2) can be written as:

$$\min \left\{ \phi(z) = \frac{1}{2} z^T P z + p^T z : Az = b, z^T q(z) = 0, z \geq 0 \right\}, \tag{3}$$

with $q(z) \geq 0, \ \forall z \geq 0$.

Consider the convex set $\mathcal{Z} = \{ z \in \mathbb{R}^n : Az = b, z \geq 0 \}$, and let be the function $\Psi : \mathbb{R}^n \to \mathbb{R}$ defined by $\Psi(z) = \sum_{i=1}^{n} \min\{q_i(z), z_i\}$.

Property 1. The function $\Psi(z)$ verifies the following properties:

- Ψ is a finite concave function on $\mathcal{Z}$;
- $\Psi(z) \geq 0, \quad \forall z \in \mathcal{Z}$;
- $\Psi(z) = 0, \quad \forall z \in \mathcal{Z}_n = \{ z \in \mathcal{Z}, \ z^t q(z) = 0 \} = \{ z \in \mathcal{Z}, \ \Psi(z) \leq 0 \}$.

From property (1), problem (3) can be rewritten in the following form:

$$\alpha = \min\{ \phi(z) \ : \ z \in \mathcal{Z}, \ \Psi(z) \leq 0 \}. \tag{4}$$

Now, we give the following proposition whose proof is based on duality in linear programming and the property of the complementary constraints.

Proposition 1. *Let $(\tilde{x}, \tilde{y}) \in RI$, then there exist vectors $\tilde{w} \in \mathbb{R}^{m_2}$, $\tilde{v} \in \mathbb{R}^{n_2}$, $\tilde{u} \in \mathbb{R}^{m_2}$, $\tilde{e} \in \mathbb{R}^{m_1}$ such that the vector $\tilde{z} = (\tilde{x}, \tilde{y}, \tilde{e}, \tilde{w}, \tilde{v}, \tilde{u}) \in \mathbb{R}^n$ verifies*

$$\tilde{z} \in \mathcal{Z} \quad and \quad \Psi(\tilde{z}) = 0.$$

4 DCA for Solving Problem (4)

This section is devoted to the resolution of problem (4) by DCA.

4.1 DC Programming

Now, we briefly review the basic properties of DC programming and DCA. We shall work with space $X = \mathbb{R}^n$ which is equipped with the canonical inner product $\langle ., . \rangle$ and the corresponding euclidean norm $\|.\|$. Thus, the dual space Y of X can be identified with X itself. Let $\Gamma_0(X)$ denotes the set of all lower semicontinuous proper convex functions on X. A general DC program has the form:

$$\alpha = \inf\{ f(x) = g(x) - h(x) \ : \ x \in X \}, \tag{5}$$

where $g, \ h \in \Gamma_0(X)$ are called DC components of the function f and $g - h$ is the DC decomposition of f.

The dual of (5) is the DC program

$$\alpha = \inf\{ h^*(y) - g^*(y) : y \in Y \}, \tag{6}$$

where g^* and h^* are respectively the conjugate function of g and h. The conjugate function of g is given by:

$$g^*(y) = \sup\{\langle x, y \rangle - g(x) : x \in X\}.$$

For problem (5), the following necessary local optimality conditions, developed in [5], have been constantly used:

$$\emptyset \neq \partial h(x^*) \subset \partial g(x^*), \tag{7}$$

where $\partial h(x^*) = \{y^* \in Y : h(x) \geq h(x^*) + < x - x^*, y^* >, \forall x \in X\}$ is the subdifferential of h at x^*.

$$\emptyset \neq \partial g(x^*) \cap \partial h(x^*). \tag{8}$$

Such a point x^* is called critical point of $g - h$.

DCA is a descent method without linesearch, consisting of the construction of the two sequences $\{x^i\}$ and $\{y^i\}$ (candidates for being primal and dual solutions, respectively), such that their corresponding limit points satisfy the local optimality conditions (7) and (8). Recall that there are two forms of DCA: the simplified DCA (or simply DCA) and the complete DCA. In practice the first is more used because it is less expensive [3]. The simplified DCA has the following form ([2] and [6]):

DCA Algorithm
1 : Let $x^0 \in \mathbb{R}^n$ given. Set $i = 0$.
2 : Compute $y^i \in \partial h(x^i)$.
3 : Compute $x^{i+1} \in \partial g^*(y^i)$.
4 : If a convergence criterion is satisfied, then **Stop; else** set $i = i + 1$ and **goto** step 2.

4.2 A DCA Scheme for Solving Problem (4)

The proposed approach for solving problem (1) is now developed. For this we have to write problem (4) in the form of a DC program. First, rewrite the function ϕ as:

$$\phi(z) = \phi_1(z) - \phi_2(z) = \frac{1}{2}z^T(P + \rho I_n)z + p^T z - \frac{1}{2}\rho\|z\|^2,$$

where ρ is a real number such that $P + \rho I_n$ is positive semidefinite.

So ϕ is a DC function with DC decomposition $\phi_1 - \phi_2$. Then, from [21] and [4] and under the assumption of boundedness of the feasible set of problem (4), there exists $k_0 \geq 0$ such that for every $k > k_0$, problem (4) is equivalent to the following penalized problem:

$$\min\{\phi(z) + k\Psi(z) : z \in \mathcal{Z}\}. \tag{9}$$

Since Ψ is a concave function on $\mathcal{Z}$, problem (9) is a DC program of the form

$$\min\{g(z) - h(z) : z \in \mathbb{R}^n\}, \tag{10}$$

with

$$g(z) = \frac{1}{2}z^T(P + \rho I_n)z + p^T z + \chi_{\mathcal{Z}}(z) \quad \text{and} \quad h(z) = \frac{1}{2}\rho\|z\|^2 - k\Psi(z), \tag{11}$$

where $\chi_{\mathcal{Z}}$ stands for the indicator function of the convex set $\mathcal{Z}$ ($\chi_{\mathcal{Z}}(z) = 0$ if $z \in \mathcal{Z}$, $+\infty$ otherwise).

The application of DCA to problem (10) consists of computing the two sequences $\{t^i\}$ and $\{z^i\}$ defined by:

$$t^i \in \partial h(z^i) \quad \text{and} \quad z^{i+1} \in \partial g^*(t^i).$$

Using the rules in convex analysis, we compute $\{t^i\}$ and $\{z^i\}$.

Computation of $t^i \in \partial h(z^i)$: we choose $t^i \in \partial h(z^i) = \left(\rho z^i - k\left(\partial \sum_{j=1}^{n} \min\{q_j(z^i), z_j^i\}\right)\right)$ as follows:

$$t^i = \rho z^i + k\theta^i, \tag{12}$$

where $\theta^i \in \sum_{j=1}^{n} \partial\big(\max\{-q_j(z^i), -z_j^i\}\big)$ and $q_j(z^i) = D_j z^i$.
Let be

$$\theta^i = -\sum_{j=1}^{n} \begin{cases} D_j^T, & \text{if } z_j^i > D_j z^i, \\ e_j, & \text{if } z_j^i < D_j z^i, \\ \gamma e_j + (1-\gamma)D_j^T, & \text{if } z_j^i = D_j z^i, \end{cases} \tag{13}$$

where D_j is the j-th row of matrix D, e_j is the j-th unit vector of $\mathbb{R}^n$ and $\gamma \in [0, 1]$.

Hence θ^i, given by (13), is an element of $\sum_{j=1}^{n} \partial\big(\max\{-D_j z^i, -z_j^i\}\big)$.

Computation of $z^{i+1} \in \partial g^*(t^i)$: following [22], we can choose z^{i+1} as the solution of the following convex problem

$$\min\{\langle(P + \rho I_n)z, z\rangle + (p - t^i)^T z : z \in \mathcal{Z}\}. \tag{14}$$

The DCA applied to problem (9) with decomposition (11) can be described as follows:

Algorithm DCABQLP

1 : Let $z^0 = (x^0, y^0, e^0, w^0, v^0, u^0)$ be an initial guess, $\epsilon > 0$, $k \in \mathbb{R}_+$, $\gamma \in [0,1]$. Set $i = 0$.

2 : Compute $t^i \in \partial h(z^i)$ using (12).

3 : Compute $z^{i+1} = (x^{i+1}, y^{i+1}, e^{i+1}, w^{i+1}, v^{i+1}, u^{i+1}) \in \partial g^*(t^i)$ by solving (14).

4 : If $y^{i+1} \in \arg\min \{f(x^{i+1}, y) : B_2 y = b_2 - A_2 x^{i+1} - w^{i+1}, \ y \geq 0\}$, **then** go to **5**; **otherwise** go to **7**.

5 : Compute the new components (v^{i+1}, u^{i+1}) of z^{i+1} by solving the following dual problem $\max\{u^t(b_2 - A_2 x^{i+1}) : B_2^t u - v = d_2, \ u \geq 0, v \geq 0\}$,

6 : If $\|z^{i+1} - z^i\|/(\|z^i\| + 1) \leq \epsilon$, **then** stop z^{i+1} is optimal for (9); and (x^{i+1}, y^{i+1}) is optimal for (1); **otherwise** go to **7**.

7 : Set $z^i = z^{i+1}$, $i = i + 1$ and go to **2**.

Remark 1. In step 4 of the algorithm, we test the feasibility of the solution (x^{i+1}, y^{i+1}) for the (BQLP). If the test in step 4 is satisfied, then we have $y^{i+1} \in R(x^{i+1})$ (*see definition 3 of the (BQLP)*). Since $(x^{i+1}, y^{i+1}) \in S$, then we have $(x^{i+1}, y^{i+1}) \in RI$ which implies that (x^{i+1}, y^{i+1}) is a feasible solution for problem (1).

The following result proves the convergence of the algorithm DCABQLP to a solution of problem (1).

Theorem 1. *Let $z^{i+1} = (x^{i+1}, y^{i+1}, e^{i+1}, w^{i+1}, v^{i+1}, u^{i+1})$ be the computed solution of problem (9) by algorithm DCABQLP. Then, (x^{i+1}, y^{i+1}) is a solution of problem (1).*

Proof. Let $k = \tilde{k}$ be the parameter value in (9) associated with the computed solution $z^{i+1} = (x^{i+1}, y^{i+1}, e^{i+1}, w^{i+1}, v^{i+1}, u^{i+1})$ of problem (10). We have

$$g(z^{i+1}) - h(z^{i+1}) \leq g(z) - h(z), \ \forall z \in \mathbb{R}^n.$$

and in particular

$$g(z^{i+1}) - h(z^{i+1}) \leq g(z) - h(z), \ \forall z \in \mathcal{Z}_n. \tag{15}$$

Then, from the definition of set $\mathcal{Z}_n$ we have

$$g(z^{i+1}) - h(z^{i+1}) \leq g(z) - h(z), \ \forall z \in \mathcal{Z}, \tag{16}$$

from which we deduce

$$\frac{1}{2}(z^{i+1})^T P z^{i+1} + p^T z^{i+1} + k\Psi(z^{i+1}) \leq \frac{1}{2} z^T P z + p^T z + k\Psi(z), \ \forall z \in \mathcal{Z}, \tag{17}$$

which means that z^{i+1} is an optimal solution for problem (9) with parameter value $k = \tilde{k}$. In addition, from remark 1, we have $y^{i+1} \in R(x^{i+1})$ which means that y^{i+1} is an optimal solution of the lower level's problem (1d)-(1f) with $x = x^{i+1}$. Moreover, the computed vector (v^{i+1}, u^{i+1}) at step 5 of the algorithm is the optimal solution of the lower level's dual problem. Then, using the property of the complementary constraint, we can prove that

$$\Psi(z^{i+1}) = 0. \tag{18}$$

Let $(\tilde{x}, \tilde{y}) \in RI$ be a feasible solution of (1). Then, by proposition 1, there exist vectors $\tilde{w} \in \mathbb{R}^{m_2}$, $\tilde{v} \in \mathbb{R}^{n_2}$, $\tilde{u} \in \mathbb{R}^{m_2}$ and $\tilde{e} \in \mathbb{R}^{m_1}$ such that

$$\tilde{z} = (\tilde{x}, \tilde{y}, \tilde{e}, \tilde{w}, \tilde{v}, \tilde{u}) \in \mathcal{Z},$$

and

$$\Psi(\tilde{z}) = 0. \tag{19}$$

From (17)-(19), we deduce

$$\frac{1}{2}(z^{i+1})^T P z^{i+1} + p^T z^{i+1} \leq \frac{1}{2}\tilde{z}^T P \tilde{z} + p^T \tilde{z} \iff F(x^{i+1}, y^{i+1}) \leq F(\tilde{x}, \tilde{y}).$$

Then (x^{i+1}, y^{i+1}) is an optimal solution of problem (1). $\qquad\square$

5   Numerical Example and Computational Experiments

In this section we give a numerical example and present some computational experiments on the performance of our algorithm. The example is taken from [24]

$$\begin{cases} \max_x F(x,y) = -x^2 - y^2 + 16x + 5xy \\ \quad s.t. \ x \leq 20 \\ \quad\quad\quad x \geq 0 \\ \quad \max_y f(x,y) = y \\ \quad s.t. \ x + y \leq 20 \\ \quad\quad\quad y \leq 10 \\ \quad\quad\quad y \geq 0 \end{cases}$$

The results of application of algorithm DCABQLP to this problem, for different values of the penalty parameter k are given in Table 1. The global solution is (11.1428, 8.8572) with objective value 469.1429. The solution in Ref. [24] is (11.14286, 8.85714) with objective value 469.14286.

Table 1. Results for the numerical example

k	$(x^*; y^*)$	F^*
1	(11.1489, 8.8511)	469.1429
10	(10, 10)	460
100	(10, 10)	460

We tested our algorithm on a collection of 20 problems randomly generated. We considered an indefinite matrix Q of the form $Q = \frac{1}{2}G^T G$, where the elements of G are random numbers in $[-10, 10]$. The elements of vectors c, d, c_2 and d_2 are random numbers in $[-10, 10]$. The constraints are generated such as the constraints set, S, of problem (1) is nonempty and bounded in the following way (see [19]):

Let $\bar{b} = \begin{pmatrix} b_1 \\ b_2 \end{pmatrix} \in \mathbb{R}^{\bar{m}}$, $\bar{A} = \begin{pmatrix} A_1 \\ A_2 \end{pmatrix} \in \mathbb{R}^{\bar{m} \times n_1}$ and $\bar{B} = \begin{pmatrix} B_1 \\ B_2 \end{pmatrix} \in \mathbb{R}^{\bar{m} \times n_2}$,

where $\bar{m} = m_1 + m_2$.

The elements of the matrices $\bar{A}$ and $\bar{B}$, $(\bar{a}_{ij})$ and $(\bar{b}_{ij})$ respectively, for $i = 1, ..., \bar{m}-1$ and $j = 1, ..., \bar{n}$ (where $\bar{n} = n_1 + n_2$) are random numbers in $[-10, 10]$. The last row of the constraints matrix is randomly generated in $[0, 10]$, and each right-hand side $\bar{b}_i$ $(i = 1, ..., \bar{m})$ is generated as follows :

$$\bar{b}_i = \sum_{j=1}^{\bar{n}} \bar{a}_{ij} + \sum_{j=1}^{\bar{n}} \bar{b}_{ij} + 2\mu, \quad i = \overline{1, m}, \quad \mu \in [0, 1]$$

The algorithm is implemented in MATLAB and run on a Intel Core CPU 1.5GHz with 2 Go of RAM. The commands *linprog* and *quadprog* of MATLAB have been used to solve the linear and quadratic programs. We take $\epsilon = 10^{-4}$ and $\gamma = 0.5$.

Initial point: to initialize the algorithm we use DCA to solve the following concave minimization problem

$$0 = \min\{\Psi(z) : \tilde{A}z = \tilde{b}, z \geq 0\}, \tag{20}$$

where

$$\tilde{A} = \begin{pmatrix} A_2 & B_2 & 0 & I_{m_2} & 0 & 0 \\ 0 & 0 & 0 & 0 & -I_{n_2} & B_2^t \end{pmatrix} \quad \text{and} \quad \tilde{b} = \begin{pmatrix} b_2 \\ d_2 \end{pmatrix}.$$

Choice of penalty parameter: the choice of the penalty parameter k is crucial for the algorithm. It must be taken large enough to have the equivalence between problem (4) and problem (9). In our experiments we take $k = 10^4$. We note that, in our several simulations (which are not reported here), the procedure of starting with a small k and increasing it led to a slow convergence of the algorithm.

The performance of DCABQLP is reported in Table 2 where #*it* stands for the average number of iterations, and *time* is the average CPU time given in seconds.

Comments: from the computational results we observe that the algorithm is very fast for problems of small dimension, but the execution time becomes higher for problems with a large number of variables. On the other hand we can see that the number of iterations required to reach the solution is not sensitive to the dimension of the tested problem and is relatively small. The average number of iterations is equal to 5.

Table 2. Computational results of DCABQLP

Problem	n_1	n_2	$\bar{m}$	#it	time
1	10	5	10	2	0.2
2	15	10	10	4	0.60
3	20	10	15	4	0.71
4	25	15	15	6	1.38
5	30	20	20	4	3.67
6	35	25	20	5	2.17
7	40	25	25	4	6.35
8	45	30	25	4	13.91
9	50	30	30	4	31.20
10	55	40	30	5	38.04
11	60	40	35	6	99.98
12	70	50	35	4	141.01
13	80	60	40	5	167.02
14	90	80	45	8	238.92
15	100	100	50	10	588.53
16	110	100	50	6	711.93
17	120	110	45	7	676.01
18	130	110	50	9	1238.1
19	140	120	40	5	819.87
20	150	130	40	7	988.53

6   Conclusion

We have presented a DC optimization approach for solving nonlinear bilevel optimization problems, in which the upper level objective function is quadratic nonconvex and the second level is a linear problem. After transformation of the problem into a DC program, via an exact penalization, we develop a DC algorithm for its resolution. Computational experiments show the efficiency of the proposed algorithm, especially for problems of small dimension.

References

1. Amouzegar, M.A., Moshirvaziri, K.: Determining optimal pollution control policies: An application of bilevel programming. European J. of Oper. Res. 119, 100–120 (1999)
2. Pham Dinh, T., Le Thi, H.A.: Convex analysis approach to DC programming: Theory, Algorithms and Applications (dedicated to Professor Hoang Tuy on the occasion of his 70th birthday). Acta Mathematica Vietnamica 22, 289–355 (1997a)
3. Le Thi, H.A., Pham Dinh, T.: Solving a class of linearly constrained indefinite quadratic problems by dc algorithms. Journal of Global Optimization 11(3), 253–285 (1997b)
4. Le Thi, H.A., Pham Dinh, T., Le, D.M.: Exact penalty in DC programming. Vietnam J. Math. 27, 169–179 (1999)

5. Le Thi, H.A., Pham Dinh, T.: A continuous approach for globally solving linearly constrained quadratic zero-one programming problems. Optimization 50, 93–120 (2001)
6. Le Thi, H.A., Pham Dinh, T.: The dc (difference of convex functions) programming and DCA revisited with DC models of real world nonconvex optimization problems. Annals of Oper. Res. 133, 23–46 (2005)
7. Le Thi, H.A., Pham Dinh, T., Nguyen, C.N., Nguyen, V.T.: DC programming techniques for solving a class of nonlinear bilevel programs. J. of Glob. Opt. 44, 313–337 (2009)
8. Le Thi, H.A., Pham Dinh, T., Huynh, V.N.: Exact penalty and error bounds in DC programming. J. of Glob. Opt. 52, 509–535 (2012)
9. Anandalingam, G., White, D.J.: A solution for the linear static Stackelberg problem using penalty functions. IEEE Trans. on Aut. Cont. 35, 1170–1173 (1990)
10. Anandalingam, G., White, D.J.: Hierarchical optimization: An introduction. Annals of Oper. Res. 34, 1–11 (1992)
11. Bard, J.F.: Convex two level optimization. Math. Program. 40, 15–27 (1988)
12. Bard, J.F.: Practical bilevel optimization: algorithms and applications. Kluwer Academic Publishers, Dordrecht (1998)
13. Cao, D., Chen, M.: Capacitated plant selection in a decentralized manufacturing environment: A bilevel optimization approach. European J. of Oper. Res. 169, 97–110 (2006)
14. Campelo, M., Dantas, S., Scheimberg, S.: A note on a penalty function approach for solving bilevel linear programs. J. of Glob. Opt. 16, 245–255 (2000)
15. Colson, B., Marcotte, P., Savard, G.: An overview of bilevel optimization. Annals of Oper. Res. 153, 235–256 (2007)
16. Dempe, S.: Foundations of Bilevel Programming. Kluwer Academic Publishers, Dordrecht (2000)
17. Dempe, S.: Annotated bibliography on bilevel programming and mathematical programs with equilibrium constraints. Optimization 52, 333–359 (2003)
18. Fortuny-Amat, J., McCarl, B.: A representation and economic interpretation of a two level programming problem. J. of Oper. Res. Soc. 321, 783–792 (1981)
19. Jacobson, E.S., Moshivaziri, K.: Computational experience using an edge search algorithm for linear reverse convex programs. J. of Glob. Opt. 9, 153–167 (1996)
20. Kalashnikov, V.V., Pérez-Valdés, G.A., Tomasgard, A., Kalashnykova, N.I.: Natural gas cash-out problem: Bilevel stochastic optimization approach. European J. of Oper. Res. 206, 18–33 (2010)
21. Luo, Z.Q., Pang, J.S., Ralph, D., Wu, S.-Q.: Exact penalization and stationarity conditions of mathematical programs with equilibrium constraints. Math. Program. 75, 19–76 (1996)
22. Rockafellar, R.T.: Convex analysis. Princeton, USA (1970)
23. Pham Dinh, T., Nguyen, C.N., Le Thi, H.A.: DC Programming and DCA for Globally Solving the Value-At-Risk, Comput. Manag. Sci. 6, 477–501 (2009)
24. Wang, G., Wang, X., Wan, Z., Lv, Y.: A globally convergent algorithm for a class of bilevel nonlinear programming problem. App. Math. and Comp. 188, 166–172 (2007)

A DC Programming Approach for Sparse Estimation of a Covariance Matrix

Duy Nhat Phan[1], Hoai An Le Thi[1], and Tao Pham Dinh[2]

[1] Laboratory of Theoretical and Applied Computer Science EA 3097
University of Lorraine, Ile de Saulcy, 57045 Metz, France
{duy-nhat.phan,hoai-an.le-thi}@univ-lorraine.fr
[2] Laboratory of Mathematics, INSA–Rouen, University of Normandie
76801 Saint-Etienne-du-Rouvray cedex, France
pham@insa-rouen.fr

Abstract. We suggest a novel approach to the sparse covariance matrix estimation (SCME) problem using the ℓ_1-norm. The resulting optimization problem is nonconvex and very hard to solve. Fortunately, it can be reformulated as DC (Difference of Convex functions) programs to which DC programming and DC Algorithms can be investigated. The main contribution of this paper is to propose a more suitable DC decomposition for solving the SCME problem. The experimental results on both simulated datasets and two real datasets in classification problem illustrate the efficiency of the proposed algorithms.

Keywords: Sparse covariance matrix, DC programming, DCA.

1   Introduction

Estimation of covariance matrix plays a major role in statistical analysis. Recently, numerous statistical methods require an estimate of a covariance matrix or its inverse, including principal component analysis (PCA), linear discriminant analysis (LDA), quadratic discriminant analysis (QDA), regression for multivariate normal data, analysis of independence and conditional independence relationships between components in graphical models, and portfolio optimization.

Suppose that we observe a sample including n observational data points $X_1, ..., X_n$ from a p-dimensional multivariate normal distribution $N(0, \Sigma)$, with a mean vector 0 and the covariance matrix Σ. Let $S = \frac{1}{n} \sum_{i=1}^{n} X_i X_i^T$ be the sample covariance matrix. The negative log-likelihood function is

$$\ell(\Sigma) = \frac{n}{2} \left[\log \det \Sigma + \text{tr}(\Sigma^{-1}S) + p \log 2\pi \right]. \tag{1}$$

The general purpose is to estimate the covariance matrix Σ. However, the problem is that with the increasing abundance of high-dimensional datasets, the sample covariance matrix S becomes an extremely noisy estimator of the covariance matrix, and besides, the number of parameters used to estimate grows

© Springer International Publishing Switzerland 2015   131
H.A. Le Thi et al. (eds.), *Model. Comput. & Optim. in Inf. Syst. & Manage. Sci.*,
Advances in Intelligent Systems and Computing 359, DOI: 10.1007/978-3-319-18161-5_12

quadratically with the number of variables. Intuitively, the most suitable approach to cope with this problem is finding an estimate of covariance matrix which is as sparse as possible, since the sparsity leads to the effective reduction in the number of parameters. Furthermore, the sparsity is visualized by the so-called covariance graph [6]. In the covariance graph, each node presents a random variable in a random vector and these nodes are connected by bidirectional edges if the covariances between the corresponding variables are nonzero. Note that two random variables are marginally independent if and only if their covariance is zero. Hence the zeros in a covariance matrix correspond to marginal independencies between variables, and a sparse estimate of the covariance matrix is a covariance graph having a small number of edges.

In the literature, there exists the number of methods that seeks the sparsity in the covariance matrix to improve the estimation accuracy and/or to explore the structure of the covariance graphical model. [1] used a diagonal estimate for the covariance matrix. [6] considered the covariance matrix estimation problem given a prespecified zero-pattern. [9],[21],[2],[5],[29] proposed some lasso regression-based methods (and/or combined with the Cholesky decomposition). [10] formulated a prior for Bayesian inference given a covariance graph structure. [22] presented the method which can be viewed as an extension of the generalized shrinkage operator [28] applied to the sample covariance matrix to achieve a sparse estimate.

In particular, [11],[3] penalized the off-diagonal elements of the covariance matrix by adding to the negative log-likelihood (1) an ℓ_1 penalty. The resulting sparse covariance matrix estimation (SCME) problem is

$$\min_{\Sigma \succ 0} \left\{ \log \det \Sigma + \mathrm{tr}(\Sigma^{-1}S) + \lambda \|W \circ \Sigma\|_1 \right\}, \tag{2}$$

where the notation $\Sigma \succ 0$ means that Σ is symmetric positive definite, W is the matrix with zero diagonal and one off-diagonal, $\circ$ denotes the Hadamard product, and λ is a nonnegative tuning parameter. If S is nonsingular, then the problem (2) is equivalent to the following problem:

$$\min_{\Sigma \succeq \delta I_p} \left\{ F(\Sigma) := \log \det \Sigma + \mathrm{tr}(\Sigma^{-1}S) + \lambda \|W \circ \Sigma\|_1 \right\}, \tag{3}$$

for some $\delta > 0$ [3]. Here, I_p denotes the $p \times p$ identity matrix, and the notation $\Sigma \succeq \delta I_p$ means that $\Sigma - \delta I_p$ is symmetric positive semidefinite. Note that if S is not full rank, we can replace S with $S + \epsilon I_p$ for some $\epsilon > 0$. Solving (3) is a formidable challenge since it is nonconvex. In [3], Bien and Tibshirani applied the minorization-maximization (MM) approach for solving this problem.

In this paper, we aim to propose an algorithm based on DC (Difference of Convex functions) programming and DCA (DC Algorithms) for solving the problem (3). DC programming and DCA were introduced by Pham Dinh Tao in their preliminary form in 1985. They have been extensively developed since 1994 by Le Thi Hoai An and Pham Dinh Tao and become now classic and increasingly popular (see e.g. [18],[26],[27]). Our motivation is based on the fact that DCA is a fast and scalable approach which has been successfully applied to many large-scale (smooth or non-smooth) nonconvex programs in various domains of applied sciences, in particular in data analysis and data mining, for which it provided quite often a global solution and proved to be more robust and efficient than standard methods (see e.g. [18],[17],[26],[27],[15],[20],[16],[19],[7] and the list of references in http://lita.sciences.univ-metz.fr/~lethi/DCA.html).

The so-called DC program is that of minimizing a DC function $F = G - H$ over a convex set with G and H being convex functions. The construction of DCA involves DC components G and H but not the function F itself. Moreover, a DC function F has infinitely many DC decompositions $G - H$ which have crucial implications on the qualities (speed of convergence, robustness, efficiency, globality of computed solutions, . . .) of DCA. Hence, the finding of an appropriate DC decomposition is important from an algorithmic point of view. In particular, for the SCME problem, the main contribution of this paper is to propose a more suitable DC decomposition, and then the corresponding DCA scheme is developed. To examine the efficiency of our proposed algorithm, we perform the experiments on both simulated datasets and two real datasets in the classification problems.

The paper is organized as follows. In Section 2, we present DC programming and DCA for general DC programs, and illustrate how to apply DCA to solve the problem (3). The numerical experiments are reported in Section 3. Finally, the conclusions are given in Section 4.

Notation. In this paper, for matrices $A, B \in \mathbb{R}^{n \times m}$, the inner and Hadamard products of A and B are defined as $\langle A, B \rangle = \text{tr}(A^T B)$, $A \circ B = [A_{ij}B_{ij}]$, respectively. The spectral and Frobenius norms are $||A||_2 = \sqrt{\lambda_{\max}(A^T A)}$, $||A||_F = \sqrt{\sum_{i,j} A_{ij}^2}$, respectively, where $\lambda_{\max}(A^T A)$ denotes the maximal eigenvalue of $A^T A$.

2 Solution Method Based on DC Programming and DCA

First, for the reader's convenience, let us give a brief introduction to DC programming and DCA.

2.1 DC Programming and DCA

A general DC program is that of the form:

$$\alpha = \inf\{F(x) := G(x) - H(x) \,|\, x \in \mathbb{R}^n\} \quad (P_{dc}),$$

where G, H are lower semi-continuous proper convex functions on $\mathbb{R}^n$. Such a function F is called a DC function, and $G - H$ a DC decomposition of F while G and H are the DC components of F. Note that, the closed convex constraint $x \in C$ can be incorporated in the objective function of (P_{dc}) by using the indicator function on C denoted by χ_C which is defined by $\chi_C(x) = 0$ if $x \in C$, and $+\infty$ otherwise.

For a convex function θ, the subdifferential of θ at $x_0 \in \text{dom}\theta := \{x \in \mathbb{R}^n : \theta(x) < +\infty\}$, denoted by $\partial\theta(x_0)$, is defined by

$$\partial\theta(x_0) := \{y \in \mathbb{R}^n : \theta(x) \geq \theta(x_0) + \langle x - x_0, y\rangle, \forall x \in \mathbb{R}^n\},$$

and the conjugate θ° of θ is

$$\theta^\circ(y) := \sup\{\langle x, y\rangle - \theta(x) : x \in \mathbb{R}^n\}, \quad y \in \mathbb{R}^n.$$

Then, the following program is called the dual program of (P_{dc}):

$$\alpha_D = \inf\{H^\circ(y) - G^\circ(y) \,|\, y \in \mathbb{R}^n\} \quad (D_{dc}).$$

One can prove (see, e.g. [26]) that $\alpha = \alpha_D$ and that there is a perfect symmetry between primal and dual DC programs: the dual to (D_{dc}) is exactly (P_{dc}).

The necessary local optimality condition for the primal DC program, (P_{dc}), is

$$\partial H(x^\circ) \subset \partial G(x^\circ). \tag{4}$$

The condition (4) is also sufficient for many important classes of DC programs, for example, for DC polyhedral programs, or when function F is locally convex at x° ([18]).

A point x° is called a *critical point* of $G - H$, or a generalized Karush-Kuhn-Tucker point (KKT) of (P_{dc})) if

$$\partial H(x^\circ) \cap \partial G(x^\circ) \neq \emptyset. \tag{5}$$

Based on local optimality conditions and duality in DC programming, the DCA consists in constructing two sequences $\{x^l\}$ and $\{y^l\}$ (candidates to be solutions of (P_{dc}) and its dual problem respectively). Each iteration l of DCA approximates the concave part $-H$ by its affine majorization (that corresponds to taking $y^l \in \partial H(x^l)$) and minimizes the resulting convex function (that is equivalent to determining $x^{l+1} \in \partial G^\circ(y^l)$).

Generic DCA scheme
Initialization: Let $x^0 \in \mathbb{R}^n$ be an initial guess, $l \leftarrow 0$.
Repeat
- Calculate $y^l \in \partial H(x^l)$
- Calculate $x^{l+1} \in \arg\min\{G(x) - \langle x, y^l\rangle : x \in \mathbb{R}^n\}$ $\quad (P_l)$
- $l \leftarrow l + 1$
Until convergence of $\{x^l\}$.

Convergences properties of DCA and its theoretical basic can be found in [18],[26]. It is worth mentioning that

- DCA is a descent method (*without linesearch*): the sequences $\{G(x^l) - H(x^l)\}$ and $\{H^\circ(y^l) - G^\circ(y^l)\}$ are decreasing.
- If $G(x^{l+1}) - H(x^{l+1}) = G(x^l) - H(x^l)$, then x^l is a critical point of $G - H$ and y^l is a critical point of $H^\circ - G^\circ$. In such a case, DCA terminates at l-th iteration.

- If the optimal value α of problem (P_{dc}) is finite and the infinite sequences $\{x^l\}$ and $\{y^l\}$ are bounded then every limit point x (resp. y) of the sequences $\{x^l\}$ (resp. $\{x^l\}$) is a critical point of $G - H$ (resp. $H^\circ - G^\circ$).
- DCA has a *linear convergence* for general DC programs, and has a finite convergence for polyhedral DC programs.

A deeper insight into DCA has been described in [18]. For instant it is crucial to note the main feature of DCA: DCA is constructed from DC components and their conjugates but not the DC function f itself which has infinitely many DC decompositions, and there are as many DCA as there are DC decompositions. Such decompositions play a crucial role in determining the speed of convergence, stability, robustness, and globality of sought solutions. Therefore, it is important to study various equivalent DC forms of a DC problem. This flexibility of DC programming and DCA is of particular interest from both a theoretical and an algorithmic point of view.

For a complete study of DC programming and DCA the reader is referred to [18],[26],[27] and the references therein.

In the last decade, a variety of works in Machine Learning based on DCA have been developed. The efficiency and the scalability of DCA have been proved in a lot of works (see e.g. [18],[17],[26],[12],[15],[20],[16],[7],[13],[14],[24],[23],[30] and the list of reference in http://lita.sciences.univ-metz.fr/~lethi/DCA.html). These successes of DCA motivated us to investigate it for solving the SCME problem.

2.2 DCA for Solving (3)

We consider a special DC formulation of the problem (3) as follows:

$$\min \{F(\Sigma) = G(\Sigma) - H(\Sigma) : \Sigma \succeq \delta I_p\}, \tag{6}$$

where

$$G(\Sigma) := \frac{\mu}{2}||\Sigma||_F^2 + \lambda||W \circ \Sigma||_1, \tag{7}$$

and

$$H(\Sigma) := \frac{\mu}{2}||\Sigma||_F^2 - \text{tr}(\Sigma^{-1}S) - \log \det \Sigma \tag{8}$$

are convex functions when μ is large enough. According to the generic DCA scheme, at each iteration l, we have to compute a subgradient V^l of H at Σ^l and then solve the convex program of the form (P_l), namely

$$\min\{G(\Sigma) - \langle V^l, \Sigma \rangle : \Sigma \succeq \delta I_p\}. \tag{9}$$

H is differentiable and $V^l = \nabla H(\Sigma^l)$ is calculated as follows:

$$V_{ij}^l = \mu\Sigma_{ij} + \left[\Sigma^{-1}S\Sigma^{-1}\right]_{ij} - \left[(\Sigma^l)^{-1}\right]_{ij}. \tag{10}$$

DCA for solving the problem (6) can be described in Algorithm 1.

Algorithm 1. (DCA applied to (6))

Initialization: Let τ be a tolerance sufficient small, set $l = 0$ and compute δ, μ. Choose $\Sigma^0 \succeq \delta I_p$.

repeat

    1. Compute V^l by $V_{ij}^l = \mu \Sigma_{ij} + \left[\Sigma^{-1} S \Sigma^{-1} \right]_{ij} - \left[(\Sigma^l)^{-1} \right]_{ij}$.

    2. Solve the following convex problem to obtain Σ^{l+1}

$$\min_{\Sigma \succeq \delta I_p} \left\{ \frac{\mu}{2} \|\Sigma\|_F^2 + \lambda \|W \circ \Sigma\|_1 - \langle V^l, \Sigma \rangle \right\} \tag{11}$$

    3. $l \leftarrow l + 1$.

until $\|\Sigma^{l+1} - \Sigma^l\|_F \leq \tau \left(\|\Sigma^l\|_F + 1 \right)$ or $|F(\Sigma^{l+1}) - F(\Sigma^l)| \leq \tau \left(|F(\Sigma^l)| + 1 \right)$.

Remark 1. For solving the convex problem (11), we use the alternating direction method of multipliers (ADMM) [4]. The augmented Lagrangian function of (11) is

$$L(\Sigma, X, Y) = \frac{\mu}{2} \|\Sigma\|_F^2 - \langle V^l, \Sigma \rangle + \lambda \|W \circ X\|_1 + \langle Y, \Sigma - X \rangle + \frac{\rho}{2} \|\Sigma - X\|_F^2. \tag{12}$$

More specifically, ADMM solves the following problems at each iteration k:

$$\Sigma^{k+1} = \arg \min_{\Sigma \succeq \delta I_p} L(\Sigma, X^k, Y^k) \tag{13}$$

$$X^{k+1} = \arg \min_{X \in \mathbb{R}^{p \times p}} L(\Sigma^{k+1}, X, Y^k) \tag{14}$$

$$Y^{k+1} = Y^k + \rho(\Sigma^{k+1} - X^{k+1}). \tag{15}$$

Finally, ADMM for solving (11) can be described as follows:

Initialization: Set $k = 0$, choose $X^0, Y^0 \in \mathbb{R}^{p \times p}$, and let $\rho > 0$.

repeat

    1. Compute $\Sigma^{k+1} = U D_\delta U^T$ where $D_\delta = \text{diag}(\max(D_{ii}, \delta))$ and $(V^l - Y^k + \rho X^k)/(\mu + \rho) = U D U^T$.

    2. Compute $X^{k+1} = S \left(\Sigma^{k+1} + Y^k/\rho, (\lambda/\rho)W \right)$.

    3. Compute $Y^{k+1} = Y^k + \rho(\Sigma^{k+1} - X^{k+1})$

    4. $k \leftarrow k + 1$.

until Converge.

Remark 2. For estimating μ, since the function $-\log \det \Sigma$ is convex and the sum of two convex functions is also convex, it is sufficient to take μ such that $\frac{\mu}{2}\|\Sigma\|_F^2 - \text{tr}(\Sigma^{-1}S)$ becomes convex. For this purpose, we can choose μ greater than the spectral radius of the Hessian matrix of $\Lambda(\Sigma) = \text{tr}(\Sigma^{-1}S)$, i.e., $\mu \geq \|\nabla^2 \Lambda(\Sigma)\|_2$ for all $\Sigma \succeq \delta I_p$. The gradient and Hessian of $\Lambda(\Sigma)$ are respectively

$$\nabla \Lambda(\Sigma) = -\Sigma^{-1} S \Sigma^{-1}, \tag{16}$$

and

$$\nabla^2 \Lambda(\Sigma) = \Sigma^{-1} S \Sigma^{-1} \otimes \Sigma^{-1} + \Sigma^{-1} \otimes \Sigma^{-1} S \Sigma^{-1}, \tag{17}$$

where $\otimes$ denotes the Kronecker product. We can deduce from (17) that

$$||\nabla^2 \Lambda(\Sigma)||_2 \le 2||S||_2 \delta^{-3},$$

thus we can assign $2||S||_2 \delta^{-3}$ to μ.

3 Numerical Experiments

3.1 Comparative Algorithms

Let W^1 be a matrix defined by $W_{ij}^1 = 0$ if $i = j$ and 1 otherwise, W^2 be a matrix defined by $W_{ij}^2 = 0$ if $i = j$ and $W_{ij}^2 = \frac{1}{|S_{ij}|}$ otherwise. DCA1 and DCA2 denote Algorithm 1 with $W = W^1$ and $W = W^2$, respectively. We will compare our proposed approaches (DCA1 and DCA2) with the methods proposed in [3] which used the MM approach (that is in fact a version of DCA) for solving the problem (3).

SPCOV1 and SPCOV2 denote the MM approach for solving the problem (3) with $W = W^1$ and $W = W^2$, respectively [3]. The R package **spcov** for SPCOV1 and SPCOV2 is available from CRAN[1].

3.2 Experimental Setups

All algorithms are implemented in the R 3.0.2, and performed on a PC Intel i7 CPU3770, 3.40 GHz of 8GB RAM.

In experiments, we set the stop tolerance $\tau = 10^{-4}$ for DCA. The starting point Σ^0 of DCA is the sample covariance matrix S. The value of parameter λ is chosen through a 5-fold cross-validation procedure on tuning or training set from a set of candidates $\{0.01, ..., 0.9\}$.

The cross-validation procedure is described as follows [3]. For $\mathcal{A} \subseteq \{1, ..., n\}$, let $S_{\mathcal{A}} = |\mathcal{A}|^{-1} \sum_{i \in \mathcal{A}} X_i X_i^T$, and $\mathcal{A}_i^c$ denotes the component of $\mathcal{A}$. We divide $\{1, ..., n\}$ into 5 subsets, $\mathcal{A}_1, ..., \mathcal{A}_5$, and then compute

$$f(\lambda) = \frac{1}{5} \sum_{i=1}^{5} \ell\left\{\hat{\Sigma}_\lambda(S_{\mathcal{A}_i^c}); S_{\mathcal{A}_i}\right\}, \tag{18}$$

where $\hat{\Sigma}_\lambda(S_{\mathcal{A}_i^c})$ is an estimate of the covariance matrix Σ with the parameter λ and $S_{\mathcal{A}_i^c}$, and $\ell\left\{\hat{\Sigma}_\lambda(S_{\mathcal{A}_i^c}); S_{\mathcal{A}_i}\right\} = -\log \det \hat{\Sigma}_\lambda(S_{\mathcal{A}_i^c}) - \text{tr}\left(\left[\hat{\Sigma}_\lambda(S_{\mathcal{A}_i^c})\right]^{-1} S_{\mathcal{A}_i}\right)$. Finally, we choose $\hat{\lambda} = \arg\max_\lambda f(\lambda)$.

---

[1] http://cran.r-project.org/web/packages/spcov/index.html

3.3   Experiments on Synthetic Datasets

We evaluate the performance of DCA1 and DCA2 on three synthetic datasets. We generate $X = [X_1, ..., X_n]$ from a multivariate normal distribution $N_p(0, \Sigma)$, where Σ is a sparse symmetric positive definite matrix. We consider two types of covariance graphs and a moving average model as follows (see [3]):

Cliques model: We generate $\Sigma = \text{diag}(\Sigma_1, ..., \Sigma_5)$, where $\Sigma_1, ..., \Sigma_5$ are dense matrices.

Random model: In this model, we take $\Sigma_{ij} = \Sigma_{ji}$ to be nonzero with the probability 0.02, independently of other elements.

First-order moving average model: We generate $\Sigma_{i,i-1} = \Sigma_{i-1,i}$ to be nonzero for $i = 2, ..., p$.

In the first two cases, the nonzero entries of matrix Σ are randomly drawn in the set $\{+1, -1\}$. In the moving average model, all nonzero values are set to be 0.4. In this experiment, for each covariance model, we generate ten training sets and one tuning set with the size $n = 200, p = 100$. The tuning set is used to choose the parameter λ.

To evaluate the performance of each method, we consider three loss functions which are the root-mean-square error (RMSE), the entropy loss (EN), and the Kullback-Leibler (KL) loss, respectively.

$$\text{RMSE} = ||\hat{\Sigma} - \Sigma||_F / p, \tag{19}$$

$$\text{EN} = -\log\det(\hat{\Sigma}\Sigma^{-1}) + \text{tr}(\hat{\Sigma}\Sigma^{-1}) - p, \tag{20}$$

$$\text{KL} = -\log\det(\hat{\Sigma}^{-1}\Sigma) + \text{tr}(\hat{\Sigma}^{-1}\Sigma) - p, \tag{21}$$

where $\hat{\Sigma}$ is a sparse estimate of the covariance matrix Σ.

The experimental results on synthetic datasets are given in Table 1. In this Table, the average of root-mean-square error (RMSE), entropy loss (EN), Kullback-Leibler (KL) loss, number of nonzero elements (NZ), CPU time in second, and their standard diviations over 10 samples are reported.

We observe from Table 1 that in terms of root-mean square error and sparsity, DCA2 gives the best results on all three models. In the random and moving average models, DCA2 also gives the best entropy loss and the best Kullback-Leibler loss. In the cliques model, DCA1 attains the lowest entropy and Kullback-Leibler losses. DCA2 and SPCOV2 perform better than DCA1 and SPCOV1, respectively because these approaches use an adaptive lasso penalty on off-diagonal elements. The training time shows that SPCOV1 and SPCOV2 are faster than DCA1 and DCA2.

3.4   Experiments on Real Datasets

We illustrate the use of sparse covariance matrix estimation problem in a real application: the classification problem of two datasets from UCI Machine Learning Repository (Ionosphere,Waveform). All the datasets are preprocessed by normalizing each dimension of the data to zero mean. The detailed information of these datasets is summarized in Table 2.

Table 1. Comparative results of DCA1, DCA2, SPCOV1 and SPCOV2 in terms of the average of root-mean-square error (RMSE), entropy loss (EN), Kullback-Leibler (KL) loss, number of nonzero elements, CPU time in second, and their standard diviations over 10 runs. Bold fonts indicate the best result in each row.

		DCA1	DCA2	SPCOV1	SPCOV2
Cliques	RMSE	0.39 ± 0.005	**0.379 ± 0.004**	0.395 ± 0.0.005	0.384 ± 0.005
	EN	**13.82 ± 0.52**	14.44 ± 2.73	22.72 ± 0.41	16.85 ± 0.51
	KL	**21.05 ± 1.46**	23.19 ± 1.84	60.46 ± 2.38	34.41 ± 1.99
	NZ	2674.4 ± 225.06	**2545.4 ± 266.38**	7620 ± 45.52	3571 ± 92
	CPU	51.41 ± 2.09	78.94 ± 48.18	**59.93 ± 12.47**	61.15 ± 18.98
Random	RMSE	0.077 ± 0.004	**0.065 ± 0.007**	0.086 ± 0.0.002	0.066 ± 0.002
	EN	3.66 ± 0.15	**2.46 ± 0.12**	3.9 ± 0.16	2.48 ± 0.15
	KL	4.88 ± 0.45	**2.78 ± 0.2**	5.15 ± 0.5	3.07 ± 0.25
	NZ	938.4 ± 80.53	**516.4 ± 44.79**	825.4 ± 54.36	527.6 ± 18.54
	CPU	78.05 ± 19.64	100.55 ± 105.91	54.5 ± 11.56	**29.78 ± 5.61**
Moving	RMSE	0.024 ± 0.001	**0.012 ± 0.001**	0.028 ± 0.0007	0.021 ± 0.0006
	EN	6.38 ± 0.7	**2.1 ± 0.26**	12.75 ± 0.46	10.78 ± 0.51
	KL	10.53 ± 1.57	**2.46 ± 0.33**	28.26 ± 1.89	22.24 ± 1.81
	NZ	1881 ± 74.78	**641.6 ± 96.65**	3834.8 ± 77.34	3004.8 ± 77.57
	CPU	88.86 ± 14.35	150.93 ± 159.9	40.98 ± 4.55	**38.84 ± 5.09**

In this experiment, we need to estimate a covariance matrix with respect to each method, and then we use the linear discriminant analysis (LDA) for these classification problems. Suppose that the samples are independent and normally distributed with a common covariance matrix Σ. The LDA classification rule is obtained by using Bayes's rule to estimate the most likely class for a test sample, i.e., the predicted class for a test sample x is

$$\arg\max_k x^T \hat{\Sigma}^{-1} \hat{\mu}_k - \frac{1}{2} \hat{\mu}_k^T \hat{\Sigma}^{-1} \hat{\mu}_k + \log n_k,$$

where $\hat{\Sigma}$ is an estimate of the covariance matrix Σ, $\hat{\mu}_k$ is the k-th class mean vector, and n_k is the number of samples in the class k. The detailed information on LDA can be found in [25],[8].

The training set is used to estimate a covariance matrix $\hat{\Sigma}$ by each approach and

$$\hat{\mu}_k = \frac{1}{n_k} \sum_{i \in \text{class} k} X_i.$$

Table 2. Real datasets used in experiments

Data	No. of features	No. of samples	No. of classes
Ionosphere	34	351	2
Waveform	40	5000	3

Table 3. Comparative results of DCA1, DCA2, SPCOV1 and SPCOV2 in terms of the average of percentage of accuracy of classifiers (ACC) and its standard deviation, the average of number of nonzero elements (NZ) its standard deviation, and the average of CPU time in second and its standard diviation over 10 training/test set splits. Bold fonts indicate the best result in each row.

		DCA1	DCA2	SPCOV1	SPCOV2
Ionosphere	ACC	87.18 ± 1.45	**87.6 ± 1**	85.12 ± 2.42	86.15 ± 02.19
	NZ	1061.2 ± 19.3	**448.8 ± 156.9**	1079 ± 3.12	534.8 ± 16.55
	CPU	**3.46 ± 1.52**	6.04 ± 4.71	4.62 ± 0.19	14.02 ± 1.27
Waveform	ACC	**85.35 ± 0.73**	85.04 ± 0.39	84.75 ± 0.7	84.82 ± 0.6
	NZ	1518.2 ± 38.85	**460 ± 10.28**	1538.2 ± 15.9	473.8 ± 9.63
	CPU	**0.09 ± 0.07**	3.39 ± 0.67	1.22 ± 0.22	1.78 ± 0.08

For the experiment, we use the cross-validation scheme to validate the performance of various approaches. The real datasets are split into a training set containing 2/3 of the samples and a test set containing 1/3 of the samples. This process is repeated 10 times, each with a random choice of training set and test set. The parameter λ is chosen via 5-fold cross-validation.

The computational results given by DCA1, DCA2, SPCOV1 and SPCOV2 were reported in Table 3. We are interested in the efficiency (the accuracy of classifiers and the sparsity in covariance matrix) as well as the rapidity of these algorithms.

We observe from computational results that in terms of accuracy of classifiers, DCA1 and DCA2 are comparable and they are better than SPCOV1 and SPCOV2 on both Ionosphere and Waveform datasets. DCA2 not only provides a high accuracy of classifiers, but also gives the best performance in terms of sparsity. In terms of CPU time, DCA1 is the fastest.

4   Conclusions

We have investigated DC programming and DCA for solving the sparse covariance matrix estimation problem using ℓ_1-norm. We proposed a more suitable DC formulation for this problem. The robustness and the effectiveness of our DCA based algorithms have been demonstrated through the computational results on both the simulated and real datasets.

As a part of future work, we plan to study more extensive applications of the sparse covariance matrix estimation problem. In particular, a natural way to deal with sparsity in machine learning is using the ℓ_0-norm in the regularization term. The resulting optimization problem is nonconvex, discontinuous, and NP-hard. We will study DC programming and DCA for solving this problem.

References

1. Bickel, P.J., Levina, E.: Some theory for Fisher's linear discriminant function, naive Bayes, and some alternatives when there are many more variables than observations. Bernoulli 10(6), 989–1010 (2004)
2. Bickel, P.J., Levina, E.: Regularized estimation of large covariance matrices. The Annals of Statistichs 36, 199–227 (2008)
3. Bien, J., Tibshirani, R.: Sparse estimation of a covariance matrix. Biometrika 98(4), 807–820 (2011)
4. Boyd, S., Parikh, N., Chu, E., Peleato, B., Eckstein, J.: Distributed optimization and statistical learning via the alternating direction method of multipliers. Foundat. Trends Mach. Learn. 3(1), 1–122 (2011)
5. Cai, T., Zhang, C., Zou, H.: Optimal rates of convergence for covariance matrix estimation. The Annals of Statistic 38, 2118–2144 (2010)
6. Chaudhuri, S., Drton, M., Richardson, T.S.: Estimation of a covariance matrix with zeros. Biometrika 94, 199–216 (2007)
7. Fawzi, A., Davies, M., Frossard, P.: Dictionary learning for fast classification based on soft-thresholding. International Journal of Computer Vision (2014), http://arxiv.org/abs/1402.1973
8. Hastie, T., Tibshirani, R., Friedman, J.: The Elements of Statistical Learning, 2nd edn. Springer, New York (2009)
9. Huang, J.Z., Liu, N., Pourahmadi, M., Liu, L.: Covariance matrix selection and estimation via penalised normal likelihood. Biometrika 93, 85–98 (2006)
10. Khare, K., Rajaratnam, B.: Wishart distributions for decomposable covariance graph models. Ann. Statist. 39, 514–555 (2011)
11. Lam, C., Fan, J.: Sparsistency and rates of convergence in large covariance matrix estimation. The Annals of Statistics 37, 4254–4278 (2009)
12. Le Thi, H.A., Huynh Van, N., Pham Dinh, T.: Exact penalty and error bounds in DC programming. Journal of Global Optimization 52(3), 509–535 (2012)
13. Le Thi, H.A., Le Hoai, M., Nguyen, V.V., Pham Dinh, T.: A DC Programming approach for feature selection in support vector machines learning. Journal of Advances in Data Analysis and Classification 2(3), 259–278 (2008)
14. Le Thi, H.A., Le Hoai, M., Pham Dinh, T.: Optimization based DC programming and DCA for hierarchical clustering. European Journal of Operational Research 183, 1067–1085 (2007)
15. Le Thi, H.A., Le Hoai, M., Pham Dinh, T.: Feature selection in machine learning: An exact penalty approachusing a difference of convex function algorithm. Machine Learning (2014), (published online July 04, 2014), doi:10.1007/s10994-014-5455-y
16. Le Thi, H.A., Nguyen, M.C.: Self-organizing maps by difference of convex functions optimization. Data Mining and Knowledge Discovery 28, 1336–1365 (2014)
17. Le Thi, H.A., Pham Dinh, T.: Solving a class of linearly constrained indefinite quadratic problems by D.C. algorithms. Journal of Global Optimization 11, 253–285 (1997)
18. Le Thi, H.A., Pham Dinh, T.: The DC (difference of convex functions) programming and DCA revisited with DC models of real world nonconvex optimization problems. Annals of Operations Research 133, 23–46 (2005)
19. Le Thi, H.A., Pham Dinh, T., Le Hoai, M., Vo Xuan, T.: DC approximation approaches for sparse optimization. To appear in European Journal of Operational Research (2014)

20. Le Thi, H.A., Vo Xuan, T., Pham Dinh, T.: Feature selection for linear SVMs under uncertain data: robust optimization based on difference of convex functions Algorithms. Neural Networks 59, 36–50 (2014)
21. Levina, E., Rothman, A., Zhu, J.: Sparse estimation of large covariance matrices via a nested lasso penalty. Ann. Appl. Stat. 2(1), 245–263 (2008)
22. Liu, H., Wang, L., Zhao, T.: Sparse covariance matrix estimation with eigenvalue contraints. Journal of Computational and Graphical Statistics 23(2), 439–459 (2014)
23. Liu, Y., Shen, X.: Multicategory ψ-Learning. Journal of the American Statistical Association 101, 500–509 (2006)
24. Liu, Y., Shen, X., Doss, H.: Multicategory ψ-Learning and Support Vector Machine: Computational Tools. Journal of Computational and Graphical Statistics 14, 219–236 (2005)
25. Mardia, K.V., Kent, J.T., Bibby, J.M.: Multivariate Analysis. Academic (1979)
26. Pham Dinh, T., Le Thi, H.A.: Convex analysis approach to D.C. programming: Theory, algorithms and applications. Acta Mathematica Vietnamica 22(1), 289–355 (1997)
27. Pham Dinh, T., Le Thi, H.A.: DC optimization algorithm for solving the trust-region subproblem. SIAM Journal of Optimization 8(1), 476–505 (1998)
28. Rothman, A.J., Levina, E., Zhu, J.: Generalized thresholding of large covariance matrices. J. Am. Statist. Assoc. 104, 177–186 (2009)
29. Rothman, A.J., Levina, E., Zhu, J.: A new approach to Cholesky-based covariance regularization in high dimensions. Biometrika 97, 539–550 (2010)
30. Thiao, M., Pham Dinh, T., Le Thi, H.A.: DC Programming. In: Le Thi, H.A., Bouvry, P., Pham Dinh, T. (eds.) MCO 2008. CCIS, vol. 14, pp. 348–357. Springer, Heidelberg (2008)

A New Approach for Optimizing Traffic Signals in Networks Considering Rerouting

Duc Quynh Tran[1], Ba Thang Phan Nguyen[2], and Quang Thuan Nguyen[2]

[1] FITA, Vietnam National University of Agriculture, Hanoi, Vietnam
[2] SAMI, Hanoi University of Science and Technology, Hanoi, Vietnam
tdquynh@vnua.edu.vn, phanbathang125692@gmail.com, thuan.nguyenquang@hust.vn

Abstract. In traffic signal control, the determination of the green time and the cycle time for optimizing the total delay time is an important problem. We investigate the problem by considering the change of the associated flows at User Equilibrium resulting from the given signal timings (rerouting). Existing models are solved by the heuristic-based solution methods that require commercial simulation softwares. In this work, we build two new formulations for the problem above and propose two methods to directly solve them. These are based on genetic algorithms (GA) and difference of convex functions algorithms (DCA).

Keywords: DC algorithm, Genetic algorithm, Traffic signal control, Bi-level optimization model.

1 Introduction

Traffic signal control plays an important role to reduce congestion, improve safety and protect environment [23]. The determination of optimal signal timings have been continuously developed. At the beginning, researchers studied isolated junctions [28]. Thus, an urban network is signalized by considering all its junctions independently. Some work study the group of junctions such as the problem of green wave in which the traffic light at a junction depends on the others [21],[29]. Normally, after finding an optimal signal timing, it is fixed. Some systems, however, use real time data to design signal timing that leads to a non-fixed time signal plan [8].

This work focuses on the fixed time plan process. Signal timings are optimized by using historical flows observed on links. This bases on the assumption that the flow rates will not change after the new optimal timing is set. Almond and Lott in 1968 showed that the assumption is not valid anymore for a wide area [1]. The signal time makes a change on journey time on a certain route and thus the users may choose another route that is better. It is theoretically explained by Wardrop user equilibrium condition [27]. To reflect the dependency of flow rates on signal timing change, when formulating optimization problem, an equilibrium model may be integrated as constraints to the problem.

The problem of determining optimum signal timing is usually formulated as a bi-level optimization problem. In the upper level, the objective function is often

© Springer International Publishing Switzerland 2015   143
H.A. Le Thi et al. (eds.), *Model. Comput. & Optim. in Inf. Syst. & Manage. Sci.*,
Advances in Intelligent Systems and Computing 359, DOI: 10.1007/978-3-319-18161-5_13

non-smooth and non-linear that optimizes some measures such as total delay, pollution, operating cost,... This upper level problem is constrained by the lower level equilibrium problem in which transport users try to alter their travel choices in order to minimize their travel costs. Such an optimization problem may has multiple optima and finding an efficient method to even get local optima is difficult [17]. Many solution methods are studied to devise an efficient technique for solving the above problem: heuristic methods ([24],[5]), linearization methods ([10], [2]), sensitivity based methods ([7],[30]), Krash-Kuhn-Tucker based methods ([26]), marginal function method ([18]), cutting plan method ([9]), stochastic search methods ([6], [4], [3]).

One of the impressive researches is of Ceylan and Bell ([3],[5]). They use a signal timings optimization method in which rerouting is taken in to account. Recall that the problem is formulated as a bi-level optimization problem in which the upper level objective is to minimize total travel time and the lower level problem is a traffic equilibrium problem. The proposed solution method was heuristic, namely, a genetic algorithm (GA) for the upper level problem and the SATURN package for the lower level one. SATURN is a simulation-assignment modeling software package [25] that gives an equilibrium solution by solving heuristically sub-routines. Since SATURN is heuristic- based and a commercial software as well, it is necessary to find a more-efficient approach to solve the problem.

In order to overcome the difficulty and to aim at getting a good equilibrium solution, we propose two new formulations that are directly solved by some efficient methods. The first formulation is then solved by genetic algorithms (GA) while the second one is done by a combination of GA and DCA (Difference of Convex functions Algorithm). As known, DCA was first introduced by Pham Dinh Tao in 1985 and has been extensively developed since 1994 by Le Thi Hoai An and Pham Dinh Tao in their common works. It has been successfully applied to many large-scale (smooth or nonsmooth) nonconvex programs in various domains of applied science, and has now become classic and popular (see [11],[12],[15] and references therein). This motivates us using DCA to improve the solution quality in GA-DCA scheme.

The paper is organized as follows. After the introduction in Section 1, the mathematical problem is described in Section 2. Section 3 is devoted to the GA-based solution method. A combined GA-DCA is presented in Section 4. Section 5 gives some conclusions.

2   Mathematical Model

In this section, we present new mathematical models for optimizing traffic signals in a network considering rerouting. The problem is first formulated as an optimization problem with complementarity constraints. The objective function is the total travel time of all vehicles in the network.

For the formulation, we use the following notations (see Table 1, Table 2). The parameters and variables are respectively defined in Table 1 and 2.

Table 1. Parameters

p	path $p = i_1^p \rightarrow i_2^p \rightarrow \dots \rightarrow i_{n(p)}^p$,
$w = (i,j)$	pair of origin i and destination j (OD pair),
P_w	set of paths from i to j,
$P = \cup P_w$	set of all paths,
d_w	demand of origin destination pair w,
$a = (u,v)$	link a,
$\delta_{a,p}$	parameter equal to 1 if link a belongs to path p, 0 otherwise,
h	junction h,
S_h	total number of stages at junction h,
$I_{r,h}$	inter-green between the end of green time for stage r and the start of the next green,
$\triangle_{h,r,p}$	parameter equal to 1 if the vehicles on path p can cross junction h at stage r,
C_{min}	minimum of cycle time,
C_{max}	maximum of cycle time,
$\phi_{h,r,min}$	minimum of duration green time of stage r at junction h,
$\phi_{h,r,max}$	maximum of duration green time of stage r at junction h,

Table 2. Variables

q_a	flow on link a,
t_a	travel time on link a,
t_p	travel time on path p,
f_p	flow on path p,
t_w	travel time for OD pair w,
$WT_{h,p}$	waiting time at junction h associated to path p,
$WT_{h,p}^0$	initial waiting time at junction h associated to path p,
$z_{h,p}$	integer variables, that is used to calculate $WT_{h,p}^0$,
$ST_{h,r}$	starting time of stage r at junction h,
θ_h	offset of junction h,
C	common cycle time,
$\phi_{h,r}$	duration of the green time for stage r at junction h

The total travel time is calculated by

$$TT = \sum_p t_p . f_p = \sum_w d_w . t_w.$$

The cycle time, the green time and the offset must satisfy the following conditions:

$$C_{min} \leq C \leq C_{max}, \tag{1}$$

$$0 \leq \theta_h \leq C - 1, \quad \forall h, \tag{2}$$

$$\phi_{h,r,min} \leq \phi_{h,r} \leq \phi_{h,r,max}. \tag{3}$$

The total of green time and inter-green time is equal to the cycle time.

$$C = \sum_{r=1}^{S_h} \phi_{h,r} + \sum_{r=1}^{S_h} I_{h,r}, \quad \forall h. \tag{4}$$

The flow on link (u, v) is the total of flows on all path p where $(u, v) \in p$.

$$q_{(u,v)} = \sum_p \delta_{u,v,p} . f_p. \tag{5}$$

The travel time on a path is the sum of the travel on links and the waiting time at junctions.

$$t_p = \sum_{k=1}^{n(p)-1} t_{(i_k^p, i_{k+1}^p)} + \sum_{k=2}^{n(p)-1} WT_{i_k^p, p} \quad \forall p. \tag{6}$$

The travel time on link (u, v), $t_{(u,v)}$, linearly depends on flow $q_{u,v}$.

$$t_{(u,v)} = t_{(u,v)}^0 + \alpha_{u,v} . q_{u,v} \quad \forall (u, v), \tag{7}$$

where $\alpha_{u,v}$ is a constant.

For each OD pair, the demand is the total of the flows on used paths

$$\sum_{p \in P_w} f_p = d_w \quad \forall w. \tag{8}$$

For each OD pair, the travel time t_w is equal to the one of all used paths and the travel time on non-used path is greater than t_w (user equilibrium).

$$t_p \geq t_w \quad \forall p \in P_w \tag{9}$$

$$f_p(t_p - t_w) = 0 \quad \forall p \in P_w \tag{10}$$

Constraints (11)-(13) are introduced to determine the starting time of stages

$$ST_{1,1} = 0 \tag{11}$$

$$ST_{h,1} = ST_{h-1,1} + \theta_h \quad \forall h \geq 2 \tag{12}$$

$$ST_{h,r} = ST_{h,r-1} + \phi_{h,r-1} + I_{h,r-1} \quad \forall h, \forall r \geq 1 \tag{13}$$

At junctions, vehicles must spend an initial waiting time that is the time from the arrival time to the beginning of the stage at which vehicle can cross the intersection to continue its journey. Constraints (14)-(15) are used to estimate the initial waiting times for the junction after the second one of a path. The integer variables $z_{i_k^p, p}$ are used in order to assure that the initial waiting time is always smaller than the common cycle time.

$$\sum_r \triangle_{i_k^p,r,p}.ST_{i_k^p,r} - \sum_r \triangle_{i_{k-1}^p,r,p}.ST_{i_{k-1}^p,r} - t_{(i_{k-1}^p,i_k^p)} - z_{i_k^p,p}.C = WT_{i_k^p,p}^0, \quad \forall p,k$$

$$(14)$$

$$0 \le WT_{i_k^p,p}^0 \le C, \quad \forall p, k = 3,..,n(p)-1. \tag{15}$$

Under the assumption that the arrival flow is under an uniform distribution, the initial waiting time at the first junction on a path is estimated by constraints (16).

$$WT_{i_2^p,p}^0 = \frac{1}{2}[C - \sum_r \triangle_{i_2^p,r,p}.\phi_{i_2^p,r}] \quad \forall p \tag{16}$$

The delay time at junctions depends on the initial waiting time and the number of vehicles crossing the junction. This relation can be expressed by constraint (17).

$$WT_{i_k^p,p} = WT_{i_k^p}^0 + \beta_{i_k^p,p}.\sum_{p_1} \triangle_{i_k^p,r,p_1}.f_{p_1}, \quad \forall p, \tag{17}$$

where $\beta_{i_k^p,p}$ is a constant.

The flows and the travel time are non-negatives, variables $z_{i_k^p,p}$ are integers.

$$f_p, t_p, t_w \ge 0 \quad \forall p, w \tag{18}$$

$$z_{i_k^p,p} \in \mathbb{Z} \quad \forall p, k \tag{19}$$

The aim of problem is to minimize the total travel time TT in the network. Therefore, it is formulated as the following optimization problem

$$(P_1) \qquad \begin{array}{c} \min\{TT = \sum_w d_w.t_w\} \\ s.t.(1)-(19) \end{array}$$

This is a mixed integer non-linear program. It is very difficult to solve due to the complementarity constraint (10) and the integer variables $z_{h,p}$. In order to overcome the difficulty above, Problem (P_1) is transformed to Problem (P_2) by using penalty techniques.

Firstly, we define set $\mathbb{D}$ as below:

$$\mathbb{D} = \{\xi = (C, \theta_h, \phi_{h,r}, f_p, t_p, t_w, t_{(u,v)}, z_{h,p})|(1)-(9),(11)-(18)\}$$

$$\mu(\xi) = \sum_p \min\{f_p, t_p - t_w\}, \quad \nu(\xi) = \sum_{h,p} \sin^2(z_{h,p}.\pi)$$

We see that constraint (10) and constraint (19) can be replaced by $\mu(\xi) \le 0$ and $\nu(\xi) \le 0$, respectively. We consider the following problem

$$(P_2) \qquad \begin{array}{c} \min\{TT(\xi) = \sum_w d_w.t_w + \lambda.\sum_p \min\{f_p, t_p - t_w\} + \lambda.\sum_{h,p} \sin^2(z_{h,p}.\pi)\} \\ s.t. \quad (1)-(9),(11)-(18) \end{array}$$

where λ is a sufficiently large number.

It is clear that if an optimal solution ξ^* to (P_2) satisfies $\mu(\xi^*) = 0, \nu(\xi^*) = 0$ then it is an optimal solution to the original problem. On the other hand, according to the general result of the penalty method (see [16], pp. 366-380), for a given large number λ, the minimizer of (P_2) should be found in a region where $\mu(\xi), \nu(\xi)$ are relatively small. Thus, we will consider in the sequel the problem (P_2) with a sufficiently large number λ. Problem (P_2) can be handled by a genetic algorithm (in the next section).

Another way, to remove the difficulty in Problem (P_1), is to transform it into an equivalent problem as below.

Since $\min\limits_{\xi \in \mathbb{D}} \nu(\xi) = 0$, Problem (P_1) is equivalent to

$$
(P_3) \quad
\begin{aligned}
&\min\{TT(\xi) = \sum_{w} d_w.t_w\} \\
&s.t \quad (1) - (4) \\
&\xi \in argmin\{\sum_{h,p} \sin^2(z_{h,p}.\pi)\} \\
&s.t. \quad (5) - (9), (11) - (18) \\
&\mu(\xi) \leq 0.
\end{aligned}
$$

In the lower level of Problem (P_3), the constraint $\mu(\xi) \leq 0$ is still hard. It is tackled by using exact penalty techniques. Theorem 1 is in order.

Theorem 1. *[13] Let Ω be a nonempty bounded polyhedral convex set, f be a finite DC function on Ω and p be a finite nonnegative concave function on Ω. Then there exists $\eta_0 \geq 0$ such that for $\eta > \eta_0$ the following problems have the same optimal value and the same solution set*

$$(P_\eta) \quad \alpha(\eta) = \min\{f(x) + \eta.p(x) : x \in \Omega\},$$

$$(P) \quad \alpha = \min\{f(x) : x \in \Omega, p(x) \leq 0\}.$$

For given $(C, \theta_h, \phi_{h,r})$, denote $\Omega = \{x = (f_p, t_p, t_w, q_{u,v}, t_{u,v}, z_{h,p}) \mid (5) - (9), (11) - (18)\}$. It is easy to see that $\mu(\xi)$ is concave and non negative on Ω. Hence, the lower problem can be rewritten as a DC program.

$$
(P_{lower}) \quad
\begin{aligned}
&\min\{\sum_{h,p} \sin^2(z_{h,p}.\pi) + \eta.\mu(\xi)\} \\
&s.t. \quad (5) - (9), (11) - (18)
\end{aligned}
$$

where $\eta > 0$ is a sufficiently large number.

The original problem is equivalent to the following one

$$
(P_4) \quad
\begin{aligned}
&\min\{TT(\xi) = \sum_{w} d_w.t_w\} \\
&s.t \quad (1) - (4) \\
&(f_p, t_p, t_w, q_{u,v}, t_{(u,v)}, z_{h,p}) \in argmin\{\sum_{h,p} \sin^2(z_{h,p}.\pi) + \eta.\mu(\xi)\} \\
&s.t. \quad (5) - (9), (11) - (18)
\end{aligned}
$$

The lower problem in (P_4) is a DC program. It can be solved by a deterministic method.

3   A GA-Based Solution Method

3.1   Introduction to Genetic Algorithm

Genetic algorithm (GA) is a branch of evolutionary computation in which one imitates the biological processes of reproduction and natural selection to solve for the fittest solutions. GA allows one to find solutions to problems that other optimization methods cannot handle due to a lack of continuity, derivatives, linearity, or other features. Although GA may not provide a global solution, but the quality of solutions obtained by GA are acceptable in practice. Moreover, GA can be easily implemented and the executable time is reasonable. Today genetic algorithms have become a classic in the field of computer science and applied successfully to solve a lot of problems in different areas. The basic steps to solve a problem using a genetic algorithm can be presented as follows:

Initialization

Coding each solution as an individual in the population. One has different ways to do this. One of the most popular way is using binary coding. In the binary coding, each individual is encoded by a sequence of bits 0 or 1.

Randomly generating an initial population.

Repeat

Step 1: Decoding and Evaluating the quality of the population by a fitness function. In reality, we can choose the objective function as the fitness function.

If stopping criteria are satisfied then STOP else goto Step 2.

Step 2: Improving the quality of population through crossover and mutation procedure (evolution). Goto Step 3.

Step 3: Selecting a new population. Go to Step 1.

In the next sub-session, we introduced a genetic algorithm for solving problem (P_2). The chromosome encoding and decoding are presented in Subsection 3.2 while the procedure of fitness function computation is described in Subsection 3.3. The crossover, mutation and selection are similar to the one in [5].

3.2   Chromosome Encoding and Decoding

Firstly, note that if the values of common cycle time C, duration of green time $\phi_{h,r}$, offset θ_h, flow f_p are given then the others variables are computed. In this study, an individual is $(C, \theta_h, \phi_{h,r}, f_p)$. We use the binary coding for variables $C, \theta_h, \phi_{h,r}, f_p$. Each variable is coded by a sequence of 8 bits. Suppose that $\overline{C}, \overline{\theta_h}, \overline{\phi_{h,r}}, \overline{f_p}$ are respectively the representations of variables $C, \theta_h, \phi_{h,r}, f_p$.

In the next paragraph, the decoding procedure is showed.

Cycle time: is the proportion of the difference $C_{max} - C_{min}$ plus C_{min}:

$$C = C_{min} + \frac{F(\overline{C})}{2^8 - 1}.(C_{max} - C_{min}),$$

where $F(X)$ is 10 base equivalent of X.

Offset: for a junction h, it is the proportion of the cycle time

$$\theta_h = \frac{F(\overline{\theta_h})}{2^8 - 1}.(C - 1)$$

Green times: for a stage at junction h, are defined as the sum of the minimum stage length and the proportion of the remaining green time, $\phi_{h,r,max} - \phi_{h,r,min}$, as follows:

$$\phi_{h,r} = \phi_{h,r,min} + \frac{F(\overline{\phi_{h,r}})}{\sum\limits_{r=1}^{S_h} F(\overline{\phi_{h,r}})} \cdot (\phi_{h,r,max} - \phi_{h,r,min}).$$

Here, $\phi_{h,r,min}$ is a given constant and $\phi_{h,r,max}$ is a parameter calculated by

$$\phi_{h,r,max} = C - \sum_{r=1}^{S_h} I_{h,r} - \sum_{y=1,y\neq r}^{S_h} \phi_{h,r,min}.$$ By this way, constraint (4) is always satisfied.

Flow on path: for a path $p \in P_w$ flow on path p is defined as the proportion of demand d_w as follows:

$$f_p = \frac{F(\overline{f_p})}{\sum\limits_{p \in P_w}} \cdot d_w.$$

By this way, constraint (8) always holds.

3.3 Computing Other Variables and the Fitness Function

Variables $q_{u,v}$ are computed via f_p by equation (5).
Variables $t_{(u,v)}$ are computed via $q_{u,v}$ by equation (7).
Variables $S_{h,r}$ are computed via $\theta_h, \phi_{h,r}, I_{h,r}$ by equations (11-13).
Variables $WT^0_{i^p_k,p}$ and $z_{i^p_k,p}$ are calculated by equation (14). Specifically, $WT^0_{i^p_k,p}$ and $z_{i^p_k,p}$ $\forall k \geq 3$ are respectively the residual and integer part of number

$$\frac{1}{C} \cdot [\sum_r \triangle_{i^p_k,r,p} \cdot ST_{i^p_k,r} - \sum_r \triangle_{i^p_{k-1},r,p} \cdot ST_{i^p_{k-1},r} - t_{(i^p_{k-1},i^p_k)}].$$

Variables $WT^0_{i^p_2,p}$, $WT_{i^p_k,p}$ and t_p are calculated via (16),(17),(6).
Variables $t_w = \min\limits_{p \in P_w} \{t_p\}$.
The fitness function FF is the objective function

$$FF = TT(\xi) = \sum_w d_w.t_w + \lambda. \sum_p \min\{f_p, t_p - t_w\} + \lambda. \sum_{h,p} \sin^2(z_{h,p}.\pi) \quad (20)$$

4   Combination of GA and DCA

In order to improve the quality of individuals in GA, we use DCA for solving the lower problem in Problem (P_4).

4.1   A Brief Presentation of DC Programming and DCA

To give the reader an easy understanding of the theory of DC programming & DCA and our motivation to use them, we briefly outline these tools in this section.

Let $\Gamma_0(\mathbb{R}^n)$ denotes the convex cone of all lower semi-continuous proper convex functions on $\mathbb{R}^n$. Consider the following primal DC program:

$$(P_{dc}) \quad \alpha = \inf\{f(x) := g(x) - h(x) \ : \ x \in \mathbb{R}^n\}, \tag{21}$$

where $g, h \in \Gamma_0(\mathbb{R}^n)$.

Let C be a nonempty closed convex set. The indicator function on C, denoted χ_C, is defined by $\chi_C(x) = 0$ if $x \in C$, ∞ otherwise. Then, the problem

$$\inf\{f(x) := g(x) - h(x) \ : \ x \in C\}, \tag{22}$$

can be transformed into an unconstrained DC program by using the indicator function of C, i.e.,

$$\inf\{f(x) := \phi(x) - h(x) \ : \ x \in \mathbb{R}^n\}, \tag{23}$$

where $\phi := g + \chi_C$ is in $\Gamma_0(\mathbb{R}^n)$.

Recall that, for $h \in \Gamma_0(\mathbb{R}^n)$ and $x_0 \in \mathrm{dom}\ h := \{x \in \mathbb{R}^n | h(x_0) < +\infty\}$, the subdifferential of h at x_0, denoted $\partial h(x_0)$, is defined as

$$\partial h(x_0) := \{y \in \mathbb{R}^n : h(x) \geq h(x_0) + \langle x - x_0, y \rangle, \forall x \in \mathbb{R}^n\}, \tag{24}$$

which is a closed convex set in $\mathbb{R}^n$. It generalizes the derivative in the sense that h is differentiable at x_0 if and only if $\partial h(x_0)$ is reduced to a singleton which is exactly $\{\nabla h(x_0)\}$.

The idea of DCA is simple: each iteration of DCA approximates the concave part $-h$ by its affine majorization (that corresponds to taking $y^k \in \partial h(x^k)$) and minimizes the resulting convex problem (P_k).

Generic DCA scheme
Initialization: Let $x^0 \in \mathbb{R}^n$ be a best guess, $0 \leftarrow k$.
Repeat
    Calculate $y^k \in \partial h(x^k)$
    Calculate $x^{k+1} \in \arg\min\{g(x) - h(x^k) - \langle x - x^k, y^k \rangle : x \in \mathbb{R}^n\}$ (P_k)
    $k + 1 \leftarrow k$
Until convergence of x^k.

Convergence properties of the DCA and its theoretical bases are described in [11,15,19,20].

4.2   DCA for Solving (P_{lower}) and GA-DCA Algorithm

In Problem (P_{lower}), the objective function $f(x) = \sum\limits_{h,p} \sin^2(z_{h,p}.\pi) + \eta.\mu(\xi)$ is a DC function. Consider function $f_{h,p}(x) = \sin^2(z_{h,p}.\pi)$, there exists a DC decomposition $f_{h,p}(x) = \tau.z_{h,p}^2 - (\tau.z_{h,p}^2 - \sin^2(z_{h,p}.\pi))$, where $\tau > 2\pi^2$. Hence, we obtain a DC decomposition of the objective function $f(x) = g(x) - h(x)$ where $g(x) = \tau. \sum_{h,p} z_{h,p}^2$ and $h(x) = \tau. \sum_{h,p} z_{h,p}^2 - \sum_{h,p} \sin^2(z_{h,p}.\pi) + \eta. \sum_{h,p} \max\{-f_p, -t_p + t_w\}$. We see that the subdifferential of $h(x)$ can be easily computed.

DCA applied to (P_{lower}) can be described as follows:

DCA

Initialization

Let ϵ be a sufficiently small positive number. Set $\ell = 0$ and x^0 is a starting point

Repeat

Calculate $y^\ell \in \partial h(x^\ell)$

Calculate $x^{\ell+1}$ by solving a convex quadratic program $\min\{g(x) \quad s.t. \quad x \in \Omega\}$

$\ell \longleftarrow \ell + 1$

Until $\|x^{\ell+1} - x^\ell\| \le \epsilon$ or $\|f(x^{\ell+1}) - f(x^\ell)\| \le \epsilon$.

In the combined GA-DCA, an individual is $(C, \theta_h, \phi_{h,r})$. The chromosome encoding and decoding are similar to GA presented in Section 3 while the values of the other variables $(f_p, t_p, t_w, q_{u,v}, t_{(u,v)}, z_{h,p})$ are the optimal solution of $P_{(lower)}$ by using DCA.

The combined GA-DCA scheme is described as follows:

GA-DCA

Initialization

Randomly generate an initial population $\mathbb{P}$.

For an individual $Id^i = (C^i, \theta_h^i, \phi_{h,r}^i) \in \mathbb{P}$, we solve problem (P_{lower}) by DCA to obtain $(f_p^i, t_p^i, t_w^i, q_{u,v}^i, t_{(u,v)}^i, z_{h,p}^i)$.

Compute the fitness of Id^i by formula (20).

Repeat

Step 1: Check the stopping criteria. If it is satisfied then STOP else go to Step 2.

Step 2: Launch crossover and mutation procedure (evolution) for improving the quality of population.

For a new individual $Id^l = (C^l, \theta_h^l, \phi_{h,r}^l) \in \mathbb{P}$, we solve problem (P_{lower}) by DCA to obtain $(f_p^l, t_p^l, t_w^l, q_{u,v}^l, t_{(u,v)}^l, z_{h,p}^l)$.

Compute the fitness of Id^l by formula (20).

Go to Step 3.

Step 3: Select a new population. Go to Step 1.

5   Conclusions

The work studied the problem of optimizing traffic signals considering rerouting. The main contribution is to build two new formulations that are probably solved by efficient methods. We also proposed two algorithms to directly solve them. GA and a combination of GA-DCA are investigated and described in detail. The effect of the parameters of the models and the algorithms on the numerical results are planned in the future work.

Acknowledgements. This work is supported by Vietnam National Foundation for Science and Technology Development (NAFOSTED) under Grant Number 101.01-2013.10.

References

1. Almond, J., Lott, R.S.: The Glasgow experiment: Implementation and assessment. Road Research Laboratory Report 142, Road Research Laboratory, Crowthorne (1968)
2. Ben Ayed, O., Boyce, D.E., Blair, C.E.: A general bi-level linear programming formulation of the network design problem. Transportation Research Part B 22(4), 311–318 (1988)
3. Ceylan, H., Bell, M.G.H.: Traffic signal timing optimisation based on genetic algorithm approach, including drivers' routing. Transportation Research Part B 38(4), 329–342 (2004)
4. Cree, N.D., Maher, M.J., Paechter, B.: The continuous equilibrium optimal network design problem: A genetic approach. In: Bell, M.G.H. (ed.) Transportation Networks: Recent Methodological Advances, pp. 163–174. Pergamon, Oxford (1998)
5. Fitsum, T., Agachai, S.: A genetic algorithm approach for optimizing traffic control signals considering routing. Computer-Aided Civil and Infrastructure Engineering 22, 31–43 (2007)
6. Friesz, T.L., Cho, H.J., Mehta, N.J., Tobin, R., Anandalingam, G.: A simulated annealing approach to the network design problem with variational inequality constraints. Transportation Science 26, 18–26 (1992)
7. Friesz, T.L., Tobin, R.L., Cho, H.J., Mehta, N.J.: Sensitivity analysis based heuristic algorithms for mathematical programs with variational inequality constraints. Mathematical Programming 48, 265–284 (1990)
8. Hunt, P.B., Robertson, D.I., Bretherton, R.D., Winton, R.I.: SCOOT - A traffic responsive method of coordinating signals. TRRL Laboratory Report 1014, TRRL, Berkshire, England (1981)
9. Lawphongpanich, S., Hearn, D.W.: An MPEC approach to second-best toll pricing. Mathematical Programming B 101(1), 33–55 (2004)
10. LeBlanc, L., Boyce, D.: A bi-level programming for exact solution of the network design problem with user-optimal. Transportation Research Part B: Methodological 20(3), 259–265 (1986)
11. Le Thi, H.A.: Contribution à l'optimisation non-convex and l'optimisation globale: Théorie, Algorithmes et Applications, Habilitation à Diriger des recherches, Université de Rouen (1997)
12. Le Thi, H.A., Pham Dinh, T.: A Continuous approach for globally solving linearly constrained quadratic zero-one programming problem. Optimization 50(1-2), 93–120 (2001)
13. Le Thi, H.A., Pham Dinh, T., Huynh, V.N.: Exact penalty and error bounds in DC programming. Journal of Global Optimization 52(3), 509–535 (2012)
14. Le Thi, H.A., Pham Dinh, T.: A Branch and Bound Method via d.c. Optimization Algorithms and Ellipsoidal Technique for Box Constrained Nonconvex Quadratic Problems. Journal of Global Optimization 13, 171–206 (1998)
15. Le Thi, H.A., Pham Dinh, T.: The DC(difference of convex functions) Programming and DCA revisited with DC models of real world non convex optimization problems. Annals of Operations Research 133, 23–46 (2005)

16. Luenberger, D.G.: Linear and Nonlinear Programming, 2nd edn. Springer (2004)
17. Luo, Z., Pang, J.S., Ralph, D.: Mathematical Programs with Equilibrium Constraints. Cambridge University Press, New York (1996)
18. Meng, Q., Yang, H., Bell, M.G.H.: An equivalent continuously differentiable model and a locally convergent algorithm for the continuous network design problem. Transportation Research Part B 35(1), 83–105 (2001)
19. Pham Dinh, T., Le Thi, H.A.: Convex analysis approach to d.c Programming: Theory, Algorithms and Applications. Acta Mathematica Vietnamica 22(1), 289–355 (1997)
20. Pham Dinh, T., Le Thi, H.A.: DC optimization algorithms for solving the trust region subproblem. SIAM J. Optimization 8, 476–505 (1998)
21. Robertson, D.I.: 'TRANSYT' method for area traffic control. Traffic Engineering and Control 10, 276–281 (1969)
22. Schaefer, R.: Foundations of Global Genetic Optimization. SCI, vol. 74. Springer, Heidelberg (2007)
23. Shepherd, S.P.: A Review of Traffic Signal Control. Publisher of University of Leeds, Institute for Transport Studies (1992)
24. Suwansirikul, C., Friesz, T.L., Tobin, R.L.: Equilibrium decomposed optimization: A heuristic for the continuous equilibrium network design problem. Transportation Science 21(4), 254–263 (1987)
25. Van Vliet, D.: SATURN - A modern assignment model. Traffic Engineering and Control 23, 578–581 (1982)
26. Verhoef, E.T.: Second-best congestion pricing in general networks: Heuristic algorithms for finding second-best optimal toll levels and toll points. Transportation Research Part B 36(8), 707–729 (2002)
27. Wardrop, J.G.: Some theoretical aspects of road traffic research. Proceedings of Institution of Civil Engineers 1(2), 325–378 (1952)
28. Webster, F.V.: Traffic Signal Settings. Road Research Technical Paper No. 39, HMSO, London (1958)
29. Wu, X., Deng, S., Du, X.: Jing MaGreen-Wave Traffic Theory Optimizationand Analysis. World Journal of Engineering and Technology 2, 14–19 (2014)
30. Yang, H.: Sensitivity analysis for the elastic-demand network equilibrium problem with with applications. Transportation Research Part B 31(1), 55–70 (1997)

Composite Convex Minimization Involving Self-concordant-Like Cost Functions

Quoc Tran-Dinh, Yen-Huan Li, and Volkan Cevher

Laboratory for Information and Inference Systems (LIONS)
EPFL, Lausanne, Switzerland

Abstract. The self-concordant-like property of a smooth convex function is a new analytical structure that generalizes the self-concordant notion. While a wide variety of important applications feature the self-concordant-like property, this concept has heretofore remained unexploited in convex optimization. To this end, we develop a variable metric framework of minimizing the sum of a "simple" convex function and a self-concordant-like function. We introduce a new analytic step-size selection procedure and prove that the basic gradient algorithm has improved convergence guarantees as compared to "fast" algorithms that rely on the Lipschitz gradient property. Our numerical tests with real-data sets show that the practice indeed follows the theory.

1 Introduction

In this paper, we consider the following composite convex minimization problem:

$$F^\star := \min_{\mathbf{x} \in \mathbb{R}^n} \left\{ F(\mathbf{x}) := f(\mathbf{x}) + g(\mathbf{x}) \right\}, \qquad (1)$$

where f is a nonlinear smooth convex function, while g is a "simple" possibly nonsmooth convex function. Such composite convex problems naturally arise in many applications of machine learning, data sciences, and imaging science. Very often, f measures a data fidelity or a loss function, and g encodes a form of low-dimensionality, such as sparsity or low-rankness.

To trade-off accuracy and computation optimally in large-scale instances of (1), existing optimization methods invariably invoke the additional assumption that the smooth function f also has an L-Lipschitz continuous gradient (cf., [11] for the definition). A highlight is the recent developments on proximal gradient methods, which feature (nearly) dimension-independent, global sublinear convergence rates [3,9,11]. When the smooth f in (1) also has strong regularity [15], the problem (1) is also within the theoretical and practical grasp of proximal-(quasi) Newton algorithms with linear, superlinear, and quadratic convergence rates [5,8,17]. These algorithms specifically exploit second order information or its principled approximations (e.g., via BFGS or L-BFGS updates [13]).

In this paper, we do away with the Lipschitz gradient assumption and instead focus on another structural assumption on f in developing an algorithmic framework for (1), which is defined below.

H.A. Le Thi et al. (eds.), *Model. Comput. & Optim. in Inf. Syst. & Manage. Sci.*,
Advances in Intelligent Systems and Computing 359, DOI: 10.1007/978-3-319-18161-5_14

Definition 1. *A convex function* $f \in \mathcal{C}^3(\mathbb{R}^n)$ *is called a self-concordant-like function* $f \in \mathcal{F}_{\text{scl}}$, *if:*

$$|\varphi'''(t)| \leq M_f \varphi''(t) \|\mathbf{u}\|_2, \qquad (2)$$

for $t \in \mathbb{R}$ *and* $M_f > 0$, *where* $\varphi(t) := f(\mathbf{x}+t\mathbf{u})$ *for any* $\mathbf{x} \in \text{dom}(f)$ *and* $\mathbf{u} \in \mathbb{R}^n$.

Definition 1 mimics the standard self-concordance concept ([10, Definition 4.1.1]) and was first discussed in [1] for model consistency in logistic regression. For composite convex minimization, self-concordant-like functions abound in machine learning, including but not limited to logistic regression, multinomial logistic regression, conditional random fields, and robust regression (cf., the references in [2]). In addition, special instances of geometric programming [6] can also be recast as (1) where $f \in \mathcal{F}_{\text{scl}}$.

The importance of the assumption $f \in \mathcal{F}_{\text{scl}}$ in (1) is twofold. First, it enables us to derive an explicit step-size selection strategy for proximal variable metric methods, enhancing backtracking-line search operations with improved theoretical convergence guarantees. For instance, we can prove that our proximal gradient method can automatically adapt to the local strong convexity of f near the optimal solution to feature linear convergence under mild conditions. This theoretical result is backed up by great empirical performance on real-life problems where the fast Lipschitz-based methods actually exhibit sublinear convergence (cf. Section 4). Second, the self-concordant-like assumption on f also helps us provide scalable numerical solutions of (1) for specific problems where f does not have Lipschitz continuous gradient, such as special forms of geometric programming problems.

Contributions. Our specific contributions can be summarized as follows:

1. We propose a new *variable metric* framework for minimizing the sum $f+g$ of a self-concordant-like function f and a convex, possibly nonsmooth function g. Our approach relies on the solution of a convex subproblem obtained by linearizing and regularizing the first term f, and uses an *analytical* step-size to achieve descent in three classes of algorithms: first order methods, second order methods, and quasi-Newton methods.

2. We establish both the global and the local convergence of different variable metric strategies. We pay particular attention to diagonal variable metrics since in this case many of the proximal subproblems can be solved exactly. We derive conditions on when and where these variants achieve locally linear convergence. When the variable metric is the Hessian of f at each iteration, we show that the resulting algorithm locally exhibits quadratic convergence without requiring any globalization strategy such as a backtracking line-search.

3. We apply our algorithms to large-scale real-world and synthetic problems to highlight the strengths and the weaknesses of our variable-metric scheme.

Relation to Prior Work. Many of the composite problems with self-concordant-like f, such as regularized logistics and multinomial logistics, also have Lipschitz continuous gradient. In those specific instances, many theoretically efficient algorithms are applicable [3,5,8,9,11,17]. Compared to these works, our framework has theoretically stronger local convergence guarantees thanks to the specific step-size strategy matched with $f \in \mathcal{F}_{\text{scl}}$. The authors of [18] consider composite problems where f is standard self-concordant and proposes a proximal Newton algorithm optimally exploiting this structure. Our structural assumptions and algorithmic emphasis here are different.

Paper Organization. We first introduce the basic definitions and optimality conditions before deriving the variable metric strategy in Section 2. Section 3 proposes our new variable metric framework, describes its step-size selection procedure, and establishes the convergence theory of its variants. Section 4 illustrates our framework in real and synthetic data.

2   Preliminaries

We adopt the notion of self-concordant functions in [10,12] to a different smooth function class. Then we present the optimality condition of problem (1).

2.1   Basic Definitions

Let $g : \mathbb{R}^n \to \mathbb{R}$ be a proper, lower semicontinuous convex function [16] and $\text{dom}(g)$ denote the domain of g. We use $\partial g(\mathbf{x})$ to denote the subdifferential of g at $\mathbf{x} \in \text{dom}(g)$ if g is nondifferentiable at $\mathbf{x}$ and $\nabla g(\mathbf{x})$ to denote its gradient, otherwise. Let $f : \mathbb{R}^n \to \mathbb{R}$ be a $\mathcal{C}^3(\text{dom}(f))$ function (i.e., f is three times continuously differentiable). We denote by $\nabla f(\mathbf{x})$ and $\nabla^2 f(\mathbf{x})$ the gradient and the Hessian of f at $\mathbf{x}$, respectively. Suppose that, for a given $\mathbf{x} \in \text{dom}(f)$, $\nabla^2 f(\mathbf{x})$ is positive definite (i.e., $\nabla^2 f(\mathbf{x}) \in \mathcal{S}_{++}^n$), we define the local norm of a given vector $\mathbf{u} \in \mathbb{R}^n$ as $\|\mathbf{u}\|_{\mathbf{x}} := [\mathbf{u}^T \nabla^2 f(\mathbf{x})\mathbf{u}]^{1/2}$. The corresponding dual norm of $\mathbf{u}$, $\|\mathbf{u}\|_{\mathbf{x}}^*$ is defined as $\|\mathbf{u}\|_{\mathbf{x}}^* := \max \left\{ \mathbf{u}^T \mathbf{v} \mid \|\mathbf{v}\|_{\mathbf{x}} \leq 1 \right\} = [\mathbf{u}^T \nabla^2 f(\mathbf{x})^{-1}\mathbf{u}]^{1/2}$.

2.2   Composite Self-Concordant-Like Minimization

Let $f \in \mathcal{F}_{\text{scl}}(\mathbb{R}^n)$ and g be proper, closed and convex. The optimality condition for (1) can be concisely written as follows:

$$0 \in \nabla f(\mathbf{x}^\star) + \partial g(\mathbf{x}^\star). \tag{3}$$

Let us denote by $\mathbf{x}^\star$ as an optimal solution of (1). Then, the condition (3) is necessary and sufficient. We also say that $\mathbf{x}^\star$ is *nonsingular* if $\nabla^2 f(\mathbf{x}^\star)$ is positive definite. We now establish the existence and uniqueness of the solution $\mathbf{x}^\star$ of (1), whose proof can be found in [19].

Lemma 1. *Suppose that $f \in \mathcal{F}_{\text{scl}}(\mathbb{R}^n)$ satisfies Definition 1 for some $M_f > 0$. Suppose further that $\nabla^2 f(\mathbf{x}) \succ 0$ for some $\mathbf{x} \in \text{dom}(f)$. Then the solution $\mathbf{x}^\star$ of (1) exists and is unique.*

For a given symmetric positive definite matrix $\mathbf{H}$, we define a generalized proximal operator $\text{prox}_{\mathbf{H}^{-1}g}$ as:

$$\text{prox}_{\mathbf{H}^{-1}g}(\mathbf{x}) := \arg\min_{\mathbf{z}} \left\{ g(\mathbf{z}) + (1/2)\|\mathbf{z} - \mathbf{x}\|^2_{\mathbf{H}^{-1}} \right\}. \tag{4}$$

Due to the convexity of g, this operator is well-defined and single-valued. If we can compute $\text{prox}_{\mathbf{H}^{-1}g}$ efficiently (e.g., by a closed form or by polynomial time algorithms), then we say that g is *proximally tractable*. Examples of proximal tractability convex functions can be found, e.g., in [14]. Using $\text{prox}_{\mathbf{H}^{-1}g}$, we can write condition (1) as:

$$\mathbf{x}^\star - \mathbf{H}^{-1}\nabla f(\mathbf{x}^\star) \in (\mathbb{I} + \mathbf{H}^{-1}\partial g)(\mathbf{x}^\star) \iff \mathbf{x}^\star = \text{prox}_{\mathbf{H}^{-1}g}(\mathbf{x}^\star - \mathbf{H}^{-1}\nabla f(\mathbf{x}^\star)).$$

This expression shows that $\mathbf{x}^\star$ is a fixed point of $\mathcal{R}_{\mathbf{H}}(\cdot) := \text{prox}_{\mathbf{H}^{-1}g}((\cdot) - \mathbf{H}^{-1}\nabla f(\cdot))$. Based on the fixed point principle, one can expect that the iterative sequence $\{\mathbf{x}^k\}_{k\geq 0}$ generated by $\mathbf{x}^{k+1} := \mathcal{R}_{\mathbf{H}}(x^k)$ converges to $\mathbf{x}^\star$. This observation is made rigorous below.

3 Our Variable Metric Framework

We first present a generic variable metric proximal framework for solving (1). Then, we specify this framework to obtain three variants: proximal gradient, proximal Newton and proximal quasi-Newton algorithms.

3.1 Generic Variable Metric Proximal Algorithmic Framework

Given $\mathbf{x}^k \in \text{dom}(F)$ and an appropriate choice $\mathbf{H}_k \in \mathcal{S}^n_{++}$, since $f \in \mathcal{F}_{\text{scl}}$, one can approximate f at $\mathbf{x}^k$ by the following quadratic model:

$$Q_{\mathbf{H}_k}(\mathbf{x}, \mathbf{x}^k) := f(\mathbf{x}^k) + \langle \nabla f(\mathbf{x}^k), \mathbf{x} - \mathbf{x}^k \rangle + \frac{1}{2}\langle \mathbf{H}_k(\mathbf{x} - \mathbf{x}^k), \mathbf{x} - \mathbf{x}_k \rangle. \tag{5}$$

Our algorithmic approach uses the variable metric forward-backward framework to generate a sequence $\{\mathbf{x}^k\}_{k\geq 0}$ starting from $\mathbf{x}^0 \in \text{dom}(F)$ and update:

$$\mathbf{x}^{k+1} := \mathbf{x}^k + \alpha_k \mathbf{d}^k \tag{6}$$

where $\alpha_k \in (0, 1]$ is a given step-size and $\mathbf{d}^k$ is a search direction defined by:

$$\mathbf{d}^k := \mathbf{s}^k - \mathbf{x}^k, \quad \text{with} \quad \mathbf{s}^k := \arg\min_{\mathbf{x}} \left\{ Q_{\mathbf{H}_k}(\mathbf{x}, \mathbf{x}^k) + g(\mathbf{x}) \right\}. \tag{7}$$

In the rest of this section, we explain how to determine the step size α_k in the iterative scheme (6) optimally for special cases of $\mathbf{H}_k$. For this, we need the following definitions:

$$\lambda_k := \|\mathbf{d}^k\|_{\mathbf{x}^k}, \quad r_k := M_f\|\mathbf{d}^k\|_2, \quad \text{and} \quad \beta_k := \|\mathbf{d}^k\|_{\mathbf{H}_k} = \langle \mathbf{H}_k\mathbf{d}^k, \mathbf{d}^k \rangle^{1/2}. \tag{8}$$

3.2   Proximal-Gradient Algorithm

When the variable matrix $\mathbf{H}_k$ is *diagonal* and g is proximally tractable, we can efficiently obtain the solution of the subproblem (7) in a distributed fashion or even in a closed form. Hence, we consider $\mathbf{H}_k = \mathbf{D}_k := \mathrm{diag}(\mathbf{D}_{k,1}, \cdots, \mathbf{D}_{k,n})$ with $\mathbf{D}_{k,i} > 0$, for $i = 1, \cdots, n$. Lemma 2, whose proof is in [19], provides a step-size selection procedure and proves the global convergence of this proximal-gradient algorithm.

Lemma 2. *Let* $\{\mathbf{x}^k\}_{k \geq 0}$ *be a sequence generated by* (6) *and* (7) *starting from* $\mathbf{x}^0 \in \mathrm{dom}\,(F)$. *For* λ_k, r_k *and* β_k *defined by* (8), *we consider the step-size* α_k *as:*

$$\alpha_k := \frac{1}{r_k} \ln \left(1 + \frac{\beta_k^2 r_k}{\lambda_k^2} \right), \tag{9}$$

If $\beta_k^2 r_k \leq (e^{r_k} - 1)\lambda_k^2$, *then* $\alpha_k \in (0, 1]$ *and:*

$$F(\mathbf{x}^{k+1}) \leq F(\mathbf{x}^k) - \frac{\beta_k^2}{r_k} \left[\left(1 + \frac{\lambda_k^2}{r_k \beta_k^2} \right) \ln \left(1 + \frac{\beta_k^2 r_k}{\lambda_k^2} \right) - 1 \right]. \tag{10}$$

Moreover, this step-size α_k *is optimal* (*w.r.t. the worst-case performance*).

By our condition, the second term on the right-hand side of (10) is always positive, establishing that the sequence $\{F(\mathbf{x}^k)\}$ is decreasing. Moreover, as $e^{r_k} - 1 \geq r_k$, the condition $\beta_k^2 r_k \leq (e^{r_k} - 1)\lambda_k^2$ can be simplified to $\beta_k \leq \lambda_k$. It is easy to verify that this is satisfied whenever $\mathbf{D}_k \preceq \nabla^2 f(\mathbf{x}^k)$. In such cases, our step-size selection ensures the best decrease of the objective value regarding the self-concordant-like structure of f (and not the actual objective instance). When $\beta_k > \lambda_k$, we scale down $\mathbf{D}_k$ until $\beta_k \leq \lambda_k$. It is easy to prove that the number of backtracking steps to find $\mathbf{D}_{k,i}$ is time constant.

Now, by using our step-size (9), we can describe the proximal-gradient algorithm as in Algorithm 1.

Algorithm 1. (Proximal-gradient algorithm with a *diagonal variable metric*)

Initialization: Given $\mathbf{x}^0 \in \mathrm{dom}(F)$, and a tolerance $\varepsilon > 0$.
for $k = 0$ **to** $k_{\max}$ **do**
  1. Choose $\mathbf{D}_k \in \mathcal{S}^n_{++}$ (e.g., using $\mathbf{D}_k := L_k \mathbb{I}$, where L_k is given by (11)).
  2. Compute the proximal-gradient search direction $\mathbf{d}^k$ as (7).
  3. Compute $\beta_k := \|\mathbf{d}^k\|_{\mathbf{D}_k}$, $r_k := M_f \|\mathbf{d}^k\|_2$ and $\lambda_k := \|\mathbf{d}^k\|_{\mathbf{x}^k}$.
  4. If $\beta_k \leq \varepsilon$ then terminate.
  5. If $\beta_k^2 r_k \leq (e^{r_k} - 1)\lambda_k^2$, then compute $\alpha_k := \frac{1}{r_k} \ln \left(1 + \frac{\beta_k^2 r_k}{\lambda_k^2} \right)$ and update $\mathbf{x}^{k+1} :=$
     $\mathbf{x}^k + \alpha_k \mathbf{d}^k$. Otherwise, set $\mathbf{x}^{k+1} := \mathbf{x}^k$ and update $\mathbf{D}_{k+1}$ from $\mathbf{D}_k$.
  end for

We combine the above analysis to obtain the following proximal gradient algorithm for solving (1). The main step in Algorithm 1 is to compute the search direction $\mathbf{d}^k$ at Step 2, which is equivalent to the solution of the convex sub-problem (7). The second main step is to compute $\lambda_k = \langle \nabla^2 f(\mathbf{x}^k)\mathbf{d}^k, \mathbf{d}^k \rangle^{1/2}$. This quantity requires the product of Hessian $\nabla^2 f(\mathbf{x}^k)$ of f and $\mathbf{d}^k$, but not the full-Hessian. It is clear that if $\beta_k = 0$ then $\mathbf{d}^k = 0$ and $\mathbf{x}^{k+1} \equiv \mathbf{x}^k$ and we obtain the solution of (1), i.e., $\mathbf{x}^k \equiv \mathbf{x}^\star$. The diagonal matrix $\mathbf{D}_k$ can be updated as $\mathbf{D}_{k+1} := c\mathbf{D}_k$ for a given factor $c > 1$.

We now explain how the new theory enhances the standard backtracking linesearch approaches. For simplicity, let us assume $\mathbf{D}_k := L_k \mathbb{I}$, where $\mathbb{I}$ is the identity matrix. By a careful inspection of (10), we see that $L_k = \sigma_{\max}(\nabla^2 f(\mathbf{x}^k))$ achieves the maximum guaranteed decrease (in the worst case sense) in the objective. There are many principled ways of approximating this constant based on the secant equation underlying the quasi-Newton methods. In Section 4, we use Barzilai-BenTal's rule:

$$L_k := \frac{\|\mathbf{y}^k\|_2^2}{\langle \mathbf{y}^k, \mathbf{s}^k \rangle}, \text{ where } \mathbf{s}^k := \mathbf{x}^k - \mathbf{x}^{k-1} \text{ and } \mathbf{y}^k := \nabla f(\mathbf{x}^k) - \nabla f(\mathbf{x}^{k-1}). \quad (11)$$

We then deviate from the standard backtracking approaches. As opposed to, for instance, checking the Armijo-Goldstein condition, we use a *new analytic condition* (i.e., Step 5 of Algorithm 1), which is computationally cheaper in many cases. Our analytic step-size then further refines the solution based on the worst-case problem structure, even if the backtracking update satisfies the Armijo-Goldstein condition.

Surprisingly, our analysis also enables us to also establish local linear convergence as described in Theorem 1 under mild assumptions. The proof can be found in [19].

Theorem 1. *Let $\{\mathbf{x}^k\}_{k \geq 0}$ be a sequence generated by Algorithm 1. Suppose that the sub-level set $\mathcal{L}_F(F(\mathbf{x}^0)) := \{\mathbf{x} \in \mathrm{dom}\,(F) : F(\mathbf{x}) \leq F(\mathbf{x}^0)\}$ is bounded and $\nabla^2 f$ is nonsingular at some $\mathbf{x} \in \mathrm{dom}\,(f)$. Suppose further that $\mathbf{D}_k := L_k \mathbb{I} \succeq \tau \mathbb{I}_n$ for given $\tau > 0$. Then, $\{\mathbf{x}^k\}$ converges to $\mathbf{x}^\star$ the solution of (1). Moreover, if $\rho_* := \max\{L_k/\sigma^*_{\min} - 1, 1 - L_k/\sigma^*_{\max}\} < \frac{1}{2}$ for k sufficiently large then the sequence $\{\mathbf{x}^k\}$ locally converges to $\mathbf{x}^\star$ at a linear rate, where $\sigma^*_{\min}$ and $\sigma^*_{\max}$ are the smallest and the largest eigenvalues of $\nabla^2 f(\mathbf{x}^\star)$, respectively.*

Linear convergence: According to Theorem 1, linear convergence is only possible when the condition number κ of the Hessian at the true solution satisfies $\kappa = \sigma^*_{\max}/\sigma^*_{\min} < 3$. While this seems too imposing, we claim that, for most f and g, this requirement is not too difficult to satisfy (see also the empirical evidence in Section 4). This is because the proof of Theorem 1 only needs the smallest and the largest eigenvalues of $\nabla^2 f(\mathbf{x}^\star)$, *restricted* to the subspaces of the union of $\mathbf{x}^\star - \mathbf{x}^k$ for k sufficiently large, to satisfy the conditions imposed by ρ_*. For instance, when g is based on the ℓ_1-norm/the nuclear norm, the differences $\mathbf{x}^\star - \mathbf{x}^k$ have at most twice the sparsity/rank of $\mathbf{x}^\star$ near convergence. Given such subspace restrictions, one can prove, via probabilistic assumptions on f (cf., [1]), that the restricted condition number is not only dramatically smaller than the full condition number κ of the Hessian $\nabla^2 f(\mathbf{x}^\star)$, but also it can even be dimension independent with high probability.

3.3   Proximal-Newton Algorithm

The case $\mathbf{H}_k \equiv \nabla^2 f(\mathbf{x}^k)$ deserves a special attention as the step-size selection rule becomes explicit and backtracking-free. The resulting method is a *proximal-Newton* method and can be computationally attractive in certain big data problems due to its low iteration count.

The main step of the proximal-Newton algorithm is to compute the proximal-Newton search direction $\mathbf{d}^k$ as:

$$\mathbf{d}^k := \mathbf{s}^k - \mathbf{x}^k, \quad \text{where } \mathbf{s}^k := \operatorname*{argmin}_{\mathbf{x}} \left\{ Q_{\nabla^2 f(\mathbf{x}^k)}(\mathbf{x}, \mathbf{x}^k) + g(\mathbf{x}) \right\}. \tag{12}$$

Then, it updates the sequence $\left\{ \mathbf{x}^k \right\}$ by:

$$\mathbf{x}^{k+1} := \mathbf{x}^k + \alpha_k \mathbf{d}^k = (1 - \alpha_k)\mathbf{x}^k + \alpha_k \mathbf{s}^k, \tag{13}$$

where $\alpha_k \in (0, 1]$ is the step size. If we set $\alpha_k = 1$ for all $k \geq 0$, then (13) is called the full-step proximal-Newton method. Otherwise, it is a damped-step proximal-Newton method.

First, we show how to compute the step size α_k in the following lemma, which is a direct consequence of Lemma 2 by taking $\mathbf{H}_k \equiv \nabla^2 f(\mathbf{x}^k)$.

Lemma 3. *Let $\left\{ \mathbf{x}^k \right\}_{k \geq 0}$ be a sequence generated by the proximal-Newton scheme (13) starting from $\mathbf{x}^0 \in \operatorname{dom}(F)$. Let λ_k and r_k be as defined by (8). If we choose the step-size $\alpha_k = r_k^{-1} \ln(1 + r_k)$ then:*

$$F(\mathbf{x}^{k+1}) \leq F(\mathbf{x}^k) - r_k^{-1} \lambda_k^2 \left[(1 + r_k^{-1}) \ln(1 + r_k) - 1 \right]. \tag{14}$$

Moreover, this step-size α_k is optimal (w.r.t. the worst-case performance).

Next, Theorem 2 proves the local quadratic convergence of the full-step proximal-Newton method, whose proof can be found in [19].

Theorem 2. *Suppose that the sequence $\left\{ \mathbf{x}^k \right\}_{k \geq 0}$ is generated by (13) with full-step, i.e., $\alpha_k = 1$ for $k \geq 0$. If $r_k \leq \ln(4/3) \approx 0.28768207$ then it holds that:*

$$\left(\lambda_{k+1} / \sqrt{\sigma_{\min}^{k+1}} \right) \leq 2M_f \left(\lambda_k / \sqrt{\sigma_{\min}^k} \right)^2, \tag{15}$$

where $\sigma_{\min}^k$ is the smallest eigenvalue of $\nabla^2 f(\mathbf{x}^k)$. Consequently, if we choose $\mathbf{x}^0$ such that $\lambda_0 \leq \sigma_{\min}(\nabla^2 f(\mathbf{x}^0)) \ln(4/3)$, then the sequence $\left\{ \lambda_k / \sqrt{\sigma_{\min}^k} \right\}$ converges to zero at a quadratic rate.

Theorem 2 rigorously establishes where we can take full steps and still have quadratic convergence. Based on this information, we propose the proximal-Newton algorithm as in Algorithm 2.

The most remarkable feature of Algorithm 2 is that it does not require any globalization strategy such as backtracking line search for global convergence.

Complexity Analysis. First, we estimate the number of iterations needed when $\lambda_k \leq \sigma$ to reach the solution $\mathbf{x}^k$ such that $\frac{\lambda_k}{\sqrt{\sigma_k}} \leq \varepsilon$ for a given tolerance $\varepsilon > 0$.

Algorithm 2. (Prototype proximal-Newton algorithm)

Initialization: Given $\mathbf{x}^0 \in \mathrm{dom}\,(F)$ and $\sigma \in (0, \sigma_{\min}(\nabla^2 f(\mathbf{x}^0)) \ln(4/3)]$.
for $k = 0$ **to** $k_{\max}$ **do**
    1. Compute $\mathbf{s}^k$ by(12). Then, define $\mathbf{d}^k := \mathbf{s}^k - \mathbf{x}^k$ and $\lambda_k := \|\mathbf{d}^k\|_{\mathbf{x}^k}$.
    2. If $\lambda_k \leq \varepsilon$, then terminate.
    3. If $\lambda_k > \sigma$, then compute $r_k := M_f \|\mathbf{d}^k\|_2$ and $\alpha_k := \frac{1}{r_k} \ln(1 + r_k)$; else $\alpha_k := 1$.
    4. Update $\mathbf{x}^{k+1} := \mathbf{x}^k + \alpha_k \mathbf{d}^k$.
end for

Based on the conclusion of Theorem 2, we can show that the number of iterations of Algorithm 2 when $\lambda_k > \sigma$ does not exceed $k_{\max} := \left\lfloor \log_2 \left(\frac{\ln(2M_f \varepsilon)}{\ln(2\sigma)} \right) \right\rfloor$. Finally, we estimate the number of iterations needed when $\lambda_k > \sigma$. From Lemma 3, we see that for all $k \geq 0$ we have $\lambda_k \geq \sigma$ and $r_k \geq \sigma$. Therefore, the number of iterations is $\left\lfloor \frac{F(\mathbf{x}^0) - F(\mathbf{x}^\star)}{\psi(\sigma)} \right\rfloor$, where $\psi(\tau) := \tau \left((1 + \tau^{-1}) \ln(1 + \tau) - 1) \right) > 0$.

3.4   Proximal Quasi-Newton Algorithm

In many applications, estimating the Hessian $\nabla^2 f(\mathbf{x}^k)$ can be costly even though the Hessian is given in a closed form (cf., Section 4). In such cases, variable metric strategies employing approximate Hessian can provide computation-accuracy tradeoffs. Among these approximations, applying quasi-Newton methods with BFGS updates for $\mathbf{H}_k$ would ensure its positive definiteness. Our analytic stepsize procedures with backtracking automatically applies to the BFGS proximal-quasi Newton method, whose algorithm details and convergence analysis are omitted here.

4   Numerical Experiments

We use a variety of different real-data problems to illustrate the performance of our variable metric framework using a MATLAB implementation. We pick two advanced solvers for comparison: TFOCS [4] and PNOPT [8]. TFOCS hosts accelerated first order methods. PNOPT provides a several proximal-(quasi) Newton implementations, which has been shown to be quite successful in logistic regression problems [8]. Both use sophisticated backtracking linesearch enhancements. We benchmark all algorithms with performance profiles [7].

A performance profile is built based on a set $\mathcal{S}$ of n_s algorithms (solvers) and a collection $\mathcal{P}$ of n_p problems. We first build a profile based on computational time. We denote by $T_{p,s} := $ *computational time required to solve problem* p *by solver* s. We compare the performance of algorithm s on problem p with the best performance of any algorithm on this problem; that is we compute the performance ratio $r_{p,s} := \frac{T_{p,s}}{\min\{T_{p,\hat{s}} : \hat{s} \in \mathcal{S}\}}$. Now, let $\tilde{\rho}_s(\tilde{\tau}) := \frac{1}{n_p} \mathrm{size}\,\{p \in \mathcal{P} : r_{p,s} \leq \tilde{\tau}\}$ for $\tilde{\tau} \in \mathbb{R}_+$. The function $\tilde{\rho}_s : \mathbb{R} \to [0, 1]$ is the probability for solver s that a performance ratio is within a factor $\tilde{\tau}$ of the best possible ratio. We use the term "performance profile" for the distribution function $\tilde{\rho}_s$ of a performance metric. In the following numerical examples, we plotted the performance profiles in $\log_2$-scale, i.e. $\rho_s(\tau) := \frac{1}{n_p}\text{size}\left\{p \in \mathcal{P} : \log_2(r_{p,s}) \leq \tau := \log_2 \tilde{\tau}\right\}$.

4.1   Sparse Logistic Regression

We consider the classical logistic regression problem of the form [20]:

$$\min_{\mathbf{x},\mu}\left\{N^{-1}\sum_{j=1}^{N}\log\left(1 + e^{-y_j(\langle \mathbf{w}^{(j)},\mathbf{x}\rangle+\mu)}\right) + \rho N^{-1/2}\|\mathbf{x}\|_1\right\}, \tag{16}$$

where $\mathbf{x} \in \mathbb{R}^p$ is an unknown vector, μ is an unknown bias, and $y^{(j)}$ and $\mathbf{w}^j$ are observations where $j = 1, \cdots, N$. The logistic term in (16) is self-concordant-like with $M_f := \max \|\mathbf{w}^{(j)}\|_2$ [1]. In this case, the smooth term in (16) has Lipschitz gradient, hence several fast algorithms are applicable.

Figure 1 illustrates the performance profiles for computational time (left) and the number of prox-operations (right) using the 36 medium size problems[1]. For comparison, we use TFOCS-N07, which is Nesterov's 2007 two prox-method; and TFOCS-AT, which is Auslender and Teboulle's accelerated method, PNOPT with L-BFGS updates, and our algorithms: proximal gradient and proximal-Newton. From these performance profiles, we can observe that our proximal gradient is the best one in terms of computational time and the number of prox-operations. In terms of time, proximal-gradient solves upto 83.3% of problems with the best performance, while these numbers in TFOCS-N07 and PNOPT-LBFGS are 2.7%. Proximal Newton algorithm solves 11.1% problems with the best performance. In prox-operations, proximal-gradient is also the best one in 75% of problems.

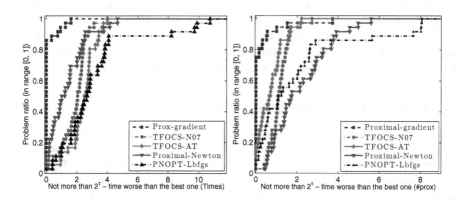

Fig. 1. Computational time (*left*) and number of prox-operations (*right*)

---

[1] Available at http://www.csie.ntu.edu.tw/~cjlin/libsvmtools/datasets/.

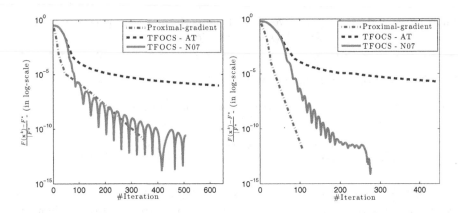

Fig. 2. *Left*: `rcv1_train.binary`, and *Right*: `real-sim`.

We now show an example convergence behavior of our proximal-gradient algorithm via two large-scale problems with $\rho = 0.1$. The first problem is `rcv1_train.binary` with the size $p = 20242$ and $N = 47236$ and the second one is `real-sim` with the size $p = 72309$ and $N = 20958$. For comparison, we use TFOCS-N07 and TFOCS-AT. For this example, PNOPT (with Newton, BFGS, and L-BFGS options) and our proximal-Newton do not scale and are omitted.

Figure 2 shows that our simple gradient algorithm locally exhibits linear convergence whereas the fast method TFOCS-AT shows a sublinear convergence rate. The variant TFOCS-N07 is the Nesterov's dual proximal algorithm, which exhibits oscillations but performs comparable to our proximal gradient method in terms of accuracy, time, and the total number of prox operations. The computational time and the number of prox-operations in these both problems are given as follows: Proximal-gradient: (15.67s, 698), (13.71s, 152); TFOCS-AT: (20.57s, 678), (33.82s, 466); TFOCS-N07: (17.09s, 1049), (22.08s, 568), respectively. For these data sets, the relative performance of the algorithms is surprisingly consistent across various regularization parameters.

4.2   Restricted Condition Number in Practice

The convergence plots in Figure 2 indicate that the linear convergence condition in Theorem 1 may be satisfied. *In fact, in all of our tests, the proximal gradient algorithm exhibits locally linear convergence.* Hence, to see if Remark 1 is grounded in practice, we perform the following test on the `a#a` dataset[1], consisting of small to medium problems. We first solve each problem with the proximal-Newton method up to 16 digits of accuracy to obtain $\mathbf{x}^\star$, and we calculate $\nabla^2 f(\mathbf{x}^\star)$. We then run our proximal gradient algorithm until convergence, and during its linear convergence, we record $\|\nabla^2 f(\mathbf{x}^\star)(\mathbf{x}^\star - \mathbf{x}^k)\|^2/\|\mathbf{x}^\star - \mathbf{x}^k\|_2^2$, and take the ratios of the maximum and the minimum to estimate the restricted condition number for each problem.

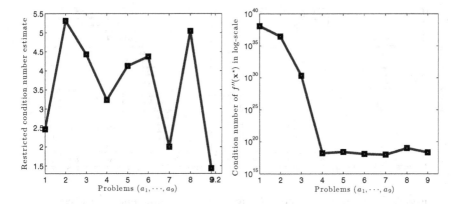

Fig. 3. Restricted condition number (*left*), and condition number (*right*) estimates

Figure 3 illustrates that while the condition number of the Hessian $\nabla^2 f(\mathbf{x}^\star)$ can be extremely large, as the algorithm navigates to the optimal solution $\mathbf{x}^\star$ through sparse subspaces, the restricted condition number estimates are in fact very close to 3. Given that algorithm still exhibit linear convergence for the cases $\# = 2, 3, 4, 5, 6, 8$ (where our condition cannot be met), we believe that the tightness of our convergence condition is an artifact of our proof and may be improved.

4.3 Sparse Multinomial Logistic Regression

For sparse multimonomial logistic regression, the underlying problem is formulated in the form of (1), which the objective function f is given as:

$$f(\mathbf{X}) := N^{-1} \sum_{j=1}^{N} \left[\log\left(1 + \sum_{i=1}^{m} e^{\langle \mathbf{w}^{(j)}, \mathbf{X}^{(i)} \rangle}\right) - \sum_{i=1}^{m} \mathbf{y}_i^{(j)} \langle \mathbf{w}^{(j)}, \mathbf{X}^{(i)} \rangle \right]. \quad (17)$$

where $\mathbf{X}$ can be considered as a matrix variable of size $m \times p$ formed from $\mathbf{X}^{(1)}, \cdots, \mathbf{X}^{(m)}$. Other vectors, $\mathbf{y}^{(j)}$ and $\mathbf{w}^{(j)}$ are given as input data for $j = 1, \ldots, N$. The function f has closed form gradient as well as Hessian. However, forming a full hessian matrix $\nabla^2 f(\mathbf{x})$ is especially costly in large scale problems when $N \gg 1$. In this case, proximal-quasi-Newton methods are more suitable. First, we show in Lemma 4 that f satisfies Definition 1, whose proof is in [19].

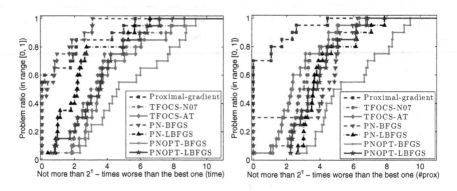

Fig. 4. Computational time (*left*), and number of prox-operations (*right*)

Lemma 4. *The function f defined by* (17) *is convex and self-concordant-like in the sense of Definition 1 with the parameter* $M_f := \sqrt{6}N^{-1} \max_{j=1,\dots,N} \|\mathbf{w}^{(j)}\|_2$.

The performance profiles of 20 small-to-medium size problems[1] are shown in Figure 4 in terms of computational time (left) as well as number of prox-operations (right), respectively. Both proximal-gradient method and proximal-Newton method with BFGS have good performance. They can solve unto 55% and 45% problems with the best time performance, respectively. These methods are also the best in terms of prox-operations (70% and 30%).

4.4 A Sytlized Example of a Non-Lipschitz Gradient Function for (1)

We consider the following convex composite minimization problem by modifying one of the canonical examples of geometric programming [6]:

$$\min_{\mathbf{x} \in \Omega} \left\{ f(\mathbf{x}) := \sum_{i=1}^{m} e^{\mathbf{a}_i^T \mathbf{x} + b_i} + \mathbf{c}^T \mathbf{x} \right\} + g(\mathbf{x}), \tag{18}$$

where Ω is a simple convex set, $\mathbf{a}_i, \mathbf{c} \in \mathbb{R}^n$ and $b_i \in \mathbb{R}$ are random, and g is the ℓ_1-norm. After some algebra, we can show that f satisfies Definition 1 with $M_f := \max\{\|\mathbf{a}_i\|_2 : 1 \le i \le m\}$. Unfortunately, f does not have Lipschitz continuous gradient in $\mathbb{R}^n$.

We implement our proximal-gradient algorithm and compare it with TFOCS and PNOPT-LBFGS. However, TFOCS breaks down in running this example due to the estimation of Lipschitz constant, while PNOPT is rather slow. Several tests on synthetic data show that our algorithm outperforms PNOPT-LBFGS. As an example, we show the convergence behavior of both these methods in Figure 5 where we plot the accuracy of the objective values w.r.t. the number of prox-operators for two cases of $\varepsilon = 10^{-6}$ and $\varepsilon = 10^{-12}$, respectively. As we can see from this figure that our prox-gradient method requires many fewer prox-operations to achieve a very high accuracy compared to PNOPT. Moreover, our method is also 20 to 40 times faster than PNOPT in this numerical test.

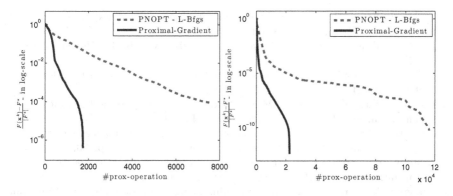

Fig. 5. Relative objective values w.r.t. #prox: *left*: $\varepsilon = 10^{-6}$, and *right*: $\varepsilon = 10^{-12}$

5 Conclusions

Convex optimization efficiency relies significantly on the structure of the objective functions. In this paper, we propose a variable metric method for minimizing the sum of a self-concordant-like convex function and a proximally tractable convex function. Our framework is applicable in several interesting machine learning problems and do not rely on the usual Lipschitz gradient assumption on the smooth part for its convergence theory. A highlight of this work is the new analytic step-size selection procedure that enhances backtracking procedures. Thanks to this new approach, we can prove that the basic gradient variant of our framework has improved local convergence guarantees under certain conditions while the tuning-free proximal Newton method has locally quadratic convergence. While our assumption on the restricted condition number in Theorem 1 is not deterministically verifiable *a priori*, we provide empirical evidence that it can hold in many practical problems. Numerical experiments on different applications that have both self-concordant-like and Lipschitz gradient properties demonstrate that the gradient algorithm based on the former assumption can be more efficient than the fast algorithms based on the latter assumption. As a result, we plan to look into fast versions of our gradient scheme as future work.

References

1. Bach, F.: Self-concordant analysis for logistic regression. Electron. J. Statist. 4, 384–414 (2010)
2. Bach, F.: Adaptivity of averaged stochastic gradient descent to local strong convexity for logistic regression (2013)
3. Beck, A., Teboulle, M.: A Fast Iterative Shrinkage-Thresholding Algorithm for Linear Inverse Problems. SIAM J. Imaging Sciences 2(1), 183–202 (2009)
4. Becker, S., Candès, E.J., Grant, M.: Templates for convex cone problems with applications to sparse signal recovery. Mathematical Programming Computation 3(3), 165–218 (2011)

5. Becker, S., Fadili, M.J.: A quasi-Newton proximal splitting method. In: Adv. Neural Information Processing Systems (2012)
6. Boyd, S., Vandenberghe, L.: Convex Optimization. Cambridge University Press (2004)
7. Dolan, E.D., Moré, J.J.: Benchmarking optimization software with performance profiles. Math. Program. 91, 201–213 (2002)
8. Lee, J.D., Sun, Y., Saunders, M.A.: Proximal Newton-type methods for convex optimization. SIAM J. Optim. 24(3), 1420–1443 (2014)
9. Mine, H., Fukushima, M.: A minimization method for the sum of a convex function and a continuously differentiable function. J. Optim. Theory Appl. 33, 9–23 (1981)
10. Nesterov, Y.: Introductory lectures on convex optimization: a basic course. Applied Optimization, vol. 87. Kluwer Academic Publishers (2004)
11. Nesterov, Y.: Gradient methods for minimizing composite objective function. Math. Program. 140(1), 125–161 (2013)
12. Nesterov, Y., Nemirovski, A.: Interior-point Polynomial Algorithms in Convex Programming. Society for Industrial Mathematics (1994)
13. Nocedal, J., Wright, S.J.: Numerical Optimization, 2nd edn. Springer Series in Operations Research and Financial Engineering. Springer (2006)
14. Parikh, N., Boyd, S.: Proximal algorithms. Foundations and Trends in Optimization 1(3), 123–231 (2013)
15. Robinson, S.M.: Strongly Regular Generalized Equations. Mathematics of Operations Research 5(1), 43–62 (1980)
16. Rockafellar, R.T.: Convex Analysis. Princeton Mathematics Series, vol. 28. Princeton University Press (1970)
17. Schmidt, M., Roux, N.L., Bach, F.: Convergence rates of inexact proximal-gradient methods for convex optimization. In: NIPS, Granada, Spain (2011)
18. Tran-Dinh, Q., Kyrillidis, A., Cevher, V.: Composite self-concordant minimization. J. Mach. Learn. Res. 15, 1–54 (2014) (accepted)
19. Tran-Dinh, Q., Li, Y.-H., Cevher, V.: Composite convex minimization involving self-concordant-like cost functions. LIONS-Tech. Report., 1–19 (2015), http://arxiv.org/abs/1502.01068
20. Yuan, Y.X.: Recent advances in numerical methods for nonlinear equations and nonlinear least squares. Numerical Algebra, Control and Optimization 1(1), 15–34 (2011)

Computational Aspects of Constrained L_1-L_2 Minimization for Compressive Sensing

Yifei Lou[1], Stanley Osher[2], and Jack Xin[3]

[1] University of Texas, Dallas, USA
[2] University of California, Los Angeles, USA
[3] University of California, Irvine, USA

Abstract. We study the computational properties of solving a constrained L_1-L_2 minimization via a difference of convex algorithm (DCA), which was proposed in our previous work [13, 19] to recover sparse signals from a under-determined linear system. We prove that the DCA converges to a stationary point of the nonconvex L_1-L_2 model. We clarify the relationship of DCA to a convex method, Bregman iteration [20] for solving a constrained L_1 minimization. Through experiments, we discover that both L_1 and L_1-L_2 obtain better recovery results from more coherent matrices, which appears unknown in theoretical analysis of exact sparse recovery. In addition, numerical studies motivate us to consider a weighted difference model L_1-αL_2 ($\alpha > 1$) to deal with ill-conditioned matrices when L_1-L_2 fails to obtain a good solution.

1 Introduction

Compressive sensing (CS) [7,9] is about acquiring or recovering a *sparse* signal (a vector with most elements being zero) from an under-determined linear system $Au = b$, where b is the data vector and A is a $M \times N$ matrix for $M < N$. Mathematically, it amounts to solving a constrained optimization problem,

$$\min \|u\|_0 \quad \text{s.t.} \quad Au = b, \tag{1}$$

where $\| \cdot \|_0$ is the L_0 norm, which counts the number of non-zero elements. Minimizing the L_0 norm is equivalent to finding the sparsest solution. Since L_0 minimization is NP-hard [14], a popular approach is to replace L_0 by a convex L_1 norm, which often gives a satisfactory sparse solution. A major step for CS was the derivation of the restricted isometry property (RIP) [3], which guarantees to recover a sparse signal by minimizing the L_1 norm. It is also proved in [3] that Gaussian random matrices satisfy the RIP with high probability. A deterministic result in [6,8,11] says that exact sparse recovery via L_1 minimization is possible if

$$\|u\|_0 < \frac{1 + 1/\mu}{2}, \tag{2}$$

where μ is the mutual coherence of A, defined as

$$\mu(A) = \max_{i \neq j} \frac{|\mathbf{a}_i^T \mathbf{a}_j|}{\|\mathbf{a}_i\| \|\mathbf{a}_j\|}, \quad \text{with } A = [\mathbf{a}_1, \cdots, \mathbf{a}_N]. \tag{3}$$

© Springer International Publishing Switzerland 2015   169
H.A. Le Thi et al. (eds.), *Model. Comput. & Optim. in Inf. Syst. & Manage. Sci.*,
Advances in Intelligent Systems and Computing 359, DOI: 10.1007/978-3-319-18161-5_15

The inequality (2) suggests that L_1 may not perform well for highly coherent matrices in that if $\mu \sim 1$, then $\|u\|_0$ is 1 at most.

Recently, there has been a surge of activities in deploying nonconvex penalties to promote sparsity while solving the linear system, because the convex L_1 norm may not perform well on some practical problems with coherent sensing matrices. In our previous studies [13,19], we advocate the use of a nonconvex functional L_1-L_2, as opposed to L_p for $p \in (0,1)$ in [5,12,18]. To minimize L_1-L_2, a difference of convex algorithm (DCA) [16] is applied. In [13], we conduct an extensive study comparing the sparse penalties, $L_0, L_1, L_p(0 < p < 1), L_1 - L_2$, and their numerical algorithms. Numerical experiments demonstrate that L_1-L_2 is always better than L_1 to promote sparsity, and using DCA for L_1-L_2 is better than iterative reweighted algorithms for L_p minimization [5, 12] when the sensing matrix exhibits high coherence.

The contributions of this work are four-fold. First, we prove that the DCA iterations for a constrained minimization converge to a stationary point (Section 2). Second, we clarify the relation of DCA to the Bregman iteration [20], which is designed for convex functional minimization (Section 3). Third, we analyze how coherence, sparsity and minimum separation (MS) contribute to exact recovery of a sparse vector from an under-determined system, and discover that both L_1 and L_1-L_2 get better recovery results towards high coherence (Section 4). Lastly, we demonstrate that exact recovery is highly correlated with DCA converging in a few steps, and then propose a weighted difference model, minimizing L_1-αL_2 for $\alpha > 1$, to improve the reconstruction accuracy when L_1-L_2 fails to find the exact solution (Section 5).

2   Constrained L_1-L_2 Minimization

Replacing L_0 in (1) by L_1-L_2, we get a constrained minimization problem,

$$\min_{u \in \mathbb{R}^N} \|u\|_1 - \|u\|_2 \quad \text{s.t.} \quad Au = b. \tag{4}$$

The idea of DCA [16] involves linearizing the second (nonconvex) term in the objective function at the current solution, and advancing to a new one by solving a L_1 type of subproblem, *i.e.*,

$$u_{n+1} = \arg\min\{\|u\|_1 - \langle q_n, u\rangle \quad \text{s.t.} \quad Au = b\}, \tag{5}$$

for $q_n = \frac{u_n}{\|u_n\|_2}$. We introduce two Lagrange multipliers y, z in an augmented Lagrangian,

$$L_{\lambda,\rho}(u,v,y,z) = \lambda\|v\|_1 - \lambda\, q_n^T u + \rho y^T(u-v) + z^T(Au-b) + \frac{\rho}{2}\|u-v\|^2 + \frac{1}{2}\|Au-b\|^2.$$

Then an alternating direction of multiplier method (ADMM) [1] is applied to solve eq. (5), by alternatively updating each variable (u, v, y and z) to minimize $L_{\lambda,\rho}$. Please refer to Figure 1 for pseudo-code. Note that the update of v can be solved efficiently by a *soft shrinkage* operator: $v = \text{shrink}(u + y, \lambda/\rho)$, where $\text{shrink}(s, \gamma) = \text{sgn}(s) \max\{|s| - \gamma, 0\}$.

Next we will prove that the DCA sequence $\{u_n\}$, defined in eq. (5), converges to a stationary point of eq. (4). The convergence proof of an unconstrained L_1-L_2 minimization can be found in [19]. For general DCA, convergence is guaranteed if the optimal value is finite and the sequence generated by DCA is bounded [16,17]. Here we can prove that the DCA sequence is bounded with probability 1, due to degree-1 homogeneity of L_1-L_2 and the fact that a nonzero signal is 1-sparse if and only if its L_1-L_2 is zero (see Lemma 1).

Lemma 1. *Suppose $u \in \mathbb{R}^N \setminus \{\mathbf{0}\}$ and $\|u\|_0 = s$, then*

$$\|u\|_1 - \|u\|_2 = 0 \quad \text{if and only if} \quad s = 1.$$

Please refer to [19] for the proof.

Lemma 2. *The objective function $E(u) = \|u\|_1 - \|u\|_2$ is monotonically decreasing for the DCA sequence $\{u_n\}$ defined in eq. (5).*

Proof. We want to show that

$$0 \leqslant E(u_n) - E(u_{n+1}) = \|u_n\|_1 - \|u_n\|_2 - \|u_{n+1}\|_1 + \|u_{n+1}\|_2. \tag{6}$$

The first-order optimality condition for a constrained problem (5) can be formulated as

$$\frac{\partial \mathcal{L}}{\partial u} = 0 \quad \text{and} \quad \frac{\partial \mathcal{L}}{\partial \nu} = 0, \tag{7}$$

where $\mathcal{L}(u, \nu) = \|u\|_1 - q_n^T u + \nu^T (Au - b)$ is the Lagrangian. Since L_1 norm is not differentiable, we consider a subgradient $p_{n+1} \in \partial \|u_{n+1}\|_1$, and hence eq. (7) is equivalent to

$$p_{n+1} - q_n + A^T \nu = 0 \quad \text{and} \quad Au_{n+1} = b. \tag{8}$$

Left multiplying the first equation in (8) by $u_n - u_{n+1}$ gives

$$\begin{aligned} 0 &= \langle p_{n+1} - q_n + A^T \nu, u_n - u_{n+1} \rangle \\ &= \langle p_{n+1}, u_n \rangle + \langle q_n, u_{n+1} \rangle - \|u_{n+1}\|_1 - \|u_n\|_2, \end{aligned} \tag{9}$$

where we use $\langle p_{n+1}, u_{n+1} \rangle = \|u_{n+1}\|_1$, $\langle q_n, u_n \rangle = \|u_n\|_2$, and $Au_n = Au_{n+1} = b$. Substituting eq. (9) into eq. (6), we get

$$E(u_n) - E(u_{n+1}) = (\|u_n\|_1 - \langle p_{n+1}, u_n \rangle) + (\|u_{n+1}\|_2 - \langle q_n, u_{n+1} \rangle). \tag{10}$$

Since any subgradient of L_1 norm has the property that $|p_{n+1}^{(i)}| \leqslant 1$ for all $1 \leq i \leq N$, we have $\|u_n\|_1 \geq \langle p_{n+1}, u_n \rangle$. The second term in eq. (10), $\|u_{n+1}\|_2 - \langle q_n, u_{n+1} \rangle \geqslant 0$ is due to Cauchy-Schwarz inequality. Therefore, we proved that $E(u_n) \geqslant E(u_{n+1})$. $\qquad \square$

Lemma 3. *The DCA sequence $\{u_n\}$ is bounded with probability 1.*

Proof. It follows from Lemma 2 that $E(u_n)$ is bounded, and hence there exists a constant C so that

$$\|u_n\|_1 - \|u_n\|_2 \leq C. \tag{11}$$

Write $u_n = \|u_n\|_2 \cdot u_n/\|u_n\|_2$, or a polar decomposition into amplitude and phase. By degree-1 homogeneity, eq. (11) is:

$$\|u_n\|_2 \left(\left\|\frac{u_n}{\|u_n\|_2}\right\|_1 - \left\|\frac{u_n}{\|u_n\|_2}\right\|_2\right) \leq C.$$

Suppose $\|u_n\|_2$ diverges (up to a sub-sequence, but denoted the same), then

$$\left\|\frac{u_n}{\|u_n\|_2}\right\|_1 - \left\|\frac{u_n}{\|u_n\|_2}\right\|_2 \to 0, \text{ as } n \to \infty.$$

Since $u_n/\|u_n\|_2$ is compact (on unit sphere), it converges to a limit point, denoted as u_*, on the unit sphere (up to a sub-sequence). Hence, $\|u_*\|_1 - \|u_*\|_2 = 0$, implying $\|u_*\|_0 = 1$ by Lemma 1.

On the other hand, the DCA sequence satisfies $Au_n = b$. Dividing by $\|u_n\|_2 \to \infty$, we find that $Au_* = 0$, so u_* is in Ker(A). Unless A has a zero column (or one component of u is absent in the constraint), Ker(A) does not contain a one-sparse vector, which is a contradiction.

So if A has no zero column (which happens with probability 1 for random matrices from continuous distribution), we conclude that u_n is bounded.

Theorem 1. *Any non-zero limit point u_* satisfies the first-order optimality condition, which means u_* is a stationary point.*

Proof. The objective function E is monotonically decreasing by Lemma 2, and bounded from below, so $E(u^n)$ converges and hence we have $\|u_{n+1}\|_2 - \langle q_n, u_{n+1}\rangle \to 0$, which implies that $u_n - u_{n+1} \to 0$ as $u_n \neq \mathbf{0}$.

As the sequence $\{u_n\}$ is bounded by Lemma 3, there exists (by definition) a subsequence of $\{u_n\}$ converging to a limit point u_*. The subsequence is denoted as $\{u_{n_k}\}$. The optimality condition at the n_k−th step of DCA is

$$q_{n_k-1} - A^T \nu \in \partial\|u_{n_k}\|_1 \quad \text{and} \quad Au_{n_k} = b. \tag{12}$$

Since $u_{n_k} \to u_*$ and subgradient of L_1 norm is a closed set, we have $\partial\|u_{n_k}\|_1 \subseteq \partial\|u_*\|_1$. Letting $n_k \to \infty$, we get

$$q_* - A^T \nu \in \partial\|u_*\|_1 \quad \text{and} \quad Au_* = b, \tag{13}$$

which means that u_* satisfies the first-order optimality condition and hence it is a stationary point. $\qquad\square$

3 Study A: DCA v.s. Bregman Iteration

We want to point out that DCA is related to Bregman iteration, which was derived from Bregman divergence [2], defined as

$$D_J^p(u, v) := J(u) - J(v) - \langle p, u - v \rangle, \tag{14}$$

where $J(\cdot)$ is a convex functional, and $p \in \partial J(v)$ is a subgradient of J at the point v. For constrained L_1 minimization,

$$\min_{u \in \mathbb{R}^N} \|u\|_1 \quad \text{s.t.} \quad Au = b, \tag{15}$$

Bregman iteration incorporates the Bregman divergence into an unconstrained formulation, and advances to a new solution u_{n+1} based on a Taylor expansion of $J(u) = \|u\|_1$ at current step, i.e., $u_{n+1} = \arg\min_u \lambda D_J^p(u, u_n) + \frac{1}{2}\|Au - b\|_2^2$. The optimality condition is

$$\lambda(p_{n+1} - p_n) + A^T(Au_{n+1} - b) = 0. \tag{16}$$

Summing from 0 to $n+1$, we have $\lambda p_{n+1} + \sum_{k=0}^{n+1} A^T(Au_k - b) = 0$, or

$$\begin{cases} \lambda p_{n+1} + A^T(Au_{n+1} - v_n) = 0 \\ v_n = \sum_{k=1}^{n}(b - Au_k), \end{cases} \tag{17}$$

for $p_0 = u_0 = 0$, which is equivalent to

$$\begin{cases} u_{n+1} = \arg\min \lambda\|u\|_1 + \frac{1}{2}\|Au - v_n\|_2^2 \\ v_{n+1} = v_n + b - Au_{n+1}. \end{cases} \tag{18}$$

We consider the same idea for L_1-L_2. In particular, we get an optimality condition by lagging the second term,

$$\lambda(p_{n+1} - p_n) - \lambda(q_n - q_{n-1}) + A^T(Au_{n+1} - b) = 0, \tag{19}$$

where p and q be subgradients of $\|u\|_1$ and $\|u\|_2$ respectively. Summing from 0 to $n+1$ and letting $p_1 = q_0 = z_1 = 0$, we obtain

$$\begin{cases} \lambda p_{n+1} - \lambda q_n + A^T(Au_{n+1} - z_n) = 0 \\ z_{n+1} = z_n + (b - Au_{n+1}), \end{cases} \tag{20}$$

for $z_n = \sum_{k=1}^{n} b - Au_k$. The first equation in (20) is equivalent to $u_{n+1} = \arg\min \lambda\|u\|_1 - \lambda\langle q_n, u\rangle + \frac{1}{2}\|Au - z_n\|_2^2$, which can be solved via ADMM.

DCA+ADMM (Alg.1) and Bregman+ADMM (Alg.2) are summarized in Fig. 1, which shows that their difference lies in the update of z and q (compare boxed lines in Figure 1). For DCA, z is updated MaxOuter iterations and then q is updated, while Bregman iteration updates z and q simultaneously. We plot relative errors of each inner solution to the ground-truth versus computational time in Fig. 2, which illustrates that DCA is more computationally efficient than Bregman iteration.

The update of z is to account for the constraint $Au = b$, which is enforced by DCA at every inner iteration. This constraint also plays an important role in proving DCA's convergence (see Theorem 1). As for Bregman iteration, Osher et. al. [15] proved that Bregman iteration (18) converges if the regularization function is convex, while convergence analysis for nonconvex formulation (20) is subject to future investigation.

Define MaxOuter, MaxInner and initialize $u \neq 0, v = y = z = 0$

Alg.1: DCA + ADMM

for 1 to MaxOuter **do**
    for 1 to MaxInner **do**
        $u = (A^T A + \rho I)^{-1}(A^T (b - z) + \rho(y + v) + \lambda q)$
        $v = shrink(u + y, \lambda/\rho)$
        $y = y + (u - v)$
        $z = z + Au - b$
    end for
    $q = u/\|u\|_2$
end for

Alg.2: Bregman + ADMM

for 1 to MaxOuter **do**
    for 1 to MaxInner **do**
        $u = (A^T A + \rho I)^{-1}(A^T (b - z) + \rho(y + v) + \lambda q)$
        $v = shrink(u + y, \lambda/\rho)$
        $y = y + (u - v)$
    end for
    $z = z + Au - b$
    $q = u/\|u\|_2$
end for

Fig. 1. Pseudo-codes for DCA+ADMM (left) and Bregman+ADMM (right)

4   Study B: Sparsity v.s. Coherence

Theoretically, the success of L_1 depends on the RIP or incoherence condition. Unfortunately, RIP is difficult to verify for a given matrix, while incoherence is not strong enough to account for exact recovery. Therefore, we are interested in non-RIP conditions to evaluate the performance of L_1 or L_1-L_2, which will contribute a better characterization of sparse solutions.

We consider a family of randomly oversampled partial discrete cosine transform (DCT) matrices of the form

$$A = [\mathbf{a}_1, \cdots, \mathbf{a}_N] \in \mathbb{R}^{M \times N} \text{ with } \mathbf{a}_j = \frac{1}{\sqrt{M}} \cos(\frac{2\pi \mathbf{w} j}{F}), \ j = 1, \cdots, N, \quad (21)$$

where $\mathbf{w}$ is a random vector uniformly distributed in $[0, 1]^M$. This matrix arises in spectral estimation [10], if the cosine function in (21) is replaced by complex exponential. The coherence of this type of matrices is controlled by F in the sense that larger F corresponds to larger coherence.

We then generate random sparse vectors as ground-truth, denoted as u_g, whose sparsity (L_0 norm) is S with nonzero elements being at least R distance apart, referred to as *minimum separation*. Let $b = Au_g$, and u_* is a reconstructed solution, from L_1 minimization using Bregman iteration (18) or L_1-L_2 minimization using DCA (5). We consider the algorithm successful, if the relative error of u_* to the ground truth u_g is less than .001, *i.e.*, $\frac{\|u_* - u_g\|}{\|u_g\|} < .001$.

We analyze whether success rates (based on 100 random realizations) are related to coherence (F), sparsity (S), and minimum separation (R) using randomly oversampled DCT matrices of size 100×2000. We include the discussion of MS here, due to the work of [4], which suggests that sparse spikes need to be further apart for more coherent matrices. However, we observe that MS seems to play a minor role in sparse recovery when it is above $2F$, a theoretical lower bound [4], as indicated by the first plot in Fig. 3 showing that success rates as

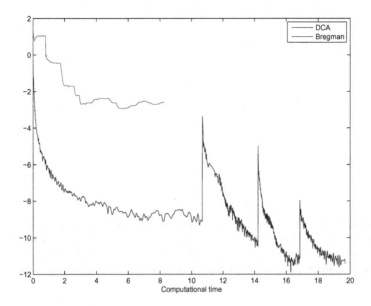

Fig. 2. Comparison between DCA and Bregman in terms of relative errors to the ground-truth solution versus computational time (sec.), which shows that DCA is more effiient than Bregman iteration. The stopping condition for outer iterations is $\|x_n - x_{n-1}\|/\|x_{n-1}\| < 1e - 4$, and Bregman stops eariler than DCA after about 8 seconds.

a function of S are almost identical for different R. The second plot in Fig. 3 is success rates as a function of F for different S while $R = 50$, which suggests that sparsity is the most important factor that contributes to exact sparse recovery, compared to MS and coherence.

In Fig. 4, we examine the success rates of using L_1 and L_1-L_2 as a function of R, while S is fixed to be 25. The two plots illustrate that the success rates of $F = 10, 15$ are higher than that of $F = 5$, which implies that more coherent matrices yield better recovery rates for both L_1 and L_1-L_2. This phenomenon appears new in L_1 sparse recovery, which is worthy of future study.

5 Study C: Exact Recovery v.s. DCA Convergence

We find that if DCA converges in a few iterations (say 3-5), the reconstructed solution coincides with the ground-truth solution with high probability. It follows from Theorem 1 that DCA sequence always converges to a stationary point. We consider a stopping condition of DCA to be $\frac{\|u_{n+1}-u_n\|_2}{\|u_n\|} < .001$. In this section, we say DCA converges if the number of iterations is less than 10. Table 1 is the confusion matrix or joint occurrence of whether DCA converges and whether the algorithm finds the exact solution, which illustrates exact recovery is highly correlated with DCA converging in a few steps.

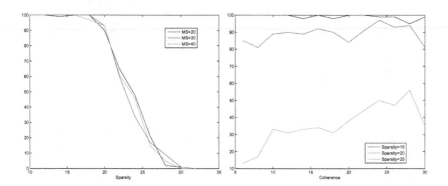

Fig. 3. Success rates of recovering sparse signals from randomly oversampled partial DCT matrices of size 100×2000 as a function of S for different R while $F = 10$ (left) and as a function of F for different S while $R = 50$ (right). Both plots suggest that sparsity is the most important factor in sparse recovery, compared to MS and coherence.

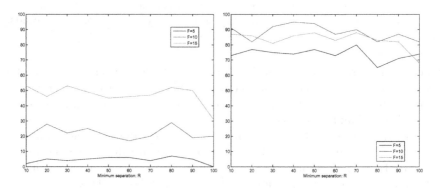

Fig. 4. Success rates of using L_1 (left) and L_1-L_2 (right) as a function of R while fixing $S = 20$ for different F showing that larger coherence yields better recovery rates

Table 1. Confusion matrix of whether DCA converges and whether the algorithm finds the exact solution.

	converge	not converge
exact	8104	17
not exact	445	5732

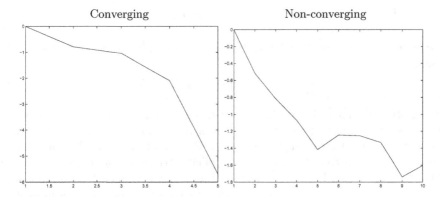

Fig. 5. Relative errors (in logarithmic scales) between two consecutive DCA iterations for converging (left) and non-converging (right) cases. If DCA converges in a few steps (3-5), it gives an exact recovery with high probability.

Table 2. Success rates (%) of L_1-L_2 and L_1-$2L_2$

$F = 10, R = 50, S =$	18	21	24	27	30	Total
L_1-L_2	97	78	34	10	0	219
L_1-$2L_2$	98	81	40	17	4	240
improve	1.03	3.85	17.65	70.00	inf	9.59
$F = 10, S = 25, R =$	20	30	40	50	60	Total
L_1-L_2	28	33	25	36	26	216
L_1-$2L_2$	37	37	34	41	32	266
improve	32.14	12.12	36.01	13.89	23.08	23.15

We examine the relative errors between two consecutive DCA iterations for both converging and non-converging cases. Fig. 5 shows that the relative errors monotonically decreases, and DCA stops in 3-5 steps. On the other hand, if the relative errors are oscillatory, it often implies that the algorithm does not converge (within 10 iterations).

We found that one reason that DCA does not converge is that L_1-L_2 of the exact solution is larger than that of some DCA iterates u_n, and hence the algorithm jumps among these local mimima. This observation suggests that L_1-L_2 is unable to promote sparsity in some degenerate cases. As a remedy, we propose a weighted difference model L_1-αL_2 with more weight on the nonconvex term ($\alpha > 1$). We find that the weighted difference model sometimes improves the recovery rate, though there is no convergence proof for $\alpha > 1$, as the objective function is not bounded from below. To illustrate this phenomenon numerically, we consider to apply the DCA for L_1-$2L_2$, if the DCA for L_1-L_2 does not converge within 10 iterations; otherwise, we use the solution of L_1-L_2 to be the one in lieu of L_1-$2L_2$. The success rates for both $\alpha = 1$ and $\alpha = 2$ are then recorded in Tables 2, which report at least 10% improvement for most testing cases.

6   Study D: Constrained V.S. Unconstrained

We now compare the constrained L_1-L_2 minimization (4) with the unconstrained version, *i.e.*,

$$\min_{u\in\mathbb{R}^N} \lambda(\|u\|_1 - \|u\|_2) + \frac{1}{2}\|Au - b\|_2^2, \tag{22}$$

which is studied thoroughly in [19]. The constrained formulation is a parameter-free model, while two auxiliary variables are introduced in the ADMM algorithm. In all experiments, we choose $\lambda = 2, \rho = 10$. For the unconstrained formulation, a small λ is chosen to enforce $Au = b$ implicitly. Here we choose $\lambda = 10^{-5}$ in (22). The comparison between constrained and unconstrained formulations for both L_1 and L_1-L_2 is given in Fig. 6, which shows that the two optimization problems yield similar performance when λ is small.

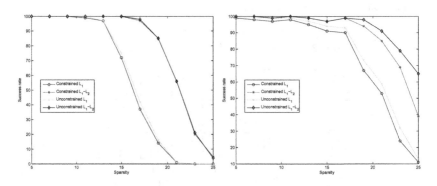

Fig. 6. The comparison between constrained and unconstrained formulations for $F = 5$ (left) and $F = 20$ (right), which shows that they give similar performance with the model parameter in unconstrained minimization problem properly chosen

7   Conclusion and Future Work

This paper studied the computational aspects of a constrained L_1-L_2 minimization, as an alternative to the conventional L_1 approach, to recover sparse signals from an under-determined system. The DCA is applied to solve this nonconvex model with guaranteed convergence to a stationary point. The relation of DCA for a nonconvex model with a convex method, Bregman iteration, was also presented. Numerical experiments demonstrated that more coherent matrices give better recovery results and we proposed a weighted difference model to improve the reconstruction results, when L_1-L_2 is not sharp enough to promote sparse solution.

As for future directions, we will analyze the convergence of Bregman iteration applied to a nonconvex model, *i.e.*, eq. (20). Furthermore, we will devote ourselves to the understanding of the peculiar phenomenon of larger coherence giving better results. As for the weighted difference model, we will further explore the possibilities of stabilizing the resulting algorithm, and adaptively choosing the weighting parameter α with respect to the matrix A in eq. (1).

Acknowledgments. We would like to thank the anonymous referee for pointing out a general convergence property of DCA and a suggestion to reorganize our previous proof of convergence. Stanley Osher was partially supported by ONR grants N000141210838, N000141410444, NSF DMS-1118971, and the Keck Foundation. Jack Xin was partially supported by NSF grant DMS-1211179.

References

1. Boyd, S., Parikh, N., Chu, E., Peleato, B., Eckstein, J.: Distributed optimization and statistical learning via the alternating direction method of multipliers. Found. Trends Mach. Learn. 3(1), 1–122 (2011)
2. Bregman, L.: The relaxation method of finding the common points of convex sets and its application to the solution of problems in convex programming. USSR Comp. Math. Math. Phys. (7), 200–217 (1967)
3. Candès, E.J., Romberg, J., Tao, T.: Stable signal recovery from incomplete and inaccurate measurements. Comm. Pure Appl. Math. 59, 1207–1223 (2006)
4. Candès, E., Fernandez-Granda, C.: Super-resolution from noisy data. J. Fourier Anal. Appl. 19(6), 1229–1254 (2013)
5. Chartrand, R., Yin, W.: Iteratively reweighted algorithms for compressive sensing. In: International Conference on Acoustics, Speech, and Signal Processing (ICASSP), pp. 3869–3872 (2008)
6. Donoho, D., Elad, M.: Optimally sparse representation in general (nonorthogonal) dictionaries via l1 minimization. Proc. Nat. Acad. Scien. 100, 2197–2202 (2003)
7. Donoho, D.L.: Compressed sensing. IEEE Trans. Inf. Theory 52(4), 1289–1306 (2006)
8. Donoho, D.L., Huo, X.: Uncertainty principles and ideal atomic decomposition. IEEE Transactions on Information Theory 47(7), 2845–2862 (2001)
9. Candès, E.J., Wakin, M.B.: An introduction to compressive sampling. IEEE Signal Process. Mag. 25(2), 21–30 (2008)
10. Fannjiang, A., Liao, W.: Coherence pattern-guided compressive sensing with unresolved grids. SIAM J. Imaging Sci. 5(1), 179–202 (2012)
11. Gribonval, R., Nielsen, M.: Sparse representations in unions of bases. IEEE Trans. Inf. Theory 49(12), 3320–3325 (2003)
12. Lai, M.J., Xu, Y., Yin, W.: Improved iteratively reweighted least squares for unconstrained smoothed lq minimization. SIAM J. Numer. Anal. 5(2), 927–957 (2013)
13. Lou, Y., Yin, P., He, Q., Xin, J.: Computing sparse representation in a highly coherent dictionary based on difference of l1 and l2. J. Sci. Comput. (2014) (to appear)
14. Natarajan, B.K.: Sparse approximate solutions to linear systems. SIAM J. Comput., 227–234 (1995)
15. Osher, S., Burger, M., Goldfarb, D., Xu, J., Yin, W.: An iterated regularization method for total variation-based image restoration. Multiscale Model. Simul. (4), 460–489 (2005)
16. Pham-Dinh, T., Le-Thi, H.A.: Convex analysis approach to d.c. programming: Theory, algorithms and applications. Acta Math. Vietnam. 22(1), 289–355 (1997)

17. Pham-Dinh, T., Le-Thi, H.A.: A d.c. optimization algorithm for solving the trust-region subproblem. SIAM J. Optim. 8(2), 476–505 (1998)
18. Xu, Z., Chang, X., Xu, F., Zhang, H.: $l_{1/2}$ regularization: A thresholding representation theory and a fast solver. IEEE Trans. Neural Netw. 23, 1013–1027 (2012)
19. Yin, P., Lou, Y., He, Q., Xin, J.: Minimization of $l_1 - l_2$ for compressed sensing. SIAM J. Sci. Comput. (2014) (to appear)
20. Yin, W., Osher, S., Goldfarb, D., Darbon, J.: Bregman iterative algorithms for l1 minimization with applications to compressed sensing. SIAM J. Imaging Sci. 1, 143–168 (2008)

Continuous Relaxation
for Discrete DC Programming

Takanori Maehara[1,*], Naoki Marumo[2,*], and Kazuo Murota[2]

[1] Department of Mathematical and Systems Engineering, Shizuoka University,
Shizuoka, Japan
maehara.takanori@shizuoka.ac.jp
[2] Department of Mathematical Informatics, University of Tokyo, Tokyo, Japan
{naoki_marumo,murota}@mist.i.u-tokyo.ac.jp

Abstract. Discrete DC programming with convex extensible functions
is studied. A natural approach for this problem is a continuous relaxation that extends the problem to a continuous domain and applies the
algorithm in continuous DC programming. By employing a special form
of continuous relaxation, which is named "lin-vex extension," the optimal solution of the continuous relaxation coincides with the original
discrete problem. The proposed method is demonstrated for the degree-
concentrated spanning tree problem.

1 Introduction

DC programming [8], [16,17], minimization of a difference of two convex functions,
is an established area in nonconvex optimization. There are many useful theorems
such as the Toland–Singer duality theorem, and practically efficient algorithms.
Moreover, most of the optimization problems can be represented as a DC programming problem [2], [16], [18]. Recently, Maehara and Murota [9] proposed a
framework of discrete DC programming. A function $f : \mathbb{Z}^n \to \mathbb{Z} \cup \{-\infty, +\infty\}$ is
defined to be a discrete DC function if it can be represented as $f = g - h$ with two
discrete convex ($M^\natural$-convex and/or $L^\natural$-convex) functions $g, h : \mathbb{Z}^n \to \mathbb{Z} \cup \{+\infty\}$.
This framework contains minimization of a difference of two submodular functions,
which often appears in machine learning [3,4], [14].

In this paper, we are dealing with a larger class of discrete convex functions,
convex extensible functions. A discrete-variable real-valued function $g : \mathbb{Z}^n \to
\mathbb{R} \cup \{+\infty\}$ is said to be *convex extensible* if it can be interpolated by a convex
function $\hat{g} : \mathbb{R}^n \to \mathbb{R} \cup \{+\infty\}$ in continuous variables. $M^\natural$-convex and $L^\natural$-convex
functions are known to be convex extensible. Convex extensibility is a natural
property required of discrete convex functions, but it is considered too weak for
a rich theory. In fact, any function g defined on a unit cube $\{0,1\}^n$ is convex
extensible. Not much theory has been developed so far for convex extensible
functions.

---

* Supported by JST, ERATO, Kawarabayashi Large Graph Project.

© Springer International Publishing Switzerland 2015   181
H.A. Le Thi et al. (eds.), *Model. Comput. & Optim. in Inf. Syst. & Manage. Sci.*,
Advances in Intelligent Systems and Computing 359, DOI: 10.1007/978-3-319-18161-5_16

Here, we study a discrete DC programming problem for $f = g - h$ with convex extensible functions g and h. This class of discrete optimization problems arise in many applications. For example, a discrete optimization problem with a continuous objective function restricted to the integer domain usually falls into this category.

A natural approach for such problem is *continuous relaxation*, which extends the discrete functions to the continuous domain and applies a continuous optimization method. As the continuous extension of $f = g - h$, we employ a special form, which we call a *lin-vex extension* $\tilde{f} = \bar{g} - \hat{h}$, where $\bar{g}$ is the convex closure (largest convex extension, usually piecewise linear) of g and $\hat{h}$ is any (often smooth) convex extension of h. Our main result (Theorem 3) shows that no integral gap exists between the discrete optimization problem for f and the continuous optimization problem for its lin-vex extension $\tilde{f}$. This approach is useful in solving a discrete optimization problem having a "nice" DC representation. If an objective function f is represented as $f = g - h$ such that the discrete optimization problem with g is efficiently solved and the subgradient of h is efficiently obtained, then the continuous DC algorithm for the lin-vex extension can be efficiently implemented and the obtained solution for the continuous relaxation is guaranteed to be an integral solution.

Use of continuous extension for discrete optimization problems is a standard technique. Integer programming problems are solved successfully via linear programing. In discrete convex analysis, in particular, we can design theoretically and practically faster algorithms by using continuous extensions and proximity theorems for M$^\natural$-convex and L$^\natural$-convex functions [10, 11].

To demonstrate the use of the proposed framework, we consider a variant of the spanning tree problem, to be called *degree-concentrated spanning tree problem*, that finds a spanning tree with the maximum variance of degrees. This problem has an application in network routing [15]. We formulate this problem as a DC programming problem, and adopt the DC algorithm. Our experiment for a real-world network shows the DC algorithm works pretty well for this problem.

2   Existing Studies of DC Programming

2.1   Continuous DC Programming

Let $g : \mathbb{R}^n \to \mathbb{R} \cup \{+\infty\}$ be a convex function. The effective domain of g is defined by $\mathrm{dom}_\mathbb{R}\, g := \{x \in \mathbb{R}^n : g(x) < +\infty\}$. Throughout the paper, we always consider functions with $\mathrm{dom}_\mathbb{R}\, g \neq \emptyset$. A vector $p \in \mathbb{R}^n$ is a *subgradient* of g at $x \in \mathrm{dom}_\mathbb{R}\, g$ if

$$g(y) \geq g(x) + \langle p, y - x \rangle \quad (y \in \mathbb{R}^n), \tag{1}$$

where $\langle p, x \rangle = \sum_{i=1}^n p_i x_i$ denotes the inner product. The set of all subgradients of g at x is called the *subdifferential* of g at x and denoted by $\partial_\mathbb{R} g(x)$. Every convex function g has a subgradient at each $x \in \mathrm{relint}(\mathrm{dom}_\mathbb{R}\, g)$, where relint denotes the relative interior.

Algorithm 1 DC algorithm

Let $x \in \operatorname{dom}_{\mathbb{R}} g$ be an initial solution.
repeat find $p \in \partial_{\mathbb{R}} h(x)$, find $x \in \partial_{\mathbb{R}} g^*(p)$ **until** convergence

---

The *Fenchel conjugate* $g^* : \mathbb{R}^n \to \mathbb{R} \cup \{+\infty\}$ of a convex function $g : \mathbb{R}^n \to \mathbb{R} \cup \{+\infty\}$ is defined by

$$g^*(p) := \sup_{x \in \mathbb{R}^n} \{\langle p, x \rangle - g(x)\}, \tag{2}$$

which is a convex function. When g is a closed proper convex function (with all level sets closed), we have $g^{**} = g$. This property is called *biconjugacy*.

A function $f : \mathbb{R}^n \to \mathbb{R} \cup \{+\infty, -\infty\}$ is called a *DC function* if it can be represented as a difference of two convex functions g and h, i.e., $f = g - h$. To guarantee $f > -\infty$, we always assume $\operatorname{dom}_{\mathbb{R}} g \subseteq \operatorname{dom}_{\mathbb{R}} h$, and define $(+\infty) - (+\infty) = +\infty$. A *DC programming problem* is a minimization problem for a DC function. The most important fact in DC programming is the Toland–Singer duality.

Theorem 1 (Toland–Singer duality). *For closed convex functions* $g, h : \mathbb{R}^n \to \mathbb{R} \cup \{+\infty\}$, *we have*

$$\inf_{x \in \mathbb{R}^n} \{g(x) - h(x)\} = \inf_{p \in \mathbb{R}^n} \{h^*(p) - g^*(p)\}. \tag{3}$$

The Toland–Singer duality can be shown by a direct calculation using biconjugacy.

DC algorithm [8], [16] is a practically efficient algorithm for finding a local optimal solution of a DC programming problem. It starts from an initial solution $x^{(0)} \in \operatorname{dom}_{\mathbb{R}} g$, and repeats the following process until convergence. Let $x^{(\nu)}$ be the ν-th solution. DC algorithm approximates the concave part h by its subgradient, $h(x) \approx h(x^{(\nu)}) + \langle p, x - x^{(\nu)} \rangle$, and minimize the convex function $g(x) - h(x^{(\nu)}) - \langle p, x - x^{(\nu)} \rangle$ to find the next solution $x^{(\nu+1)}$. Since $x \in \operatorname{argmin}_{y \in \mathbb{R}^n} \left(g(y) - h(x^{(\nu)}) - \langle p, y - x^{(\nu)} \rangle \right)$ is equivalent to $x \in \partial_{\mathbb{R}} g^*(p)$, the algorithm is simply expressed as in Algorithm 1. When the algorithm terminates, we obtain a pair of vectors (x, p) such that $p \in \partial_{\mathbb{R}} g(x) \cap \partial_{\mathbb{R}} h(x)$. If both g and h are differentiable, this condition is equivalent to $p = \nabla g(x) = \nabla h(x)$, which implies $\nabla f(x) = \nabla g(x) - \nabla h(x) = 0$. Thus the DC algorithm terminates at a stationary point.

It should be emphasized that the theory of DC programming relies on the biconjugacy (for the Toland–Singer duality) and the existence of subgradient (for DC algorithm). See [16,17] for more details of continuous DC programming.

2.2 Discrete DC Programming

Extending DC programming to discrete setting is a natural idea to conceive. But we must specify what we mean by "convex functions" in a discrete space.

Table 1. Complexity of discrete DC programming $\min\{g(x) - h(x)\}$ [9]

(a) $x \in \mathbb{Z}^n$		
$g\backslash h$	$\mathrm{M}^{\natural}$	$\mathrm{L}^{\natural}$
$\mathrm{M}^{\natural}$	NP-hard	NP-hard
$\mathrm{L}^{\natural}$	open	NP-hard

(b) $x \in \{0,1\}^n$		
$g\backslash h$	$\mathrm{M}^{\natural}$	$\mathrm{L}^{\natural}$
$\mathrm{M}^{\natural}$	NP-hard [5]	NP-hard
$\mathrm{L}^{\natural}$	P	NP-hard

Moreover, as mentioned above, such discrete convex functions should satisfy "existence of subgradient" and "biconjugacy." Here, discrete versions of the subgradient and Fenchel conjugate are defined similarly to (1) and (2) with "$\mathbb{R}$" replaced by "$\mathbb{Z}$."

A theory of discrete DC programming has been proposed recently by Maehara and Murota [9] using discrete convex analysis [1], [12,13]. A function $g : \mathbb{Z}^n \to \mathbb{Z} \cup \{+\infty\}$ is called $M^{\natural}$-*convex* if it satisfies a certain exchange axiom. A linear function on a (poly)matroid is a typical example of $\mathrm{M}^{\natural}$-convex functions, and a matroid rank function is an $\mathrm{M}^{\natural}$-concave function. A function $g : \mathbb{Z}^n \to \mathbb{Z} \cup \{+\infty\}$ is called $L^{\natural}$-*convex* if it satisfies the translation submodularity. A submodular set function is a typical example of $\mathrm{L}^{\natural}$-convex functions. $\mathrm{M}^{\natural}$-convex and $\mathrm{L}^{\natural}$-convex functions are endowed with nice properties related to subgradient and biconjugacy. A *discrete DC function* means a function $f : \mathbb{Z}^n \to \mathbb{Z} \cup \{+\infty\}$ that can be represented as a difference of two discrete convex functions g and h, i.e., $f = g - h$. Since there are two classes of discrete convex functions ($\mathrm{M}^{\natural}$-convex functions and $\mathrm{L}^{\natural}$-convex functions), there are four types of discrete DC functions (an $\mathrm{M}^{\natural}$-convex function minus an $\mathrm{M}^{\natural}$-convex function, an $\mathrm{M}^{\natural}$-convex function minus an $\mathrm{L}^{\natural}$-convex function, and so on).

Minimization problems of discrete DC functions are referred to as *discrete DC programming problems*. According to the four classes of discrete DC functions, we have four classes of discrete DC programming problems. The computational complexity of these four classes is summarized in Table 1. It is noted that the NP-hardness of $\mathrm{M}^{\natural}$–$\mathrm{M}^{\natural}$ DC programming has been shown recently [5] through a reduction from the maximum clique problem.

The Toland–Singer duality is extended to the discrete case.

Theorem 2 (Discrete Toland–Singer duality [9]). *For $M^{\natural}$- and/or $L^{\natural}$-convex functions $g, h : \mathbb{Z}^n \to \mathbb{Z} \cup \{+\infty\}$, we have*

$$\inf_{x \in \mathbb{Z}^n} \{g(x) - h(x)\} = \inf_{p \in \mathbb{Z}^n} \{h^*(p) - g^*(p)\}. \tag{4}$$

A discrete version of the DC algorithm can also be defined similarly with $\mathbb{R}$ in Algorithm 1 replaced by "$\mathbb{Z}$." Each step of the algorithm, $p \in \partial_{\mathbb{Z}} h(x)$ and $x \in \partial_{\mathbb{Z}} g^*(p)$, can be executed efficiently by using the existing algorithms in discrete convex analysis. Moreover, by exploiting polyhedral properties of $\mathrm{M}^{\natural}$-convex and $\mathrm{L}^{\natural}$-convex functions, we can guarantee a stronger local optimality condition. See [9] for more details of discrete DC programming.

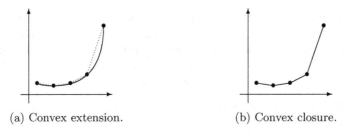

(a) Convex extension.   (b) Convex closure.

Fig. 1. Convex extension and convex closure

3 Continuous Relaxation for Discrete DC Programming

A framework of DC programming for the difference of two convex extensible discrete functions is presented in this section.

3.1 Lin-vex Extension of Discrete DC Function

Let $g : \mathbb{Z}^n \to \mathbb{R} \cup \{+\infty\}$ be a real-valued function in discrete variables. A *convex extension* $\hat{g} : \mathbb{R}^n \to \mathbb{R} \cup \{+\infty\}$ of g is a convex function (in continuous variables) that satisfies

$$\hat{g}(x) = g(x) \quad (x \in \mathbb{Z}^n). \tag{5}$$

Note that there are (possibly) many convex extensions for a discrete function. The *convex closure* $\overline{g} : \mathbb{R}^n \to \mathbb{R} \cup \{+\infty\}$ of g is the point-wise maximum of all affine functions that are global underestimators of g, i.e.,

$$\overline{g}(x) := \sup\{\ell(x) : \ell \text{ affine}, \ell(y) \le g(y) \ (y \in \mathbb{Z}^n)\}. \tag{6}$$

Under mild assumptions (e.g., if the effective domain is bounded), the convex closure of a function on $\mathbb{Z}^n$ is a piecewise linear function. In this paper, we say that g is *convex extensible* if $g(x) = \overline{g}(x)$ for $x \in \mathbb{Z}^n$, although it really means that g is extensible to a closed convex function. Fig. 1 illustrates the difference between the convex extension and the convex closure.

We consider a discrete DC programming problem that is represented in terms of two convex extensible functions g and h: $\text{minimize}_{x \in \mathbb{Z}^n} f(x) = g(x) - h(x)$. A natural approach to this problem is continuous relaxation that extends the objective function to the continuous domain and solves the continuous optimization problem by some existing method in continuous optimization.

A special form of continuous relaxation plays a crucial role. Let $\overline{g}$ be the convex closure of g and $\hat{h}$ be any convex extension of h. A *lin-vex extension* of $f = g - h$ is defined as a function $\tilde{f}$ given by

$$\tilde{f}(x) = \overline{g}(x) - \hat{h}(x), \tag{7}$$

where "lin-vex" is intended to mean "piecewise linear for g and general convex for h." By definition, we have $f(x) = \tilde{f}(x)$ for all $x \in \mathbb{Z}^n$; therefore $\inf_{x \in \mathbb{Z}^n} f(x) \ge$

$\inf_{x \in \mathbb{R}^n} \tilde{f}(x)$. In discrete (or integer) optimization, in general, the optimal values of the original problem and that of the continuous relaxation are different, and the discrepancy between these optimal values are referred to as the *integrality gap*. Fortunately, however, our continuous relaxation based on lin-vex extension does not suffer from integrality gap. This is our main result.

Theorem 3. *For convex extensible functions* $g, h : \mathbb{Z}^n \to \mathbb{R} \cup \{+\infty\}$ *with* $\mathrm{dom}_{\mathbb{Z}}\, g$ *bounded and* $\mathrm{dom}_{\mathbb{Z}}\, g \subseteq \mathrm{dom}_{\mathbb{Z}}\, h$, *we have*

$$\inf_{x \in \mathbb{Z}^n} \{g(x) - h(x)\} = \inf_{x \in \mathbb{R}^n} \{\overline{g}(x) - \hat{h}(x)\}. \tag{8}$$

Proof. Let $x^* \in \mathbb{R}^n$ be a minimizer of $\overline{g} - \hat{h}$. Since $\overline{g}$ is a piecewise linear function, we can take a convex polyhedron R such that $\overline{g}$ is linear on R and $x^* \in R$. Since $\overline{g}$ is linear on R, $\overline{g} - \hat{h}$ is concave on R; therefore its minimum is attained at an extreme point of R, which is integral.   ∎

The lin-vex extension of f has two kinds of freedoms. First, it depends on the DC representation $f = g - h$. For an arbitrary convex extensible function $k : \mathbb{Z}^n \to \mathbb{R}$, we can obtain another DC representation $f = (g + k) - (h + k)$, and the corresponding lin-vex extension may change. Second, it depends on the choice of convex extension $\hat{h}$ of h. The convex closure $\overline{h}$ is eligible for $\hat{h}$, but in some cases, there can be a more suitable choice for $\hat{h}$. For example, if h is defined by the restriction of a continuous (smooth) convex function $\varphi : \mathbb{R}^n \to \mathbb{R} \cup \{+\infty\}$, i.e., $h(x) = \varphi(x)$ for $x \in \mathbb{Z}^n$, then φ is a reasonable candidate $\hat{h}$. We intend to make use of these freedoms to design an efficient algorithm.

Remark 1. Theorem 3 does not hold for a continuous extension of the form $\hat{g} - \hat{h}$. For example, let us consider $\hat{g}(x) = (x - 1/2)^2$ and $\hat{h}(x) = 0$ for $x \in \mathbb{R}$, and $g(x) = \hat{g}(x)$ and $h(x) = \hat{h}(x)$ for $x \in \mathbb{Z}$. Then we have $\inf_{x \in \mathbb{Z}} \{g(x) - h(x)\} = 1/4 \neq 0 = \inf_{x \in \mathbb{R}} \{\hat{g}(x) - \hat{h}(x)\}$.

Remark 2. In convex analysis, the *Legendre–Fenchel duality*

$$\inf_{x \in \mathbb{R}^n} \{g(x) + h(x)\} = - \inf_{p \in \mathbb{R}^n} \{g^*(p) + h^*(-p)\} \tag{9}$$

is frequently used and a discrete version of (9) is also known in discrete convex analysis. It should be clear that the Toland–Singer duality (3) deals with the infimum of $g - h$, but the Legendre–Fenchel duality (9) deals with the infimum of $g + h$. For the Legendre–Fenchel duality, there is an integrality gap, i.e.,

$$\inf_{x \in \mathbb{Z}^n} \{g(x) + h(x)\} \neq \inf_{x \in \mathbb{R}^n} \{\overline{g}(x) + \hat{h}(x)\}, \tag{10}$$

in general. For example, let $g(x_1, x_2) = |x_1 + x_2 - 1|$ and $h(x_1, x_2) = |x_1 - x_2|$ for $(x_1, x_2) \in \mathbb{Z}^2$, and $\overline{g}(x_1, x_2) = |x_1 + x_2 - 1|$ and $\hat{h}(x_1, x_2) = |x_1 - x_2|$ for $(x_1, x_2) \in \mathbb{R}^2$. Then, the left-hand side of (10) is $\inf\{g(x_1, x_2) + h(x_1, x_2)\} = 1$ and the right-hand side is $\inf\{\overline{g}(x_1, x_2) + \hat{h}(x_1, x_2)\} = 0$ with $(x_1, x_2) = (1/2, 1/2)$.   ∎

3.2 DC Algorithm for lin-vex Extension

By Theorem 3, solving a convex extensible DC problem (left-hand side of (8)) is equivalent to solving its lin-vex relaxation problem (right-hand side of (8)), which is a continuous DC programming problem. Here, we consider an implementation of the DC algorithm for this continuous DC programming problem.

As shown in Algorithm 1, the DC algorithm for $\inf_{x \in \mathbb{R}^n} \{\overline{g}(x) - \hat{h}(x)\}$ repeats the dual step $p \in \partial_{\mathbb{R}} \hat{h}(x)$ and the primal step $x \in \partial_{\mathbb{R}} \overline{g}^*(p)$. We consider situations where these two steps can be done efficiently. For the dual step, we assume that a subgradient p of $\hat{h}$ at x can be computed efficiently. For example, if $h : \mathbb{Z}^n \to \mathbb{R}$ is given by $h(x) = x^\top A x$ $(x \in \mathbb{Z}^n)$ for some positive-definite A, we can take the convex extension $\hat{h}(x) = x^\top A x$ $(x \in \mathbb{R}^n)$, whose subgradient is explicitly obtained as $\partial_{\mathbb{R}} \hat{h}(x) = \{2Ax\}$. For the primal step, recall that $x \in \partial_{\mathbb{R}} \overline{g}^*(p)$ is equivalent to

$$x \in \operatorname*{argmin}_{y \in \mathbb{R}^n} \{\overline{g}(y) - \langle p, y \rangle\}. \tag{11}$$

Since $\overline{g}$ is a piecewise linear function and its linearity domain is an integral polytope, the problem (11) has an integral optimal solution; therefore it is essentially equivalent to the discrete optimization problem

$$x \in \operatorname*{argmin}_{y \in \mathbb{Z}^n} \{g(y) - \langle p, y \rangle\}. \tag{12}$$

We assume that the minimization problem (12) can be solved efficiently. This is the case, for example, if g is an $M^\natural$-convex or $L^\natural$-convex function, or if g corresponds to some efficiently-solvable discrete optimization problem, such as matching on a graph, maximum independent set on a tree, knapsack problem, etc.

4 Application to Degree-Concentrated Minimum Spanning Tree Problem

We consider a problem in network routing. Let $G = (V, E)$ be an undirected graph that represents a computer network. Here, V denotes a set of computers and E represents the connection of the computers, i.e., $(i, j) \in E$ if computer i communicates with computer j. The *spanning tree routing* [15] is a routing system determined by a spanning tree T such that all packets are sent along the spanning tree.

Here, we consider monitoring of network communications. If a routing system, which is specified by a spanning tree, has large-degree vertices, we can obtain much information by observing packets on these vertices. Thus, for efficient monitoring, we want to construct a spanning tree with some high-degree vertices. To construct such spanning tree, we solve the following problem, to be named *degree-concentrated spanning tree problem*:

$$\max_{T:\text{spanning tree}} \sum_{v \in V} \deg_T(v)^2, \tag{13}$$

where $\deg_T(v)$ is the degree of a vertex v in the tree T. This problem is NP-hard because, for a cubic graph, it can be reduced from the maximum leaf spanning tree problem, which is known to be NP-hard in a cubic graph [7].

The above problem can be formulated in a DC programming problem as follows. Let $B \in \mathbb{R}^{|V| \times |E|}$ be the incidence matrix of graph G, i.e., $B_{ie} = 1$ if an edge e is incident to a vertex i. Then, for $x \in \{0,1\}^{|E|}$ and $T = \{e \in E : x_e = 1\}$, we have

$$Bx = (\deg_T(v_1), \ldots, \deg_T(v_{|V|}))^\top. \tag{14}$$

Therefore, for $A = B^\top B \in \mathbb{R}^{|E| \times |E|}$, we have

$$x^\top A x = \sum_{v \in V} \deg_T(v)^2. \tag{15}$$

We define the concave part by $h(x) = x^\top A x$ and the convex part by

$$g(x) = \begin{cases} 0, & \{e \in E : x_e = 1\} \text{ is a spanning tree,} \\ +\infty, & \text{otherwise.} \end{cases} \tag{16}$$

Then we have

$$g(x) - h(x) = \begin{cases} -\sum_{v \in V} \deg_T(v)^2, & T = \{e \in E : x_e = 1\} \text{ is a spanning tree,} \\ +\infty, & \text{otherwise.} \end{cases} \tag{17}$$

Thus the minimization problem for $f(x) = g(x) - h(x)$ coincides with the degree-concentrated spanning tree problem.

In this representation, both g and h are convex extensible. The gradient of $\hat{h}(x) = x^\top A x$ is explicitly obtained as $\partial_{\mathbb{R}} \hat{h}(x) = \{p\} = \{2Ax\}$; here $(Ax)_{(u,v)} = \deg_T(u) + \deg_T(v)$. The minimization of $\overline{g}(x) - \langle p, x \rangle$ is performed by solving a maximum spanning tree problem with edge weight p_e for $e \in E$; see Figure 2 for the illustration of the algorithm. If there are two or more optimal solutions in this minimization problem, we randomly choose one of them. Thus, the local optimal solution for the degree-concentrated spanning tree problem can be found efficiently by the DC algorithm. The complexity of the algorithm is $O(|E| \log |E|)$ for each iteration.

To evaluate the performance of the above DC algorithm and the quality of solutions, we conduct the following experiment. We use a real-world network, p2p-Gnutella08, obtained from Stanford Large Network Dataset Collection,[1] representing a network of a peer-to-peer communication network. For comparison, we also implemented the greedy algorithm that iteratively selects an edge e^* randomly from $\mathrm{argmin}_{e \notin T} f(T \cup \{e\})$ and updates the solution T to $T \cup \{e^*\}$.

Figure 3a shows the objective values in the first 10-iterations of the DC algorithm. Here, each plot corresponds to a single run of the algorithm, and

---

[1] http://snap.stanford.edu/data/ For other real-world networks in this collection, we performed the same experiments, to obtain similar results.

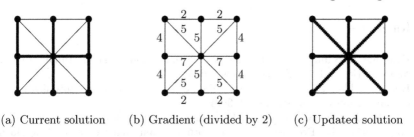

(a) Current solution   (b) Gradient (divided by 2)   (c) Updated solution

Fig. 2. Illustration of the DC algorithm for the degree-concentrated spanning tree problem. For a current solution shown in (a), the gradient p is given as (b), and the updated solution (c) is a maximum spanning tree with respect to p.

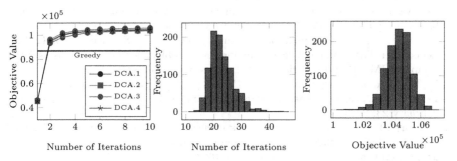

(a) Objective values vs iterations

(b) Distribution of the number of iterations

(c) Distribution of the obtained solutions

Fig. 3. DC algorithm for the degree-concentrated spanning tree problem

the horizontal line shows the best greedy solution from among 1000 runs. In the DC algorithm, a solution quickly approaches the optimal solution in the first few iterations. The greedy solution is outperformed in the second iteration. Since the converged solutions are different in different runs, the algorithm only finds a local optimal solution. We adopted the rule that the algorithm should terminate if $f(x^{(\nu+1)}) = f(x^{(\nu)})$. The number of iterations for convergence and the obtained objective values are shown in Figures 3b and 3c, respectively. Since the number of iterations is small, the DC algorithm is efficient and scales to large instances. Moreover, the quality of the obtained solutions is not much diverged. Thus the DC algorithm works pretty well for the degree-concentrated spanning tree problem.

Acknowledgement. This work is supported by JSPS/MEXT KAKENHI Grant Number 26280004, and JST, CREST.

References

1. Fujishige, S.: Submodular Functions and Optimization, 2nd edn. Annals of Discrete Mathematics, vol. 58. Elsevier, Amsterdam (2005)
2. Horst, R., Thoai, N.V.: DC Programming: Overview. Journal of Optimization Theory and Applications 103, 1–43 (1999)
3. Iyer, R., Jegelka, S., Bilmes, J.: Fast semidifferential-based submodular function optimization. In: Proceedings of the 30th International Conference on Machine Learning, pp. 855–863 (2013)
4. Kawahara, Y., Washio, T.: Prismatic algorithm for discrete D.C. programming problem. In: Proceedings of the 25th Annual Conference on Neural Information Processing Systems, pp. 2106–2114 (2011)
5. Kobayashi, Y.: The complexity of maximizing the difference of two matroid rank functions. METR2014-42, University of Tokyo (2014)
6. Korte, B., Vygen, J.: Combinatorial Optimization, 5th edn. Springer, Berlin (2012)
7. Lemke, P.: The maximum leaf spanning tree problem for cubic graphs is NP-complete. IMA Preprint Series #428, University of Minnesota (1988)
8. Le-Thi, H.A.: DC Programming and DCA, http://www.lita.univ-lorraine.fr/~lethi/index.php/en/research/dc-programming-and-dca.html (retrieved at February 23, 2015)
9. Maehara, T., Murota, K.: A framework of discrete DC programming by discrete convex analysis. Mathematical Programming (2014), http://link.springer.com/article/10.1007
10. Moriguchi, S., Shioura, A., Tsuchimura, N.: M-convex function minimization by continuous relaxation approach—Proximity theorem and algorithm. SIAM Journal on Optimization 21, 633–668 (2011)
11. Moriguchi, S., Tsuchimura, N.: Discrete L-convex function minimization based on continuous relaxation. Pacific Journal of Optimization 5, 227–236 (2009)
12. Murota, K.: Discrete Convex Analysis. Society for Industrial and Applied Mathematics, Philadelphia (2003)
13. Murota, K.: Recent developments in discrete convex analysis. In: Cook, W., Lovász, L., Vygen, J. (eds.) Research Trends in Combinatorial Optimization, ch. 11, pp. 219–260. Springer, Berlin (2009)
14. Narasimhan, M., Bilmes, J.: A submodular-supermodular procedure with applications to discriminative structure learning. In: Proceedings of the 21st Conference on Uncertainty in Artificial Intelligence, pp. 404–412 (2005)
15. Tanenbaum, A.S.: Computer Networks, 5th edn. Prentice Hall, Upper Saddle River (2010)
16. Pham Dinh, T., Le Thi, H.A.: Convex analysis approach to D.C. programming: Theory, algorithms and applications. Acta Mathematica Vietnamica 22, 289–355 (1997)
17. Hoang, T.: D.C. optimization: Theory, methods and algorithms. In: Horst, R., Pardalos, P.M. (eds.) Handbook of Global Optimization, pp. 149–216. Kluwer Academic Publishers, Dordrecht (1995)
18. Yuille, A.L., Rangarajan, A.: The concave-convex procedure. Neural Computation 15, 915–936 (2003)

Solving Relaxation Orienteering Problem Using DCA-CUT

Anh Son Ta[1], Hoai An Le Thi[2], and Trong Sy Ha[1]

[1] SAMI, Hanoi University of Sicence and Technology, Vietnam
[2] LITA, Lorraine University, Site Metz, France

Abstract. Orienteering problem is well-known as a NP-hard problem in transportation with many applications. This problem aims to find a path between a given set of control points, where the source and destination points are specified with respect to maximize the total score of collected points and satisfy the distance constraint. In this paper, we first analyze the structure of a generalized orienting problem and a new solution method, based on DC programming, DCA and Cutting plane method, is introduced. Preliminary numerical experiments are reported to show the efficiency of the proposed algorithm.

Keywords: Orienteering problem, DC programming, DC Algorithm, Binary Linear Integer Program.

1 Introduction

Given a set of control points with associated scores along with the start and end points, the orienteering problem (OP) deals with finding a path between the start and end points, visiting each points at most one in order to maximize the total score subject to a given distance budget, denoted by D_{max}. Due to the fact that distance is limited, tours may not include all points. It should be noted that the OP is equivalent to the Traveling Salesman Problem (TSP) when the time is relaxed just enough to cover all points and the start and end points are not specified.

The selective traveling salesperson problem ([1]), the maximum collection problem ([2]), and the bank robber problem ([3]) are introduced in the form of the OP.

The OP also has applications in vehicle routing and production scheduling. Golden et al. (1984) ([4]) discussed certain applications of the OP to customer/vehicle assignment and inventory/routing problems. Golden et al. (1987) ([5]) also applied the OP to a vehicle routing problem in which oil tankers are routed to service stations in different locations. The total score of the route is maximized while minimizing the route distance without violating the D_{max} constraint. Balas (1989) ([9]) modeled certain types of production scheduling problems as the OP. These problems are concerned with product-mix planning of production to maximize the total profit without violating the production time constraints. Keller (1989) ([6]) modified his multi-objective vending problem as

© Springer International Publishing Switzerland 2015
H.A. Le Thi et al. (eds.), *Model. Comput. & Optim. in Inf. Syst. & Manage. Sci.,*
Advances in Intelligent Systems and Computing 359, DOI: 10.1007/978-3-319-18161-5_17

the OP in which there is a trade off between maximizing reward potential and minimizing the travel cost. Kantor and Rosenwein (1992) ([7]) presented the OP with time windows in which a point can only be visited within a specified time interval. This approach seems to be promising for potential applications such as bank and postal delivery, industrial refuse collection, dial-a-ride services and school bus routing. Golden et al. (1987) ([5]) have shown that the OP is NP-hard.

In recent years, in the survey ([8]), Vansteenwegen et al. (2011) summarizes many practical applications of orienteering problems and many exact and heuristic solution approaches were proposed.

In this paper, we consider a relaxation of orienteering problem. It is the OP without subtour elimination constraints. This is also a hard problem and our proposed method, based on DC programming, DCA and cutting method, can be applied to general OP in a similar way.

The paper is organized as follows. Section 2 introduces the problem statement and mathematical model. The solution method based on DC programming and DCA, DCA combine with cutting plane method and how to choosing initial point for DCA are presented in Section 3 while the numerical simulation is reported in Section 4. Finally, the conclusion is presented in Section 5.

2   Mathematical Model

To model the orienteering problem, we denote V as the set of control points and, E as the set of edges between points in V. Then, the complete graph can be defined by $G = (V, E)$. Each control point, i, in V has an associated score $s_i \geq 0$ whereas the start point 1 and the end point n have no scores. The distance d_{ij}, between the points i and j is the nonnegative cost for each edge in E or the cost of traveling between points i and j. So, the objective is to find a path P from the start point 1 to the end point n through a subset of control points, visiting each points at most one such that the total score collected from the visited points will be maximized without violating the given distance constraint.

Assumption that t_{ij} is the travel time from node i to node j and a_i is the arrival time at node i ($a_i = 0$ if the optimal path do not visit node i)

We define binary variable as follows,

$$x_{ij} = \begin{cases} 1 \text{ if } (i, j) \in P, (\text{ direction } i \to j) \\ 0 \text{ otherwise.} \end{cases}$$

The mathematical model of the OP is given as follows

$$\max \sum_{i=1}^{n} \sum_{j=1}^{n} s_i x_{ij} \tag{1}$$

subject to

$$\sum_{i=2}^{n} x_{1j} = 1, \tag{2}$$

$$\sum_{i=1}^{n-1} x_{in} = 1, \tag{3}$$

$$\sum_{i=2}^{n-1} x_{ik} - \sum_{j=2}^{n-1} x_{kj} = 0, \quad k = 2, \cdots, n-1, \tag{4}$$

$$\sum_{i=2}^{n-1} x_{ik} \leq 1, \quad k = 2, \cdots, n-1 \tag{5}$$

$$\sum_{i=1}^{n} \sum_{j=1}^{n} d_{ij} x_{ij} \leq D_{max} \tag{6}$$

$$x_{ij} \in \{0,1\}\ i, j = 1, \cdots, n. \tag{7}$$

We defined a directed graph $G^* = (V, E^*)$ as follows, the vertex set V is the same with graph G, the arc set $E^* = \{(1, i) : i = 2, \cdots, n$, and $(j, n) : j = 1, \cdots, n-1$, and $(i, j), (j, i) : \forall i, j = 2, \cdots, n-1, i \neq j\}$ and set $|E^*| = m$.

Let us set $A_1 \in \mathbb{R}^{n \times m}$ be the vertices-arcs incidence matrix of G^*,

$$a_{i,(u,v)} = \begin{cases} 1 \text{ if } i \equiv u \text{ and } (i, v) \in E^*, \\ -1 \text{ if } i \equiv v \text{ and } (u, i) \in E^*, \\ 0 \text{ otherwise.} \end{cases} \quad \forall i = 1, \cdots, n, \ (u, v) \in E^* \tag{8}$$

We define vector $b_1 = (1, 0, \cdots, 0, -1) \in \mathbb{R}^n$ and variable vector $x \in \mathbb{R}^m$ such that the constraints (2),(3) and (4) can be presented as $A_1 x = b_1$.

Let us define A_2 be a $(n-2) \times m$ matrix and vector $b_2 = (1, 1, \cdots, 1) \in \mathbb{R}^{n-2}$ such that the constraints (5) is presented by $A_2 x \leq b_2$.

Define vector $d = (d_{ij}) \in \mathbb{R}^m$ such that the constraints (6) is rewritten as $\langle d, x \rangle \leq D_{max}$.

The problem (1)-(7) is a Binary Integer Linear Program (BILP), DC programming and DC Algorithm for solving this problem is presented as follows.

3   Solving Orienteering Problem by DCA CUT

3.1   Solving Orienteering Problem by DCA

By using an exact penalty result, we can reformulate the problem (1)-(7) in the form of a concave minimization program. The exact penalty technique aims at transforming the original problem into a more tractable equivalent DC program. Let $K := \{x \in \mathbb{R}^m : A_1 x = b_1,\ A_2 x \leq b_2,\ \langle d, x \rangle \leq D_{max},\ x \in [0, 1]^m\}$. The feasible set of the original problem is then $S = \{x : x \in K,\ x \in \{0, 1\}^m\}$. The original program is rewritten as problem (P),

$$\min\{-\langle s, x \rangle : x \in S\}. \quad (P) \tag{9}$$

Let us consider the function $p : \mathbb{R}^n \to \mathbb{R}$ defined by: $p(x) = \sum_{i=1}^{m} \min\{x_i, 1-x_i\}$. It is clear that $p(x)$ is concave and finite on K, $p(x) \geq 0$ $\forall x \in K$ and that:

$$\{x : x \in S\} = \{x : x \in K, p(x) \leq 0\}. \tag{10}$$

Hence the problem OR can be rewritten as:

$$\min\left\{\langle s, x \rangle : x \in K,\ p(x) \leq 0\right\}. \tag{11}$$

The following theorem can then be formulated.

Theorem 1. *Let K be a nonempty bounded polyhedral convex set, f be a finite concave function on K and p be a finite nonnegative concave function on K. Then there exists $\eta_0 \geq 0$ such that for $\eta > \eta_0$ the following problems have the same optimal value and the same solution set:*

$$(P_\eta) \qquad \alpha(\eta) = \min\{f(y) + \eta p(y) : y \in K\},$$
$$(P) \qquad \alpha = \min\{f(y) : y \in K, p(y) \leq 0\}.$$

Furthermore

- *If the vertex set of K, denoted by $V(K)$, is contained in $x \in K : p(y) \leq 0$, then $\eta_0 = 0$.*
- *If $p(y) > 0$ for some y in $V(K)$, then $\eta_0 = \min\left\{\frac{f(y)-\alpha(0)}{S_0} : y \in K, p(y) \leq 0\right\}$, where $S_0 = \min\left\{p(y) : y \in V(K), p(y) > 0\right\} > 0$.*

Proof. The proof for the general case can be found in [10].

From Theorem 1 we get, for a sufficiently large number η ($\eta > \eta_0$), the equivalent concave minimization problem:

$$\min\{f_\eta(x) := \langle s, x \rangle + \eta p(x)\ : x \in K\},$$

which is a DC program of the form:

$$\min\{g(x) - h(x) : x \in \mathbb{R}^m\}, \tag{12}$$

where: $g(x) = \chi_K(x)$ and $h(x) = -f_\eta(x) = -\langle s, x \rangle - \eta p(x)$.

We have successfully transformed an optimization problem with integer variables into its equivalent form with continuous variables. Notice that (12) is a polyhedral DC program where g is a polyhedral convex function (i.e., the pointwise supremum of a finite collection of affine functions). DCA applied to the DC program (12) consists of computing, at each iteration k, the two sequences $\{x^k\}$ and $\{y^k\}$ such that $y^k \in \partial h(x^k)$ and x^{k+1} solves the next linear program of the form (P_k).

$$\min\left\{g(x) - \langle x - x^k, y^k \rangle : x \in \mathbb{R}^m\right\} \Leftrightarrow \min\{-\langle x, y^k \rangle : x \in K\}. \tag{13}$$

From the definition of h, a sub-gradient $y^k \in \partial h(x^k)$ can be computed as follows:

$$y^k = \begin{cases} -s_i - \eta & \text{if } x_i \geq 1/2 \\ -s_i + \eta & \text{if } x_i < 1/2 \end{cases} \tag{14}$$

The DCA scheme applied to (12) can be summarized as follows:

Algorithm 1.

Initialization:
   Choose a initial point x^0, set $k = 0$;
   Let ϵ_1, ϵ_2 be sufficiently small positive numbers;
Repeat
   Compute y^k via (14);
   Solve the linear program (13) to obtain z^{k+1};
   $k \leftarrow k + 1$;
Until either $\|x^{k+1} - x^k\| \leq \epsilon_1(\|x^k\| + 1)$ or $|f_\eta(x^{k+1}) - f_\eta(x^k)| \leq \epsilon_2(|f_\eta(x^k)| + 1)$.

Theorem 2. *(Convergence properties of Algorithm DCA)*

– DCA generates the sequence $\{x^k\}$ contained in $V(K)$ such that the sequence $\{f_\eta(x^k)\}$ is decreasing.
– The sequence $\{x^k\}$ converges to $x^* \in V(K)$ after a finite number of iterations.
– The point x^* is a critical point of Problem (12). Moreover if $x_i^* \neq \frac{1}{2}$ for all $i \in \{0, \ldots, m\}$, then x^* is a local solution to (12).
– For a number η sufficiently large, if at iteration r we have $x^r \in \{0, 1\}^m$, then $x^k \in \{0, 1\}^m$ for all $k \geq r$.

Proof. Immediate consequences of the DCA applied to concave quadratic zero-one programming whose proof can be found in [10].

3.2   Initial Point for DCA

Totally unimodular matrices
A matrix A is called *totally unimodular matrix* if each subdeterminant of A is 0, +1 or -1. So, each element of A is 0, +1 or -1. The relation of integer linear program and totally unimodular matrix is presented as the following theorem:

Theorem 3. *Let A be a totally unimodular matrix and let b be an integral vector. Then the polyhedron $P := \{x : Ax \leq b\}$ is integral.*

Proof. See Theorem 19.1 in [11].

The fundamental characteristic of totally unimodular matrix is introduced as follows.

Theorem 4. *Let A be the matrix with entries 0, +1 or -1. The following are equivalent:*

(i) *A is totally unimodular, i.e;, each square submatrix of A has determinant 0, +1 or -1;*
(ii) *for each integral vector b the polyhedron $\{x : x \geq 0, Ax \leq b\}$ has only integral vertices;*
(iii) *for all integral vector a, b, c, d the polyhedron $\{x : c \leq x \leq d, a \leq Ax \leq b\}$ has only integral vertices;*

(iv) *each collection of columns of A can be split into two parts so that the sum of the columns in one part minus the sum of the column in the other part is a vector with entries only 0, +1 or -1;*

(v) *each nonsingular submatrix of A has a row with an odd number of non-zero components;*

(vi) *the sum of the entries in any square submatrix with even row and column sums is divisible by four;*

(vii) *no square of A has determinant +2 or -2.*

Proof. See Theorem 19.3 in [11].

As a result in [11], the vertices-arcs incidence matrix A_1 (8) is totally unimodular. Moreover, we will prove that matrix $A_0 = \begin{bmatrix} A_1 \\ A_2 \end{bmatrix}$ is totally unimodular.

Theorem 5. *Let matrix* $A_0 = \begin{bmatrix} A_1 \\ A_2 \end{bmatrix}$ *be totally unimodular where* A_1, A_2 *defined as in previous section.*

Proof. As the definition of matrix A_1 and A_2, the matrix A_1 has n row corresponding with vertices form 1 to n. Each column of A_1 there are exact two non-zero elements, one +1 and one -1. In row 1, all the no-zero elements equal to +1. In row n, all the non-zero elements equal to -1. Matrix A_2 is presented for the constraints (5), then A_2 is $(n-2) \times m$-matrix presents the flow condition at $n-2$ vertices $2, 3, \cdots, n-1$, a row of A_2 (for instance, row i-associated with vertex $i+1$) is created by copied row $i+1$ at matrix A_1 (associated with vertex $i+1$) and then replaced all -1 elements in this rows by 0.

We consider now matrix $A_0 = \begin{bmatrix} A_1 \\ A_2 \end{bmatrix}$, obviously each column of A_0 there are exact two or three non-zero elements. If there are exact two non-zero elements then one +1 and one -1. If there are exact three non-zero elements, then two elements +1, and one -1. Let us set R be a collection of rows of A_0, we do partition R into two part R_1 and R_2 by the following rule:

1. all the rows in R which are selected from A_1 are belong to R_1;
2. for one row i (associated with vertex j) in R which selected from A_2, we consider that if the row associated with vertex j is in R_1 then row i belong to R_2, if not row i belong to R_1.

By doing this way, we have a partition R_1, R_2 of R, which satisfies sum of all the rows in R_1 minus sum of all the rows in R_2 is a vector with entries only 0, +1 or -1. In the other hand, matrix A is totally unimodular if and only if its transpose matrix A^T is totally unimodular. Therefore, from property iv) in Theorem 4, the role of column is equivalent with the role of row, then this implies A_0 is totally unimodular.

Initial point for DCA

We consider a subproblem (P1) as follows,

$$D = \min\{\langle d, x \rangle : A_0 x \leq b_0, \ x \in [0, 1]^m\}, \qquad (P1) \qquad (15)$$

where $b_0 = \begin{bmatrix} b_1 \\ b_2 \end{bmatrix}$.

By Theorem 4 and Theorem 5, the solution x_{P1} of linear problem (P1) is integral (0-1) solution. Thus, if $D > D_{max}$ then the feasible set of the original problem is empty set, *we STOP the algorithm*. If $D \leq D_{max}$, so the solution x_{P1} of problem (P1) is a feasible point of the original problem and we use x_{P1} as an initial point of DCA. Thanks to a nice property of DCA in Theorem 2, we see that when we use x_{P1} as an initial point of DCA, *the result of DCA is always a local optimal solution*, and recall that this problem is a polyhedral DC then DCA is finite convergence.

3.3   DCA-Cut for Global Solution

Let us consider the following problem:

$$\min\{\chi_K(z) + c^T x + d^T y + tp(x) \ : \ z = (x,y) \in \mathbb{R}^n \times \mathbb{R}^p\}, \tag{16}$$

where

$$\chi_K(z) := \begin{cases} 0 & \text{si } z \in K, \\ +\infty & \text{otherwise} \end{cases} \tag{17}$$

is the indicator function on K.

We set $g(z) := \chi_K(z)$ and

$$h(z) := -c^T x - d^T y + t(-p)(z) = -c^T x - d^T y + t \sum_{j=1}^{n} \max\{-x_j, x_j - 1\}. \tag{18}$$

Thus, the problem is equivalent with a DC program:

$$\min\{g(z) - h(z) \ : \ z = (x,y) \in \mathbb{R}^n \times \mathbb{R}^p\}. \tag{19}$$

We define a valid inequality for all point of S from a solution of penalty function p on K. Let $z^* \in K$, we define:

$$I_0(z^*) = \{j \in \{1,\ldots,n\} : x_j^* \leq 1/2 \} \quad , \quad I_1(z^*) = \{1,\ldots,n\} \setminus I_0(z^*).$$

and

$$l_{z^*}(z) \equiv l_{z^*}(x) = \sum_{i \in I_0(z^*)} x_i + \sum_{i \in I_1(z^*)} (1 - x_i).$$

Lemma 1. *(see [12]) Let $z^* \in K$, we have*

(i) $l_{z^*}(x) \geq p(x) \ \forall x \in \mathbb{R}^n$.
(ii) $l_{z^*}(x) = p(x)$ *if and only if*

$$(x,y) \in R(z^*) := \{(x,y) \in K \ : \ x_i \leq 1/2, i \in I_0(x^*); \ x_i \geq 1/2, i \in I_1(x^*) \}.$$

Lemma 2. *(see [12]) Let $z^* = (x^*, y^*)$ be a local minimum of function p on K, then the inequality*

$$l_{z^*}(x) \geq l_{z^*}(x^*) \tag{20}$$

is valid for all $(x, y) \in K$.

Theorem 6. *(see [13], [12]) There exists a finite number $t_1 \geq 0$ such that, for all $t > t_1$, if $z^* = (x^*, y^*) \in V(K) \setminus S$ is a local minimum of problem (19) then*

$$l_{z^*}(x) \geq l_{z^*}(x^*), \ \forall(x, y) \in K. \tag{21}$$

Construction of a cut from an infeasible solution

Let z^* be a solution that is not a feasible point of S such that $l_{z^*}(z) \geq l_{z^*}(z^*) \ \forall z \in K$. In this case, there exists at least one index $j_0 \in \{1, \ldots, n\}$ such that $x_{j_0}^*$ is non binary. We consider two following cases:

Case 1: The value of $l_{z^*}(z^*)$ is not integer.
As $l_{z^*}(z)$ is integer for all $z \in S$, we have immediately:

$$\begin{cases} l_{z^*}(z) \geq \rho := \lfloor l_{z^*}(z^*) \rfloor + 1, \forall z \in S \\ l_{z^*}(z^*) \leq \rho. \end{cases} \tag{22}$$

In other words, the inequality

$$l_{z^*}(z) \geq \rho \tag{23}$$

is a strictly separate cut z^* of S

Case 2: the value of $l_{z^*}(z^*)$ is integer. It is possible that there are feasible points z' such that $l_{z^*}(z') = l_{z^*}(z^*)$. If such a point exists, we could update the best solution (PLM01) and also improve the upper bound of the optimal value.

Otherwise, for all $z \in S$, we have $l_{z^*}(z) > l_{z^*}(z^*)$. That is to say,

$$l_{z^*}(z) \geq l_{z^*}(z^*) + 1 \tag{24}$$

is a separate cut z^* of S.

We consider below a procedure (called *Procedure P*) is providing a cut, or a feasible point, or a potential point.

Let us set $I_F(z^*) := \{i \in I : x_i^* \notin \{0, 1\}\}$.
Let us set $I_{\overline{F}}(z^*) := \{i \in I : x_i^* \in \{0, 1\}\}$, then $I = I_{\overline{F}}(z^*) \cup I_F(z^*)$ and $I_{\overline{F}}(z^*) \cap I_F(z^*) = \emptyset$.

We observe that if we can find z^1 such that $l_{x^*}(x^1) = l_{x^*}(x^*)$ and there exists $i_1 \in I_{\overline{F}}$ and $x_{i_1}^1 = 1 - x^*_{i_1}$ (there are only two possibilities $x_{i_1}^1 = 1 - x^*_{i_1}$ or $x_{i_1}^1 = x^*_{i_1}$) then $z^1 \notin R(z^*)$.

By Lemma 1 we have:

$$p(x^\star) = l_{x^\star}(x^\star) = l_{x^\star}(x^1) > p(x^1).$$

Step 1. Let us set $K_1 = \{z = (x, y) \in K; \ x_i = x^\star_i \ \forall i \in I_{\overline{F}}\}$.

Step 2. Choose $i_s \in I_F(z^\star)$.

Step 2.1. If $i_s \in I_0(x^\star)$ then we solve the linear program:

$$(Pmax1) \quad \overline{x}_{i_s} = \max\{x_{i_s} \ : \ z = (x, y) \in K_1; \ l_{x^\star}(x) = l_{x^\star}(x^\star)\}. \tag{25}$$

– If $\overline{x}_{i_s} = 1$ then $\overline{z} \notin R(z^\star)$ and by Lemma 1 we have:

$$p(x^\star) = l_{x^\star}(x^\star) = l_{x^\star}(\overline{x}) > p(\overline{x}).$$

– If $\overline{x}_{i_s} < 1$ then $x_{i_s} = 0$ and we update the indices set $I_{\overline{F}}(z^\star) = I_{\overline{F}}(z^\star) \cup \{i_s\}$, $I_F(z^\star) = I_F(z^\star) \backslash \{i_s\}$ and

$$K_1 = K_1 \cap \{z = (x, y) \in K; \ x_{i_s} = 0\}.$$

– If Problem $(Pmax1)$ is infeasible then added a cut $l_{x^\star}(x) \geq l_{x^\star}(x^\star) + 1$ in our problem.

Step 2.2. If $i_s \in I_1(x^\star)$ then we solve the linear program:

$$(Pmin2) \quad \overline{x}_{i_s} = \min\{x_{i_s} \ : \ z = (x, y) \in K_1; \ l_{x^\star}(x) = l_{x^\star}(x^\star)\}. \tag{26}$$

– If $\overline{x}_{i_s} = 0$ then $\overline{z} \notin R(z^\star)$ and by Lemma 1 we have:

$$p(x^\star) = l_{x^\star}(x^\star) = l_{x^\star}(\overline{x}) > p(\overline{x}).$$

– If $\overline{x}_{i_s} > 0$ then $x_{i_s} = 1$ and we update the indices set $I_{\overline{F}}(z^\star) = I_{\overline{F}}(z^\star) \cup \{i_s\}$, $I_F(z^\star) = I_F(z^\star) \backslash \{i_s\}$ and

$$K_1 = K_1 \cap \{z = (x, y) \in K; \ x_{i_s} = 1\}.$$

– If Problem $(Pmin2)$ is infeasible then added a cut $l_{x^\star}(x) \geq l_{x^\star}(x^\star) + 1$ in our problem.

4   Numerical Result

To globally solution, we combine DCA and Cutting plane method which presented in Section 3. The initial point of DCA procedure is used for first iteration of DCA. The results are compared with CPLEX 12.2. The algorithm has been coded in VC++ and implemented on a Intel Core i3 CPU 2.3 Ghz, RAM 4GB. The benchmark instances are presented in [16], [15] and [14]. All these benchmark instances are available via www.mech.kuleuven.be/cib/op.

Table 1. Results of DCA - CUT of data in Tsiligirides 1984

Data	DBCut	TimeD (s)	Gap %	Cplex12.2	Data	DBCut	TimeD (s)	Gap%	Cplex12.2
T101	**45**	0.34	0.05	**45**	T110	**240**	1.00	0.54	**240**
T102	**70**	5.64	0.17	**70**	T111	**250**	2.92	1.38	**250**
T103	**95**	63.52	1.91	**95**	T112	**265**	2.15	0.44	**265**
T104	**120**	16.84	1.89	**120**	T113	270	1.91	2.43	275
T105	**140**	106.04	1.99	**140**	T114	**280**	2.51	1.59	**280**
T106	**165**	0.89	0.47	**165**	T115	**285**	3.57	0.0	**285**
T107	**180**	1.22	3.49	**180**	T116	**285**	4.75	0.0	**285**
T108	**200**	0.54	2.67	**200**	T117	**285**	4.75	0.0	**285**
T109	**225**	1.12	3.45	**225**	T118	**285**	0.22	0.0	**285**
T201	265	1.37	4.52	270	T207	420	0.72	1.60	430
T202	**330**	2.83	2.70	**330**	T208	**450**	0.11	0.0	**450**
T203	**360**	0.49	2.27	**360**	T209	**450**	0.12	0.0	**450**
T204	**370**	1.94	4.95	**370**	T210	**450**	1.09	0.0	**450**
T205	**400**	0.70	1.89	**400**	T211	**450**	0.13	0.0	**450**
T206	**420**	0.73	1.60	**420**					

Table 2. Results of DCA - CUT of data in Tsiligirides 1984

Data	DBCut	TimeD (s)	Gap %	Cplex12.2	Data	DBCut	TimeD (s)	Gap%	Cplex12.2
T301	**260**	16.85	2.87	**260**	T311	700	1.16	2.25	710
T302	**330**	1.57	3.92	**330**	T312	**740**	0.57	0.93	**740**
T303	380	0.49	2.27	390	T313	**760**	6.63	1.40	**760**
T304	**440**	1.57	0.55	**440**	T314	**790**	0.38	0.06	**790**
T305	470	0.62	4.03	480	T315	**790**	2.72	1.21	**790**
T306	**530**	1.61	1.84	**530**	T316	**800**	1.58	0.0	**800**
T307	550	0.50	4.41	560	T317	**800**	2.22	0.0	**800**
T308	**600**	1.79	2.02	**600**	T318	**800**	0.21	0.0	**800**
T309	630	0.86	2.95	640	T319	**800**	0.21	0.0	**800**
T310	**670**	1.57	1.91	**670**	T320	**800**	0.21	0.0	**800**

Reference	N. of test	N. of vertices
Tsiligirides (1984) ([14])	18	32
	11	21
	20	33
Chao (1993) ([16]) and Chao et al. (1996) ([15])	26	66
	14	64

In Tables of results, Data, DBCut, TimeD, Gap and Cplex 12.2 stands for name of data, objective value of DCA-CUT, running time by DCA-CUT, Gap and objective value of Cplex 12.2 (the global optimal value), respectively, where

$$Gap = \frac{\text{Upper bound - Lower bound}}{\text{Upper bound}}.$$

Table 3. Results of DCA - CUT of data in Chao 1983 and Chao et al. 1996

Data	DBCut	TimeD (s)	Gap %	Cplex12.2	Data	DBCut	TimeD (s)	Gap%	Cplex12.2
C101	**360**	1.72	5.05	**360**	C108	**972**	11.96	2.51	**972**
C102	**480**	1.59	0.88	**480**	C109	**1044**	2.07	1.54	**1044**
C103	540	9.11	8.39	570	C110	**1116**	1.66	0.69	**1116**
C104	**648**	7.25	4.49	**648**	C111	**1170**	49.21	0.98	**1170**
C105	**744**	2.17	2.41	**744**	C112	1212	13.89	1.48	1224
C106	**840**	2.05	0.81	**840**	C113	**1260**	32.63	0.98	**1260**
C107	**912**	26.92	4.91	**912**	C114	**1308**	2.12	0.26	**1308**
C201	**70**	235.35	3.60	**70**	C214	**1130**	2.15	2.21	**1130**
C202	**140**	1263.10	2.5	**140**	C215	**1205**	33.98	3.18	**1205**
C203	210	4940.92	25	245	C216	**1280**	0.53	0.0	**1280**
C204	**350**	3.66	0.0	**350**	C217	1330	2.09	2.17	1340
C205	**420**	1.89	4.16	**420**	C218	**1380**	5.9	1.81	**1380**
C206	**490**	1.43	7.14	**490**	C219	1430	3.48	2.62	1455
C207	560	35.21	9.37	595	C220	**1510**	0.52	0.0	**1510**
C208	**700**	9.18	0.0	**700**	C221	**1540**	1.69	0.49	**1540**
C209	**770**	1.01	2.22	**770**	C222	1470	1.84	0.96	1570
C210	**840**	18.86	4.16	**840**	C223	1600	6.68	1.40	1615
C211	910	17.41	5.76	945	C224	**1660**	0.50	0.0	**1660**
C212	**1030**	0.64	0.0	**1030**	C225	**1670**	1.81	0.15	**1670**
C213	**1080**	2.29	1.16	**1080**	C226	**1680**	0.88	0.0	**1680**

From the numerical results, we observe that:

- DCA-CUT always provides an integer solution and it converges after a few number of iterations.
- The GAPs are small. It means that the objective value obtained by DCA-CUT are rather close to the global optimal value. In all of experiments
  - in Table 1 and Table 2, almost of GAP is not larger than 3.0%,
  - in Table 3, almost of GAP is less than 4.5%.

5   Conclusion and Future Work

In this paper, we consider the BILP formulation of a relaxation orienteering problem. An efficient approach based on DC algorithm (DCA) and Cutting plane method is proposed for solving this problem. The computational results obtained show that this approach is efficient and original as it can give integer solutions while working in a continuous domain. From the promising result, in a future work we plan to combine DCA, Branch-and-Bound and Cutting plane method to globally solve the general orienteering problem.

Acknowledgment. This research is funded by Vietnam National Foudation for Science and Technology Development (NAFOSTED) under grant number 101.01-2013.19.

References

1. Laporte, G., Martello, S.: The selective travelling salesman problem. Discrete Applied Mathematics 26, 193–207 (1990)
2. Steven, E.B., Tom, M.C.: A heuristic for the multiple tour maximum collection problem. Computers & Operations Research 21(1), 101–111 (1994)
3. Arkin, E.M., Mitchell, J.S.B., Narasimhan, G.: Resource-constrained Geometric Network Optimization. In: Proceedings of the Fourteenth Annual Symposium on Computational Geometry, pp. 307–316 (1998)
4. Golden, B.L., Assad, A., Dahl, R.: Analysis of a large-scale vehicle routing problem with inventory component. Large Scale Systems 7, 181–190 (1984)
5. Golden, B.L., Levy, L., Vohra, R.: The orienteering problem. Naval Research Logistics 34, 307–318 (1987)
6. Keller, P.: Algorithms to solve the orienteering problem: a comparison. European Journal of Operational of Research 41, 224–231 (1989)
7. Kantor, M., Rosenwein, M.: The orienteering problem with time windows. Journal of Operational Research Society 43(6), 629–635 (1992)
8. Pieter, V., Wouter, S., Dirk, V.O.: The orienteering problem: A survey. European Journal of Operational Research 209(1), 1–10 (2011)
9. Balas, E.: The prize collecting traveling salesman problem. Networks 19, 621–636 (1989)
10. Le Thi, H.A., Pham Dinh, T.: A Continuous Approach for Globally Solving Linearly Constrained Quadratic Zero - One Programming Problems. Optimization 50, 93–120 (2001)
11. Alexander, S.: Theory of Linear and Integer Programming. Wiley-Interscience Series in Discrete Mathematics and Optimization (1998)
12. Nguyen Quang, T.: Approches locales et globales basées sur la programmation DC et DCA pour des problèmes combinatoires en variables mixtes 0-1: applications à la planification opérationnelle, Thèse de doctorat dirigée par Le Thi H.A Informatique Metz (2010)
13. Nguyen, V.V.: Methodes exactes basées sur la programmation DC et nouvelles coupes pour l'optimisation en variables mixtes zero-un, Thèse de doctorat dirigée par Pham Dinh T. and Le Thi H.A, LMI, INSA Rouen (2006)
14. Tsiligirides, T.: Heuristic Methods Applied to Orienteering. Journal of the Operational Research Society 35, 797–809 (1984)
15. Chao, I., Golden, B., Wasil, E.: Theory and methodology: a fast and effective heuristic for the orienteering problem. European Journal of Operational Research 88, 475–489 (1996)
16. Chao, I.: Algorithms and Solutions to Multi-level Vehicle Routing Problems. PhD thesis, University of Maryland at College Park (1993)

Solving the Quadratic Eigenvalue Complementarity Problem by DC Programming

Yi-Shuai Niu[1], Joaquim Júdice[2], Hoai An Le Thi[3], and Tao Pham Dinh[4]

[1] Shanghai JiaoTong University, Maths Departement and SJTU-Paristech, Shanghai, China
niuyishuai@sjtu.edu.cn
[2] Instituto de Telecomunicações, Lisbon, Portugal
judice@co.it.pt
[3] University of Lorraine, Metz, France
hoai-an.le-thi@univ-lorraine.fr
[4] National Institute of Applied Sciences, Rouen, France
pham@insa-rouen.fr

Abstract. We present in this paper some results for solving the Quadratic Eigenvalue Complementarity Problem (QEiCP) by using DC(Difference of Convex functions) programming approaches. Two equivalent Nonconvex Polynomial Programming (NLP) formulations of QEiCP are introduced. We focus on the construction of the DC programming formulations of the QEiCP from these NLPs. The corresponding numerical solution algorithms based on the classical DC Algorithm (DCA) are also discussed.

Keywords: Eigenvalue Problem, Complementarity Problem, Nonconvex Polynomial Programming, DC Programming, DCA.

1 Introduction

Given three matrices $A, B, C \in \mathbb{R}^{n \times n}$, the *Quadratic Eigenvalue Complementarity Problem (QEiCP)* consists of finding a $\lambda \in \mathbb{R}$ and an associated nonzero vector $x \in \mathbb{R}^n$ such that

$$w = \lambda^2 Ax + \lambda Bx + Cx$$
$$x^T w = 0, x \geq 0, w \geq 0 \tag{1}$$

This problem and some applications have been firstly introduced in [19] and is usually denoted by QEiCP(A, B, C). In any solution (λ, x) of QEiCP(A, B, C), the λ-component is called a *quadratic complementary eigenvalue*, and the vector x-component is a *quadratic complementary eigenvector* associated to λ.

QEiCP is an extension of the well-known Eigenvalue Complementarity Problem (EiCP) [18], which consists of finding a complementary eigenvalue $\lambda \in \mathbb{R}$ and an associated complementary eigenvector $x \in \mathbb{R}^n \setminus \{0\}$ such that

$$w = \lambda Bx - Cx$$
$$x^T w = 0, x \geq 0, w \geq 0 \tag{2}$$

where $B, C \in \mathbb{R}^{n \times n}$ are two given matrices.

© Springer International Publishing Switzerland 2015   203
H.A. Le Thi et al. (eds.), *Model. Comput. & Optim. in Inf. Syst. & Manage. Sci.*,
Advances in Intelligent Systems and Computing 359, DOI: 10.1007/978-3-319-18161-5_18

Clearly, EiCP is a special case of QEiCP where the matrix A is null. During the past several years, many applications of EiCP have been discussed and a number of algorithms have been proposed for the solution of this problem and some extensions [1,2,3,6,7,8,9,10,11,15,16].

EiCP has at least one solution if the matrix B of the leading λ-term is positive definite (PD) [9,18]. Contrary to the EiCP, the QEiCP may have no solution even when the matrix A of leading λ-term is PD. For instance, if $B = 0$, A, C are PD matrices, there is no solution for QEiCP since $x^T w = \lambda^2 x^T A x + x^T C x > 0, \forall \lambda \in \mathbb{R}, x \in \mathbb{R}^n \setminus \{0\}$.

The existence of a solution to QEiCP depends on the given (A, B, C). If the matrix A is PD, QEiCP has at least a solution if one of the two following conditions holds:

(i) $C \notin S_0$ [4], where S_0 is the class of matrices defined by

$$C \in S_0 \Leftrightarrow \exists x \geq 0, x \neq 0, Cx \geq 0.$$

(ii) co-hyperbolicity[19] : $(x^T B x)^2 \geq 4(x^T A x)(x^T C x)$ for all $x \geq 0, x \neq 0$.

In practice, investigating whether $C \in S_0$ reduces to solving a special linear program [4]. On the other hand, it is relatively hard to prove that co-hyperbolicity holds. However, there are some sufficient conditions which imply the co-hyperbolicity. For instance, this occurs if A and $-C$ are both PD matrices.

A number of algorithms have been proposed for the solution of QEiCP when $A \in$ PD and one of the conditions $C \notin S_0$ or co-hyperbolicity holds [1,4,5,6,19]. As discussed in [4,5,6], some of these methods are based on nonlinear programming (NLP) formulations of QEiCP such that (λ, x) is a solution of QEiCP if and only if (λ, x) is a global minimum of NLP with an optimal value equal to zero. In this paper, we introduce two nonlinear programming formulations and their corresponding DC programming formulations when co-hyperbolicity holds, and we briefly discuss the DC Algorithm for the solution of these DC programs.

The paper is organized as follows. Section 2 contains the nonlinear programming formulations of QEiCP, and the corresponding dc formulations mentioned before. A new result on lower and upper bounds estimation of the quadratic complementary eigenvalue is given in section 3. The numerical solution algorithms for solving these DC programming formulations are discussed in section 4. Some conclusions are presented in the last section.

2   DC Programming Formulations for QEiCP

In this section, we introduce two DC programming formulations of QEiCP when $A \in$ PD and the co-hyperbolic property holds. These DC programs are based on two NLP formulations of QEiCP. The construction of the DC programming problem requires lower and upper bounds on the λ-variable which can be computed by the procedures discussed in [6]. We will also present a new procedure for such a goal in the section 3.

2.1 Nonlinear Programming Formulations

As discussed in [6], QEiCP is equivalent to the following NLP:

$$(P) \qquad 0 = \min f(x,y,z,w,\lambda) := \|y - \lambda x\|^2 + \|z - \lambda y\|^2 + x^T w$$
$$\text{s.t. } w = Az + By + Cx \tag{3}$$
$$e^T x = 1, e^T y = \lambda$$
$$x \geq 0, w \geq 0, z \geq 0.$$

As (x,y,z,w,λ) is an optimal solution of the problem (P) if and only if (λ, x) is a solution of QEiCP. In fact, for any solution of QEiCP (λ, x) that does not satisfy $e^T x = 1$, we can always construct a solution $(\lambda, \frac{x}{e^T x})$ of QEiCP satisfying such a constraint.

The problem (P) is a polynomial programming problem where a nonconvex polynomial function $f(x,y,z,w,\lambda)$ is minimized subject to linear constraints. Due to the fact that any polynomial function is a dc function, we can reformulate the problem (P) as a dc program.

On the other hand, observing that the complementarity constraint $w^T x = 0, x \geq 0, w \geq 0$ holds if and only if $w^T x = \sum_{i=1}^{n} \min(x_i, w_i) = 0$, we have the following equivalent nonlinear programming formulation of (P):

$$(P') \quad 0 = \min f'(x,y,z,w,\lambda) = \|y - \lambda x\|^2 + \|z - \lambda y\|^2 + \sum_{i=1}^{n} \min(x_i, w_i)$$
$$\text{s.t } w = Az + By + Cx$$
$$e^T x = 1, e^T y = \lambda$$
$$x \geq 0, w \geq 0, z \geq 0.$$

The problems (P) and (P') have the same set of linear constraints. The difficulty for solving (P) and (P') relies on the non-convexity on their objective functions.

2.2 DC Programming Formulations

The polynomial function f in (P) can be decomposed into four parts:

$$f(x,y,z,w,\lambda) = \|y\|^2 + \|z\|^2 - 2\lambda y^T(x+z) + \lambda^2(\|x\|^2 + \|y\|^2) + x^T w$$
$$= f_0(y,z) + f_1(x,y,z,\lambda) + f_2(x,y,\lambda) + f_3(x,w)$$

with

$$\begin{cases} f_0(y,z) = \|y\|^2 + \|z\|^2 \\ f_1(x,y,z,\lambda) = -2\lambda y^T(x+z) \\ f_2(x,y,\lambda) = \lambda^2(\|x\|^2 + \|y\|^2) \\ f_3(x,w) = x^T w \end{cases}$$

The function f_0 is convex quadratic function, while f_1, f_2, f_3 are nonconvex polynomial functions. Similarly, the objective function f' in (P') is also decomposed into the following four terms as:

$$f'(x,y,z,w,\lambda) = f_0(y,z) + f_1(x,y,z,\lambda) + f_2(x,y,\lambda) + \tilde{f}_3(x,w)$$

where $\tilde{f}_3(x,w)$ defined by $\sum_{i=1}^{n} \min(x_i, w_i)$ is a polyhedral concave function.

Both the bilinear function f_3 and the polyhedral concave function $\tilde{f}_3$ are classical dc functions whose dc decompositions are as follows:

1. DC decomposition of bilinear function f_3:

$$f_3(x, w) = \frac{\|x + w\|^2}{4} - \frac{\|x - w\|^2}{4} \tag{4}$$

in which $\frac{\|x+w\|^2}{4}$ and $\frac{\|x-w\|^2}{4}$ are both convex quadratic functions.

2. DC decomposition of polyhedral function $\tilde{f}_3$:

$$\tilde{f}_3(x, w) = \sum_{i=1}^{n} \min(x_i, w_i) = (0) - \left(-\sum_{i=1}^{n} \min(x_i, w_i)\right) \tag{5}$$

where $-\sum_{i=1}^{n} \min(x_i, w_i)$ is a convex polyhedral function.

To obtain a dc decompositions of the nonconvex polynomial functions f_1 and f_2, we first obtain the expressions of their gradients and hessians:

1. Gradient and Hessian of f_1:

$$\nabla f_1(x, y, z, \lambda) = \begin{bmatrix} \nabla_x f_1(x, y, z, \lambda) \\ \nabla_y f_1(x, y, z, \lambda) \\ \nabla_z f_1(x, y, z, \lambda) \\ \nabla_\lambda f_1(x, y, z, \lambda) \end{bmatrix} = \begin{bmatrix} -2\lambda y \\ -2\lambda(x + z) \\ -2\lambda y \\ -2y^T(x + z) \end{bmatrix}.$$

$$\nabla^2 f_1(x, y, z, \lambda) = \begin{bmatrix} 0 & -2\lambda I & 0 & -2y \\ -2\lambda I & 0 & -2\lambda I & -2(x+z) \\ 0 & -2\lambda I & 0 & -2y \\ -2y^T & -2(x+z)^T & -2y^T & 0 \end{bmatrix}.$$

2. Gradient and Hessian of f_2:

$$\nabla f_2(x, y, \lambda) = \begin{bmatrix} \nabla_x f_2(x, y, \lambda) \\ \nabla_y f_2(x, y, \lambda) \\ \nabla_\lambda f_2(x, y, \lambda) \end{bmatrix} = \begin{bmatrix} 2\lambda^2 x \\ 2\lambda^2 y \\ 2\lambda(\|x\|^2 + \|y\|^2) \end{bmatrix}.$$

$$\nabla^2 f_2(x, y, z, \lambda) = \begin{bmatrix} 2\lambda^2 I & 0 & 4\lambda x \\ 0 & 2\lambda^2 I & 4\lambda y \\ 4\lambda x^T & 4\lambda y^T & 2(\|x\|^2 + \|y\|^2) \end{bmatrix}.$$

The spectral radius of the hessian matrices $\nabla^2 f_1$ and $\nabla^2 f_2$ (denoted by $\rho(\nabla^2 f_1)$ and $\rho(\nabla^2 f_2)$) can be bounded above by the induced 1-norm as follows:

$$\rho(\nabla^2 f_1) \le \|\nabla^2 f_1\|_1 = 2\max\{|\lambda| + |y_i|, |x_i + z_i| + 2|\lambda|, \sum_i (2|y_i| + |x_i + z_i|)\}$$

$$\rho(\nabla^2 f_2) \le \|\nabla^2 f_2\|_1 = 2\max\{\lambda^2 + 2|\lambda||x_i|, \lambda^2 + 2|\lambda||y_i|, \|x\|^2 + \|y\|^2 + 2|\lambda|\sum_i (|x_i| + |y_i|)\}$$

Thus $\rho(\nabla^2 f_1)$ and $\rho(\nabla^2 f_2)$) are bounded when the variables (x, y, z, w, λ) of (P) and (P') are bounded.

The next proposition shows that if the quadratic complementary eigenvalue λ of QEiCP is bounded, then the variables x, y, z, w are bounded with respect to the bounds of λ.

Proposition 1. *If the quadratic complementary eigenvalue λ of QEiCP is bounded in an interval $[l, u]$, then any optimal solution of (P) and (P') satisfies:*

$$x \in [0,1]^n; y \in [\min\{0,l\}, \max\{0,u\}]^n; z \in [0, \max\{u^2, l^2\}]^n;$$

$$0 \leq w \leq \begin{bmatrix} \max\{u^2, l^2\} \sum_j |A_{1j}| + \max\{|l|, |u|\} \sum_j |B_{1j}| + \sum_j |C_{1j}| \\ \vdots \\ \max\{u^2, l^2\} \sum_j |A_{nj}| + \max\{|l|, |u|\} \sum_j |B_{nj}| + \sum_j |C_{nj}| \end{bmatrix}.$$

Proof. Suppose that we could determine some values l and u such that λ-component of QEiCP is located in the interval $[l, u]$.

1. $e^T x = 1, x \geq 0$ implies $x \in [0,1]^n$.
2. $y = \lambda x, x \in [0,1]^n$ and $\lambda \in [l, u]$ imply $y \in [\min\{0,l\}, \max\{0,u\}]^n$.
3. $z = \lambda y, y = \lambda x \Rightarrow z = \lambda^2 x$, with $x \in [0,1]^n, \lambda \in [l, u]$, leads to $z \in [0, \max\{u^2, l^2\}]^n$.
4. Since $w \geq 0$, the upper bound of w is obtained from the definition of w as $Az + By + Cx$. As $x \in [0,1]^n, y \in [\min\{0,l\}, \max\{0,u\}]^n, z \in [0, \max\{u^2, l^2\}]^n$, then w is also bounded:

$$|w| \leq \begin{bmatrix} \max\{u^2, l^2\} \sum_j |A_{1j}| + \max\{|l|, |u|\} \sum_j |B_{1j}| + \sum_j |C_{1j}| \\ \vdots \\ \max\{u^2, l^2\} \sum_j |A_{nj}| + \max\{|l|, |u|\} \sum_j |B_{nj}| + \sum_j |C_{nj}| \end{bmatrix}.$$

$\square$

Let us define the convex polyhedral set:

$$\mathcal{C} := \{(x, y, z, w, \lambda) : w = Az + By + Cx, e^T x = 1, e^T y = \lambda, x \in [0,1]^n,$$
$$y \in [\min\{0,l\}, \max\{0,u\}]^n, z \in [0, \max\{u^2, l^2\}]^n, w \geq 0, l \leq \lambda \leq u\}.$$

The problems (P) and (P') defined on $\mathcal{C}$ have the same set of optimal solutions, and $\rho(\nabla^2 f_1)$ and $\rho(\nabla^2 f_2)$ are bounded. In fact, the following proposition holds:

Proposition 2. *For $(x, y, z, w, \lambda) \in \mathcal{C}$,*

$$\rho(\nabla^2 f_1) \leq 2 + 2n(p^2 + 2p) = \rho_1$$

$$\rho(\nabla^2 f_2) \leq 2(3np^2 + 2p + 1) = \rho_2$$

where $p = \max\{|l|, |u|\}$.

Proof. Since $\lambda \in [l, u]$, then $|\lambda| \leq \max\{|l|, |u|\} = p$. Hence,

$$\rho(\nabla^2 f_1) \leq 2 \max\{|\lambda| + |y_i|, |x_i + z_i| + 2|\lambda|, \sum_i (2|y_i| + |x_i + z_i|)\}.$$

But,

$$\sum_i |y_i| \leq np.$$

$$\sum_i |x_i + z_i| \le \sum_i |x_i| + \sum_i |z_i| \le 1 + np^2.$$

Hence,

$$\rho(\nabla^2 f_1) \le 2 \max\{2p, 1 + p^2 + 2p, 1 + n(p^2 + 2p)\} = 2 + 2n(p^2 + 2p) = \rho_1$$

Similarly,

$$\rho(\nabla^2 f_2) \le 2 \max\{\lambda^2 + 2|\lambda||x_i|, \lambda^2 + 2|\lambda||y_i|, \|x\|^2 + \|y\|^2 + 2|\lambda| \sum_i (|x_i| + |y_i|)\}$$

$$\le 2 \max\{p^2 + 2p, 3p^2, 3np^2 + 2p + 1\} = 2(3np^2 + 2p + 1) = \rho_2.$$

$\square$

Thus, we get a dc decomposition for f_1 and f_2 as follows:

$$f_1(x, y, z, \lambda) = \frac{\rho_1}{2} \|(x, y, z, \lambda)\|^2 - (\frac{\rho_1}{2} \|(x, y, z, \lambda)\|^2 - f_1(x, y, z, \lambda))$$

$$f_2(x, y, \lambda) = \frac{\rho_2}{2} \|(x, y, \lambda)\|^2 - (\frac{\rho_2}{2} \|(x, y, \lambda)\|^2 - f_2(x, y, \lambda))$$

where $\frac{\rho_1}{2} \|(x, y, z, \lambda)\|^2$ and $\frac{\rho_2}{2} \|(x, y, \lambda)\|^2$ are quadratic convex functions. While $\frac{\rho_1}{2} \|(x, y, z, \lambda)\|^2 - f_1(x, y, z, \lambda)$ and $\frac{\rho_2}{2} \|(x, y, \lambda)\|^2 - f_2(x, y, \lambda)$ are locally convex restricted on $\mathcal{C}$.

Using the dc decompositions of f_1, f_2, f_3 and $\tilde{f}_3$ derived in this section, we get the following dc decomposition for the objective functions f and f'.

1. A dc decomposition for f is given by:

$$f(x, y, z, w, \lambda) = g(x, y, z, w, \lambda) - h(x, y, z, w, \lambda)$$

where

$$g(x, y, z, w, \lambda) = \frac{\|x + w\|^2}{4} + \frac{\rho_1 + \rho_2}{2} \|x\|^2 + (\frac{\rho_1 + \rho_2}{2} + 1)\|y\|^2 + (\frac{\rho_1}{2} + 1)\|z\|^2 + \frac{\rho_1 + \rho_2}{2} \lambda^2,$$

$$h(x, y, z, w, \lambda) = g(x, y, z, w, \lambda) - f(x, y, z, w, \lambda).$$

2. A dc decomposition for f' is given by:

$$f'(x, y, z, w, \lambda) = g'(x, y, z, w, \lambda) - h'(x, y, z, w, \lambda)$$

where

$$g'(x, y, z, \lambda) = \frac{\rho_1 + \rho_2}{2} \|x\|^2 + (\frac{\rho_1 + \rho_2}{2} + 1)\|y\|^2 + (\frac{\rho_1}{2} + 1)\|z\|^2 + \frac{\rho_1 + \rho_2}{2} \lambda^2,$$

$$h'(x, y, z, w, \lambda) = g'(x, y, z, \lambda) - f'(x, y, z, w, \lambda).$$

The functions g and g' are both convex quadratic functions, while h and h' are locally convex functions restricted on the convex polyhedral set $\mathcal{C}$.

Finally, we get the following equivalent DC programs of (P) and (P') as below:

$$(P_{DC}) \qquad 0 = \min g(x, y, z, w, \lambda) - h(x, y, z, w, \lambda)$$
$$\text{s.t. } (x, y, z, w, \lambda) \in \mathcal{C}. \tag{6}$$

$$(P'_{DC}) \qquad 0 = \min g'(x, y, z, w, \lambda) - h'(x, y, z, w, \lambda)$$
$$\text{s.t. } (x, y, z, w, \lambda) \in \mathcal{C}. \tag{7}$$

3   Lower and Upper Bounds for the Quadratic Complementary Eigenvalue λ

Since the bounds of the variables x, y, z, w in $\mathcal{C}$, as well as the dc decompositions given in the previous section depend on the bounds of λ, we need to estimate its upper and lower bounds. The following theorem gives these values.

Proposition 3. *If $A \in PD$ and the co-hyperbolic condition holds, the λ-component of any solution of QEiCP satisfies*

$$l = \beta - \sqrt{\alpha} \leq \lambda \leq \gamma + \sqrt{\alpha} = u$$

with $s = \min\{x^T A x : e^T x = 1, x \geq 0\}$, $\alpha = \max\{\gamma^2, \beta^2\} + \frac{\max_{i,j}\{-C_{ij}\}}{s}$,

$$\beta = \begin{cases} \frac{\min\{-B_{ij}\}}{2\max\{A_{ij}\}}, & \text{if } \min\{-B_{ij}\} > 0; \\ \frac{\min\{-B_{ij}\}}{2s}, & \text{if } \min\{-B_{ij}\} \leq 0. \end{cases}$$

$$\gamma = \begin{cases} \frac{\max\{-B_{ij}\}}{2s}, & \text{if } \max\{-B_{ij}\} > 0; \\ \frac{\max\{-B_{ij}\}}{2\max\{A_{ij}\}}, & \text{if } \max\{-B_{ij}\} \leq 0. \end{cases}$$

Proof. Since $A \in PD$ and the co-hyperbolic condition holds, the λ-component of any solution of QEiCP satisfies

$$\lambda = \frac{-x^T B x \pm \sqrt{(x^T B x)^2 - 4(x^T A x)(x^T C x)}}{2 x^T A x}.$$

Let $U = \{e^T x = 1, x \geq 0\}$. For a given matrix $M \in \mathbb{R}^{n \times n}$ and for any $x \in U$, we next prove that:

$$\min_{i,j} M_{ij} \leq x^T M x \leq \max_{i,j} M_{ij}, \forall x \in U. \tag{8}$$

If fact, let $(Mx)_i$ denote the i-th element of the vector Mx. Then

$$x^T M x = \sum_{i=1}^{n} x_i (Mx)_i$$

But, $(Mx)_i$ is bounded by

$$\min\{\sum_{j=1}^{n} M_{ij} x_j : x \in U\} \leq (Mx)_i \leq \max\{\sum_{j=1}^{n} M_{ij} x_j : x \in U\}, \forall x \in U.$$

Since the linear programs $\min\{\sum_{j=1}^{n} M_{ij} x_j : x \in U\}$ and $\max\{\sum_{j=1}^{n} M_{ij} x_j : x \in U\}$ have optimal solutions on vertices, the optimal values of the above linear programs are exactly $\min_j\{M_{ij}\}$ and $\max_j\{M_{ij}\}$. Hence, we can compute bounds for $x^T M x$ on U as follows:

$$\min_{i,j}\{M_{ij}\} = \min\{\sum_i x_i \min_j\{M_{ij}\} : x \in U\} \leq \sum_i x_i \min_j\{M_{ij}\} \leq \sum_i x_i (Mx)_i$$

$$= x^T M x \le \sum_i x_i \max_j \{M_{ij}\} \le \max\{\sum_i x_i \max_j \{M_{ij}\} : x \in U\} = \max_{i,j} \{M_{i,j}\}.$$

Hence, (8) is true.

Using the bounds (8) for the matrices B and C, we have:

$$\min_{i,j} \{-B_{ij}\} \le -x^T B x \le \max_{i,j} \{-B_{ij}\},$$

$$\min_{i,j} \{-C_{ij}\} \le -x^T C x \le \max_{i,j} \{-C_{ij}\}$$

Since $A \in$ PD, we have

$$0 < s = \min\{x^T A x : x \in U\} \le x^T A x \le \max_{i,j} \{A_{ij}\}, \forall x \in U.$$

Accordingly, $\frac{-x^T B x}{2x^T A x}$ is bounded by:

$$\frac{\min\{-B_{ij}\}}{2x^T A x} \le \frac{-x^T B x}{2x^T A x} \le \frac{\max\{-B_{ij}\}}{2x^T A x} \le \gamma = \begin{cases} \frac{\max\{-B_{ij}\}}{2s}, & \text{if } \max\{-B_{ij}\} > 0; \\ \frac{\max\{-B_{ij}\}}{2\max\{A_{ij}\}}, & \text{if } \max\{-B_{ij}\} \le 0. \end{cases}$$

and

$$\frac{\min\{-B_{ij}\}}{2x^T A x} \ge \beta = \begin{cases} \frac{\min\{-B_{ij}\}}{2\max\{A_{ij}\}}, & \text{if } \min\{-B_{ij}\} > 0; \\ \frac{\min\{-B_{ij}\}}{2s}, & \text{if } \min\{-B_{ij}\} \le 0. \end{cases}$$

Then

$$(\frac{-x^T B x}{2x^T A x})^2 + \frac{-x^T C x}{x^T A x} \le \max\{\gamma^2, \beta^2\} + \frac{\max\{-C_{ij}\}}{s} = \alpha.$$

Finally, we can compute bounds for λ as follows:

$$\beta - \sqrt{\alpha} \le \frac{-x^T B x}{2x^T A x} - \sqrt{(\frac{-x^T B x}{2x^T A x})^2 + \frac{-x^T C x}{x^T A x}} \le$$

$$\lambda \le \frac{-x^T B x}{2x^T A x} + \sqrt{(\frac{-x^T B x}{2x^T A x})^2 + \frac{-x^T C x}{x^T A x}} \le \gamma + \sqrt{\alpha}.$$

$\square$

In practice, it is interesting to compare in the future the bound proposed here with the one given in [6]. The bounds given in this paper have been designed such that they can be computed in a small amount of effort, even for large-scale problems.

4   DC Algorithms for Solving P_{DC} and P'_{DC}

In this section, we investigate how to solve the DC programming formulations (P_{DC}) and (P'_{DC}).

Given a general DC program:

$$\min\{g(x) - h(x) : x \in C\},$$

where C is a non-empty convex set, the general DC algorithm (DCA) consists of constructing two sequences $\{x^k\}$ and $\{y^k\}$ via the following scheme[12,13,14]:

$$x^k \to y^k \in \partial h(x^k)$$
$$x^{k+1} \in \partial g^*(y^k) = \mathrm{argmin}\{g(x) - \langle x, y^k \rangle : x \in C\}.$$

The symbol ∂h stands for the sub-differential of the convex function h, and g^* is the conjugate function of g. These definitions are fundamental and can be found in any textbook of the convex analysis (see for example [17]).

The sequence $\{x^k\}$ and $\{y^k\}$ are respectively candidates for optimal solutions of the primal and dual DC programs.

In DCA, two major computations should be considered:

1. Computing $\partial h(x^k)$ to get y^k.
2. Solving the convex program $\mathrm{argmin}\{g(x) - \langle x, y^k \rangle : x \in C\}$ to obtain x^{k+1}.

Now, we investigate the use of DCA to solve the DC programs (P_{DC}) and (P'_{DC}). Concerning to (P_{DC}), since the function h is differentiable, $\partial h(x, y, z, w, \lambda)$ is reduced to a singleton $\{\nabla h(x, y, z, w, \lambda)\}$, where

$$
\nabla h(x, y, z, w, \lambda) = \nabla g(x, y, z, w, \lambda) - \nabla f(x, y, z, w, \lambda)
$$
$$
= \begin{bmatrix} \frac{x+w}{2} + (\rho_1 + \rho_2)x + 2\lambda y - 2\lambda^2 x - w \\ (\rho_1 + \rho_2 - 2\lambda^2)y + 2\lambda(x + z) \\ \rho_1 z + 2\lambda y \\ \frac{w-x}{2} \\ (\rho_1 + \rho_2 - 2(\|x\|^2 + \|y\|^2))\lambda + 2y^T(x + z) \end{bmatrix}. \tag{9}
$$

For (P'_{DC}), since the function h' is non-differentiable, we compute the convex set $\partial h'(x, y, z, w, \lambda)$ as follows:

$$
\partial h'(x, y, z, w, \lambda) = \left\{ \begin{bmatrix} (\rho_1 + \rho_2)x + 2\lambda y - 2\lambda^2 x - u \\ (\rho_1 + \rho_2 - 2\lambda^2)y + 2\lambda(x + z) \\ \rho_1 z + 2\lambda y \\ -v \\ (\rho_1 + \rho_2 - 2(\|x\|^2 + \|y\|^2))\lambda + 2y^T(x + z) \end{bmatrix} \right\} \tag{10}
$$

where

$$
u = (u_i)_{i=1,\dots,n}, u_i = \begin{cases} 1, & x_i < w_i; \\ \{0,1\}, & x_i = w_i; \\ 0, & x_i > w_i. \end{cases}
$$

$$
v = (v_i)_{i=1,\dots,n}, v_i = \begin{cases} 0, & x_i < w_i; \\ \{0,1\}, & x_i = w_i; \\ 1, & x_i > w_i. \end{cases}
$$

Finally, DCA applied to (P_{DC}) and (P'_{DC}) requires solving respectively one convex quadratic program over a polyhedral convex set in each iteration.

The following two fixed-point schemes describe our dc algorithms:

$$(x^{k+1}, y^{k+1}, z^{k+1}, w^{k+1}, \lambda^{k+1}) = \operatorname{argmin}\{g(x, y, z, w, \lambda) \\ - \langle (x, y, z, w, \lambda), \nabla h(x^k, y^k, z^k, w^k, \lambda^k) \rangle : (x, y, z, w, \lambda) \in \mathcal{C}\} \tag{11}$$

with $g(x, y, z, w, \lambda) = \frac{\|x+w\|^2}{4} + \frac{\rho_1+\rho_2}{2}\|x\|^2 + (\frac{\rho_1+\rho_2}{2}+1)\|y\|^2 + (\frac{\rho_1}{2}+1)\|z\|^2 + \frac{\rho_1+\rho_2}{2}\lambda^2$.

$$(x^{k+1}, y^{k+1}, z^{k+1}, w^{k+1}, \lambda^{k+1}) = \operatorname{argmin}\{g'(x, y, z, \lambda) \\ - \langle (x, y, z, w, \lambda), Y^k \rangle : (x, y, z, w, \lambda) \in \mathcal{C}\} \tag{12}$$

with $Y^k \in \partial h'(x^k, y^k, z^k, w^k, \lambda^k)$ and $g'(x, y, z, \lambda) = \frac{\rho_1+\rho_2}{2}\|x\|^2 + (\frac{\rho_1+\rho_2}{2}+1)\|y\|^2 + (\frac{\rho_1}{2}+1)\|z\|^2 + \frac{\rho_1+\rho_2}{2}\lambda^2$.

These convex quadratic programs can be efficiently solved via a quadratic programming solver such as CPLEX, Gurobi, XPress, etc.

DCA should terminate if one of the following stopping criteria is satisfied for given tolerances ϵ_1, ϵ_2 and ϵ_3.

(1) The sequence $\{(x^k, y^k, z^k, w^k, \lambda^k)\}$ converges, i.e.,

$$\|(x^{k+1}, y^{k+1}, z^{k+1}, w^{k+1}, \lambda^{k+1}) - (x^k, y^k, z^k, w^k, \lambda^k)\| \leq \epsilon_1$$

(2) The sequence $\{f(x^k, y^k, z^k, w^k, \lambda^k)\}$ (resp. $\{f'(x^k, y^k, z^k, w^k, \lambda^k)\}$) converges, i.e.,

$$\|f(x^{k+1}, y^{k+1}, z^{k+1}, w^{k+1}, \lambda^{k+1}) - f(x^k, y^k, z^k, w^k, \lambda^k)\| \leq \epsilon_2$$

(resp. $\|f'(x^{k+1}, y^{k+1}, z^{k+1}, w^{k+1}, \lambda^{k+1}) - f'(x^k, y^k, z^k, w^k, \lambda^k)\| \leq \epsilon_2$).

(3) *The sufficient global ϵ-optimality condition holds, i.e.,*

$$f(x^k, y^k, z^k, w^k, \lambda^k) \leq \epsilon_3 \quad (\text{resp. } f'(x^k, y^k, z^k, w^k, \lambda^k) \leq \epsilon_3).$$

The following theorem indicates the convergence of DCA:

Theorem 1 (Convergence theorem of DCA). *DCA applied to QEiCP generates convergence sequences $\{(x^k, y^k, z^k, w^k, \lambda^k)\}$ and $\{f(x^k, y^k, z^k, w^k, \lambda^k)\}$ (resp. $\{f'(x^k, y^k, z^k, w^k, \lambda^k)\}$) such that:*

- *The sequence $\{f(x^k, y^k, z^k, w^k, \lambda^k)\}$ (resp. $\{f'(x^k, y^k, z^k, w^k, \lambda^k)\}$) is decreasing and bounded below.*
- *The sequence $\{(x^k, y^k, z^k, w^k, \lambda^k)\}$ converges either to a solution of QEiCP when the third stopping condition is satisfied or to a general KKT point of (P_{DC}) (resp. (P'_{DC})).*

Proof. The proof of the theorem is an obvious consequence of the general convergence theorem of DCA [12,13,14]. The sufficient global optimality condition is due to the fact that the optimal value of the dc program is equal to zero. □

5   Conclusions

In this paper, we have presented two DC programming formulations of the Quadratic Eigenvalue Complementarity Problem. The corresponding numerical solution algorithms based on the classical DCA for solving these dc programs were briefly discussed.

The numerical results and the analysis of the performance of DCA for solving QEiCP will be given in a future paper. We will discuss a new *local dc decomposition algorithm* that is designed to speed up the convergence of DCA. Furthermore, that paper will also be devoted to the solution of QEiCP when the condition $A \in$PD and $C \notin S_0$ holds. A new DC formulation of QEiCP based on the reformulation of an equivalent extended EiCP will be introduced to deal with this case and the corresponding DC Algorithm will be discussed.

Acknowledgements. The research of Yi-Shuai Niu in this project is partially supported and financed by the Innovative Research Fund of Shanghai Jiao Tong University 985 Program. Joaquim Júdice was partially supported in the scope of R&D Unit UID/EEA/5008/2013, financed by the applicable financial framework (FCT/MEC through national funds and the applicable co-funded by FEDER-PT2002 partnership agreement).

References

1. Adly, S., Seeger, A.: A non-smooth algorithm for cone constrained eigenvalue problems. Computational Optimization and Applications 49, 299–318 (2011)
2. Adly, S., Rammal, H.: A new method for solving second-order cone eigenvalue complementarity problem. Journal of Optimization Theory and Applications (2014), doi:10.1007/s10957-014-0645-0
3. Brás, C., Fukushima, M., Júdice, J., Rosa, S.: Variational inequality formulation for the asymmetric eigenvalue complementarity problem and its solution by means of a gap function. Pacific Journal of Optimization 8, 197–215 (2012)
4. Brás, C., Iusem, A.N., Júdice, J.: On the quadratic eigenvalue complementarity problem. To appear in Journal of Global Optimization
5. Fernandes, L.M., Júdice, J., Fukushima, M., Iusem, A.: On the symmetric quadratic eigenvalue complementarity problem. Optimization Methods and Software 29, 751–770 (2014)
6. Fernandes, L.M., Júdice, J., Sherali, H.D., Forjaz, M.A.: On an enumerative algorithm for solving eigenvalue complementarity problems. Computational Optimization and Applications 59, 113–134 (2014)
7. Júdice, J., Sherali, H.D., Ribeiro, I.: The eigenvalue complementarity problem. Computational Optimization and Applications 37, 139–156 (2007)
8. Júdice, J., Raydan, M., Rosa, S., Santos, S.: On the solution of the symmetric complementarity problem by the spectral projected gradient method. Numerical Algorithms 44, 391–407 (2008)
9. Júdice, J., Sherali, H.D., Ribeiro, I., Rosa, S.: On the asymmetric eigenvalue complementarity problem. Optimization Methods and Software 24, 549–586 (2009)

10. Le Thi, H.A., Moeini, M., Pham Dinh, T., Júdice, J.: A DC programming approach for solving the symmetric Eigenvalue Complementarity Problem. Computational Optimization and Applications 51, 1097–1117 (2012)
11. Niu, Y.S., Le Thi, H.A., Pham Dinh, T., Júdice, J.: Efficient dc programming approaches for the asymmetric eigenvalue complementarity problem. Optimization Methods and Software 28, 812–829 (2013)
12. Pham Dinh, T., Le Thi, H.A.: DC optimization algorithms for solving the trust region subproblem. SIAM Journal of Optimization 8, 476–507 (1998)
13. Pham Dinh, T., Le Thi, H.A.: DC Programming. Theory, Algorithms, Applications: The State of the Art. In: First International Workshop on Global Constrained Optimization and Constraint Satisfaction, Nice, October 2-4 (2002)
14. Pham Dinh, T., Le Thi, H.A.: The DC programming and DCA Revisited with DC Models of Real World Nonconvex Optimization Problems. Annals of Operations Research 133, 23–46 (2005)
15. Pinto da Costa, A., Seeger, A.: Cone constrained eigenvalue problems, theory and algorithms. Computational Optimization and Applications 45, 25–57 (2010)
16. Queiroz, M., Júdice, J., Humes, C.: The symmetric eigenvalue complementarity problem. Mathematics of Computation 73, 1849–1863 (2003)
17. Rockafellar, R.T.: Convex Analysis. Princeton University Press, Princeton (1970)
18. Seeger, A.: Eigenvalue analysis of equilibrium processes defined by linear complementarity conditions. Linear Algebra and Its Applications 294, 1–14 (1999)
19. Seeger, A.: Quadratic eigenvalue problems under conic constraints. SIAM Journal on Matrix Analysis and Applications 32, 700–721 (2011)

The Maximum Ratio Clique Problem: A Continuous Optimization Approach and Some New Results

Mahdi Moeini

Chair of Business Information Systems and Operations Research (BISOR),
Technical University of Kaiserslautern, Postfach 3049,
Erwin-Schrödinger-Str., D-67653 Kaiserslautern, Germany
moeini@wiwi.uni-kl.de

Abstract. In this paper, we are interested in studying the maximum ratio clique problem (MRCP) that is a variant of the classical maximum weight clique problem. For a given graph, we suppose that each vertex of the graph is weighted by a pair of rational numbers. The objective of MRCP consists in finding a maximal clique with the largest ratio between two sets of weights that are assigned to its vertices. It has been proven that the decision version of this problem is NP-complete and it is hard to solve MRCP for large instances. Hence, this paper looks for introducing an efficient approach based on Difference of Convex functions (DC) programming and DC Algorithm (DCA) for solving MRCP. Then, we verify the performance of the proposed method. For this purpose, we compare the solutions of DCA with the previously published results. As a second objective of this paper, we identify some valid inequalities and evaluate empirically their influence in solving MRCP. According to the numerical experiments, DCA provides promising and competitive results. Furthermore, the introduction of the valid inequalities improves the computational time of the classical approaches.

Keywords: Maximum Ratio Clique Problem, Fractional Programming, DC Programming, DCA.

1   Introduction

In this paper, we are given a simple undirected graph. We denote this graph by $G = (V, E)$, where V is the set of vertices and E is the set of edges. We denote the vertices of G by i such that $i \in \{1, \ldots, n\}$. In such a graph, a subset C of V defines a *clique* if C induces a complete subgraph of G. The concept of clique is very important in graph theory and its applications (see e.g., [1,16,21] and references therein). There are several variants of cliques: a *maximal clique* is defined as a clique that cannot be extended to another clique by adding new vertices (and consequently, adding new edges). For a given graph, a clique that has the maximum cardinality of vertices, is called a *maximum clique*. Finding a maximum clique of a graph is a classical combinatorial optimization problem.

© Springer International Publishing Switzerland 2015   215
H.A. Le Thi et al. (eds.), *Model. Comput. & Optim. in Inf. Syst. & Manage. Sci.*,
Advances in Intelligent Systems and Computing 359, DOI: 10.1007/978-3-319-18161-5_19

There are other optimization problems related to the concept of clique. For example, if we associate a weight $w_i \geq 0$ with each vertex i of the graph G, then we have the maximum weight clique problem (MWCP) that looks for a clique for which the sum of vertex weights is maximized.

In this paper, we suppose that we are given a graph G such that rational weights $a_i \geq 0$ and $b_i \geq 0$ are assigned to each vertex $i \in V$. We are interested in finding a maximal clique C in the graph G, such that the fractional quantity $\frac{\sum_{i \in C} a_i}{\sum_{i \in C} b_i}$ is maximized. This problem is called the *Maximum Ratio Clique Problem (MRCP)* [21]. The Maximum Ratio Clique Problem (MRCP) has various applications: e.g., portfolio optimization, social networks, etc. In these applications the interactions between different members of a set or society may be measured by means of a fractional function. This function focuses on the influence of the each member of the set or society on its neighbors. The objective consists in measuring the overall outcome of the influences.

Sethuraman et al. [21] formulated MRCP as an integer fractional programming problem. We know that unconstrained fractional $0 - 1$ programming problem is NP-hard [20] and Sethuraman et al. proved that the decision version of MRCP is NP-complete [21]. For solving MRCP, three solution approaches have been proposed: the first one is based on linearizing the fractional programming problem and the other methods are binary search and Newton's method. These methods have already been introduced for solving fractional programming problems [2,3,5].

In this paper, we are interested in studying MRCP. For this purpose, the objective of this paper is twofold: at first, in a similar way as Le Thi et al. [7], we investigate a novel approach based on techniques of non-convex programming. More precisely, for solving MRCP, we propose an approach based on Difference of Convex functions (DC) programming and DC Algorithm (DCA) (see [7,8,9]). This approach has a rich history of applications for solving a wide variety of problems and it proved to be efficient and robust; particularly, in solving large scale optimization problems (see e.g., [4,6,8,10,11,12,13,14,15,17,18,19]). As the second objective, in this paper, we study some mathematical properties of MRCP and propose some valid inequalities. Finally, we investigate the role of the valid inequalities in reducing computational time for solving the linearized MRCP.

We organize this paper as follows: In Section 2, the mathematical formulation of MRCP as well as some mathematical properties of the model are presented. This section is completed by linearizing MRCP. Section 3 is devoted to the basic concepts of DC programming, a DC formulation of MRCP, and the DC Algorithm (DCA) for solving MRCP. The computational experiments and numerical results are presented in Section 4. Finally, some conclusions are drown in Section 5.

2 Formulation of the Maximum Ratio Clique Problem (MRCP)

In order to present the mathematical model of MRCP, we define the decision variables $x_i \in \{0, 1\}$ ($\forall i \in V$), such that x_i is 1 if i belongs to the solution clique, otherwise $x_i = 0$. Furthermore, we suppose that $A = (a_{ij})$ is the adjacency matrix of the graph $G = (V, E)$. More precisely, $a_{ij} \in \{0, 1\}$ and $a_{ij} = 1$ iff the vertices i and j are connected ($i \neq j$). Assuming this notation, MRCP can be formulated as the following integer fractional programming model:

$$(MRCP): \quad \max \quad \frac{\sum_{i=1}^{n} a_i x_i}{\sum_{i=1}^{n} b_i x_i} \tag{1}$$

$$\text{Such that: } x_i + x_j \leq 1 \quad : \forall (i, j) \notin E, i \neq j, \tag{2}$$

$$\sum_{i=1}^{n} (1 - a_{ij}) x_i \geq 1 \quad : \forall j \in V, \tag{3}$$

$$x_i \in \{0, 1\} \quad : \forall i \in V. \tag{4}$$

In this model, the objective consists in maximizing the ratio $\frac{\sum_{i=1}^{n} a_i x_i}{\sum_{i=1}^{n} b_i x_i}$. Conforming to any specific application, this ratio can have different interpretations. It is important to note that for the classical maximum weight clique problem (MWCP), it is sufficient to consider only the constraints (2). Indeed, in the case of MWCP, by the non-negativity of the weights and the variables, any optimal solution will be a maximal clique. However, for MRCP, we need to add the constraints (3). In fact, if we ignore these constraints, it will be sufficient to take any single vertex k satisfying the following condition:

$$\frac{a_k}{b_k} = \max_{i \in V} \{ \frac{a_i}{b_i} \}.$$

because such a vertex can be considered as an optimal solution for the fractional program [21]. But the constraints (3) ensure that the optimal solution is a maximal clique. More precisely, for any optimal solution C of MRCP, the constraints (3) guarantee that:

$$j \in V \setminus C \Rightarrow \exists i \in C : a_{ij} = 0,$$

that is, such a vertex j cannot be added to C.

2.1 Some Mathematical Properties of MRCP

In this section, we explore, briefly, the mathematical structure of MRCP and we introduce some valid inequalities.

Property 1: Suppose that there are three different vertices i, j, and k in V such that any of edges (i, j), (i, k), and (j, k) satisfy the constraints (2), then the following inequality is valid for MRCP:

$$x_i + x_j + x_k \leq 1. \tag{5}$$

Proof: Since the conditions of the constraints (2) are met, in the optimal solution, just one of the variables x_i, x_j, x_k can be equal to 1. Hence, $x_i + x_j + x_k$ can be at most equal to 1. ∎

Clearly, the statement of *Property 1* can be extended for a larger number of vertices. However, from computational point of view, such a set of valid inequalities (with larger number of vertices) is challenging and impractical.

Property 2: Let $\{i, j, k\} \subset V$ be any 3-tuple of different vertices. Suppose that $\{i, j, k\}$ does not make a 3-clique (triangle) in G, then the following inequality is valid for MRCP:

$$x_i + x_j + x_k \leq 2. \tag{6}$$

Proof: Since $\{i, j, k\} \subseteq V$ do not make a 3-clique (triangle) in G; hence, at least, one of the edges $(i, j), (i, k), (j, k)$ do not belong to E. This is equivalent to that just one of the following cases may be true

$(x_i + x_j \leq 2$ and $x_k = 0)$ or $(x_i + x_k \leq 2$ and $x_j = 0)$ or $(x_j + x_k \leq 2$ and $x_i = 0)$.

If we consider any possible case and sum up the terms, we obtain (6). ∎

2.2   Linearization of MRCP

The formulation (1)-(4) is an integer fractional programming model that can be linearized by using the following classical method [21,22]: the first step of this linearization method consists in introducing supplementary variables as follows:

$$y = \frac{1}{\sum_{i=1}^{n} b_i x_i} \quad \text{and} \quad z_i = y x_i : \forall i \in V.$$

Since z_i is described by a quadratic term, we need the following linear constraints that replace $z_i = y x_i$

$$z_i \geq L x_i \text{ and } z_i \leq U x_i \qquad : \forall i \in V, \tag{7}$$
$$z_i \leq y - L(1 - x_i) \text{ and } z_i \geq y - U(1 - x_i) \qquad : \forall i \in V, \tag{8}$$

where L and U are some constants defining, respectively, lower and upper bounds on y. These bounds can be easily obtained through the following formulas:

$$L = \frac{1}{\sum_{i=1}^{n} b_i} \quad \text{and} \quad U = \frac{1}{\min_{i \in V} b_i}.$$

By gathering these materials, we obtain the following mixed integer linear programming (MILP) formulation for MRCP:

$$(MRCP - MILP): \quad \max \sum_{i=1}^{n} a_i z_i \tag{9}$$

$$\text{Such that: } x_i + x_j \leq 1 \qquad : \forall (i,j) \notin E, i \neq j, \qquad (10)$$

$$\sum_{i=1}^{n}(1 - a_{ij})x_i \geq 1 \qquad : \forall j \in V, \qquad (11)$$

$$\sum_{i=1}^{n} b_i z_i = 1 \qquad : \forall j \in V, \qquad (12)$$

$$z_j \leq U x_j \qquad : \forall j \in V, \qquad (13)$$

$$z_j \geq L x_j \qquad : \forall j \in V, \qquad (14)$$

$$z_j \leq y - L(1 - x_j) \qquad : \forall j \in V, \qquad (15)$$

$$z_j \geq y - U(1 - x_j) \qquad : \forall j \in V, \qquad (16)$$

$$x_j \in \{0,1\}; L \leq y \leq U; z_j \geq 0 \qquad : \forall j \in V. \qquad (17)$$

Any standard MILP-solver can be used for solving (9)-(17). In this paper, we are interested in investigating a non-convex programming approach for solving (9)-(17). The detailed description of the basic materials as well as the proposed algorithm are presented in the next section.

3   DC Programming and DC Formulation for MRCP

3.1   DC Programming: A Short Introduction

In this section, we review some of the main definitions and properties of DC programming and DC Algorithms (DCA); where, "DC" stands for "difference of convex functions".

Consider the following primal DC program

$$(P_{dc}) \qquad \beta_p := \inf\{F(x) := g(x) - h(x) \ : \ x \in \mathbb{R}^n\},$$

where g and h are convex and differentiable functions. F is a *DC function*, g and h are *DC components* of F, and $g - h$ is called a *DC decomposition* of F.

Let C be a nonempty closed convex set and χ_C be the indicator function of C, i.e., $\chi_C(x) = 0$ if $x \in C$ and $+\infty$ otherwise. Then, by using χ_C, one can transform the constrained problem

$$\inf\{g(x) - h(x) \ : \ x \in C\}, \qquad (18)$$

into the following unconstrained DC program

$$\inf\{f(x) := \phi(x) - h(x) \ : \ x \in \mathbb{R}^n\}, \qquad (19)$$

where $\phi(x)$ is a convex function defined by $\phi(x) := g(x) + \chi_C(x)$.

Hence, without loss of generality, we can suppose that the primal DC program is unconstrained and in the form of (P_{dc}).

For any convex function g, its conjugate is defined by $g^*(y) := \sup\{\langle x, y \rangle - g(x) : x \in \mathbb{R}^n\}$ and the dual program of (P_{dc}) is defined as follows

$$(D_{dc}) \qquad \beta_d := \inf\{h^*(y) - g^*(y) \ : \ y \in \mathbb{R}^n\}. \qquad (20)$$

One can prove that $\beta_p = \beta_d$ [19].

For a convex function θ and $x_0 \in \text{dom }\theta := \{x \in \mathbb{R}^n | \theta(x) < +\infty\}$, the subdifferential of θ at x_0 is denoted by $\partial\theta(x_0)$ and is defined by

$$\partial\theta(x_0) := \{y \in \mathbb{R}^n : \theta(x) \geq \theta(x_0) + \langle x - x_0, y \rangle, \forall x \in \mathbb{R}^n\}. \tag{21}$$

We note that $\partial\theta(x_0)$ is a closed convex set in $\mathbb{R}^n$ and is a generalization of the concept of derivative.

For the primal DC program (P_{dc}) and $x^* \in \mathbb{R}^n$, the necessary local optimality condition is described as follows

$$\partial h(x^*) \subset \partial g(x^*). \tag{22}$$

This condition is also sufficient for many important classes of DC programs, for example, for the polyhedral DC programs [18] (in order to have a *polyhedral* DC program, at least one of the functions g and h must be a polyhedral convex function; i.e., the point-wise supremum of a finite collection of affine functions).

We are now ready to present the main scheme of the DC Algorithms (DCA) [18,19] that are used for solving the DC programming problems. The DC Algorithms (DCA) are based on local optimality conditions and duality in DC programming, and consist of constructing two sequences $\{x^l\}$ and $\{y^l\}$. The elements of these sequences are trial solutions for the primal and dual programs, respectively. In fact, x^{l+1} and y^{l+1} are solutions of the following convex primal program (P_l) and dual program (D_{l+1}), respectively:

$$(P_l) \quad \inf\{g(x) - h(x^l) - \langle x - x^l, y^l \rangle : x \in \mathbb{R}^n\}, \tag{23}$$

$$(D_{l+1}) \quad \inf\{h^*(y) - g^*(y^l) - \langle y - y^l, x^{l+1} \rangle : y \in \mathbb{R}^n\}. \tag{24}$$

One must note that, (P_l) and (D_{l+1}) are convexifications of (P_{dc}) and (D_{dc}), respectively, in which h and g^* are replaced by their corresponding affine minorizations. By using this approach, the solution sets of (P_{dc}) and (D_{dc}) are $\partial g^*(y^l)$ and $\partial h(x^{l+1})$, respectively. To sum up, in an iterative scheme, DCA takes the following simple form

$$y^l \in \partial h(x^l); \quad x^{l+1} \in \partial g^*(y^l). \tag{25}$$

One can prove that the sequences $\{g(x^l) - h(x^l)\}$ and $\{h^*(y^l) - g^*(y^l)\}$ are decreasing, and $\{x^l\}$ (respectively, $\{y^l\}$) converges to a primal feasible solution (respectively, a dual feasible solution) satisfying the local optimality conditions. For a complete study of DC programming and DCA, readers are referred to [8,18,19] and the list of references on http://lita.sciences.univ-metz.fr/ lethi/ DCA.html.

3.2   DC Programming For Solving MRCP

In order to solve the maximum ratio clique problem, we investigate a novel approach based on DC programming and DCA. For this purpose, we need a reformulation of MRCP with a DC objective function that is minimized over a convex set. In this section, we explain the mathematical operations that we need for transforming MRCP to an *equivalent* DC programming model. Then, we present a DCA for solving the proposed DC program. More precisely, in a similar way as Le Thi et al. [7], by using an exact penalty result (presented in [7,9]) we can formulate (MRCP-MILP) as a DC minimization problem subject to linear constraints, which is consequently a DC program. At the first step, in order to simplify the notations, we define:

$A := \{(\mathbf{x}, y, \mathbf{z}) \in [0,1]^n \times [L, U] \times \mathbb{R}^n : (\mathbf{x}, y, \mathbf{z}) \text{ satisfies } (10) - (16)\}.$

Let $p(\mathbf{x}, y, \mathbf{z})$ be the concave function defined as follows

$$p(\mathbf{x}, y, \mathbf{z}) := \sum_{i=1}^n x_i(1 - x_i).$$

Since $p(\mathbf{x}, y, \mathbf{z})$ is non-negative on A, (MRCP-MILP) can be re-written as follows

$$\min\left\{ -\sum_{i=1}^n a_i z_i : p(\mathbf{x}, y, \mathbf{z}) \le 0, (\mathbf{x}, y, \mathbf{z}) \in A \right\}. \tag{26}$$

Moreover, the objective function of (26) is linear, A is a bounded polyhedral convex set, and the concave function $p(\mathbf{x}, y, \mathbf{z})$ is non-negative on A; consequently, we can use the exact penalty result presented in [9] and we obtain the following equivalent formulation of MRCP

$$\min\left\{ F(\mathbf{x}, y, \mathbf{z}) := -\sum_{i=1}^n a_i z_i + t p(\mathbf{x}, y, \mathbf{z}) : (\mathbf{x}, y, \mathbf{z}) \in A \right\}, \tag{27}$$

where $t > t_0$ and $t_0 \in \mathbb{R}_+$ is a sufficiently large positive number. Furthermore, the function F is concave in variables $\mathbf{x}$ and linear in variables y and $\mathbf{z}$; hence, F is a DC function. A natural DC formulation of the problem (27) is

(MRCP-DC): $\min\left\{ F(\mathbf{x}, y, \mathbf{z}) := g(\mathbf{x}, y, \mathbf{z}) - h(\mathbf{x}, y, \mathbf{z}) : (\mathbf{x}, y, \mathbf{z}) \in \mathbb{R}^{2n+1} \right\},$

where

$$g(\mathbf{x}, y, \mathbf{z}) = -\sum_{i=1}^n a_i z_i + \chi_A(\mathbf{x}, y, \mathbf{z})$$

and

$$h(\mathbf{x}, y, \mathbf{z}) = t \sum_{i=1}^n x_i(x_i - 1).$$

Here, χ_A is the indicator function on A, i.e. $\chi_A(\mathbf{x}, y, \mathbf{z}) = 0$ if $(\mathbf{x}, y, \mathbf{z}) \in A$ and $+\infty$ otherwise.

3.3   DCA for solving (MRCP-DC)

According to the general scheme of DCA, firstly, we require a point in the sub-differential of the function $h(\mathbf{x}, y, \mathbf{z})$ defined by $h(\mathbf{x}, y, \mathbf{z}) = t \sum_{i=1}^{n} x_i(x_i - 1)$. From the definition of $h(\mathbf{x}, y, \mathbf{z})$ we have

$$(\mathbf{u}^k, s^k, \mathbf{w}^k) \in \partial h(\mathbf{x}^k, y^k, \mathbf{z}^k) \Leftrightarrow s^k = 0, w_i^k = 0, u_i^k := t(2x_i^k - 1) : i = 1, \ldots, n. \tag{28}$$

Secondly, we need to find $(\mathbf{x}^{k+1}, y^{k+1}, \mathbf{z}^{k+1})$ in $\partial g^*(\mathbf{u}^k, s^k, \mathbf{w}^k)$. Such a point can be an optimal solution of the following linear program:

$$\min \left\{ -\sum_{i=1}^{n} a_i z_i - \langle (\mathbf{x}, y, \mathbf{z}), (\mathbf{u}^k, s^k, \mathbf{w}^k) \rangle : (\mathbf{x}, y, \mathbf{z}) \in A \right\} \tag{29}$$

To sum up, the DCA applied to (MRCP-DC) can be summarized as follows:

Algorithm DCA for solving (MRCP-DC)

---

1. **Initialization**: Choose $(\mathbf{x}^0, y^0, \mathbf{z}^0) \in \mathbb{R}^{2n+1}$, $\epsilon > 0$, $t > 0$, and set $k = 0$.
2. **Iteration**:
   - Set $s^k = 0$, $w_i^k = 0$, and $u_i^k := t(2x_i^k - 1)$ for $i = 1, \ldots, n$.
   - Solve the linear program (29) to obtain $(\mathbf{x}^{k+1}, y^{k+1}, \mathbf{z}^{k+1})$.
3. If $\left\| (\mathbf{x}^{k+1}, y^{k+1}, \mathbf{z}^{k+1}) - (\mathbf{x}^k, y^k, \mathbf{z}^k) \right\| \leq \epsilon$ then stop the algorithm and take the vector $(\mathbf{x}^{k+1}, y^{k+1}, \mathbf{z}^{k+1})$ as an optimal solution, otherwise set $k \longleftarrow k+1$ and go to step **2**.

Finding a suitable initial point for DCA:
One of the key questions in DCA consists in finding a good initial point for starting DCA. In this work, for a given graph $G = (V, E)$, we took a maximal clique $\widetilde{C}$ in G [21]. Such a clique can be found as follows:

1. Select a vertex having the maximum value for the ratio of its weights,
2. Add its neighbors in decreasing order of the ratio of their weights,
3. After adding a new vertex, make sure that the new set of vertices is still a clique.

Once the maximal clique $\widetilde{C}$ is formed, we construct $(\mathbf{x}^0, y^0, \mathbf{z}^0)$ as follows:

- $x_i^0 = 1 \Longleftrightarrow i \in \widetilde{C}$ for $i = 1 \ldots, n$.
- $y^0 = \frac{1}{\sum_{i=1}^{n} b_i x_i^0}$.
- $z_i^0 = y^0 x_i^0$ for $i = 1 \ldots, n$.

In fact, we tested different initial points for starting DCA, some of them are:

- The point obtained by the above procedure;
- $(\mathbf{x}, y, \mathbf{z}) = (0, \ldots, 0) \in \mathbb{R}^{2n+1}$;

- $(\mathbf{x}, y, \mathbf{z}) = (1, \dots, 1) \in \mathbb{R}^{2n+1}$;
- The optimal solution of the relaxed (MRCP-MILP) problem obtained by replacing the binary constraints $x_i \in \{0, 1\}$ by $0 \leq x_i \leq 1$ for all $i = 1, \dots, n$.

According to our experiments, the initial point provided by the first procedure gives the best results.

4   Computational Experiments

This section is devoted to the computational experiments and the numerical results. Through the experiments, we are interested in:

- evaluating the performance of DCA in solving (MRCP-DC),
- investigating the influence of valid inequalities (5) and (6) in solving ($MRCP-MILP$) by means of the standard MILP-Solver *IBM CPLEX*.

The experiments have been carried out on two types of data sets: randomly generated instances and real-world data related to construction of wind turbines. A more detailed description of the test instances can be found in [21].

We compared our solutions with the results from earlier studies. Indeed, Sethuraman et al. [21] proposed three approaches for solving MRCP: solving (MRCP-MILP) by the standard MILP-Solver *IBM CPLEX*, an adaptation of the Binary search [2,5], and an adaptation of the Newton's method [3]. We did our experiments (under same conditions) by using these methods as well as DCA. More precisely, we implemented all of the algorithms by $C++$ and ran the codes on a DELL laptop equipped with Linux operating system, Intel Core 2 Duo CPU of 2.53GHz and 3.8 GB of memory. The standard solver IBM CPLEX 12.5.1 has been used as the MILP/LP solver.

Concerning the parameters that we need to set for DCA, we chose $\epsilon = 10^{-6}$ as the precision of the solutions, and for a test instance of size n, the penalty parameter (i.e., t) is set to $n/4$.

Table 1 shows some information about each of the test instances: number of vertices ($|V|$) and number of edges ($|E|$). Also, for each instance (*instance*), the best optimal values (*best val.*) and the size of their corresponding maximum ratio clique (*C.Size*) are reported. These values correspond to the solutions of the exact methods.

4.1   Numerical Results of DCA for Solving MRCP

The first set of results concerns the assessment of DCA in solving (MRCP-DC) versus the other methods (i.e., IP-Solver, Binary Search, and Newton's method). The results are presented in Table 1. In this table, the computational CPU time (in seconds) of all solving methods are shown. A separate section of Table 1 is dedicated to the results of DCA algorithm: for each instance, the objective value provided by DCA (*dc val.*), the size of the maximum ratio clique found by DCA ((*C.Size*)), the computational time of DCA (*CPU*) in seconds, and its number of iterations (*iter.*) are presented.

Table 1. The results of the DCA in comparison to the other methods: IBM CPLEX, Binary Search, and Newton's method

instance	$|V|$	$|E|$	best val.	C.Size	Computational Time(s.)			DCA			
					CPLEX	Binary	Newton	dc Val.	C.Size	CPU	iter.
random-1	100	2266	3.28	10	1.69	1.00	0.64	1.85	6	0.06	2
random-2	150	5212	4.69	8	2.12	1.79	1.79	1.31	11	0.30	4
random-3	200	10008	4.21	5	11.14	4.10	3.48	0.96	7	0.17	2
random-4	400	40786	4.83	7	2348.76	57.91	41.04	0.98	11	1.06	2
random-5	500	63789	3.65	9	14674.00	235.80	488.826	0.91	13	2.37	2
random-6	100	2655	1.15	13	1.50	0.71	0.47	1.04	10	0.05	2
random-7	150	5767	1.20	9	16.02	4.05	3.53	0.90	14	0.10	2
random-8	200	10220	1.19	10	103.14	12.67	8.96	0.84	14	0.17	2
random-9	400	38942	1.32	7	1076.10	29.43	29.77	0.95	13	1.11	2
wind-2004	500	10277	92736.30	3	87.24	39.87	17.97	92736.30	3	8.78	4
wind-2005	500	10516	94686.60	2	65.93	18.36	9.13	91999.40	3	2.16	2
wind-2006	500	9681	98471.00	2	45.15	16.59	15.48	93666.00	3	2.22	2

The presented DCA algorithm has produced satisfactory results in comparison to the other methods. The exact methods are efficient in solving some small/medium sized instances; however, they need longer time for solving the other instances. Among the tested instances, we observe that DCA has a very good performance in solving *random-6, ..., random-9, wind-2004, wind-2005,* and *wind-2006.* The results are particularly interesting for the test instance *wind-2004,* for which the proposed DCA method gives the same results as the exact methods, but in a significantly shorter CPU time.

4.2   Numerical Results of the Valid Inequalities

In the second part of experiments, we assess the influence of the valid inequalities (5) and (6) in improving the performance of the IP-Solver. More precisely, we add (5) and (6) to (MRCP-MILP) and solve the augmented models. Depending on the size of instance, the number of these inequalities can be huge (indeed, $O(n^3)$). Consequently, if we include all of the constraints (5) and/or (6), the model becomes intractable. Hence, we use a simple heuristic in order to include a smaller number of them. For this purpose, we define a *"size limit"* and for $\{i, j, k\} \subset V$, we add (5) and/or (6) iff $i, j, k \leq size\ limit$. In our experiments, *size limit* is set to 10 and 5 for (5) and (6), respectively. The results are shown in Table 2. In this table, the performance of CPLEX is assessed, in terms of CPU time, against the exclusion/inclusion of each set of the valid inequalities (5) and (6). The column "only (5)" (respectively, "only (6)") concerns the model (MRCP-MILP) after adding the valid inequalities (5) (respectively, (6)). The number of added inequalities are represented by #(5) (respectively, #(6)). The last column shows the results for the case of including both types of inequalities.

According to the results, we observe that the valid inequalities (5) have more positive influence in reducing computational time of CPLEX for solving (MRCP-

MILP). Furthermore, when we add both types of inequalities (5) and (6), the computational time is relatively improved but adding only clique inequalities (5) gives a better performance to the exact method.

Table 2. The computational time (in seconds) of IBM CPLEX in solving (MRCP-MILP) with/without valid inequalities (5) & (6).

instance	$|V|$	$|E|$	(MRCP-MILP)	only (5)	#(5)	only (6)	#(6)	both (5) & (6)
random-1	100	2266	1.69	1.75	180	1.68	60	1.74
random-2	150	5212	2.12	2.01	90	2.47	96	2.03
random-3	200	10008	11.14	8.51	114	14.60	60	8.52
random-4	400	40786	2348.76	2240.64	294	1897.80	60	1743.53
random-5	500	63789	14674.00	10820.70	186	24585.80	60	20348.30
random-6	100	2655	1.50	1.31	72	1.47	60	1.30
random-7	150	5767	16.02	16.08	258	57.89	60	12.59
random-8	200	10220	103.14	83.91	66	103.73	84	177.93
random-9	400	38942	1076.10	834.26	60	1461.94	60	1326.18
wind-2004	500	10277	87.24	66.55	519	113.71	60	53.23
wind-2005	500	10516	65.93	48.09	504	66.46	60	48.44
wind-2006	500	9681	45.15	33.75	720	35.84	60	37.58

The header spans: Computational Time(s.) of IP-Solver IBM CPLEX

5 Conclusion

In this paper, we presented a new approach based on DC programming and DCA for solving the maximum ratio clique problem (MRCP). We saw that DCA provides competitive results in comparison to the other methods and shows to be computationally quick and efficient in giving high quality solutions. Furthermore, we investigated the mathematical properties of MRCP and we proposed two sets of valid inequalities. Finally, we presented the numerical experiments on some data sets and described our observations. Our experiments confirms that adding clique inequalities can have significant improvement in computational time of exact methods.

The computational results suggest to us extending the numerical experiments in higher dimensions and combining the proposed approach as well as the most promising valid inequalities in the framework of an exact approach (such as Branch-and-Bound algorithms) for globally solving MRCP. Works in these directions are currently in progress and the results will be reported in future.

Acknowledgements. Mahdi Moeini acknowledges the chair of Business Information Systems and Operations Research (BISOR) at the TU-Kaiserslautern (Germany) for the financial support through the research program "CoVaCo". Mahdi Moeini has also been supported by both CNRS and OSEO within the

ISI project "Pajero" (France). The research was started while the author was still affiliated with the *Center of Research in Computer Science (CNRS-CRIL)*, *France.*

References

1. Boginski, V., Butenko, S., Pardalos, P.: Mining market data: a network approach. Comput. Oper. Res. 33, 3171–3184 (2006)
2. Ibaraki, T.: Parametric approaches to fractional programs. Math. Prog. 26, 345–362 (1983)
3. Isbell, J.R., Marlow, W.H.: Attrition games. Naval Res. Logist. Q 3(1-2), 71–94 (1956)
4. Kröller, A., Moeini, M., Schmidt, C.: A Novel Efficient Approach for Solving the Art Gallery Problem. In: Ghosh, S.K., Tokuyama, T. (eds.) WALCOM 2013. LNCS, vol. 7748, pp. 5–16. Springer, Heidelberg (2013)
5. Lawler, E.L.: Combinatorial optimization: networks and matroids. Holt, Rinehart and Winston, New York (1976)
6. Le Thi, H.A.: Contribution à l'optimisation non convexe et l'optimisation globale: Théorie, Algorithmes et Applications. Habilitation à Diriger des Recherches, Université de Rouen (1997)
7. Le Thi, H.A., Pham Dinh, T.: A continuous approach for globally solving linearly constrained quadratic zero-one programming problems. Optimization 50(1-2), 93–120 (2001)
8. Le Thi, H.A., Pham Dinh, T.: The DC (difference of convex functions) Programming and DCA revisited with DC models of real world non convex optimization problems. Annals of Operations Research 133, 23–46 (2005)
9. Le Thi, H.A., Pham Dinh, T., Huynh, V.N.: Exact Penalty Techniques in DC Programming. Research Report, LMI, INSA-Rouen, France (2005)
10. Le Thi, H.A., Moeini, M.: Portfolio selection under buy-in threshold constraints using DC programming and DCA. In: International Conference on Service Systems and Service Management (IEEE/SSSM 2006), pp. 296–300 (2006)
11. Le Thi, H.A., Moeini, M., Pham Dinh, T.: Portfolio Selection under Downside Risk Measures and Cardinality Constraints based on DC Programming and DCA. Computational Management Science 6(4), 477–501 (2009)
12. Le Thi, H.A., Moeini, M., Pham Dinh, T.: DC Programming Approach for Portfolio Optimization under Step Increasing Transaction Costs. Optimization 58(3), 267–289 (2009)
13. Le Thi, H.A., Moeini, M., Pham Dinh, T., Judice, J.: A DC Programming Approach for Solving the Symmetric Eigenvalue Complementarity Problem. Computational Optimization and Applications 51, 1097–1117 (2012)
14. Le Thi, H.A., Moeini, M.: Long-Short Portfolio Optimization under Cardinality Constraints by Difference of Convex Functions Algorithm. Journal of Optimization Theory and Applications 161(1), 199–224 (2014)
15. Liu, Y., Shen, X., Doss, H.: Multicategory ψ-Learning and Support Vector Machine: Computational Tools. Journal of Computational and Graphical Statistics 14, 219–236 (2005)
16. Luce, R., Perry, A.: A method of matrix analysis of group structure. Psychometrika 14, 95–116 (1949)

17. Nalan, G., Le Thi, H.A., Moeini, M.: Robust investment strategies with discrete asset choice constraints using DC programming. Optimization 59(1), 45–62 (2010)
18. Pham Dinh, T., Le Thi, H.A.: Convex analysis approach to d.c. programming: Theory, Algorithms and Applications. Acta Mathematica Vietnamica, dedicated to Professor Hoang Tuy on the occasion of his 70th birthday 22(1), 289–355 (1997)
19. Pham Dinh, T., Le Thi, H.A.: DC optimization algorithms for solving the trust region subproblem. SIAM J. Optimization 8, 476–505 (1998)
20. Prokopyev, O.A., Huang, H., Pardalos, P.M.: On complexity of unconstrained hyperbolic 0-1 programming problems. Oper. Res. Lett. 3(3), 312–318 (2005)
21. Sethuraman, S., Butenko, S.: The Maximum Ratio Clique Problem. Comp. Man. Sci. 12(1), 197–218 (2015)
22. Wu, T.: A note on a global approach for general 0–1 fractional programming. Eur. J. Oper. Res. 101(1), 220–223 (1997)

Part III
Dynamic Optimization

Dynamic Adaptive Large Neighborhood Search for Inventory Routing Problem

Viacheslav A. Shirokikh and Victor V. Zakharov

St.Petersburg State University,
Faculty of Applied Mathematics and Control Processes,
Universitetsky Prospect 35, St.Petersburg, Peterhof, 198504, Russia
shva.erivel@gmail.com, mcvictor@mail.ru

Abstract. This paper is devoted to new approach to increase level of time consistency of heuristics and propose Dynamic Adaptive Large Neighborhood Search (DALNS) algorithm to improve solutions generated by ALNS.

To evaluate effectiveness of DALNS implementation computational experiments were performed on benchmark instances. It was shown that the number of tests in which solution was improved equals 5236 (46% of total amount).

Keywords: time consistency, inventory routing problem (IRP), heuristic algorithms, adaptive large neighborhood search (ALNS), dynamic adaptive large neighborhood search (DALNS).

1   Introduction

The main purpose of this paper is to describe new method which could help to improve performance of heuristics used for Inventory Routing Problems (IRP). Heuristic methods are very popular for solving this type of problems, since they are NP-hard and arise in large-scale systems.

Heuristics do not guarantee the obtained solution to be optimal. If the solution is not optimal, then there exist at least one period such that continuation of this solution is not optimal in corresponding subproblem. While reducing the scale of subproblem compare to initial one, the probability to get better route (if the algorithm includes randomization, of course) could increase, in general. Thus, there is a possibility to improve already obtained solution using the same heuristic algorithm.

We describe in this paper general model of IRP and implementation of Adaptive Large Neighborhood Search (ALNS) heuristic algorithm. Then, we discuss the idea of time consistency and suggest its possible application for improving performance of used heuristics, and present results of computing experiments.

2   Literature Review

Inventory routing problem arises as generalization of VRP, more exactly its capacitated version (CVRP), by the way of including in consideration multiple periods and holding costs.

© Springer International Publishing Switzerland 2015
H.A. Le Thi et al. (eds.), *Model. Comput. & Optim. in Inf. Syst. & Manage. Sci.*,
Advances in Intelligent Systems and Computing 359, DOI: 10.1007/978-3-319-18161-5_20

First papers on this problem are practical research studies: propane distribution in study by Golden, Dror et al. in 1982 [1] and following theoretical paper in 1985 [2]; distribution of industrial gases in study by Bell, Dalberto, Fisher et al. in 1983 [3]; reducing logistics costs at General Motors in study by Blumenfeld, Burns et al., theoretical paper in 1985 [4] and practical research report in 1987 [5].

2.1   Typology of the Problem

We consider one-to-many case of IRP. Namely, we examine joint routing and distribution planning over T periods from one depot to N customers, using a single vehicle. Optimization objective is to minimize total transportation and holding costs. Such model is also considered in papers by Bertazzi et al. in [6, 7].

Different modifications of the problem have been studied over the years. Natural modifications are generalization for several vehicles [3], [8] or several products [3], [9], [10].

Depending on considered demand model, there are deterministic (with constant demand), stochastic (when probability distribution or its parameters are known [11, 12, 13]) and dynamic (when demand is known only at current time [11]) versions of problem.

Along with the basic version, in which stock-outs at customers are not allowed, there are some studies, dealing with shortage [12, 13].

Versions of the problem differ by objective function. Most of the researches consider total transportation and holding costs as objective function, but some examine the problem of multicriteria optimization [14].

Some researchers modify the form of feasible solution in order to simplify solution search procedure. Such modifications are IRP with direct deliveries [4, 5], with transshipments [15] and order-up-to level policy [7, 15, 17].

Aside from others, maritime IRP is being considered. Specifications of this problem include relatively small amount of ports (customers/depots), travel and operations time accounting, consideration of discharging and waiting costs, heterogeneous fleet [10], [16].

2.2   Algorithms

First studies on the problem utilized exact methods combined with existing algorithms for VRP [1, 2, 3], and analytical study possibilities also were considered [4, 5].

Among used exact algorithms, the most popular are brunch-and-cut algorithm modifications [7, 8, 9], but other algorithms also used (like column generation or Lagrangian relaxation), along with special packages, like CPLEX. Exact algorithms often used for solving subproblem, combining with heuristic [3], [17].

Because of large scale and complexity of the problem, most studies use heuristic methods. At this point, almost every popular global optimization method was applied, such as genetic algorithm [18], particle swarm algorithm [19] or Monte Carlo simulation [13].

Various complicated heuristic search algorithms have been developed for the IRP. Popular examples are tabu search based hybrid algorithm [17] or adaptive large neighborhood search [11], [15].

2.3   Recent Papers

Recent studies are focused mostly on exploring of new formulations for the problem. Used algorithms are mostly adaptations of well-known local search methods, but some researchers examine exact methods [22].

Several studies modify the model by tightening restrictions in order to accelerate search procedure [23, 24].

Some new formulations appears with practical applications, such as ATMs refilling [25], bikes transportation [26] and waste oil collection [27].

Other notable problem formulations in recent studies are cyclic IRP [28, 29], IRP with pick-ups [25, 26], IRP with fixed routing decisions [29, 30, 31], "green" approach [32] and multi-objective IRP [33].

More detailed and wide literature review can be found in recent review papers "thirty years of inventory routing" [34] and "formulations for an inventory routing problem" [35].

3   Mathematical Problem Definition

Let $G = (N, A)$ be the graph, where $N = \{0,1,2, \dots, n\}$ is set of nodes (points), 0 for depot and $1,2, \dots, n$ for customers, and $A = \{(i,j) \mid i \in N, j \in N\}$ is set of arcs (direct paths). Each arc has its travel cost c_{ij}. There is also the Euclidean version of the problem in which N is given with corresponding coordinates set $\{(x_i, y_i), i \in N\}$ and travel cost is defined as Euclidean distance: $c_{ij} = \sqrt{(x_i - x_j)^2 + (y_i - y_j)^2}$.

At each node $i \in N$ one can store some amount I_i of the product which is integer value. Holding cost h_i is known for each node. Stored amount of product is bounded: $0 \leq I_i \leq U_i$.

Time horizon consists of T periods. At each period vehicle can perform one route which starts and ends in the depot. At the beginning of each period r units of product is being produced at the depot and d_i units is being consumed at each customer i .

Routes and deliveries are variables in the problem.

Product amount loaded in the vehicle is restricted by its capacity Q.

The problem objective function is total transportation and holding costs during time horizon.

Suppose x_{ijt} is a binary variable which is equal to one if route at period t passes the arc (i, j), and zero otherwise, and q_{it} is an integer variable for amount of delivery to customer i at period t. Initial inventory level is given by $I_{i0} = I_i^0 = const, \forall i \in N$. Then the problem can be defined in the form of linear program as follows:

$$f = \sum_{t=1}^{T}\sum_{i=0}^{n}\sum_{j=0}^{n} c_{ij} * x_{ijt} + \sum_{t=1}^{T}\sum_{i=0}^{n} h_i * I_{it} \to min$$

subject to:

$$I_{0t} = I_{0,t-1} + r - \sum_{i=1}^{n} q_{it}, \qquad t = 1, \dots, T$$

$$I_{it} = I_{i,t-1} - d_i + q_{it}, \qquad t = 1, \dots, T, i = 1, \dots, n$$

$$0 \le I_{it} \le U_i, \qquad \forall i \in N, t = 1, \dots, T$$

$$\sum_{i=1}^{n} q_{it} \le Q, \qquad t = 1, \dots, T$$

$$\sum_{i=0}^{n} x_{ijt} = \sum_{i=0}^{n} x_{jit}, \qquad \forall j \in N, t = 1, \dots, T$$

$$\sum_{j=1}^{n} x_{0jt} \le 1, \qquad \forall t = 1, \dots, T$$

Where the restrictions are, respectively: inventory dynamic at depot and at customers, inventory restrictions, vehicle capacity restrictions, route continuity restrictions and single vehicle restrictions.

4   Adaptive Large Neighborhood Search Algorithm

This algorithm has been described in details in [15].

The algorithm is adaptive randomized local search in large solution neighborhood, including elements of simulated annealing.

Solution neighborhood here is defined as all solutions, obtained from applying one of elementary moves, such as random insertion/removal customers in the route, best insertion/removal, cluster insertion/removal or routes re-assign.

Because of largeness of such neighborhood, the move to apply is chosen randomly, respectively to their weights, which accounts the successiveness of recent applications.

Simulated annealing is used to include possibility of move to worse solution with probability, depends on current "temperature".

In this implementation, a system of penalties has been added to the objective function in order to exclude unfeasible solution in result.

Initial solution is obtained using "greedy" algorithm to construct route which passes all customers at each period, all y_{it} is set to d_i.

Following values of algorithm parameters were used in this implementation: $\tau_{start} = 30000$, $\varphi = 0{,}9994$, $\sigma_1 = 10$, $\sigma_2 = 5$, $\sigma_3 = 2$, $k = 200$.

We present general scheme of ALNS algorithm below.

Algorithm 1: ALNS

1: Initialization: set $weight = 1$ and $score = 0$ for each move
2: Set initial solution: $s = s_{best} = s_0$
3: Set initial temperature: $\tau = \tau_{start}$
4: **while** $\tau > 0.01$ **do**
5:     $s' = s$
6:     Randomly select move m and apply it to s'
8:     Fix routing decisions for s' and solve the remaining problem
9:     **if** $f(s') < f(s)$ **then**
10:         $s = s'$
11:         **if** $f(s) < f(s_{best})$ **then**
12:             $s_{best} = s$
13:             $score[m] = score[m] + \sigma_1$
14:         **else**
15:             $score[m] = score[m] + \sigma_2$
16:     **else if** $p = random(0;1) \leq exp\left(\frac{f(s)-f(s')}{\tau}\right)$ **then**
18:         $s = s'$
19:         $score[m] = score[m] + \sigma_3$
20:     $\tau = \varphi * \tau$
21:     **if** the iteration count is a multiple of k **then**
22:             update the weights and reset the scores
23:             perform 2-opt procedure for each route

Remaining problem in the algorithm consists in choosing of delivery volumes, while routes are fixed. To solve it a simple heuristic based on inventory function analysis was developed.

We consider inventory level function $I(t)$ on the interval $[t_1; t_2]$ between two vehicle visits. This function decreases monotonically on such interval. Delivering y units of product to the customer at t_1 means that $I(t)$ increases by this value in all successive periods. Therefore, maximum delivery value is leftover capacity at period t_1 and minimum is deficit (negative inventory level) at period t_2, regarding the capacity.

Algorithm 2: Heuristic for the remaining problem

0: Initialization: nullify deliveries: $\forall n, t: q_n(t) = 0$, recalculate $I_n(\cdot)$
1: **for each** $n \in N$ **do**
2:     **if** $h_n < h_0$ **do**
3:         **for** $t = 1, \ldots, T$, **if** n is visited at t **then**
4:             $q_n(t) = U_n - I_n(t)$, recalculate $I_n(\cdot)$
5:     **if** $h_n \geq h_0$ **then**
6:         **for each** interval between visits $[t_1; t_2]$, incuding $t_2 = T$ **do**
7:             **if** $I_n(t_2) < 0$ **then**
8:                 $q_n(t_1) = min\{-I_n(t_2 - 1), U_n - I_n(t_1)\}$, recalculate $I_n(\cdot)$
9: Fix the solution to satisfy capacity restrictions: $\sum_n q_n(t) \leq Q$, $I_n(t) \leq U_n$

---

5   Time Consistency and Heuristic Algorithms

Bellman in 1957 [20] formulated the optimality principle. According to it, "an optimal policy has the property that whatever the initial state and initial decision are, the remaining decisions must constitute an optimal policy with regard to the state resulting from the first decision". Therefore, we can conclude that for any step restriction of optimal solution for remaining time interval is optimal for corresponding current subproblem.

If the solution is obtained with a heuristic algorithm, its optimality for the main problem is not guaranteed, and the same is for restricted solutions and corresponding subproblems. However, subproblems are smaller in scale, therefore algorithm working time decreases and the probability of obtaining better solution could increase.

5.1   Algorithm Quality Criterion: Time Consistency

This criterion can be used for general routing problems, because their solutions (routes) can be represented as consequence of nodes visited during time horizon. The same representation also can be used for IRP.

Let us consider set of benchmark instances P for the IRP class. Suppose that for each instance $p \in P$ we can obtain a set of different solutions $s(p) \in N(p)$ generated by an algorithm. Length of the consequence of nodes corresponding to $s(p)$ is denoted by $K(s(p))$. Define $s(k, p)$ as remaining consequence of nodes after step $k = 1, ..., K(s(p))$. Suppose k to be executed in period $t(k)$. At step k we construct the subproblem $p'(k)$ on interval $[t(k); T]$ where nodes already visited at period $t(k)$ are excluded for visit in current period, their initial inventory level is equal to $I_{i,t(k)}$ and $I_{i,t(k)-1}$ for others. Then we obtain solution $s(p'(k))$ for the subproblem using the same algorithm.

Definition. Solution $s(p)$ is *time consistent*, if for each $k = 1, ..., K(s(p))$ the following inequality holds: $f(s(k, p)) \leq f(s(p'(k)))$, where f is the objective function in the current problem.

Let us perform $M(s)$ experiments for each solution $s \in N(p)$. Each experiment consists in checking whether solution is time consistent or not.   Let $b(s, k)$ be the number of experiments, in which time consistency is violated for the solution s at step k. If s is optimal, then regarding Bellman's principle of optimality we would have the following quality: $\sum_{k=1}^{K(s)} b(s, k) = 0$. Since heuristics do not guarantee optimality of the solution, we can only have $\sum_{k=1}^{K(s)} b(s, k) \leq M(s)$.

Definition. Assume *experimental level of heuristic time consistency* ($conL$) to be the value calculated as follows

$$conL = 1 - \frac{1}{|P|} \times \sum_{p \in P} \frac{1}{|N(p)|} \sum_{s \in N(p)} \frac{1}{M(s)} \sum_{k=1}^{K(s)} b(s, k)$$

One can note that $0 \leq conL \leq 1$. Higher value of experimental level of time consistency for heuristic forms expectation that this heuristic generates solutions which appear to be "more" time consistent compare to other heuristics having less value of the defined criterion.

5.2   Dynamic Adaptive Large Neighborhood Search (DALNS)

Iterative method for improving solution of VRP (more exactly, CVRG – cooperative vehicle routing game) was proposed in paper [21]. Using its main idea, we upgrade this method for the class of inventory routing problems.

We introduce the following notation. Suppose $ALNS(\cdot)$ is the ALNS algorithm function, s is the solution obtained with the algorithm, S_1 is the consequence of passed nodes of the solution s, S_2 is the consequence of remaining nodes, C is constants of the problem, f is the objective function of the problem. Then, we can present the general scheme of the Dynamic Adaptive Large Neighborhood Search (DALNS).

Algorithm 3: DALNS

0:   Initialization: initial solution s_0, $S_1 = \emptyset$, $S_2 = s_0$
1:   **while** $S_2 \neq \emptyset$ **do**
2:       $s = ALNS\,(\,S_2,\ C\,)$
3:       **if** $f(s) < f(S_2)$ **then** $S_2 = s$
4:       $C \leftarrow S_2[1]$, $S_1 \leftarrow S_2[1]$, $S_2 = S_2 \setminus S_2[1]$

---

6   Computational Results

All algorithms were programmed in C++. Experiments were performed at HPC-cluster of Electromechanical and Computer Systems Modelling department, faculty of Applied Mathematics and Control Processes, SPbU. Cluster specifications: operating system Linux SLES 11 SP1, 12 nodes × 2 processors Intel Xeon 5335 × 4 cores, 16 GB RAM per node.

To perform experiments, 20 instances were taken from resource http://www.leandro-coelho.com/instances/inventory-routing/. They originate from [6], and have been considered in number of IRP studies. Chosen instances differ in level of holding cost (h: low or high), number of periods (T: 3 or 6) and number of customers (n: 10/20/30/40/50/100). These characteristics are also presented in Table 1 in column "instance".

We have calculated the experimental level of time consistency for ALNS algorithm:

$$conL = 0{,}545424$$

For each instance, 100 solutions have been obtained, among them $N(p)$ were different. For each solution DALNS has been implemented M times (one for large instances: 6×50 and 6×100; ten for others).

We use following additional notation. $N1$ is the number of violations of time consistency among all $N(p) \times M$ tests. $N2$ is the number of time inconsistent solutions, i.e. solutions that were time inconsistent in all M tests. In the same sense, $N3$ is the number of time consistent solutions. Note that $N2 + N3 \leq N(p)$. This values are presented in Table 1.

In the last columns of this table, we compare ALNS and DALNS performance. We present average (Avg. f) and standard deviation (σ) of objective function vaules of generated solutions set for each instance.

Table 1. Results of DALNS implementation

Instance			N(p)	M	N1	N2	N3	ALNS		DALNS	
h	T	n						Avg. f	σ	Avg. f	σ
low	3	10	7	10	0	0	7	1625,04	56,5964	1625,04	56,5964
		20	26	10	0	0	26	2695,07	351,649	2695,07	351,649
		30	72	10	18	1	67	3770,14	406,636	3770,07	406,12
		40	74	10	73	5	61	4284,57	483,59	4267,39	448,406
		50	92	10	38	2	85	4647,59	375,56	4638,27	347,246
	6	10	51	10	323	25	10	3787,11	172,105	3716,21	152,237
		20	100	10	783	68	13	6143,56	459,988	5922,44	384,808
		30	100	10	827	61	4	8113,19	597,666	7807,11	471,044
		50	100	1	90	90	10	10526,8	624,891	10209,2	709,574
		100	100	1	75	75	25	16873	938,445	16507,2	912,191
high	3	10	7	10	0	0	7	3663,97	55,3075	3663,97	55,3075
		20	30	10	17	0	24	6040,8	266,576	6039,99	265,39
		30	87	10	125	4	64	10140,3	361,031	10129,3	330,997
		40	87	10	180	14	62	11398,7	445,25	11377	392,563
		50	91	10	197	13	67	12360,4	305,425	12348,1	290,239
	6	10	72	10	585	48	6	7574,81	191,467	7484,25	193,026
		20	100	10	833	70	8	13045,5	361,751	12900,8	333,288
		30	100	10	884	81	5	20357,3	527,622	20165,1	422,808
		50	100	1	94	94	6	27358,4	656,647	27109,8	635,106
		100	100	1	94	94	6	52188,2	780,861	51821,9	698,469

7   Conclusions

The main purpose of the study was to explore the idea of heuristic solution time consistency. One can notice that level of time consistency of the solution generated by ALNS is little bit more than 0.5, that is about half of generated solutions appears to be time inconsistent. We have suggested in the paper a new approach to find level of time consistency of heuristics and proposed the Dynamic Adaptive Large Neighborhood Search (DALNS) algorithm which can generate better solutions than ALNS.

DALNS shows itself well in computational experiments: the value of improvement reaches 22% of primary solution and the number of tests in which solutions generated by ALNS were improved equals 5236 (46% of total amount). Note, that effectiveness of the method increases with the scale of problem.

References

1. Assad, A., Golden, B., Dahl, R., Dror, M.: Design of an inventory routing system for a large propane-distribution firm. In: Gooding, C. (ed.) Proceedings of the 1982 Southeast TIMS Conference, pp. 315–320 (1982)
2. Dror, M., Ball, M., Golden, B.: A computational comparison of algorithms for the inventory routing problem. Annals of Operations Research 4(1), 1–23 (1985)
3. Bell, W.J., Dalberto, L.M., Fisher, M.L., Greenfield, A.J., Jaikumar, R., Kedia, P., Mack, R.G., Prutzman, P.J.: Improving the distribution of industrial gases with an on-line computerized routing and scheduling optimizer. Interfaces 13(6), 4–23 (1983)
4. Blumenfeld, D.E., Burns, L.D., Diltz, J.D., Daganzo, C.F.: Analyzing trade-offs between transportation, inventory and production costs on freight networks. Transportation Research Part B: Methodological 19(5), 361–380 (1985)
5. Blumenfeld, D.E., Burns, L.D., Daganzo, C.F., Frick, M.C., Hall, R.W.: Reducing logistics costs at General Motors. Interfaces 17(1), 26–47 (1987)
6. Bertazzi, L., Paletta, G., Speranza, M.G.: Deterministic order-up-to level policies in an inventory routing problem. Transportation Science 36(1), 119–132 (2002)
7. Archetti, C., Bertazzi, L., Laporte, G., Speranza, M.G.: A branch-and-cut algorithm for a vendor-managed inventory-routing problem. Transportation Science 41(3), 382–391 (2007)
8. Coelho, L.C., Laporte, G.: The exact solution of several classes of inventory-routing problems. Computers and Operations Research 40(2), 558–565 (2013)
9. Coelho, L.C., Laporte, G.: A branch-and-cut algorithm for the multi-product multi-vehicle inventory-routing problem. International Journal of Production Research 51(23-24), 7156–7169 (2013)
10. Ronen, D.: Marine inventory routing: Shipments planning. Journal of the Operational Research Society 53(1), 108–114 (2002)
11. Coelho, L.C., Cordeau, J.-F., Laporte, G.: Heuristics for dynamic and stochastic inventory-routing. Computers and Operations Research 52(Part A), 55–67 (2014)
12. Bertazzi, L., Bosco, A., Guerriero, F., Laganà, D.: A stochastic inventory routing problem with stock-out. Transportation Research Part C: Emerging Technologies 27, 89–107 (2013)
13. Cáceres-Cruz, J., Juan, A.A., Bektas, T., Grasman, S.E., Faulin, J.: Combining Monte Carlo simulation with heuristics for solving the inventory routing problem with stochastic demands. In: Laroque, C., Himmelspach, J., Pasupathy, R., Rose, O., Uhrmacher, A. (eds.) Proceedings of the 2012 Winter Simulation Conference, pp. 274–274. IEEE Press, Piscataway (2012)

14. Geiger, M.J., Sevaux, M.: The biobjective inventory routing problem – problem solution and decision support. In: Pahl, J., Reiners, T., Voß, S. (eds.) INOC 2011. LNCS, vol. 6701, pp. 365–378. Springer, Heidelberg (2011)
15. Coelho, L.C., Cordeau, J.F., Laporte, G.: The inventory-routing problem with transshipment. Computers and Operations Research 39(11), 2537–2548 (2012)
16. Papageorgiou, D.J., Nemhauser, G.L., Sokol, J., Cheon, M.S., Keha, A.B.: MIRPLib–A library of maritime inventory routing problem instances: Survey, core model, and benchmark results. European Journal of Operational Research 235(2), 350–366 (2014)
17. Archetti, C., Bertazzi, L., Hertz, A., Speranza, M.G.: A hybrid heuristic for an inventory routing problem. INFORMS Journal on Computing 24(1), 101–116 (2012)
18. Moin, N.H., Salhi, S., Aziz, N.A.B.: An efficient hybrid genetic algorithm for the multi-product multi-period inventory routing problem. International Journal of Production Economics 133(1), 334–343 (2011)
19. Marinakis, Y., Marinaki, M.: A particle swarm optimization algorithm with path relinking for the location routing problem. Journal of Mathematical Modelling and Algorithms 7(1), 59–78 (2008)
20. Bellman, R.: Dynamic Programming. Princeton University Press, Princeton (1957)
21. Zakharov, V.V., Schegryaev, A.N.: Multi-period cooperative vehicle routing games. Contributions to Game Theory and Management 7, 349–359 (2014)
22. Desaulniers, G., Rakke, J.G., Coelho, L.C.: A branch-price-and-cut algorithm for the inventory-routing problem (2015), http://www.leandro-coelho.com/publications/
23. Coelho, L.C., Laporte, G.: Improved solutions for inventory-routing problems through valid inequalities and input ordering. International Journal of Production Economics 155(1), 391–397 (2014)
24. Agra, A., Christiansen, M., Delgado, A., Simonetti, L.: Hybrid heuristics for a short sea inventory routing problem. European Journal of Operational Research 236(3), 924–935 (2014)
25. Van Anholt, R.G., Coelho, L.C., Laporte, G., Vis, I.F.A.: An Inventory-Routing Problem with Pickups and Deliveries Arising in the Replenishment of Automated Teller Machines (2015), http://www.leandro-coelho.com/publications/
26. Brinkmann, J., Ulmer, M.W., Mattfeld, D.C.: Inventory Routing for Bike Sharing Systems (2015), https://www.tu-braunschweig.de/winfo/team/brinkmann
27. Aksen, D., Kaya, O., Salman, F.S., Tüncel, Ö.: An adaptive large neighborhood search algorithm for a selective and periodic inventory routing problem. European Journal of Operational Research 239(2), 413–426 (2014)
28. Vansteenwegen, P., Mateo, M.: An iterated local search algorithm for the single-vehicle cyclic inventory routing problem. European Journal of Operational Research 237(3), 802–813 (2014)
29. Raa, B.: Fleet optimization for cyclic inventory routing problems. International Journal of Production Economics 160, 172–181 (2015)
30. Bertazzi, L., Bosco, A., Laganà, D.: Managing stochastic demand in an Inventory Routing Problem with transportation procurement. Omega (October 16, 2014), http://dx.doi.org/10.1016/j.omega.2014.09.010
31. Cordeau, J.F., Laganà, D., Musmanno, R., Vocaturo, F.: A decomposition-based heuristic for the multiple-product inventory-routing problem. Computers and Operations Research 55, 153–166 (2015)

32. Mirzapour Al-e-hashem, S.M.J., Rekik, Y.: Multi-product multi-period Inventory Routing Problem with a transshipment option: A green approach. International Journal of Production Economics 157, 80–88 (2014)
33. Huber, S., Geiger, M.J., Sevaux, M.: Interactive approach to the inventory routing problem: computational speedup through focused search. In: Dethloff, J., Haasis, H.-D., Kopfer, H., Kotzab, H., Schönberger, J. (eds.) Logistics Management. Lecture Notes in Logistics, pp. 339–353. Springer International Publishing (2015)
34. Coelho, L.C., Cordeau, J.F., Laporte, G.: Thirty years of inventory routing. Transportation Science 48(1), 1–19 (2013)
35. Archetti, C., Bianchessi, N., Irnich, S., Speranza, M.G.: Formulations for an inventory routing problem. International Transactions in Operational Research 21(3), 353–374 (2014)

Reactive Multiobjective Local Search Schedule Adaptation and Repair in Flexible Job-Shop Problems

Hélène Soubaras

Thales Research & Technology France
Decision & Optimization Laboratory
Campus Polytechnique, Palaiseau, France

Abstract. This paper deals with the flexible job-shop scheduling problem (FJSP): an amount of jobs have to be executed by a limited number of resources that can be exchanged for some tasks. Solving such a schedule consists in allocating a resource for each task in the jobs. But one must be able to cope with unexpected changes in the model, i.e. uncertainties such as a modification of the duration of some tasks, or an additional job, or a resource that is added or removed... Yet, for operational reasons, the change in the schedule must remain little. We propose a domain-independent plan adaptation algorithm satisfying those requirements, which principle is to move tasks within the plan like sliding puzzle pieces. This algorithm is also able to cope with uncertainties on the tasks duration. It does not need the initial solver. This local search approach is compared to another, a classical tabu search [7] in which we introduced several criteria.

1 Introduction

We are interested in temporal planning problems of resource allocation such as in the crisis management context [1,2], where one must organize the rescuers intervention after a disaster. There are several independent goals (to rescue each victim). So there is a job (ordered list of tasks) to execute for each victim. And one has several resources that cannot be shared at the same time (e.g. different types of ambulances, vehicles...). There are resources that can be exchanged (e.g. there are several ambulances). Such a problem where a set of jobs must be scheduled on resources (or machines) is classically called a job-shop problem. In our case, there are several alternative resources that can be allocated to some tasks: it is a flexible job-shop problem (FJSP) [7].

We suppose that a schedule, i.e. is a timed list of tasks, has already been computed by an external solver. It is being executed, but a problem occurs suddenly: a task is delayed or ending at an unexpected time, or a new estimation is provided for a future task duration. The disruption corresponds to a change in the FJSP model. The schedule is thus no more valid. It must be repaired or adapted.

© Springer International Publishing Switzerland 2015
H.A. Le Thi et al. (eds.), *Model. Comput. & Optim. in Inf. Syst. & Manage. Sci.*,
Advances in Intelligent Systems and Computing 359, DOI: 10.1007/978-3-319-18161-5_21

The aim of this work is to propose a plan repair method. The ongoing tasks must not be stopped. The initial planning solver is supposed not to be reasonably available, so one cannot replan entirely from scratch. One must propose a solution which is executable and reaches the goals minimizing the makespan (overall plan duration). There must also be few changes between the original plan and the repaired one, since the agents, who are humans, must not be disrupted. We will thus consider multiobjective approaches to cope with these two criteria (makespan and distance to the original schedule).

Local search approaches are well suited to this kind of problem since their principle is to choose a solution which is close to a previous one. In this paper we will present an approach derived from a classical tabu search that was proposed by Gambardella et al.[7], and a new approach using the computation of potentials: the 15 puzzle potential algorithm. It has been proposed in order to improve the time computation.

2   Problem Modelling

2.1   Model for the Flexible Job-Shop Problem (FJSP)

A job-shop problem is a very classical scheduling problem consisting of jobs that have to be executed on a set of resources (or machines). Each job is composed of tasks. Each task has a given duration and must be executed by one resource, and each resource can perform only one task at a time. Solving the job-shop problem consists in determining a start time of each task by optimizing a criterion – which is generally the makespan (overall duration).

A FJSP is a job-shop problem where the resource that can be used to perform a given task can be chosen in a set of possible alternative resources. Formally, the FJSP is defined by

- a finite set of resources $\mathcal{R}$,
- a set $\{R_1, R_2, ...R_k, ...\}$ of subsets of $\mathcal{R}$ (classes of resources),
- a set of jobs J_i, each one being defined by a ordered sequence of tasks,
- all the tasks of all the jobs are denoted as $a = a_p$ with $1 \leq p \leq N_p$; each p represents the task belonging to a job J_i at a rank j,
- $class(a)$ the class of resource needed for each task,
- $\tau(a)$ the time duration of each task.

To solve a FJSP, one must find for each task a a start time $t_{start}(a)$ and a resource $r(a) = r \in R_k$ such that $class(a) = k$. A solution to the FJSP is a schedule (plan). It can be denoted as $\Pi = (t_{start}, r)$.

The subplan (i.e. a subset of tasks of a plan) using one given resource is called a *queue*. Each task a of the plan belongs to at most one job and at most one queue.

In the job shop scheduling problem formalism, the tasks a_p are vertices of the associated disjunctive graph $G(V, C \cup D)$ ([7]) with the set of vertices $V = \Pi \cup \{a_0, a_{N_p+1}\}$ (one adds dummy vertices with 0 duration at the beginning and the end of the plan). The *precedence constraints* represented by the directed edges C correspond to the order that must be respected in each job. Since the resources cannot be used by more than one task at the same time, one has additional constraints in each queue. So we can express all of them by a direct dependence relationship between tasks: $\mathcal{D}(a_p, a_q) = 1$ if and only if a_p precedes a_q in a job or in a queue. $\mathcal{D}(a_p, a_q) = 0$ otherwise.

2.2   Extension of the FJSP Model to Duration Uncertainty Intervals

We will consider also an extension of the FJSP by introducing an uncertainty about the duration of the tasks. In that case, each task a will have a duration uncertainty interval $[\tau_{min}(a), \tau_{max}(a)]$.

2.3   Existing Techniques

In the literature, there are several plan repair strategies that do not use the initial solver; in [12] the dilemma of repairing a plan is discussed; in [3,10], a theoretical formalism of causality is studied. Some plan adaptation with search techniques use the LPG (Local search for Planning Graphs) algorithm [8,6], which is a local search-based on a graph. Many other approaches use partial plans, such as *refinement planning* [11], which proceeds in two phases: first removing the tasks that prevent the plan from reaching its goal, then planning to extend the partial plans.

Apart from the plan repair issue, the problem of handling uncertainties on task durations has also been studied in the literature [4,5]. These approaches are stochastic (they are interested in the probability of satisfying a goal). In these approaches time is a constraint in models involving concurrency between tasks.

Gambardella [7] proposed a local search approach for FJSP scheduling. It is based on an intuitive heuristics that we will apply to plan repair and plan adaptation in this paper. It has also been associated to genetic approaches ([14]). Huang [9] proposed a repair method for FJSP by formulating as a constraint satisfaction problem.

The proposed domain-independent plan repair approach, where the overall duration of the plan is a criterion to minimize and not a constraint, handles uncertainties about tasks durations within a simple theoretical model of the plan – a set of jobs. It is an empirical sub-optimal strategy including an original method to shortlist solutions. The merit of our new approach is to reduce the computation time.

2.4   Expression of the "Few Changes" Criterion

In order to evaluate the changes between two plans, one must define a distance criterion. There exists a criterion for the distance between two plans Π and Π' that was proposed by Fox *et al.* [6], which is the number of tasks that are in Π' and not in Π plus the number of tasks that are in Π and not in Π':

$$D(\Pi, \Pi') = \sharp(\Pi' \setminus \Pi) + \sharp(\Pi \setminus \Pi')$$

This criterion can be used locally and it expresses what the authors call the *plan stability*. It is adapted to plans that are repaired by local modifications, e.g. where a local subplan has been replaced by another. Thus this distance criterion is equal to the sum of the lengths of the two subplans.

In our approach the modifications of the plan will not be local. The changes will affect the parameters of some tasks. So, it is thus more relevant to consider the distances between jobs and to average them on the jobs. We thus chose the quadratic distance between times in order to fulfill the mathematical properties of a distance:

$$D_t(\Pi, \Pi') = \sqrt{\sum_i d_t(a_p, a'_p)}$$

where a_p are the tasks of Π, a'_p are the tasks of Π', and

$$d_t(a_p, a'_p) = \left(t_{start}(a_p) - t_{start}(a'_p)\right)^2 + \left(t_{end}(a_p) - t_{end}(a'_p)\right)^2$$

is the square distance between the initial time $t_{start}(a_p)$ and final times $t_{end}(a_p)$ of corresponding tasks a_p and a'_p before and after a plan modification. In practice, we used the minimum duration: $t_{end}(a_p) = t_{start}(a_p) + \tau_{min}(a_p)$.

3   The Two Multiobjective Local Search Approaches

In this section we describe the two approaches to be compared: the Gambardella's tabu search [7] extended to multiple criteria, and the 15 puzzle potential algorithm, which is orignial. Both methods use some common issues that will be described first.

3.1   Common Issues for the Two Approaches

Plan Analysis. Before running the algorithm, an initial plan analysis phase extracts the resource it uses and the job it belongs to (if any). The initial plan is supposed to be compressed, i.e. each task is dated as soon as possible according to the previous tasks it depends on. Then the plan is decomposed into jobs and queues (see Section 2.1). One computes the durations T_i of each job J_i. The maximum among them, T_{max}, is the duration of the plan.

Moving a Task. The idea is to apply so-called *elementary modifications* in the plan and to choose good candidates iteratively. They consist in replacing the resource r of a task by another one r' belonging to the same class. Each elementary modification can be seen as "moving" a task from one queue to another within the plan like moving a piece in a 15 puzzle[1]. An elementary modification will be denoted as $d = (r, n, r', n')$. The moved task was formerly located at rank n on the queue of resource r, and it is moved onto the queue of resource r' at rank n'.

---

[1] The most famous sliding puzzle.

Critical Path. A critical path of a plan Π is a subplan Π' of Π of tasks blocking each other and ending at T_{max}. In other words, $\Pi' = \{a'_p, p = 1...N'_p \leq N_p\}$ in which a'_{p+1} cannot be performed before a'_p is finished and $t_{end}(a'_p) = t_{start}(a'_{p+1})$ for $1 \leq p \leq N'_p - 1$ and $t_{end}(a'_{N'_p}) = T_{max}$. An example is illustrated in Figure 1.

In the two approaches examined in this paper, one computes all the elementary modifications obtained by moving a task from the critical paths only. It is obvious that moving a task from a non-critical path is less relevant.

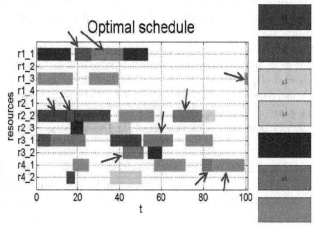

Fig. 1. Optimal schedule for the jobs of Table 1. An example of critical path is indicated (tasks pointed by arrows).

Hierarchical Multi-criteria Optimization. In both studied approaches, we want to take into account two criteria: the makespan T_{max} and the distance to the original plan, which is will be represented by $D_t(\Pi, \Pi')$ but can be replaced by another distance criterion or the total number of elementary modifications. To introduce these two criteria in the algorithms, we will use hierarchy: the makespan is optimized first. In there are several ex-aequo solutions, the second criterion is optimized.

There is also another criterion: the computation time (CPU). It will be evaluated afterwards and of course it is not an optimization criterion.

3.2   The Multicriteria Tabu Search Approach

The idea of tabu search is to start from one solution and find successive new solutions. The solutions that have already been examined are eliminated and listed in a so-called *tabu list*. Each new solution is chosen as being the best one in a neighborhood of the previous solution out of the tabu list.

We will use Gambardella's neighborhood ([7]), which is the set of solutions obtained by moving one task (see Paragraph 3.1).

And we use also his tabu list, i.e.: when moving a task a onto the queue of resource r' behind task a_{prev}, one adds the triple $=a_{prev}, a, r')$ to the tabu list. If the tabu list is too long its size is limited by forgetting the oldest triples.

3.3   The 15 Puzzle Potential Algorithm

Principle. The idea of this new approach is to use an approximate criterion – the potential – which is faster to compute than the actual makespan, in order to reduce the computational time. Some solutions are shortlisted on the basis of the potentials, and then the overall computation is performed on the shortlist only (and not on all the possible solutions).

The algorithm examines successively a growing number N_{modif} of elementary modifications. At each iteration, we have a set of selected plans containing $N_{modif} - 1$ modifications.

For each of them, we examine all possible additional elementary modification and reject it if it does not satisfy the precedence constraints; otherwise, we compute a function called potential. Then we shortlist the $N_{shortlist}$ modifications who have the best potentials; then we compute and store their associated plans.

When all the plans containing $N_{modif} - 1$ modifications have been examined, we select the N_{select} best solutions among all the shortlists.

Computation of the Potentials. We are considering here a possible elementary modification, which consists in moving a given task a from its original queue to queue $\Pi(r)$ at rank k. First one must eliminate some cases: already seen cases, impossible cases due to cycles, or impossible cases due to the fact that the user may have imposed that some tasks cannot move.

In an elementary modification, one must:

- delay if necessary the task a_2 following a_p in its new queue;
- delay if necessary the task a_{next} following a_p in its job J_i;
- move forward if possible the task a_1 following a_p in its former queue;
- move forward a_{next} (if it has not been delayed).

Is is possible to compute a lower bound for the new start time : $t'_{start}(a_1)$, $t'_{start}(a_2)$ and $t'_{start}(a_{next})$. We thus introduce the following variables:

$$\delta_1 = t'_{start}(a_1) - t_{start}(a_1) \qquad \delta_2 = t'_{start}(a_2) - t_{start}(a_2)$$

$$\delta_{next} = t'_{start}(a_{next}) - t_{start}(a_{next})$$

and to obtain the potential of the elementary modification, we compute the uncertainty interval I_i for the durations of all the jobs J_i that are depending on a_1, a_2 or a_{next}. The overall duration will be inside the interval defined by

$$\underline{T}_{max} = max_i\left(inf(I_i)\right) \qquad \text{and} \qquad \overline{T}_{max} = max_i\left(sup(I_i)\right)$$

and the potential, which is our criterion including the risk aversion α, is

$$V = \alpha \underline{T}_{max} + (1 - \alpha)\overline{T}_{max} \tag{1}$$

Thus, for a given number of elementary modifications, we select the $N_{shortlist}$ candidate plans who have the best value for $\mathcal{V}$. Then, we store each of them, and compute exactly their T_{max}.

When this has been done for all the numbers of elementary modifications, we choose the minimal T_{max} among all the stored plans. If several plans have the same minimal value, we choose the plan with least temporal distance to the initial plan.

This expression of the potentials has been chosen such that it is representative of the makespan of a schedule, which can be seen in Figure 2 and however it is faster to compute than the real makespan (since one must not compute all the values of t_{start}).

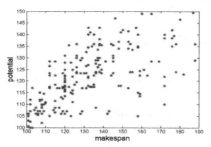

Fig. 2. Relationship between the potential ane the corresponding makespan (computed in one of the test plans of Section 4.1

Structure of the Algorithm. So, there are two levels for selection: the elementary modifications are shortlisted from their potential; the best plan is chosen among the shortlists according to its computed criteria. Here below is the pseudo-code (Algorithm 1).

4   Results

4.1   Comparison of the Two Approaches

The proposed tabu search approach and the 15 puzzle potential algorithm are suited to repair plans for various types of disruptions. One idea to test its capacity to repair plans is to compare its results on various random disruptions of a same plan. To do that, we took as a reference a given set of jobs (that was generated randomly), and a given set of resources.

The Test Sample. The performances of the studied plan adaptation approaches have been studied by Monte-Carlo on randomly generated schedules. This test sample was built in the same way than the first test sample used in [7]. It consists in N_s jobs that were generated randomly for a total number M_p of tasks, for each task a class of resource is determined randomly. The total number of resources is N_r.

Algorithm 1. 15 Puzzle Potential Algorithm

1. $\Sigma_0 = \{\Pi_0\}$	// Initial schedule
2. $\Sigma' = \Sigma_0$	// Shortlist of solutions
3. **for** $1 \leq k \leq k_{max}$ **do**	// number of modifications
4. $\Sigma_k = \emptyset$	// set of solutions with k modifications
5. **for** $\Pi \in \Sigma_{k-1}$ **do**	
6. **for** $d = (r, n, r', n')$ **do**	
7. **if** $\Pi' = f(\Pi, d)$ is possible **then**	
8. $\mathcal{D}(\Pi) = \mathcal{D}(\Pi) \cup \{d\}$	// set of possible modifications
9. COMPUTE $V(\Pi, d)$	
10. keep the $N_{shortlist}$ elements of $\mathcal{D}(\Pi)$ that have best $V(\Pi, d)$	
11. **for** $d \in \mathcal{D}(\Pi)$ **do**	
12. COMPUTE $\Pi' = f(\Pi, d)$	
13. COMPUTE $\Sigma_k = \Sigma_k \cup \{\Pi'\}$	
14. keep the N_{select} best elements of Σ_k	
15. $\Sigma' = \Sigma' \cup \Sigma_k$	
16. keep the best $\Pi \in \Sigma'$	

For each job, we generated randomly:

- a number of tasks;
- a resource class needed for each task;
- a duration τ for each task, with average $\tau_0 = 10$ and max variation rate $max(\tau - \tau_0)/\tau_0 = 0.8$.

The set of jobs is then fixed but we generate randomly a great number of allocations (picking a resource in each class). The time of each task is chosen as soon as possible such that the plan is feasible.

It this series of tests we had $N_s = 7$ jobs that were generated randomly for a total number $M_p = 50$ of tasks. The total number of resources is $N_r = 11$, with 4 classes of resources, each one containing respectively 4, 3, 2 and 2 resources. The jobs and the classes of resources are given in Table 1.

Table 1. The jobs used in the tests. The numbers are the classes of resources to be used in the tasks. One supposes that two consecutive tasks may be different however they use the same class of resource.

J1: 3 4 2 3	J5: 1 2 1 3
J2: 3 2 1	J6: 3 1 3 3 2 4
J3: 2 2	J7: 1 1 4 1 2 4 3 4 1
J4: 2 4	

The optimal solution was computed using the CPT solver [13]. It is shown in Figure 1. Its makespan is 101. To use CPT, we had to generate the description of the jobs in PDDL language.

To do the Monte-Carlo evaluation, we generated a set of 10 schedules by random allocations on these jobs. We ran the algorithms on each of them and compared the averaged performances the obtained schedules.

Obtained Performance. For the 15 puzzle algorithm, we adjusted experimentally the parameters: $\alpha = 0.5$ and $N_{select} = 1$. To study the influence of the shortlist size $N_{shortlist}$, we obtained the results shown in Table 2, and the corresponding curves are shown in Figure 3. When $N_{shortlist}$ is too small, Table 2 even shows that there are failures (the algorithm does not find a solution). We conclude that we must choose $N_{shortlist} < 10$.

Finally, if we consider independently the 3 performance criteria (the makespan, the temporal distance, and the CPU), the comparison of the tabu search and the 15 puzzle approaches is summarized in Table 3:

Table 2. Obtained performances with various values for $N_{shortlist}$

	Tabu	15puzzle					
$N_{shortlist}$	1	1	2	3	5	10	15
makespan	101	112.5	108.6	104.9	105.3	103.7	103.7
CPU	2.53	0.85	1.03	1.36	1.80	2.82	3.89
D_t	98.55	75.14	74.33	80.02	73.72	76.04	74.15
N_{modifs}	4.7	4.3	5.1	4.6	3.4	4.0	4.4
optimal	100 %	30 %	50 %	60 %	60 %	80 %	80 %
failure	0 %	10 %	10 %	0 %	0 %	0 %	0 %

Table 3. Best approach for each criterion

	Tabu	15puzzle
makespan	X	
temporal distance		x
CPU		X

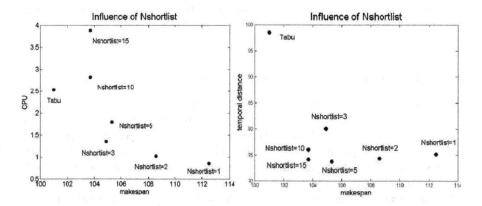

Fig. 3. (*Left*) Average CPU versus makespan for various values of $N_{shortlist}$. (*Right*) Average temporal distance versus makespan for various values of $N_{shortlist}$

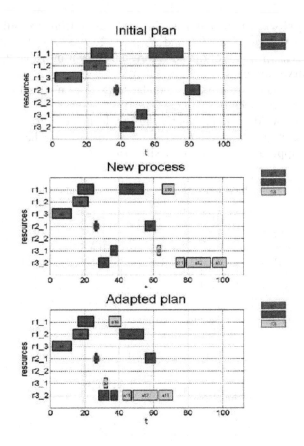

Fig. 4. (*Top*) Initial plan with 2 ships allocated on 7 resources of the harbor. (*Middle*) New process (job) for a new ship arriving at the harbor at $t = 30$. (*Bottom*) Adapted plan for the 3 ships, after running the 15-puzzle algorithm.

4.2 Application to a Flow of Arriving Jobs

The 15 puzzle potential algorithm can be applied to assess a flow of jobs that arrive randomly and must be dispatched to a limited number of resources, as for example ships arriving in a harbor. When a ship comes, it needs to be processed with an ordered sequence of tasks : disembark passengers, unload, oil supply, maintenance... Each task uses a resource: platform, crane, manpower... Each ship is represented by a job.

In the following numerical example, the jobs were generated randomly as in the tests below but one at a time. We also used the same model for the 7 resources in the harbor. At a given time, one has an initial plan with 2 ships. At time $t = 30$, a 3rd ship arrives: it is randomly allocated and simply added to the previous plan without optimizing the overall plan duration. After running the 15 puzzle algorithm, the overall duration has been shortened (see Figure 4).

5   Conclusion

We have examined two approaches for plan repair and adaptation for FJSP. They are based on a local search consisting in moving tasks from one resource to another. One is derived from a classical tabu search, the other one – the 15 puzzle algorithm – is an original approach based on potentials. Both work from a given initial plan without using the initial solver. One of the main operational advantages is that these algorithms handle a criterion to maintain little distance to the initial plan. The 15 puzzle is less optimal than the tabu search but it often obtains a solution which is close to the optimum in a shorter time and with less distance to the initial plan. The 15 puzzle can also a larger class of problems since it can address uncertainties about the durations of the tasks. Note that the approaches were implemented with an additional operational constraint: some tasks were fixed.

Implementation and validation tests show that this new algorithm works, providing repaired plans that are shorter in time duration than the single compression of a disrupted plan. It often finds an optimal solution with few modifications. And if is well suited to manage a flow of demands.

Its capability to involve uncertainty about tasks duration can be exploited more in further works. We just showed theoretically in this paper that an uncertainty about task durations can be taken into account in the potential function. But one could do more testing, and also introduce a fuzzy model for the tasks durations.

References

1. Aligne, F.: Which information and decision support system for crisis management? In: Proc. of Information Syst. Technology Panel Symp (IST-086/RSY-019), C3I for Crisis, Emergency and Consequence Management (May 2009)
2. Aligne, F., Savéant, P.: Automated planning in evolving contexts: an emergency planning model with traffic prediction and control. In: Proc. Future Security Conf., Bonn, Germany, September 4-6 (2011)
3. Alterman, R.: Adaptive Planning. Cognitive Science 12(3), 393–421 (1988)
4. Beaudry, E., Kabanza, F., Michaud, F.: Planning for concurrent action executions under action duration uncertainty using dynamically generated bayesian networks. In: ICAPS, pp. 10–17 (2010)
5. Coles, A.J., Coles, A.I., Clark, A., Gilmore, S.T.: Cost-sensitive concurrent planning under duration uncertainty for service level agreements. In: Proceedings of the Twenty First International Conference on Automated Planning and Scheduling (ICAPS 2011) (June 2011)
6. Fox, M., Gerevini, A., Long, D., Serina, I.: Plan stability: Replanning Versus Plan Repair. In: 16th Int. Conf. on Automated Planning and Scheduling (ICAPS 2006), pp. 212–221. AAAI Press (2006)
7. Gambardella, L., Mastrolilli, M.: Effective neighborhood functions for the flexible job shop problem. Journal of Scheduling 3(3) (1996)
8. Gerevini, A., Serina, I.: Fast Plan Adaptation through Planning Graphs: Local and Systematic Search Techniques. In: 5th Int. Conf. on AI Planning and Scheduling (AIPS 2000), pp. 112–121. AAAI Press, Menlo Park (2000)

9. Huang, Y., Zheng, L., Williams, B.C., Tang, L., Yang, H.: Incremental temporal reasoning in job shop scheduling repair. In: 2010 IEEE International Conference on Industrial Engineering and Engineering Management (IEEM), pp. 1276–1280. IEEE (2010)

10. Kambhampati, S., Hendler, J.A.: A Validation-Structure-Based Theory of Plan Modification and Reuse. Artificial Intelligence 55(2-3), 193–258 (1992)

11. van der Krogt, R., de Weerdt, M.: Plan Repair as an Extension of Planning. In: 15th International Conference on Automated Planning and Scheduling (ICAPS 2005), pp. 161–170. AAAI Press (2005)

12. Nebel, B., Koehler, J.: Plan Reuse versus Plan Generation: A Theoretical and Empirical Analysis. Artificial Intelligence 76, 427–454 (1995)

13. Vidal, V., Geffner, H.: Branching and Pruning: An Optimal Temporal POCL Planner based on Constraint Programming. Artificial Intelligence 170(3), 298–335 (2006)

14. Zhang, G., Shi, Y., Gao, L.: A genetic algorithm and tabu search for solving flexible job shop schedules. In: International Symposium on Computational Intelligence and Design, ISCID 2008, vol. 1, pp. 369–372. IEEE (2008)

Part IV
Modelling and Optimization in Financial Engineering

An Approach to Forming and Managing a Portfolio of Financial Securities by Small and Medium Price-Taking Traders in a Stock Exchange

Alexander S. Belenky[1,2] and Lyudmila G. Egorova[1]

[1] National Research University Higher School of Economics, Moscow, Russian Federation
legorova@hse.ru
[2] Center for Engineering Systems Fundamentals, MIT, Cambridge, MA, USA
abelenky@mit.edu

Abstract. The paper discusses a new approach to developing tools for quantitatively analyzing the financial behavior of small and medium price-taking traders each possessing abilities to predict share price values for a set of financial securities traded in a stock exchange. Tools for forming and managing a trader's portfolio of securities from this set are proposed. Particularly, it is shown that when the trader can treat share price values from the portfolio as random variables with known (to her) distributions, an optimal portfolio composition is found by solving a linear programming problem. Otherwise, this optimal composition is found as the trader's equilibrium strategy in an antagonistic two-person game with the stock exchange being the other player. In this game on polyhedra of disjoint player strategies, described by systems of linear equations and inequalities of a balance kind, calculating saddle points is reduced to solving linear programming problems forming a dual pair.

Keywords: dynamics of financial securities, linear programming, portfolio, price-taking traders, random variable distribution, saddle points, two-person games on polyhedral sets of disjoint player strategies.

1 Introduction

Stock exchanges are an economic phenomenon affecting the economy in every country, and this fact contributes to a great deal of attention to studying the stock exchange behavior, which has been displayed for years by a wide spectrum of specialists (experts), especially by financiers, economists, sociologists, psychologists, politicians, and mathematicians. What becomes known as a result of their studies, what these experts can (and wish to) explain and interpret from findings of the studies to both interested individuals and society as a whole to help them understand how the stock exchanges work, and how good (or bad) these explanations are make a difference. Indeed, economic issues and policies, the financial stability and the financial security of every country and the personal financial status of millions of individuals in the world who invest their personal money in sets of financial instruments traded in stock

© Springer International Publishing Switzerland 2015
H.A. Le Thi et al. (eds.), *Model. Comput. & Optim. in Inf. Syst. & Manage. Sci.*,
Advances in Intelligent Systems and Computing 359, DOI: 10.1007/978-3-319-18161-5_22

exchanges are affected by the stock exchange behavior. The existing dependency of the above "customers" on the ability (or inability) of the experts to provide trustworthy explanations of this behavior makes advances in developing tools for quantitatively analyzing the work of stock exchanges important for both the financial practice and economic science.

These tools seem indispensible, first of all, for specialists in economics and finance, since they let them a) receive, process, analyze, and interpret available information on the financial behavior of both stock exchanges and their participants, b) form, test, and analyze both scientific and experience-based hypotheses on the stock exchange behavior, along with mathematical models for its description, and c) study, evaluate, and generalize the experience of successful traders. So developing the tools, especially those being easy to operate, widely available, and producing results of their work that are easy to understand presents interest for a sizable number of individuals.

The present paper discusses a new approach to developing such mathematical tools. The idea of the approach is to propose models for forming and managing a trader's portfolio of securities that would let one use linear programming techniques for finding an optimal composition of the portfolio, and two models are proposed to this end. The first model is applied when at the time of making a decision on forming the portfolio, the trader possesses information that allows her to treat changes of each security as those of a random variable with a known (to her) distribution, and an optimal portfolio composition is found by solving a linear programming problem. The second model covers the case in which no information on the probability distribution of the above random variable is available to or can be obtained by the trader. In this case, the trader's decision is sought as that of a player in a two-person game on polyhedra of disjoint player strategies, described by systems of linear equations and inequalities of a balance kind, where the stock exchange is the other player. In this game, an optimal trader's strategy is a vector component of a saddle point of the game, which can be found by solving linear programming problems forming a dual pair.

2   Detecting the Ability of a Price-Taking Trader to Predict Future Prices of a Financial Security and to Succeed in Using this Ability in a Standard and in a Margin Trading

The ability of a trader to predict (or to divine) the price dynamics for a set of particular financial securities traded in a stock exchange is a determining factor in forming her optimal portfolio. However, even for a person gifted in predicting either values of any time series in general or only those describing the dynamics of particular financial securities with a probability exceeding 50%, this ability as such may turn out to be insufficient for successfully trading securities in a long run or even in a short period of time. Thus, tools for detecting the ability of a potential trader to predict share price values for a set of financial securities with a probability exceeding 50% and those for testing this ability from the viewpoint of the final financial result (that the trader may expect to achieve by trading particular financial securities within any particular period of time) are needed. These tools may give the potential trader the impression of what she should expect by embarking the gamble of trading in stock exchanges and may warn those who do not display the ability to success in trading financial securities either in general or in a margin trading with a particular leverage to abstain from participating in these activities.

A software complex for estimating the probability with which a potential trader can predict the upward and downward directions of changing the share price values of a financial security works as follows: The trader is offered a set of segments of a time series describing these values for a financial security of her choice, and the result of a trial (consisting of the trader's prediction made at the end point of a chosen segment of the time series) is compared with the real value of the share price. The set contains segments of the same length to secure the same conditions of conducting the trials, and the segments are chosen in a manner securing the independence of each trial outcome of those of the other trials. Meeting both conditions lets one consider the ratio of the number of the correct predictions to the total number of trials an estimate of the above probability [1].

Another software complex lets a potential trader estimate possible final financial results of trading a particular security with a detected probability to predict the directions of changing the share price values of this security. The trader predicts the direction of the share price values of the security at a consequent set of moments (at which the real values of the security share prices constitute a time series) and chooses the number of shares that she wishes to trade (to buy or to sell) at each moment from the set. The financial result of both activities is compared with that calculated at the real share price values for the sets (time series segments) of various lengths of the trader's choice. Also, the trader can see financial results of simulations conducted for a sizable number of "artificial traders," each of who predicts share price values of a particular security with the same probability as does the trader, along with the average final financial result for the whole group [2, 3].

3   Two Mathematical Models for Analyzing the Interaction of a Price-Taking Trader with a Stock Exchange

Consider a price-taking trader who buys and sells financial securities on a stock exchange, whose ability to predict the direction of changing the share price values of the securities with the probability exceeding 50% for each of them has been tested and confirmed for a set of financial securities traded there. Let a) S be this set for which $n = |S|$ elements, where for any set $K, |K|$ is the number of elements in K, b) $p_i > 0.5, i \in \overline{1, n}$ be the above probabilities and c) $H(t)$ be the amount of financial resources that the trader can spend at the moment t both on buying additional shares of financial securities from the set S and on margin trading. Let the trader know share price values for financial securities from the set S for a period of time up to the moment t, and let us assume that the trader is 'careful,' meaning that this trader trades financial securities from the set S at the moment t only if the inequality $H(\tau) \geq H(0)/2$ has been held for all $0 < \tau \leq t$, where $H(0)$ is the amount of money that the trader had at the time of forming her portfolio. Finally, let us assume that at any moment t, the trader can implement her decisions on changing the structure of her portfolio of financial securities from S without a delay, either directly or via a broker.

3.1 Analyzing Financial Strategies of a Price-Taking Trader under a Known Probability Distribution of Share Price Values for Each Financial Security from the Portfolio

Let $I^+(t) \subseteq S$ be a subset of securities from the set S such that for each security from $I^+(t)$, the trader expects its share price value to change upwards at the moment $t + 1$, $I^-(t) \subseteq S \backslash I^+(t)$ be a subset of securities from the set S such that for each security from $I^-(t)$, the trader expects its share price value to change downwards at the moment $t + 1$, and $I^0(t) = S \backslash (I^+(t) \cup I^-(t))$ be a subset of S such that for each security from $I^0(t)$, the trader is not certain about the direction of changing its share price value at the moment $t + 1$.

Let $a_i(t)$ be the number of shares of security i that the trader possesses at the moment t, $i \in \overline{1, n}$, $c_i(t)$ is the share price value for security i at the moment t, $i \in S$, $c_i^{max}(t)$ is the maximum share price value for security i that the trader has seen up to the moment t, $i \in S$, $c_i^{min}(t)$ is the minimum share price value for security i that the trader has seen up to the moment t, $i \in S$, $c_i(t + 1)$ is the share price value for security i that the trader expects to be at the moment t, $i \in S$, $x_i(t)$ be the number of shares of security i that the trader intends to buy at the moment t, $i \in I^+(t) \subseteq S$, $y_i(t)$ be the number of shares of security i that the trader intends to sell at the moment t, $i \in I^-(t) \subseteq S$, $z_i(t)$ be the number of shares of security i that the trader intends to receive from brokers to open a short position based upon the size of her collateral, $i \in I^-(t) \subseteq S$, and let the following assumptions on the dynamics of changing the share price values of securities from the set S hold:

a) at the moment t, the value of the share price of security i changes upwards (compared with its value $c_i(t)$) as a continuous random variable u that is uniformly distributed on the segment $[c_i(t), c_i^{max}(t)]$ with the probability density

$$f_1(u) = \begin{cases} \dfrac{1}{c_i^{max}(t) - c_i(t)}, if\ u \in [c_i(t), c_i^{max}(t)], \\ 0, if\ u \notin [c_i(t), c_i^{max}(t)], \end{cases}$$

b) at the moment t, the values of the share price of security i changes downwards (compared with its value $c_i(t)$) as a continuous random variable v that is uniformly distributed on the segment $[c_i^{min}(t), c_i(t)]$ with the probability density

$$f_2(v) = \begin{cases} \dfrac{1}{c_i(t) - c_i^{min}(t)}, if\ v \in [c_i^{min}(t), c_i(t)], \\ 0, if\ v \notin [c_i^{min}(t), c_i(t)], \end{cases}$$

and c) u and v are independent random variables.

Further, let $A_i(t)$ be the event consisting of the trader's prediction that the share price value of security i will change upwards at the moment $t + 1$ compared with the value $c_i(t), i \in S$, i.e. the inequality $c_i(t + 1) > c_i(t)$ will hold, $B_i(t)$ be the event consisting of the trader's prediction that the share price value of security i will change downwards at the moment $t + 1$ compared with the value $c_i(t), i \in S$, i.e. the inequality $c_i(t + 1) < c_i(t)$ will hold.

It is clear that if the event $A_i(t)$ takes place at the moment t for security $i \in S$, then the expectation of the share price value of security i at the moment $t + 1$ equals $Mc_i(t + 1) = \frac{c_i(t)+ c_i^{max}(t)}{2}, i \in S$, whereas if the event $B_i(t)$ takes place at the moment t instead, then $Mc_i(t + 1) = \frac{c_i^{min}(t)+c_i(t)}{2}, i \in S$.

If the event $A_i(t)$ occurs with the probability p_i^+, then the opposite event $\bar{A}_i(t)$ occurs with the probability $1 - p_i^+$. The event $\bar{A}_i(t)$ is a sum of two incompatible events $\bar{A}_i^1(t)$ and $\bar{A}_i^2(t)$, where $\bar{A}_i^1(t)$ is the event consisting of changing the share price value for security i at the moment $t + 1$ downwards, whereas $\bar{A}_i^2(t)$ is the event consisting of holding the equality $c_i(t) = c_i(t + 1)$. Assuming the events $\bar{A}_i^1(t)$ and $\bar{A}_i^2(t)$ are equally possible, one can conclude that the expectation of the share price value for security $i \in I^+(t)$ is a discrete random variable with the values $\frac{c_i(t)+ c_i^{max}(t)}{2}, \frac{c_i^{min}(t)+c_i(t)}{2}$ and $c_i(t)$ and with the probabilities of these values equaling $p_i^+, \frac{1-p_i^+}{2}$ and $\frac{1-p_i^+}{2}$, respectively. Analogously, if the event $B_i(t)$ occurs with the probability p_i^- for security i from the set $I^-(t)$, the probability distribution of the share price value for security $i \in I^-(t)$ is a discrete random variable with the values $\frac{c_i^{min}(t)+c_i(t)}{2}, \frac{c_i(t)+ c_i^{max}(t)}{2}$ and $c_i(t)$ and with the probabilities $p_i^-, \frac{1-p_i^-}{2}$ and $\frac{1-p_i^-}{2}$, respectively. Finally, the expectation of the share price value for security $i \in I^0(t)$ is a discrete variable with the probability distribution described by the values $c_i(t)$, $\frac{c_i^{min}(t)+c_i(t)}{2}, \frac{c_i(t)+ c_i^{max}(t)}{2}$ that are assumed with the probabilities $p_i^0, \frac{1-p_i^0}{2}$ and $\frac{1-p_i^0}{2}$, respectively.

At every moment t the trader a) divides all the securities from the set S into three groups so that $S = I^+(t) \cup I^-(t) \cup I^0(t)$, where $I^+(t) \cap I^-(t) = \emptyset, I^+(t) \cap I^0(t) = \emptyset, I^-(t) \cap I^0(t) = \emptyset$, b) buys shares of securities from the set $I^+(t)$, sells shares of securities from the set $I^-(t)$ and invests the revenue from this sell into buying shares of securities from the set $I^+(t)$, c) holds all the shares of securities from the set $I^0(t)$, and d) may to decide to borrow shares of securities from the set $I^-(t)$ (using her total financial resources as a collateral) to open short positions. Thus, her financial strategy at the moment t is determined by the numbers of shares of securities from the set S that she intends to buy, to sell, to borrow, and to hold. Let $x_i(t), i \in I^+(t), y_i(t), i \in I^-(t)$, and $z_i(t), i \in I^-(t)$ be the numbers of shares from the first three of the above four categories, and let $k(t)$ be the value of the leverage in the margin trading of securities from the ser $I^-(t)$ offered by a broker to the trader at the moment t. One can easily be certain that the expected share price values for securities from the sets $I^+(t), I^-(t)$, and $I^0(t)$ at the moment $t + 1$ are

$$c_i^+(t + 1) = p_i^+ \frac{c_i(t) + c_i^{max}(t)}{2} + \frac{1 - p_i^+}{2} \frac{c_i^{min}(t)+c_i(t)}{2} + \frac{1 - p_i^+}{2} c_i(t), i \in I^+(t),$$

$$c_i^-(t+1) = p_i^- \frac{c_i^{min}(t)+c_i(t)}{2} + \frac{1-p_i^-}{2} \frac{c_i(t) + c_i^{max}(t)}{2} + \frac{1-p_i^-}{2}c_i(t), i \in I^-(t),$$

$$c_i^0(t+1) = p_i^0 c_i(t) + \frac{1-p_i^0}{2} \frac{c_i^{min}(t)+c_i(t)}{2} + \frac{1-p_i^0}{2} \frac{c_i(t) + c_i^{max}(t)}{2}, i \in I^0(t).$$

If the optimal financial strategy of the trader at the moment t is understood in the sense of her expected amount of total financial resources (associated with trading securities) at the moment $t+1$, this strategy can be found by solving the following linear programming problem:

$$\sum_{i \in I^+(t)} x_i(t)c_i^+(t+1) + \sum_{i \in I^-(t)} z_i(t)(c_i(t) - c_i^-(t+1)) \to \max$$

$$\sum_{i \in I^+(t)} a_i(t)c_i^+(t+1) + \sum_{i \in I^+(t)} x_i(t)c_i^+(t+1) + \sum_{i \in I^-(t)} z_i(t)(c_i(t) - c_i^-(t+1))$$
$$+ \sum_{i \in I^0(t)} a_i(t)c_i^0(t+1) \geq H(t) + \sum_{i=1}^{n} a_i(t)c_i(t), (1)$$

$$\sum_{i \in I^+(t)} x_i(t)c_i^+(t+1) \leq H(t) + \sum_{i \in I^-(t)} y_i(t)c_i(t),$$

$$k(t)\left(H(t) + \sum_{i=1}^{n} a_i(t)c_i(t) \right) \geq \sum_{i \in I^-(t)} z_i(t)c_i(t),$$

$$x_i(t) \geq 0, i \in I^+(t), y_i(t) \geq 0, i \in I^-(t), z_i(t) \geq 0, i \in I^-(t).$$

Let $\left(x_i^*(t), i \in I^+(t), y_i^*(t), i \in I^-(t), z_i^*(t), i \in I^-(t)\right)$ be a solution to problem (1). Then the number

$$\sum_{i \in I^+(t)} a_i(t)c_i^+(t+1) + \sum_{i \in I^+(t)} x_i^*(t)c_i^+(t+1) + \sum_{i \in I^-(t)} z_i^*(t)(c_i(t) - c_i^-(t+1))$$
$$+ \sum_{i \in I^0(t)} a_i(t)c_i^0(t+1),$$

determines the expected total amount of the financial resources of the trader at the moment $t+1$.

To verify whether the system of constraints of problem (1) is compatible, one can use a technique, proposed in [4], whose application requires solving an auxiliary linear programming problem with a wittingly compatible system of constraints. Moreover, if system (1) is incompatible, the use of the technique proposed in [4] also allows the trader to find out a) whether under her expectations on the probability values for the share price values $c_i(t+1), i \in S$, she can increase the total amount of her financial resources (associated with trading securities) at the moment $t+1$

(compared with $H(t)$), and b) how much she should expect to lose by making the transactions at the moment t as she planned.

One should bear in mind that, generally, problem (1) should be formulated as an integer programming problem, since the numbers $x_i(t), i \in I^+(t), y_i(t), i \in I^-(t)$ and $z_i(t), i \in I^-(t)$ are integers. However, since usually, all these numbers are substantially larger than unit, problem (1), which is a relaxation of the corresponding integer programming problem, can be solved instead, and non-integer values of the variables in the solution can be rounded-off. Certainly, for a relatively small number n, problem (1) can be solved as an integer programming problem (with the additional constraints $x_i(t) \in N \cup \{0\}, i \in I^+(t), y_i(t) \in N \cup \{0\}, i \in I^-(t), z_i(t) \in N \cup \{0\}, i \in I^-(t),$ where N is the set of all natural numbers).

Remark 1. The uniform distribution for changing share price values of the securities from the set S is the basic assumption on analyzing the financial behavior of the price-taking traders who rely on their own intuition regarding the dynamics of these share price values more than on any statistical regularity of the dynamics of their changes. However, any other trustworthy probability distributions can be used for calculating the numbers $c_i^+(t + 1), i \in I^+(t), c_i^-(t + 1), i \in I^-(t), c_i^0(t + 1), i \in I^0(t)$, as long as these probability distributions are known to the trader.

Remark 2. The consideration of the share price values of security i at the moment t as those of two random variables (one for the upward changes and the other for the downward ones) reflects the fact that the regularities underlying these changes are, generally, different and usually correspond to the prevailing of a particular type of the market (such as, for instance, the 'bull market' and the 'bear market,' respectively).

3.2   Analyzing Financial Strategies of a Price-Taking Trader in the Absence of Assumptions on the Probability Distributions of Share Price Values of Financial Securities from the Trader's Portfolio

The assumption on the uniformity of the distribution of the share price values for securities from the trader's portfolio reflects a kind of uncertainty that the trader may have on the regularities of changes of these prices. Another kind of uncertainty corresponds to the case in which no information on the probability distributions of the above share price values becomes available to the trader, and constraints of the balance kind describing the 'ranges' of these price changes are all the trader can assume about them, mostly from her personal trading experience and observations. In this case, a game-theoretic approach to choosing trader's strategies on managing her portfolio seems reasonable to apply. Under this approach, at the moment t, the trader plays against the stock exchange by choosing the volumes of shares of securities from the portfolio that she would like to have at the moment $t + 1$, whereas as a player, the stock exchange 'chooses' the share price values for these securities. Such a game resembles games with the nature in which the trader should consider the worst case scenario for her and choose a strategy aimed at obtaining a certain guaranteed result (should this scenario be the case). As is known, in games with the nature, the latter is, generally, insensitive to the person's move (if the person is a 'small" player with respect to her ability to affect the nature, and the same is the case in the game between a small trader and the stock exchange). However, both the person (in the game with the nature) and the trader (in the game with the stock exchange) try to develop their best strategies to counteract the worst case scenario that the nature and the stock exchange can "offer" (according to the person's and to the trader's knowledge and expectations), respectively.

For some classes of such games in industry, agriculture, transportation, brokerage, investment, public administration, and energy, which are games on polyhedral sets of either disjoint or connected player strategies and the payoff functions being a sum of two linear and a bilinear function of vector variables, the possibility to use linear programming techniques for finding the best players' strategies has been proven [4, 5]. The aim of this section of the paper is to demonstrate the possibility to extend this result for the game of the interaction of a price-taking trader with a stock exchange under some natural assumptions.

Assumptions. At each moment t, the trader expects all the share price values of securities from the group $I^+(t)$ to change in the same direction (i.e., either upwards or downwards). The same expectation holds for all the share price values of securities from the group $I^-(t)$, whereas the trader expects share price values of securities from the group $I^0(t)$ to change in different directions. Depending on whether these expectations hold for all the securities from each group, their share price values are components of one of two polyhedra whose descriptions in the form of linear equations and inequalities are known to the trader, and parts of these linear inequalities describe parallelepipeds in the corresponding spaces (so that all these polyhedra are subsets of the corresponding parallelepipeds).

To simplify the notation to follow and to bring it closer to traditional descriptions of the game to be considered, in particular, to the description of the games in [4,5], some variables and parameters in this notation (being different from the one used in the previous section of this paper) are introduced.

Let $x^{I^+(t)}(t) \in R^{|I^+(t)|}$ be the vector whose component $x_i^{I^+(t)}(t)$ is the number of shares of security $i \in I^+(t)$ that the trader intends to buy at the moment t, $y^{I^+(t)}(t+1) \in R^{|I^+(t)|}$ be the vector whose component $y_i^{I^+(t)}(t+1)$ is the (expected by the trader) share price value for security $i \in I^+(t)$ at the moment $t+1$, $x^{I^-(t)}(t) \in R^{|I^-(t)|}$ be the vector whose component $x_i^{I^-(t)}(t)$ is the number of shares of security $i \in I^-(t)$ that the trader intends to sell at the moment t, $y^{I^-(t)}(t+1) \in R^{|I^-(t)|}$ be the vector whose component $y_i^{I^-(t)}(t+1)$ is the (expected by the trader) share price value for security $i \in I^-(t)$ at the moment $t+1$, $z^{I^+(t)}(t+1) \in R^{|I^+(t)|}$ be the vector whose component $z_i^{I^+(t)}(t+1)$ is the share price value for security $i \in I^+(t)$ that would go downwards at the moment $t+1$, contradictory to the trader's (mistaken) expectations, $z^{I^-(t)}(t+1) \in R^{|I^-(t)|}$ be the vector whose component $z_i^{I^-(t)}(t+1)$ is the share price value for security $i \in I^-(t)$ that would go upwards at the moment $t+1$ contradictory, to the trader's (mistaken) expectations, $x^{I^0(t)}(t) \in R^{|I^0(t)|}$ be the vector whose component $x_i^{I^0(t)}(t)$ is the number of shares of security $i \in I^0(t)$ about which the trader is not certain on whether its share price value will go upwards or downwards at the moment $t + 1$, $\overline{y}^{I^0(t)}(t + 1) \in R^{|I^0(t)|}$ be the vector whose component $\overline{y}_i^{I^0(t)}(t + 1)$ is the share price value for security $i \in I^0(t)$ at the moment $t + 1$ if this price value goes upwards (with the probability ½), and $\underline{y}^{I^0(t)}(t + 1) \in R^{|I^0(t)|}$ be the vector whose component $\underline{y}_i^{I^0(t)}(t + 1)$ is the share price value for security $i \in I^0(t)$ at the moment $t + 1$ if this share price value goes downwards (with the probability ½).

Further, let $M^{I^+(t)}(t), \Omega^{I^+(t)}(t + 1), \Lambda^{I^+(t)}(t + 1) \in R^{|I^+(t)|}$, $M^{I^-(t)}(t), \Omega^{I^-(t)}(t + 1), \Lambda^{I^-(t)}(t + 1) \in R^{|I^-(t)|}$, and $M^{I^0(t)}(t), \overline{\Omega}^{I^0(t)}(t + 1), \underline{\Omega}^{I^0(t)}(t + 1) \in R^{|I^0(t)|}$ be polyhedra for which the inclusions $x^{I^+(t)}(t) \in M^{I^+(t)}(t), x^{I^-(t)}(t) \in M^{I^-(t)}(t), x^{I^0(t)}(t) \in M^{I^0(t)}(t), y^{I^+(t)}(t + 1) \in \Omega^{I^+(t)}(t + 1), y^{I^-(t)}(t + 1) \in \Omega^{I^-(t)}(t + 1), \overline{y}^{I^0(t)}(t + 1) \in \overline{\Omega}^{I^0(t)}(t + 1), \underline{y}^{I^0(t)}(t + 1) \in \underline{\Omega}^{I^0(t)}(t + 1), z^{I^+(t)}(t + 1) \in \Lambda^{I^+(t)}(t + 1), z^{I^-(t)}(t + 1) \in \Lambda^{I^-(t)}(t + 1)$ hold.

Theorem. At each moment t, the above interaction between the trader and the stock exchange can be described by the game on polyhedra $\tilde{M}(t)$ and $\theta(t + 1)$ of disjoint player strategies, where $\tilde{M}(t)$ is the set of trader's strategies and $\theta(t + 1)$ is that of the stock exchange, with the payoff function $\langle x(t), D(t)w(t + 1)\rangle$, where

$$x(t) = \left(x^{I^+(t)}(t), x^{I^-(t)}(t), x^{I^0(t)}(t)\right) \in \tilde{M}(t) = M^{I^+(t)}(t) \times M^{I^-(t)}(t) \times M^{I^0(t)}(t),$$

$$w(t + 1) = \left(w^{I^+(t)}(t + 1), w^{I^-(t)}(t + 1), w^{I^0(t)}(t + 1)\right) \in \theta(t + 1) =$$
$$= \theta^{I^+(t)}(t + 1) \times \theta^{I^-(t)}(t + 1) \times \theta^{I^0(t)}(t + 1),$$

and $D(t)$ is a matrix of corresponding dimensions, $\theta^{I^+(t)}(t + 1) = \Omega^{I^+(t)}(t + 1) \times \Lambda^{I^+(t)}(t + 1)$, $\theta^{I^-(t)}(t + 1) = \Omega^{I^-(t)}(t + 1) \times \Lambda^{I^-(t)}(t + 1)$, $\theta^{I^0(t)}(t + 1) = \overline{\Omega}^{I^0(t)}(t + 1) \times \underline{\Omega}^{I^0(t)}(t + 1)$, are polyhedra and $w^{I^+(t)}(t + 1) = \left(y^{I^+(t)}(t + 1), z^{I^+(t)}(t + 1)\right) \in \theta^{I^+(t)}(t + 1)$, $w^{I^-(t)}(t + 1) = \left(y^{I^-(t)}(t + 1), z^{I^-(t)}(t + 1)\right) \in \theta^{I^-(t)}(t + 1), w^{I^0(t)}(t + 1) = \left(\overline{y}^{I^0(t)}(t + 1), \underline{y}^{I^0(t)}(t + 1)\right) \in \theta^{I^0(t)}(t + 1)$ are vectors, and the saddle point in this game is formed by vector components of solutions to linear programming problems forming a dual pair.

Proof

1. From the definition of the sets $I^+(t)$ and $I^-(t)$ it follows that at the moment $t + 1$, the trader expects the share price values for all the securities from the set $I^+(t)$ to change upwards and the share price values for all the securities from the set $I^-(t)$ to change downwards, whereas share price values for the securities from the set $I^0(t)$ can change either way.

Let us consider first the set of securities $I^+(t)$ from the trader's portfolio at the moment t. If the trader predicted the directions of changing the share price values for all the securities from this set correctly, then her best strategy of choosing the numbers of shares of securities to buy would be determined by a solution to the problem

$$\min_{y^{I^+(t)}(t+1)\in\Omega^{I^+(t)}(t+1)} \langle x^{I^+(t)}(t), y^{I^+(t)}(t+1)\rangle \rightarrow \max_{x^{I^+(t)}(t)\in M^{I^+(t)}(t)}$$

If the trader did not predict the direction of changing the share price values for all the securities from the set $I^+(t)$ correctly, then her best strategy would be determined by a solution to the problem

$$\min_{z^{I^+(t)}(t+1)\in\Lambda^{I^+(t)}(t+1)} \langle x^{I^+(t)}(t), z^{I^+(t)}(t+1)\rangle \rightarrow \max_{x^{I^+(t)}(t)\in M^{I^+(t)}(t)}$$

Since the trader correctly predicts the direction of changes of the share price values of securities from the set $I^+(t)$ only with a certain probability, say, p^+, the expectation of the financial result of her prediction (which is a discrete random variable) can be written as

$$\max_{x^{I^+(t)}(t)\in M^{I^+(t)}(t)} \min_{w^{I^+(t)}(t+1)\in\theta^{I^+(t)}(t+1)} \langle x^{I^+(t)}(t), D^{2|I^+(t)|}(p^+,(1-p^+))w^{I^+(t)}(t+1)\rangle$$

where $D^{|I^+(t)|}(p^+)$ and $D^{|I^+(t)|}(1-p^+)$ are $|I^+(t)| \times |I^+(t)|$–diagonal matrices whose all elements on the main diagonal equal p^+ and $1-p^+$, respectively, and $D^{2|I^+(t)|}(p^+,(1-p^+))$ is a $|I^+(t)| \times 2|I^+(t)|$ – matrix that is formed by writing the matrix $D^{|I^+(t)|}(1-p^+)$ next to the matrix $D^{|I^+(t)|}(p^+)$ from the right.

Analogously, for securities from the set $I^-(t)$, the expectation of the financial results of the trader's prediction can be written as

$$\max_{x^{I^-(t)}(t)\in M^{I^-(t)}(t)} \min_{w^{I^-(t)}(t+1)\in\theta^{I^-(t)}(t+1)} \langle x^{I^-(t)}(t), D^{2|I^-(t)|}(p^-,(1-p^-))w^{I^-(t)}(t+1)\rangle$$

where $D^{2|I^-(t)|}(p^-,(1-p^-))$ is a $|I^-(t)| \times 2|I^-(t)|$ –matrix of the same structure as is the matrix $D^{2|I^+(t)|}$, and all the elements on the main diagonals of the matrices $D^{|I^-(t)|}(p^-)$ and $D^{|I^-(t)|}(1-p^-)$ equal p^- and $1-p^-$, respectively.

2. The same reasoning for securities from the set $I^0(t)$ allows one to assert that the expectation of the financial result of the trader's prediction at the moment t (regarding the changes of the share price values for securities from $I^0(t)$) can be written as

$$\max_{x^{I^0(t)}(t)\in M^{I^0(t)}(t)} \min_{w^{I^0(t)}(t+1)\in\theta^{I^0(t)}(t+1)} \langle x^{I^0(t)}(t), D^{2|I^0(t)|}\left(\frac{1}{2},\frac{1}{2}\right)w^{I^0(t)}(t+1)\rangle,$$

where $D^{2|I^0(t)|}\left(\frac{1}{2},\frac{1}{2}\right) = D^{|I^0(t)|}\left(\frac{1}{2}\right)D^{|I^0(t)|}\left(\frac{1}{2}\right)$, where $D^{|I^0(t)|}\left(\frac{1}{2}\right)$ is a $|I^0(t)| \times |I^0(t)|$- diagonal matrix, all whose elements on the main diagonal equal ½.

Further, let

$$D(t) = \begin{pmatrix} D^{2|I^+(t)|}(p^+,(1-p^+)) & 0 & 0 \\ 0 & D^{2|I^-(t)|}(p^-,(1-p^-)) & 0 \\ 0 & 0 & D^{2|I^0(t)|}\left(\frac{1}{2},\frac{1}{2}\right) \end{pmatrix}.$$

3. Finally, let $\tilde{M}(t) = \{x(t) \in R_+^n: A(t)x(t) \geq b(t)\}$, $\theta(t+1) = \{w(t+1) \in R_+^{3n}: B(t+1)w(t+1) \geq d(t+1)\}$, where $A(t)$, $B(t+1)$, and $b(t), d(t+1)$ are matrices and vectors of corresponding dimensions.

Taking into account the notation introduced before, the antagonistic game between the trader and the stock exchange that is played at the moment t can be written as that with the payoff function $\langle x(t), D(t)w(t+1)\rangle$ on the polyhedra $\tilde{M}(t)$ and $\theta(t+1)$, which are sets of disjoint player strategies (of the trader and of the stock exchange), respectively.

4. Saddle points in games on (generally unbounded) polyhedral sets $M \times \Omega$ [4] of disjoint player strategies with the payoff function $< p, x > + < x, Dy > + < q, y >$, where $x \in M, y \in \Omega$, D is a matrix, and p, q are vectors of corresponding dimensions, are formed by vector components of solutions to a dual pair of linear programming problems [4], which for the game under consideration with the payoff function $\langle x(t), D(t)w(t+1)\rangle$ on the set $\tilde{M}(t) \times \theta(t+1)$ take the form

$$\langle d(t+1), h(t)\rangle \rightarrow \max_{(h(t),x(t))\in Q(t,t+1)} \quad (2)$$

$$\langle -b(t), s(t+1)\rangle \rightarrow \min_{(s(t+1),w(t+1))\in P(t,t+1)}$$

Where $Q(t, t+1) = \{(h(t), x(t)) \geq 0 : h(t)B(t+1) \leq x(t)D(t), A(t)x(t) \geq b(t)\}$, $P(t, t+1) = \{(s(t+1), w(t+1)) \geq 0 : s(t+1)A(t) \leq -D(t)w(t+1), B(t+1)w(t+1) \geq d(t+1)\}$.

The Theorem is proved.

Let $(x^*(t), h^*(t), w^*(t+1), s^*(t+1))$ be a solution be a solution to the pair of linear programming problems (2). Then the vectors $\left(x^{I^+(t)}(t)\right)^*$, $\left(x^{I^-(t)}(t)\right)^*$ and $\left(x^{I^0(t)}(t)\right)^*$ from the supervector $x^*(t) = \left(\left(x^{I^+(t)}(t)\right)^*, \left(x^{I^-(t)}(t)\right)^*, \left(x^{I^0(t)}(t)\right)^*\right)$ determine the optimal trader's strategy in the game, which is attained at the saddle point of the game and delivers

$$\max_{x(t)\in\tilde{M}(t)}\left[\min_{w(t+1)\in\theta(t+1)}\langle x(t), D(t)w(t+1)\rangle\right].$$

This number determines how many shares of each security from the sets $I^+(t)$ and $I^-(t)$ the trader should buy and sell at the moment t, respectively, as well as what is the financial result of her decision to change the portfolio at the moment t according to her predictions of the direction of changing share price values of all the three sets of the securities forming her portfolio at the moment t.

4   Concluding Remarks

1. A new approach to modeling the financial behavior of small and medium price-taking traders with respect to managing their portfolios of securities for each of which they possess a tested and confirmed ability to predict the direction of changing the share price values is proposed. This approach is based on considering two models reflecting what information on the dynamics of the above share price values the trader

possesses, and the use of each model allows one to determine an optimal trader's strategy by using linear programming techniques. As is known, these techniques have enormous computational potential, which is critical for studying large-scale systems, stock exchanges being one of them. Both models, along with the techniques for detecting the probability with which the trader can predict the direction of changing the share price values of each security from the portfolio at any moment, constitute elements of the tools for quantitatively analyzing the financial behavior of small and medium price-taking traders and that of the stock exchange as a whole.

2. The proposed tools seem to be helpful for quantitatively analyzing the particular markets that the stock exchange often exhibits, particularly, the 'bull market' and the 'bear market,' as well as financial bubbles, by conducting corresponding simulations.

3. While the proposed approach opens new research opportunities for studying stock exchanges, first of all, from the viewpoint of the financial behavior of small and medium price-taking traders, the question on to what extent this approach can help study the financial behavior of large traders remains open and will be the subject of further research of the authors.

Acknowledgements. The financial support from the Russian Federation Government within the framework of the implementation of the 5-100 Programme Roadmap of the National Research University Higher School of Economics is acknowledged. The authors are grateful for financial support of their work to DeCAn Laboratory at the National Research University Higher School of Economics, headed by Prof. Fuad Aleskerov, with whom the authors have had fruitful discussions on problems of developing tools for quantitatively analyzing large-scale systems in general and those for stock exchanges in particular. L. Egorova expresses her gratitude to LATNA Laboratory, NRU HSE, RF government grant, ag.11.G34.31.0057 and A. Belenky expresses his gratitude to the MIT Center for Engineering Systems Fundamentals.

References

1. Feller, W.: An Introduction to Probability Theory and Its Applications, vol. 1, 2. Wiley (1991)
2. Egorova, L.G.: Effectiveness of Different Trading Strategies for Price-takers. Procedia Computer Science 31, 133–142 (2014)
3. Egorova, L.G.: Agent-Based Models of Stock Exchange: Analysis via Computational Simulation. In: Network Models in Economics and Finance, pp. 147–158. Springer (2014)
4. Belenky, A.S.: Minimax planning problems with linear constraints and methods for their solutions. Automation and Remote Control 42(10), 1409–1419 (1981)
5. Belenky, A.S.: Two Classes of Games on Polyhedral Sets in Systems Economic Studies. In: Network Models in Economics and Finance, pp. 35–84. Springer (2014)

Behavioral Portfolio Optimization with Social Reference Point*

Yun Shi, Duan Li, and Xiangyu Cui

School of Management, Shanghai University, Shanghai, China
y_shi@shu.edu.cn
Department of Systems Engineering and Engineering Management,
The Chinese University of Hong Kong, Hong Kong
dli@se.cuhk.edu.hk
School of Statistics and Management,
Shanghai University of Finance and Economics, Shanghai, China
cui.xiangyu@mail.shufe.edu.cn

Abstract. In this study, we address the social interaction process in which PT (Prospect Theory) preferences are influenced by other market participants, e.g., regular CRRA (Constant Relative Risk Averse) investors or other PT investors, and study then the long run wealth convergence of the two trading parties: one PT agent vs. one CRRA agent or both agents of PT types. In the model with one PT agent vs. one CRRA agent, the PT agent knows the CRRA agent's optimal terminal wealth and takes it as his/her reference point. If the PT agent starts with an initial wealth level higher than that of the CRRA agent, he/she will always do better than the CRRA agent by imitating the CRRA agent's policy. On the other hand, if the PT agent starts with a wealth level lower than that of the CRRA agent, he/she can still do better than the CRRA agent by adopting a "gambling policy". When both trading parties are of PT type, we consider two types of reference points: either both PT agents take their average wealth as their reference point or they are mutually reference dependent. Under both situations, we give sufficient conditions on the long run wealth convergence.

Keywords: Portfolio optimization, reference point, prospect theory, social comparison, relative wealth concern.

1 Introduction

In daily life, our choices and decisions are inevitably influenced by our friends and neighbors in our social networks. For example, the households will put more effort in energy saving when knowing that their neighbors are more efficient

---

* This research work was partially supported by Hong Kong Research Grants Council under grant CUHK 419511. The second author is also grateful to the support from Patrick Huen Wing Ming Chair Professorship of Systems Engineering & Engineering Management.

(see [1]). High school students will performance better when they are provided relative performance feedback (see [2]). A fund manager has incentives to out-perform her peers to increase fund flows and in turn her compensation (see [3]).

Prospect theory (PT, see [4] and [5]) provides a proper framework to incorpo-rate such features of social interactions and social comparison into the formation and updating of the reference point, thus enabling us to investigate whether (and how) these social features affect the behavior of the investor.[1] More specifically, different people may have different reference points. How do these differences interact and evolve? This paper represents a preliminary work aiming to answer this question.

We start with a simple model, where a PT investor takes the CRRA (Constant Relative Risk Averse) agent's optimal terminal wealth level as his/her reference point. By carrying out theoretical analysis and providing numerical examples, we show that if the PT agent starts with an initial wealth level higher than that of the CRRA agent, he/she will take a "conservative policy" and always be better than the CRRA agent; conversely, if the PT agent starts with an initial wealth level lower than the CRRA agent, he/she will take a "gambling policy" and keep gambling until he/she is better than the CRRA agent. In the second model, both agents are of PT types, and they either take their average wealth as a reference point or mutually take the other one's wealth as his/her reference point. Under both situations, we provide sufficient conditions under which their wealth levels converge.

A recent paper by Jin and Zhou [7] solves a behavioral portfolio choice problem in a continuous-time complete market, with a general utility function and a general non-linear transformations in probability distortion. We start in this paper by reviewing the method in [7]. We then apply their method in our study to consider interactions between a rational (CRRA) investor and a PT investor or between two PT investors.

2   Preliminary

Assume that the PT investor has an S-shaped value function and an inverse-S-shape probability distortion with the following elements,

$$u_+(x) = x^\gamma, u_-(x) = \lambda \cdot x^\gamma, x \geq 0, \tag{1}$$

$$T(p) = \frac{p^\alpha}{(p^\alpha + (1-p)^\alpha)^{1/\alpha}}, \tag{2}$$

where loss aversion coefficient satisfies $\lambda > 1$, risk aversion coefficient satisfies $0 < \gamma < 1$, and distortion coefficient α takes value α_+ in the gain region and value α_- in the loss region with $0 < \alpha_+ < 1$ and $0 < \alpha_- < 1$. Let the reference point be 0, and the investor's initial wealth be x_0. Under this PT framework, the

---

[1] The dynamics of the reference point and its impact on investors' behavior have been studied in [6].

behavioral portfolio selection problem is to find the most preferable portfolio in terms of maximizing the value $V(X_T)$ as shown below,

$$(P) \quad \max \ E[V(X_T)]$$
$$\text{s.t. } E[\xi_T X_T] = x_0, \quad x_0 \text{ is given,}$$

where ξ_T is the state price density, and

$$V(X) := V_+(X^+) - V_-(X^-)$$

$$V_+(Y) := \int_0^{+\infty} T_+(P\{u_+(Y) > y\})\, dy, \ \ V_-(Y) := \int_0^{+\infty} T_-(P\{u_-(Y) > y\})\, dy.$$

Because the behavioral type model is easy to be ill-posed, we need to add the following two assumptions to make the problem itself well-posed:

- Condition on gain part distortion $T_+(\cdot)$:

$$\frac{F^{-1}(z)}{T'_+(z)} \text{ is nondecreasing.} \tag{3}$$

- Condition on loss part distortion $T_-(\cdot)$:

$$\inf_{c>0} k(c) \geq 1. \tag{4}$$

where

$$\phi(c) = \mathbb{E}\left[\left(\frac{T'_+(F(\xi_T))}{\xi_T} \right)^{1/(1-\gamma)} \xi_T \mathbf{1}_{\xi_T \leq c} \right], \quad k(c) = \frac{\lambda \cdot T_-(1 - F(c))}{\phi(c)^{1-\gamma} \left(\mathbb{E}[\xi_T \mathbf{1}_{\xi_T \geq c}] \right)^{\gamma}},$$

with $F(\cdot)$ being its distribution function.

Remark 1. *The second condition has an intuitive interpretation: the variable $k(c)$ is a "benefit/cost" measure, showing the cost of short selling "bad state" to finance "good state" investment, and balancing the trade-off between the gain-part problem and the loss-part problem. When $k(c) \geq 1$, it is not worth short selling "bad state" to finance "good state". Suppose that there exists some c such that $k(c) < 1$, and there is no bankruptcy constraint, what would the PT investor do? The investor will keep short selling "bad state", without any fear about big loss. So, in this situation, there will be an infinite solution, and then the problem is ill-posed.*

Theorem 1. *(Jin and Zhou 2008) Assume that $x_0 \geq 0$ and condition (3) holds.*

- *If condition (4) also holds, then the optimal portfolio is the replicating portfolio for the contingent claim*

$$X^* = \frac{x_0}{\phi(+\infty)} \left(\frac{T'_+(F(\xi_T))}{\xi_T} \right)^{1/(1-\gamma)}. \tag{5}$$

- *If condition (4) does not hold, then the problem is ill-posed.*

Theorem 2. *(Jin and Zhou 2008) Assume that $x_0 < 0$ and condition (3) holds.*

– *If condition (4) also holds, then the optimal portfolio is the replicating portfolio for the contingent claim*

$$X^* = \frac{x_+^*}{\phi(c^*)} \left(\frac{T'_+(F(\xi_T))}{\xi_T} \right)^{1/(1-\gamma)} 1_{\xi_T \leq c^*} - \frac{x_+^* - x_0}{\mathbb{E}\left[\xi_T 1_{\xi_T \geq c^*} \right]} 1_{\xi_T \geq c^*}, \qquad (6)$$

where $x_+^* = \frac{-x_0}{k(c^*)^{1/(1-\gamma)} - 1}$ and c^* solves

$$\min_{c \geq 0} \quad v(c) = \left[k(c)^{1/(1-\gamma)} - 1 \right] \cdot \phi(c).$$

– *If condition (4) does not hold, then the problem is ill-posed.*

3   CRRA Agent vs. PT Agent

3.1   Theoretical Model

We first apply the method in [7] to a game model with one CRRA agent, whose utility is of a power form $u(x) = \frac{(x)^\gamma}{\gamma}$, and one PT agent. For their different utility functions $u(\cdot)$ and $V(\cdot)$, both investors solve the following respective *myopic* problem in each round,

$$\begin{array}{ll} (Myopic^{CRRA}) \; \max \; E[u(W_T)] & (Myopic^{PC}) \; \max \; E[V(W_T)] \\ \qquad\qquad \text{s.t. } E[\xi_T W_T] = w_0, & \qquad\qquad \text{s.t. } E[\xi_T W_T] = w_0, \end{array}$$

where w_0 is given. The CRRA agent starts with initial wealth w_0^{CRRA}, and solves the above problem with the power utility, resulting in the following optimal terminal wealth,

$$W_T^{CRRA} = \frac{w_0^{CRRA}}{E[\xi_T^{-\gamma/(1-\gamma)}]} \left(\frac{1}{\xi_T} \right)^{1/(1-\gamma)}.$$

Note that the CRRA agent always behaves his/her own way, i.e., his/her decision is not affected by other participants in the market.

The PT agent starts with initial wealth w_0^{PT}, knows the distribution of the CRRA agent's optimal terminal wealth and takes this random terminal wealth as his/her reference point, i.e. $\theta_T = W_T^{CRRA}$. Then he/she solves the above myopic problem with an S-shaped value function and an inverse S-shape distortion function. Denote $X_T = W_T^{PT} - \theta_T$ and $x_0 = w_0^{PT} - w_0^{CRRA}$. For X_T, we can directly use the method suggested in [7]. Based on different initial wealth levels, there are two different situations for our PT investor:

– If PT agent's initial wealth satisfies $w_0^{PT} \geq w_0^{CRRA}$, the optimal terminal wealth W_T^{PT} takes the form in (5):

$$W_T^{PT} = \theta_T + \frac{x_0}{\phi(+\infty)} \left(\frac{T'_+(F(\xi_T))}{\xi_T} \right)^{1/(1-\gamma)} \tag{7}$$

$$= W_T^{CRRA} + \frac{w_0^{PT} - w_0^{CRRA}}{\phi(+\infty)} \left(\frac{T'_+(F(\xi_T))}{\xi_T} \right)^{1/(1-\gamma)}.$$

– If PT agent's initial wealth satisfies $w_0^{PT} < w_0^{CRRA}$, the optimal terminal wealth W_T^{PT} takes the form in (6):

$$W_T^{PT} = \theta_T + \frac{x_+^*}{\phi(c^*)} \left(\frac{T'_+(F(\xi_T))}{\xi_T} \right)^{\frac{1}{1-\gamma}} 1_{\xi_T \leq c^*} - \frac{x_+^* - x_0}{\mathbb{E}\left[\xi_T 1_{\xi_T \geq c^*}\right]} 1_{\xi_T \geq c^*} \tag{8}$$

$$= W_T^{CRRA} + \frac{x_+^*}{\phi(c^*)} \left(\frac{T'_+(F(\xi_T))}{\xi_T} \right)^{\frac{1}{1-\gamma}} 1_{\xi_T \leq c^*} - \frac{x_+^* - x_0}{\mathbb{E}\left[\xi_T 1_{\xi_T \geq c^*}\right]} 1_{\xi_T \geq c^*},$$

where $x_+^* = \frac{-(w_0^{PT} - w_0^{CRRA})}{k(c^*)^{1/(1-\gamma)} - 1}$ and c^* solves

$$\min_{c \geq 0} \quad v(c) = \left[k(c)^{1/(1-\gamma)} - 1 \right] \cdot \phi(c).$$

We can summarize these two different results as follows:

– If PT agent starts with an initial wealth larger than the reference point, then he/she would set a "conservative" target and take a "conservative policy". More specifically, he/she would split the initial wealth w_0^{PT} into two parts: using the first part, w_0^{CRRA}, to replicate CRRA's optimal terminal wealth and investing the remaining part, $w_0^{PT} - w_0^{CRRA}$, in the market in a hope of gaining more.
– If PT agent starts with an initial wealth smaller than the reference point, then he/she would aspire for an "aggressive" target and thus take a "gambling policy". More specifically, he/she needs to short sell "bad state" contingent claim, generating $(x_+^* - x_0)$, to finance the "good state" investments.

3.2 Toy Example-1: One-Step Success?

When a PT agent takes the CRRA's optimal terminal wealth as his/her reference point, what the PT agent aspires is to do better than the CRRA agent. A natural question is: can the PT agent achieve his/her target in one round? If not, how about the situation in long run? We will use the following numerical example to answer these questions.

Assume that the state price density ξ_T follows a lognormal distribution, i.e., $\ln \xi_T \sim N(\mu, \sigma)$. We take S&P 500 yearly return as our market setting, $\mu = 0.1, \sigma = 0.2, r = 0.03$, and $T = 1$ yr, where r is the risk-free rate. Considering the two key assumptions in [7], we set the preference and distortion parameters of investors as follows.

– With respect to condition (3), we set:

$$\alpha_+ = 1, \text{ i.e. } T_+(x) = x, \text{ no probability distortion on gain part.}$$

– With respect to condition (4), we set[2]:

$$\alpha_- = 0.69, \gamma = 0.88, \lambda = 9 \text{ or } \alpha_- = 0.6, \gamma = 0.75, \lambda = 10.$$

If PT agent starts with an initial wealth larger than the initial wealth of the CRRA agent, he/she will take the "conservative" policy in (7). After some simple calculations, we have

$$W_T^{PT} = \theta_T + \frac{x_0}{\mathbb{E}[\xi_T^{-\gamma/(1-\gamma)}]}\left(\frac{1}{\xi_T}\right)^{1/(1-\gamma)} = \frac{w_0^{PT}}{\mathbb{E}[\xi_T^{-\gamma/(1-\gamma)}]}\left(\frac{1}{\xi_T}\right)^{1/(1-\gamma)},$$

which essentially follows the same conservative policy as the CRRA agent. Furthermore, the PT agent will always have a higher wealth level than the CRRA agent at time T, as a result of positive ξ_T. In plain language, the PT agent, who starts with an initial wealth higher than that of the CRRA agent, can always do better than the CRRA agent by imitating what the CRRA agent does.

Between the two situations presented above, the latter situation with the PT agent starting with an initial wealth level smaller than that of the CRRA agent is more interesting. In such a situation, he/she must take some gambling policy according to (8), and keep gambling until the situation reverses when his/her wealth is larger than the CRRA's. Assume that $\xi_n, n = T, 2T, \cdots$, are i.i.d. for each round. We can interpret every gambling of PT agent in each round as a Bernoulli trial with a success probability p^s given by

$$p^s = P(W_n^{PT} \geq W_n^{CRRA}) = P(\xi_n \leq c^*), n = T, 2T, \cdots.$$

As c^* is independent of the initial wealth, the success probability p^s is also independent of the initial wealth level. From our numerical experiments, p^s is usually a large number, for example, $1 - 1.6 \times 10^{-5}$ in this example. Denote N as the first "success" time when the wealth of the PT agent surpasses the wealth of the CRRA agent. Then the random number N follows the following geometrical distribution,

$$P(\text{ succeed after m years }) = P(N = m) = (1 - p^s)^{m-1}p^s.$$

Note also that once the PT agent attains his/her target, i.e., he/she attains a wealth level which is higher than the wealth level of the CRRA agent, he/she will then switch to the conservative policy in (7) immediately. Figure 1 shows one such an instance: the PT agent starts with an initial wealth of $50, which is smaller than the initial wealth of the CRRA agent, $100, but he/she can succeed

---

[2] Parameters $\alpha_- = 0.69$ and $\gamma = 0.88$ are the suggested numbers from Tversky and Kahneman (1992). But we need to set λ as larger as 9 to satisfy condition (4) when ξ_T follows a lognormal distribution.

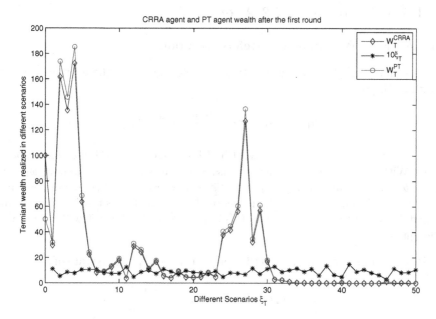

Fig. 1. One Step Success: CRRA $100 vs. PT $50

in one step. The black star stands for the different scenarios of ξ_T (scaled by 10 times). The blue diamond represents different realizations of W_T^{CRRA} under different scenarios of ξ_T; and the red circle represents different realizations of W_T^{PT} under different scenarios of ξ_T. The PT agent always has a higher wealth than the CRRA agent after the first round, regardless of the different scenarios. Furthermore, good scenarios (small ξ_T) correspond to high wealth levels for both the CRRA agent and PT agent.

From the above results, it seems that the optimal policy is always to gamble. With a high probability, the PT agent can do better than the CRRA agent by "gambling", no matter what initial wealth he/she starts with. Why do we have such results? The first reason comes from the fact that our result is for a model of a complete market. "Completeness" means that every policy can be carried out, no matter how crazy the contingent claim is, such as in (8), you want to short sell. The second reason may be due to our two-agent model setting. In our setting, the CRRA agent is in some passive position. The PT agent knows the CRRA's optimal terminal wealth and takes it as his/her reference point, and then he/she takes an advantagious position. In contrast, the CRRA agent can only passively accept his/her position. A more interesting problem would be to consider a situation where the CRRA agent modifies his/her behavior style with a consideration of market interaction too. In the following text we will consider a two-agent model, in which both agents are of PT-type.

4   PT-1 Agent vs. PT-2 Agent

4.1   Average Wealth as the Reference Point

Suppose that both investors are of PT-type and both take the average of their wealth as the reference point:

$$\theta_{1T} = \theta_{2T} = \theta_T = \frac{w_{10} + w_{20}}{2} e^{rT},$$

where w_{10} and w_{20} are the initial wealth of PT-1 agent and PT-2 agent, respectively. Without loss of generality, we assume that $w_{10} \geq w_{20}$. Denote $x_{1T} = w_{1T} - \theta_T$, $x_{2T} = w_{2T} - \theta_T$, $x_{10} = \frac{w_{10}-w_{20}}{2}$ and $x_{20} = \frac{w_{20}-w_{10}}{2}$. Once again, we can directly apply the method proposed in [7] in this situation. Based on their different initial wealth levels, PT-1 agent and PT-2 agent will adopt two different policies.

- PT-1 agent starts with a larger initial wealth (or a positive state of $x_{10} \geq 0$):

$$w_{10} \Rightarrow w_{1T} = \theta_T + A_1 \cdot \xi_T^{\frac{-1}{1-\gamma}} \cdot \frac{w_{10} - w_{20}}{2}.$$

- PT-2 agent starts with a smaller initial wealth (or a negative state of $x_{20} \leq 0$):

$$w_{20} \Rightarrow w_{2T} = \theta_T + B_1 \cdot \xi_T^{\frac{-1}{1-\gamma}} \cdot \frac{w_{10} - w_{20}}{2} 1_{\xi_T \leq c^*} - B_2 \cdot \frac{w_{10} - w_{20}}{2} 1_{\xi_T \geq c^*},$$

where $A_1 = \frac{1}{\phi(+\infty)}$, $B_1 = \frac{1}{v(c^*)}$, $B_2 = \left[\frac{1}{k(c^*)^{\frac{1}{1-\gamma}} - 1} + 1 \right] \frac{1}{\mathbb{E}\left[\xi_T 1_{\xi_T > c^*} \right]}$.

Remark 2. *Someone may criticize the rationality of taking "average wealth" as a reference point. Question could be raised as: In a two-agent model, the poorer one takes average wealth as his/her reference point, which means that he/she strives to be better and it seems to be a reasonable behavior. But, why does the richer person want to take average wealth (a lower wealth level than that of his/her status quo) as his/her reference point? The reason could be as follows: humans rarely choose things in their absolute terms; rather, they focus on the relative advantage of one thing over another, and estimate value accordingly. Here, we are considering neighborhood influence when people make decisions. Some people at the bottom of society, may set high wealth levels as their targets with a hope to have a better tomorrow. At the same time, the others, who are already in the upper class, take average wealth as their reference point in the sense of that relative wealth concern makes them happier.*

When both PT agents set the average wealth as their targets, a natural question is: whether their long-run wealth levels converge or not. In the first round, we denote the gap of the initial wealth levels of the two PT agents by $\Delta w_0 := w_{10} - w_{20}$ and the gap of their terminal wealth levels by $\Delta w_1 := w_{1T} - w_{2T}$.

In the second round, taking the previous terminal wealth gap as the initial wealth gap $\Delta w_1 := w_{1T} - w_{2T}$, and then their terminal wealth gap becomes $\Delta w_2 := w_{1,2T} - w_{2,2T}$. Generally, the dynamics of their wealth gap is governed by

$$\Delta w_n := \eta_n \cdot \Delta w_{n-1}, n = 1, 2, \cdots,$$

where $\Delta w_n := w_{1,nT} - w_{2,nT}$ and η_n, $n = 1, 2, \cdots$, are i.i.d. random variables,

$$\eta = \frac{A_1 - B_1}{2} \cdot \xi^{\frac{-1}{1-\gamma}} \cdot \mathbf{1}_{\xi \leq c^*} + \frac{1}{2}\left(A_1\xi^{\frac{-1}{1-\gamma}} + B_2\right) \cdot \mathbf{1}_{\xi \geq c^*}.$$

In particular, we have

$$\Delta w_1 = \eta_1 \cdot \Delta w_0, \Delta w_2 = \eta_2 \cdot \Delta w_1, \cdots \Delta w_n = \eta_n \cdot \Delta w_{n-1}, \cdots$$

If the random variables η_n, $n = 1, 2, \cdots$, satisfy the following sufficient convergence condition, we can claim that the wealth levels of the two PT investors converge.

Theorem 3. *(Sufficient Convergence Condition) Assume that Condition 3 and Condition 4 in [7] hold. If both PT investors take their average wealth as their reference points, then we have*

$$\mathbb{E}[|\eta|] < 1 \Longrightarrow w_{1,nT} - w_{2,nT} \xrightarrow{L_1} 0 \Longleftrightarrow w_{1,nT} - w_{2,nT} \xrightarrow{a.s.} 0.$$

Proof. Please refer to [8].

4.2   Toy Example-2: Converge or Diverge?

We still use S&P 500 yearly return as our market setting:

$$\mu = 0.1, \sigma = 0.2, r = 0.03, T = 1.$$

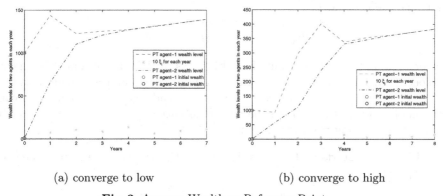

(a) converge to low   (b) converge to high

Fig. 2. Average Wealth as Reference Point

In order to satisfy the two key assumptions for well-posedness, we set investors' preference parameters as

$$T_+(x) = x, \alpha_- = 0.6, \gamma = 0.75, \lambda = 10.$$

Under these parameters, we have $\mathbb{E}[|\eta|] = 0.7194 < 1$, which satisfies the sufficient condition in Theorem 3. Figures 2(a) and 2(b) illustrate two convergent instances. In both figures, PT-1 agent starts with \$100 and PT-2 agent starts with \$1. They take their average wealth as their reference points and their wealth levels quickly converge to some common level, which depends on the market realizations of each year ξ_i. Generally, a better market (such as smaller ξ_i in Figure 2(b)) makes the two agents ending up with a higher common wealth level, while a worse market (such as larger ξ_i in Figure 2(a)) makes the society ending up with a lower common wealth level. Actually, what Theorem 3 really means is that: as long as the random variable η satisfies the convergence condition, their wealth will converge, no matter how large their initial wealth gap is. It seems that we find a "communist world". In this "communist world", investors' wealth will converge, when they all hope so (by setting the average wealth level as the reference point).

As you probably imagine, this sufficient convergence condition heavily depends on the characteristics of participants (preference parameters). In fact, if we consider some "aggressive" investors instead, the convergence result will change. "Aggressiveness" means less distortion (larger α_-) on the loss part, less risk aversion (larger γ) and less loss aversion (smaller λ). For example, if we set

$$\alpha_- = 0.69, \gamma = 0.88, \lambda = 9 \Rightarrow \mathbb{E}[|\eta|] = 4.6711 > 1,$$

which violates the sufficient condition in Theorem 3. Their wealth levels may diverge.

4.3   Mutual Reference Dependence

In this subsection, both investors are still of PT type, but they are mutually reference dependent, i.e.,

$$\theta_{1T} = w_{20} \cdot e^{rT}, \theta_{2T} = w_{10} \cdot e^{rT},$$

where w_{10} and w_{20} are the initial wealth levels of PT-1 agent and PT-2 agent, respectively. Once again, we assume that $w_{10} \geq w_{20}$, and also denote $x_{1T} = w_{1T} - \theta_{1T}$, $x_{2T} = w_{2T} - \theta_{2T}$, $x_{10} = w_{10} - w_{20}$ and $x_{20} = w_{20} - w_{10}$. Applying the method proposed in [7] to the current situation, our two PT investors will apply the following two different policies:

- PT-1 agent starts with a larger initial wealth (or a positive state variable of $x_{10} \geq 0$),

$$w_{10} \Rightarrow w_{1T} = \theta_{1T} + A_1 \cdot \xi_T^{\frac{-1}{1-\gamma}} \cdot (w_{10} - w_{20}).$$

– PT-2 agent starts with a smaller initial wealth (or a negative state variable of $x_{20} \leq 0$)

$$w_{20} \Rightarrow w_{2T} = \theta_{2T} + B_1 \cdot \xi_T^{\frac{-1}{1-\gamma}} \cdot (w_{10} - w_{20})1_{\xi_T \leq c^*} - B_2 \cdot (w_{10} - w_{20})1_{\xi_T \geq c^*},$$

where $A_1 = \frac{1}{\phi(+\infty)}$, $B_1 = \frac{1}{v(c^*)}$, $B_2 = \left[\frac{1}{k(c^*)^{\frac{1}{1-\gamma}} - 1} + 1\right] \frac{1}{\mathbb{E}\left[\xi_T 1_{\xi_T > c^*}\right]}$.

Similarly, we denote the dynamics of their wealth gap by $\Delta w_n := \widetilde{\eta}_n \cdot \Delta w_{n-1}$, $n = 1, 2, \cdots$, where $\widetilde{\eta} = -e^{rT} + (A_1 - B_1) \cdot \xi^{\frac{-1}{1-\gamma}} \cdot 1_{\xi \leq c^*} + \left(A_1 \xi^{\frac{-1}{1-\gamma}} + B_2\right) \cdot 1_{\xi \geq c^*}$. With similar arguments, we have the following sufficient condition for convergence.

Theorem 4. *(Sufficient Convergence Condition) Assume that Condition 3 and Condition 4 in [7] hold. If both PT investors are mutually reference dependent, then we have*

$$\mathbb{E}[|\widetilde{\eta}|] < 1 \Longrightarrow w_{1,nT} - w_{2,nT} \xrightarrow{L_1} 0 \Longleftrightarrow w_{1,nT} - w_{2,nT} \xrightarrow{a.s.} 0.$$

Proof. Please refer to [8].

4.4   Toy Example-3: Converge or Diverge?

We still use S&P 500 yearly return as our market setting:

$$\mu = 0.1, \sigma = 0.2, r = 0.03, T = 1.$$

In order to satisfy the two assumptions for well-posedness, we set investors' preference parameters as

$$T_+(x) = x, \alpha_- = 0.6, \gamma = 0.72, \lambda = 10 \Rightarrow E[|\widetilde{\eta}|] = 0.8364 < 1.$$

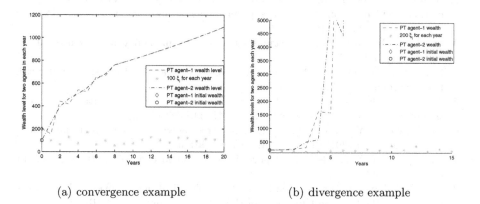

(a) convergence example   (b) divergence example

Fig. 3. Mutual Reference Dependence

By Theorem 4, the wealth levels of these two investors will converge. Figure 3(a) shows one instance of such a convergence: PT-1 agent starts with \$200 and PT-2 agent starts with \$100. From Figure 3(a), we can see that each of the two investors takes the other's wealth level as his/her reference point and their wealth levels indeed converge. It seems that we find another "communist world" here. But if we change the parameters a little as follows,

$$\alpha_- = 0.6, \gamma = 0.71, \lambda = 10 \Rightarrow E[\|\tilde{\eta}\|] = 5.5873 > 1,$$

we can see that their wealth levels diverge as shown in Figure 3(b). The PT-1 agent starts with \$200 and PT-2 agent starts with \$199. Even with such a small initial wealth gap, their wealth levels eventually diverge.

5   Conclusion

An investor's reference point is not only affected by his/her own investment experience, but also affected by other investors' wealth levels, especially his/her "neighbors" in his/her "social network" who are close to him/her. By carrying out theoretical analysis and providing numerical examples, we show that the comparison of the initial wealth levels between a PT agent and a CRRA agent matters greatly to the PT agent, as its sign dictates whether the PT agent will take a "conservative policy" or a "gambling policy". When both agents are of PT-type, we provide sufficient convergence conditions for two types of reference points: average wealth as their reference points and mutual reference dependence.

References

1. Schultz, P.W., Nolan, J.M., Cialdini, R.B., Goldstein, N.J., Griskevicius, V.: The constructive, destructive, and reconstructive power of social norms. J. Public Econ. 94(7), 435–452 (2007)
2. Azmat, G., Iriberri, N.: The importance of relative performance feedback information: Evidence from a natural experiment using high school students. Psychol. Sci. 18(5), 429–434 (2010)
3. Basak, S., Makarov, D.: Strategic asset allocation in money management. J. Financ. 69(1), 179–217 (2014)
4. Kahneman, D., Tversky, A.: Prospect theory: An analysis of decisions under risk. Econometrica 47(2), 263–292 (1979)
5. Tversky, A., Kahneman, D.: Advances in prospect theory: Cumulative representation of uncertainty. J. Risk Uncertainty 5(4), 297–323 (1992)
6. Shi, Y., Cui, X.Y., Yao, J., Li, D.: Dynamic trading with reference point adaptation and loss aversion (2013) (submitted for publication)
7. Jin, H., Zhou, X.Y.: Behavioral portfolio selection in continuous time. Math. Financ. 18(3), 385–426 (2008)
8. Shi, Y.: Behavioral portfolio models and their implications in investors' behaviors, PhD thesis, the Chinese University of Hong Kong, Hong Kong (2013)

Finding Equilibrium in a Financial Model by Solving a Variational Inequality Problem

Vyacheslav V. Kalashnikov[1], Nataliya I. Kalashnykova[2],
and Felipe J. Castillo-Pérez[1]

[1] Department of Systems and Industrial Engineering
Tecnológico de Monterrey (ITESM), Campus Monterrey
Ave. Eugenio Garza Sada 2501 Sur
Monterrey, Nuevo León, Mexico 64849
kalash@itesm.mx
[2] Department of Physics and Mathematics
Universidad Autónoma de Nuevo León
San Nicolás de los Garza, Nuevo León, Mexico 66450
nkalash2009@gmail.com

Abstract. In this paper, a general multi-sector, multi-instrument model of financial flows and prices is developed, in which the utility function for each sector is assumed to be quadratic and constraints satisfy a certain accounting identity that appears in flow-of-funds accounts. Each sector uses conjectures of its influence upon the prices of instruments. The equilibrium conditions are first derived, and then the governing variational inequality is presented. Subsequently, a qualitative analysis of the model is conducted, and a concept of consistent conjectures is introduced and examined.

Keywords: Consistent conjectures, consistent equilibrium, financial models.

1 Introduction

Consider an economy consisting of m sectors, with a typical sector denoted by i, and with n instruments, with a typical instrument denoted by j. Denote the volume of instrument j held in sector i's portfolio as an asset, by x_{ij}, and the volume of instrument j held in sector's i's portfolio as a liability, by y_{ij}. The assets in sector i's portfolio are grouped into a column vector $x_i \in R^n$, and the liabilities are grouped into the column vector $y_i \in R^n$. Further group the sector asset vectors into the column vector $x \in R^{mn}$, and the sector liability vectors into the column vector $y \in R^{mn}$.

Each sector's utility can be defined as a function of the expected future portfolio value. The expected value of the future portfolio may be described by two characteristics: the expected mean value and the uncertainty surrounding the expected mean. In this model, the expected mean portfolio value of the next period is assumed to be equal to the market value of the current period portfolio. Each sector's uncertainty, or assessment of risk, with respect to the future value of the portfolio is based on a variance-covariance matrix denoting the sector's assessment of the standard deviation of prices for each instrument. The $2n \times 2n$ variance-covariance matrix associated with sector i's assets and liabilities is denoted by Q^i.

Since each sector's expectations are formed by reference to current market activity, sector utility maximization can be written in terms of optimizing the current portfolio. Sectors may trade, issue, or liquidate holdings in order to optimize their portfolio compositions.

In this model, it is assumed that the total volume of each balance sheet side is exogenous. Let r_j denote the price of instrument j, and group the prices into the column vector $r \in R^n$. In contrast to the model by A.Nagurney [1] that makes use of the assumption of perfect competition, i.e., supposes that each sector will behave as if its actions cannot affect the instruments' prices and thus the behaviour of the other sectors, we examine the simplest oligopoly model. In other words, we assume that each sector i expects the price of instrument j to grow up together with the (positive) gap $y_{ij} - x_{ij}$ between his liability and asset holdings, and the rate of this grow is $w_{ij} \geq 0$ that will be referred to as sector i's *coeficient of influence* upon the price of instrument j. This assumption is the main novelty of our model compared to that studied in [1].

The rest of the paper is arranged as follows. The quadratic-type portfolio optimization problem is specified and studied in Section 2. Section 3 deals with the exterior conjectural variations equilibrium (CVE) in the financial model with general utility functions of the sector, and the exterior CVE properties are examined in Section 4. Finally, a new concept of consistent (interior) conjectural variations equilibrium is introduced and discussed in Section 5. Concluding remarks, acknowledgments, and a list of references complete the paper.

2   Model Specification

Define each sector's portfolio optimization problem as follows. Sector i seeks to determine its optimal composition of instruments held as assets and as liabilities, so as to minimize the risk while at the same time maximizing the value of its asset holdings and minimizing the value of its liabilities. The portfolio optimization problem for sector i is, hence, given by:

$$\text{Minimize} \qquad \begin{pmatrix} x_i \\ y_i \end{pmatrix}^T Q^i \begin{pmatrix} x_i \\ y_i \end{pmatrix} - \sum_{j=1}^{n} r_{ij}(x_{ij} - y_{ij})$$

subject to

$$\sum_{j=1}^{n} x_{ij} = s_i, \qquad \sum_{j=1}^{n} y_{ij} = s_i, \tag{1}$$

$$x_{ij} \geq 0, \quad y_{ij} \geq 0, \quad j = 1, \ldots, n, \tag{2}$$

where

$$r_{ij} = r_{ij}(x_{ij}, y_{ij}) = r_j - w_{ij}(x_{ij} - y_{ij}). \tag{3}$$

Here, the instrument price vector $r \in R^n$ is exogenous to the individual sector optimization problem, whereas the quotient $w_{ij} \geq 0$ reflects the degree of influence of sector i on the price of instrument j conjectured by this sector itself. That is, sector i conjectures the expected price for its liability to equal r_{ij} determined by (3).

Constraints (1) represent the accounting identity reflecting that the accounts for sector i must balance, where s_i is the total financial volume held by sector i. Constraints (2) are the usual non-negativity restrictions. Let Ω_i denote the closed convex subset of (x_i, y_i) formed by constraints (1) and (2). Since Q^i is a variance-covariance matrix, it will be assumed that this matrix is positive definite and, therefore, the objective function for each sector i's portfolio optimization problem is strictly convex.

Necessary and sufficient conditions for a portfolio $(x_i^*, y_i^*) \in \Omega_i$ to be optimal is that it satisfy the following system of inequalities and equalities called the *linear complementarity problem* (LCP).

For each instrument j $(j = 1, \ldots, n)$:

$$
\begin{aligned}
\varphi_{ij}^1(x^*, y^*) &\equiv 2Q_{(11)j}^{i}{}^T x_i^* + 2Q_{(21)j}^{i}{}^T y_i^* - r_j^* + \\
&\quad + 2w_{ij}(x_{ij}^* - y_{ij}^*) - \mu_i^1 \geq 0, \\
\varphi_{ij}^2(x^*, y^*) &\equiv 2Q_{(22)j}^{i}{}^T y_i^* + 2Q_{(12)j}^{i}{}^T x_i^* + r_j^* - \\
&\quad - 2w_{ij}(x_{ij}^* - y_{ij}^*) - \mu_i^2 \geq 0, \\
x_{ij}^* \cdot \varphi_{ij}^1(x^*, y^*) &= 0, \\
y_{ij}^* \cdot \varphi_{ij}^2(x^*, y^*) &= 0,
\end{aligned}
\tag{4}
$$

where r_j^* denotes the price for instrument j, which is assumed to be fixed from the perspective of all the sectors. Note that Q^i has been partitioned as $Q^i = \begin{pmatrix} Q_{11}^i & Q_{12}^i \\ Q_{21}^i & Q_{22}^i \end{pmatrix}$, and is symmetric. Furthermore, $Q_{(\alpha\beta)j}^i$ denotes the j-th column of $Q_{(\alpha\beta)}^i$, with $\alpha = 1, 2; \beta = 1, 2$. The terms μ_i^1 and μ_i^2 are the Lagrange multipliers of constraints (1).

A corresponding LCP will be solved by each of the m sectors.

2.1 Definition of Equilibrium

The inequalities governing the instrument prices in the economy are now described. These prices provide the feedback from the economic system to the sectors regarding the equilibration of the total assets and total liabilities of each instrument. Here, it is assumed that there is free disposal and, hence, the instrument prices will be nonnegative. The economic system conditions ensuring market clearance then take on the following form.

For each instrument j; $j = 1, \ldots, n$:

$$
\sum_{i=1}^{m} (x_{ij}^* - y_{ij}^*) \begin{cases} = 0, & \text{if } r_j^* > 0; \\ \geq 0, & \text{if } r_j^* = 0. \end{cases}
\tag{5}
$$

In other words, if the price is positive, then the market must clear for the instrument; if there is an excess supply of an instrument in the economy, then its price must be zero. Combining the above sector and market inequalities and equalities yields the following.

Definition 1. (Exterior equilibrium in the financial model)
For a fixed set of conjectures described by the $m \times n$ matrix $W = (w_{ij})_{i=1,j=1}^{m\ \ n}$, a vector $(x^*, y^*, r^*) \in \prod_{i=1}^{m} \Omega_i \times R_+^n$ is called the exterior equilibrium of the financial model if and only if it satisfies the system of equalities and inequalities (4) and (5), for all sectors $1 \leq i \leq m$, and for all instruments $1 \leq j \leq n$, simultaneously.

Now we are ready to deduce the variational inequality problem governing the exterior equilibrium conditions of our financial model.

Theorem 1. (Variational Inequality Formulation of Exterior Financial Equilibrium)
For a fixed set of conjectures described by the $m \times n$ matrix $W = (w_{ij})_{i=1,j=1}^{m\ \ n}$, a vector $(x^*, y^*, r^*) \in \prod_{i=1}^{m} \Omega_i \times R_+^n$ of sector assets, liabilities, and instrumental prices is the exterior financial equilibrium if and only if it solves the following variational inequality problem: Determine an $(x^*, y^*, r^*) \in \prod_{i=1}^{m} \Omega_i \times R_+^n$, satisfying:

$$\sum_{i=1}^{m} \sum_{j=1}^{n} \left[2Q_{(11)j}^{i}{}^{T} x_i^* + 2Q_{(21)j}^{i}{}^{T} y_i^* - r_j^* + 2w_{ij}(x_{ij}^* - y_{ij}^*) \right] \times$$

$$\times \left(x_{ij} - x_{ij}^* \right) +$$

$$+ \sum_{i=1}^{m} \sum_{j=1}^{n} \left[2Q_{(22)j}^{i}{}^{T} y_i^* + 2Q_{(12)j}^{i}{}^{T} x_i^* + r_j^* - 2w_{ij}(x_{ij}^* - y_{ij}^*) \right] \times$$

$$\times \left(y_{ij} - y_{ij}^* \right) + \sum_{j=1}^{n} \left[\sum_{i=1}^{m} (x_{ij}^* - y_{ij}^*) \right] \times \left[r_j - r_j^* \right] \geq 0,$$

$$\forall (x, y, r) \in \prod_{i=1}^{m} P_i \times R_+^n. \tag{6}$$

Proof. The proof being quite long will be published elsewhere.

The qualitative analysis of the variational inequality (6) governing the financial equilibrium model introduced in this section is presented in Section 2 in the framework of a more general model, of which the quadratic model introduced here is a special case.

3   General Utility Functions

In this section, the quadratic financial model is extended, and a variational inequality formulation of the (exterior) equilibrium conditions presented.

Assume that each sector seeks to maximize its utility, where the utility function, $U_i(x_i, y_i, r_i)$, is given by:

$$U_i(x_i, y_i, r_i) = u_i(x_i, y_i) + \langle r_i^T, x_i - y_i \rangle, \tag{7}$$

where $u_i : R^n \times R^n \to R$ is a differentiable function, and $r_i = r_i(x_i, y_i) \in R^n$ is the price vector with its components r_{ij}, $j = 1, \ldots, n$, defined in (3). Taking that into account, we can rewrite (13) as follows:

$$U_i(x_i, y_i, r_i) = u_i(x_i, y_i) + \langle r^T, x_i - y_i \rangle - \sum_{j=1}^n w_{ij}(x_{ij} - y_{ij})^2. \tag{8}$$

Then the optimization problem for sector i can be specified as:

$$\text{Maximize}_{(x_i, y_i) \in \Omega_i} \quad U_i(x_i, y_i, r_i), \tag{9}$$

where Ω_i is a closed, convex, nonempty, and bounded subset of R^{2n}, denoting the feasible set of asset and liability choices. Note that in this model, we no longer require the constraint set Ω_i to be of the form defined by equalities (1) and inequalities (2). Nevertheless, the model introduced in this section captures the general financial equilibrium model of Section 1 as a special case, where

$$u_i(x_i, y_i) = - \begin{pmatrix} x_i \\ y_i \end{pmatrix}^T Q^i \begin{pmatrix} x_i \\ y_i \end{pmatrix}.$$

Assuming that each sector's utility function is concave, necessary and sufficient conditions for an optimal portfolio (x_i^*, y_i^*), given a fixed vector of instrument prices r^*, are that $(x_i^*, y_i^*) \in \Omega_i$, and satisfy the inequality:

$$-\langle \nabla_{x_i} U_i(x_i^*, y_i^*, r_i^*)^T, x_i - x_i^* \rangle -$$

$$-\langle \nabla_{y_i} U_i(x_i^*, y_i^*, r_i^*)^T, y_i - y_i^* \rangle \geq 0, \quad \forall (x_i, y_i) \in \Omega_i, \tag{10}$$

where $\nabla_{x_i} U_i(\cdot)$ denotes the gradient of $U_i(\cdot)$ with respect to x_i, or, equivalently, in view of (7)–(8),

$$-\langle (\nabla_{x_i} u_i(x_i^*, y_i^*) + r^*)^T, x_i - x_i^* \rangle +$$

$$+2 \sum_{j=1}^n w_{ij}(x_{ij}^* - y_{ij}^*)(x_{ij} - x_{ij}^*) -$$

$$-\langle (\nabla_{y_i} u_i(x_i^*, y_i^*) - r^*)^T, x_i - x_i^* \rangle -$$

$$-2 \sum_{j=1}^n w_{ij}(x_{ij}^* - y_{ij}^*)(y_{ij} - y_{ij}^*) \geq 0, \quad \forall (x_i, y_i) \in \Omega_i. \tag{11}$$

A respective variational inequality must hold for each of the m sectors.

The system of equalities and inequalities governing the instrument prices in the economy as in (5) is still valid. Hence, one can immediately write down the following economic system conditions.

For each instrument j; $j = 1, \ldots, n$:

$$\sum_{i=1}^{m} x_{ij}^* - \sum_{i=1}^{m} y_{ij}^* \begin{cases} = 0, & \text{if } r_j^* > 0, \\ \geq 0, & \text{if } r_j^* = 0. \end{cases} \tag{12}$$

In other words, as before, if there is an excess supply of an instrument in the economy, then its price must be zero; if the price of an instrument is positive, then the market for this instrument must clear.

Combining the above sector and market inequalities and equalities yields the following.

Definition 2. (Exterior Financial Equilibrium with General Utility Functions)
For a fixed set of conjectures described by the $m \times n$ matrix $W = (w_{ij})_{i=1,j=1}^{m \quad n}$, a vector $(x^*, y^*, r^*) \in \prod_{i=1}^{m} \Omega_i \times R_+^n$ is the exterior conjectural variations equilibrium (CVE) in the financial model developed above if and only if it satisfies inequalities (11) and (12), for all sectors $1 \leq i \leq m$, and for all instruments $1 \leq j \leq n$, simultaneously.

The variational inequality formulation of the equilibrium conditions of the model is now presented. The proof of this theorem is similar to that of Theorem 1.

Theorem 2. (Variational Inequality Formulation of Exterior Financial Conjectural Variations Equilibrium with General Utility Functions)
For a fixed set of conjectures described by the $m \times n$ matrix $W = (w_{ij})_{i=1,j=1}^{m \quad n}$, a vector of assets and liabilities of the sectors, and instrument prices $(x^*, y^*, r^*) \in \prod_{i=1}^{m} \Omega_i \times R_+^n$ is the exterior financial conjectural variations equilibrium if and only if it solves the variational inequality problem:

$$-\sum_{i=1}^{m} \langle (\nabla_{x_i} u_i(x_i^*, y_i^*) + r^*)^T, x_i - x_i^* \rangle +$$

$$+2 \sum_{i=1}^{m} \sum_{j=1}^{n} w_{ij}(x_{ij}^* - y_{ij}^*)(x_{ij} - x_{ij}^*) -$$

$$-\sum_{i=1}^{m} \langle (\nabla_{y_i} u_i(x_i^*, y_i^*) - r^*)^T, y_i - y_i^* \rangle -$$

$$-2 \sum_{i=1}^{m} \sum_{j=1}^{n} w_{ij}(x_{ij}^* - y_{ij}^*)(y_{ij} - y_{ij}^*) +$$

$$+\sum_{j=1}^{n} \left[\sum_{i=1}^{m} x_{ij}^* - \sum_{i=1}^{m} y_{ij}^* \right] \times [r_j - r_j^*] \geq 0,$$

$$\forall (x, y, r) \in \prod_{i=1}^{m} \Omega_i \times R_+^n. \tag{13}$$

4   Qualitative Properties

In this section, certain qualitative properties of the exterior CVE in the model outlined in Section 2 are examined.

Theorem 3. (Existence)
For any fixed set of conjectures described by the $m \times n$ matrix $W = (w_{ij})_{i=1,j=1}^{m\ \ n}$, if $(x^*, y^*, r^*) \in \prod_{i=1}^{m} \Omega_i \times R_+^n$ is the exterior CVE in the model, that is, it solves the variational inequality (13), then the equilibrium asset and liability vector (x^*, y^*) is a solution to the variational inequality:

$$-\sum_{i=1}^{m}\langle(\nabla_{x_i}u_i(x_i^*, y_i^*))^T, x_i - x_i^*\rangle +$$

$$+2\sum_{i=1}^{m}\sum_{j=1}^{n}w_{ij}(x_{ij}^* - y_{ij}^*)(x_{ij} - x_{ij}^*) -$$

$$-\sum_{i=1}^{m}\langle(\nabla_{y_i}u_i(x_i^*, y_i^*))^T, y_i - y_i^*\rangle -$$

$$-2\sum_{i=1}^{m}\sum_{j=1}^{n}w_{ij}(x_{ij}^* - y_{ij}^*)(y_{ij} - y_{ij}^*) \geq 0, \quad \forall(x, y) \in S, \tag{14}$$

where the subset

$$S \equiv \left\{(x, y) \in \prod_{i=1}^{m}\Omega_i : \sum_{i=1}^{m}(x_{ij} - y_{ij}) \geq 0; j = 1, \ldots, n\right\}$$

is nonempty.
  Conversely, if (x^*, y^*) is a solution of (14), there exists an $r^* \in R_+^n$, such that (x^*, y^*, r^*) is a solution of (13), and, thus, the exterior CVE in the financial model.

Proof. The proof being quite long will be published elsewhere.

At last, we show that if the utility functions u_i are strictly concave for all i, then the exterior equilibrium asset and liability pattern (x^*, y^*) is also unique for any fixed conjectures W.
  It is clear that if the functions u_i are strictly concave then the functions U_i defined by (8) are also strictly concave with respect to the variables (x_i, y_i).
  Assume now that for the same fixed conjectures W, there are two distinct exterior equilibrium solutions (x^1, y^1, r^1) and (x^2, y^2, r^2). Then

$$-\sum_{i=1}^{m}\langle(\nabla_{x_i}u_i(x_i^1, y_i^1) + r^1)^T, x_i' - x_i^1\rangle +$$

$$+2\sum_{i=1}^{m}\sum_{j=1}^{n}w_{ij}(x_{ij}^1 - y_{ij}^1)(x_{ij}' - x_{ij}^1)-$$

$$-\sum_{i=1}^{m}\langle(\nabla_{y_i}u_i(x_i^1, y_i^1) - r^1)^T, y_i' - y_i^1\rangle-$$

$$-2\sum_{i=1}^{m}\sum_{j=1}^{n}w_{ij}(x_{ij}^1 - y_{ij}^1)(y_{ij}' - y_{ij}^1)+$$

$$+\sum_{j=1}^{n}\left[\sum_{i=1}^{m}x_{ij}^1 - \sum_{i=1}^{m}y_{ij}^1\right](r_j' - r_j^1) \geq 0,$$

$$\forall(x', y', r') \in \prod_{i=1}^{m}\Omega_i \times R_+^n, \tag{15}$$

and

$$-\sum_{i=1}^{m}\langle(\nabla_{x_i}u_i(x_i^2, y_i^2) + r^2)^T, x_i - x_i^2\rangle+$$

$$+2\sum_{i=1}^{m}\sum_{j=1}^{n}w_{ij}(x_{ij}^2 - y_{ij}^2)(x_{ij} - x_{ij}^2)-$$

$$-\sum_{i=1}^{m}\langle(\nabla_{y_i}u_i(x_i^2, y_i^2) - r^2)^T, y_i - y_i^2\rangle-$$

$$-2\sum_{i=1}^{m}\sum_{j=1}^{n}w_{ij}(x_{ij}^2 - y_{ij}^2)(y_{ij} - y_{ij}^2)+$$

$$+\sum_{j=1}^{n}\left[\sum_{i=1}^{m}x_{ij}^2 - \sum_{i=1}^{m}y_{ij}^2\right](r_j - r_j^2) \geq 0,$$

$$\forall(x, y, r) \in \prod_{i=1}^{m}\Omega_i \times R_+^n. \tag{16}$$

Now select $(x', y', r') = (x^2, y^2, r^2)$ and substitute it to (15); symmetrically, set $(x, y, r) = (x^1, y^1, r^1)$ and put it into inequality (16). Summing up the resulting inequalities, we come to

$$-\sum_{i=1}^{m}\langle(\nabla_{x_i}u_i(x_i^1, y_i^1) - \nabla_{x_i}u_i(x_i^2, y_i^2))^T, x_i^2 - x_i^1\rangle-$$

$$-\sum_{i=1}^{m}\langle(\nabla_{y_i}u_i(x_i^1, y_i^1) - \nabla_{y_i}u_i(x_i^2, y_i^2))^T, y_i^2 - y_i^1\rangle-$$

$$-2\sum_{i=1}^{m}\sum_{j=1}^{n} w_{ij}\left[(x_{ij}^1 - x_{ij}^2) - (y_{ij}^1 - y_{ij}^2)\right]^2 \geq 0, \qquad (17)$$

which contradict (23). Hence, we have thus established the following result.

Theorem 4. (Uniqueness of Exterior Equilibrium Asset and Liability Pattern)
If the utility functions u_i are strictly concave for all sectors i, then for any fixed conjectures matrix W, the exterior CVE asset and liability pattern (x^*, y^*) exists uniquely.

Remark 1. Theorem 4 clearly implies that under its assumptions, the exterior CVE involving the asset and liability pattern (x^*, y^*) also exists uniquely for any fixed conjectures matrix W.

Remark 2. Observe that in the above analysis, if the utility function U_i had been assumed to be concave, rather than strictly concave with respect to the variables (x_i, y_i), then existence would still have been established, but one would no longer be able to guarantee the uniqueness of the (exterior) conjectural variations equilibrium (CVE) asset and liability pattern for a fixed conjectures matrix W.

5   Consistent Conjectures

In all the previous sections, we implicitly assumed that the conjectures W are given exogenously for the model. However, in a series of recent publications by the authors [4], [5], [6], a concept of *consistent* conjectures has been proposed and justified. Although this concept is impossible to apply to our financial model immediately, Theorem 4 allows one to construct an *upper level game* and define consistent conjectures W indirectly. This procedure works as follows.

Under assumptions of Theorem 4, define a many-person game $\Gamma = (N, W, V)$ by the following rules:
(*i*) $N = \{1, \ldots, m\}$ is the set of the same sectors as in our financial model;
(*ii*) the set of feasible conjectures $W = (w_{ij})_{i=1,j=1}^{m\ \ n} \in R_+^{m\times n}$ is the set of possible strategies in the upper level game;
(*iii*) $V = V(W) = (V_1, \ldots, V_m)$ are payoff functions used by the sectors $i = 1, \ldots, m$, in order to estimate their payoffs as the result of being stuck to their strategies $w_i = (w_{i1}, \ldots, w_{in})$. These functions are well-defined via Theorems 3 and 4 as follows: For each matrix W, according to those theorems, there exists uniquely an exterior CVE (x^*, y^*, r^*). Now the payoff function $V_i = V_i(W)$ is well-defined as the optimal value of the utility function $U_i(x^*, y^*, r^*)$, introduced in (14), i.e.,

$$V_i(W) = U_i(x^*, y^*, r^*), \quad i = 1, \ldots, m. \qquad (18)$$

Indeed, the payoff values (18) are defined by formula (8), where the (equilibrium) assets and liabilities holdings (x^*, y^*), as well as the equilibrium price r^*, are the elements of the exterior CVE whose existence and uniqueness is guaranteed by Theorem 4 of Section III.

Now if the strategies set W in the upper level game is a compact (i.e., closed and bounded) subset of $R_+^{m \times n}$, one can guarantee the existence of the classical (Cournot-Nash) equilibrium in this game [3]. Then we name this Cournot-Nash equilibrium in the upper level game a *consistent*, or *interior* CVE in the original financial model.

Definition 3. (Consistent, or Interior Financial CVE with General Utility Functions)
The asset, liabilities, and price vector (x^*, y^*, r^*) generated in the Cournot-Nash equilibrium of the upper level game is called the interior equilibrium in the financial model, and the corresponding conjectures W^* involved in the upper level game Cournot-Nash equilibrium, are named consistent.
The section is finished with the following result.

Theorem 5. (Existence of Interior CVE in Our Financial Model)
Under assumptions of Theorem 4, and for a compact feasible set of conjectures W, there exists the consistent (interior) CVE in the financial model (9).

Proof. It is straightforward (*cf.* [3]).

6   Conclusion

In the paper, a general multi-sector, multi-instrument model of financial flows and prices is developed, in which the utility function for each sector is assumed to be quadratic and constraints satisfy a certain accounting identity that appears in flow-of-funds accounts. Each sector uses conjectures of its influence upon the prices of instruments. The equilibrium conditions are first derived, and then the governing variational inequality is presented. Subsequently, a qualitative analysis of the model is conducted.

Finally, a criterion of consistency of the influence coefficients w_{ij} is introduced, and the existence of at least one interior (consistent) CVE in the financial model for a compact feasible conjectures set W, is established.

Acknowledgments. The research activity of the authors was financially supported by the Research Department of the Tecnológico de Monterrey (ITESM), Campus Monterrey, and by the SEP-CONACYT project CB-2013-01-221676, Mexico. The second author was also supported by the PAICYT project No. CE250-09.
We also wish to thank the anonymous reviewers for helpful suggestions.

References

1. Nagurney, A.: Network Economics: A Variational Inequality Approach. Kluwer Academic Publishers, Dordrecht (1999)
2. Isac, G., Bulavsky, V.A., Kalashnikov, V.V.: Complementarity, Equilibrium, Efficiency and Economics. Kluwer Academic Publishers, Dordrecht (2002)

3. Ferris, M.C., Pang, J.-C.: Engineering and economic applications of complementarity problems. SIAM Review 39, 669–713 (1997)
4. Kalashnikov, V.V., Bulavsky, V.A., Kalashnykova, N.I., Castillo, F.J.: Mixed oligopoly with consistent conjectures. European J. Oper. Res. 210, 729–735 (2011)
5. Bulavsky, V.A., Kalashnikov, V.V.: Games with linear conjectures about system parameters. J. Optim. Theory and Appl. 152, 152–170 (2012)
6. Kalashnikov, V.V., Bulavsky, V.A., Kalashnikov Jr., V.V., Kalashnykova, N.I.: Structure of demand and consistent conjectural variations equilibrium (CCVE) in a mixed oligopoly model. Ann. Oper. Res. 2014, 281–297 (2014)

Multiperiod Mean-CVaR Portfolio Selection[*]

Xiangyu Cui[1] and Yun Shi[2,**]

[1] School of Statistics and Management,
Shanghai University of Finance and Economics, Shanghai, China
cui.xiangyu@mail.shufe.edu.cn
[2] School of Management, Shanghai University, Shanghai, China
y_shi@shu.edu.cn

Abstract. Due to the time inconsistency issue of multiperiod mean-CVaR model, two important policies of the model with finite states, the pre-committed policy and the time consistent policy, are derived and discussed. The pre-committed policy, which is global optimal for the model, is solved through linear programming. A detailed analysis shows that the pre-committed policy doesn't satisfy time consistency in efficiency either, i.e., the truncated pre-committed policy is not efficient for the remaining short term mean-CVaR problem. The time consistent policy, which is the subgame Nash equilibrium policy of the multiperson game reformulation of the model, takes a piecewise linear form of the current wealth level and the coefficients can be derived by a series of integer programming problems and two linear programming problems. The difference between two polices indicates the degree of time inconsistency.

Keywords: mean-CVaR, pre-committed policy, time consistency in efficiency, time consistent policy, linear programming, integer programming.

1 Introduction

Conditional value at risk (CVaR) is a widely used risk measure in financial institutions. Different from value at risk (VaR), CVaR is a coherent risk measure (see [1]). Mathematically, CVaR is defined as the conditional expected loss,

$$\nu_C(\xi, \alpha) := \mathbb{E}[\ \xi|\ \xi \geq \nu(\xi, \alpha)],$$

where ξ is the loss, α is the confidential level and $\nu(\xi, \alpha)$ is the VaR of ξ (see [9]). In [9], Rockafellar and Uryasev showed the CVaR, $\nu_C(\xi, \alpha)$, has the following representation

$$\nu_C(\xi, \alpha) = \min_z \left[z + \frac{1}{1 - \alpha} \mathbb{E}\left[\ (\xi - z)^+\right] \right], \tag{1}$$

---

[*] This research work was partially supported by the National Natural Science Foundation of China under grant 71201094.
[**] Corresponding author.

    293
H.A. Le Thi et al. (eds.), *Model. Comput. & Optim. in Inf. Syst. & Manage. Sci.*,
Advances in Intelligent Systems and Computing 359, DOI: 10.1007/978-3-319-18161-5_25

where $y^+ = \max\{y, 0\}$ is the positive part of $y \in \mathbb{R}$. With the help of the representation, CVaR has been successfully applied in static (single period) portfolio selection models.

However, CVaR has been seldom discussed in dynamic portfolio selection problem. As the dynamic mean-CVaR model does not satisfy the Bellman's principle of optimality of dynamic programming, the global optimal policy for the entire investment time horizon is NOT consistent with the local optimal policies at different time instants for the corresponding tail parts of the time horizon, which results in *time inconsistency* (see [6], [2]).

If the investor only concerns the global interests, he/she may adopt the global optimal policy, which is called pre-committed policy following the term used in Basak and Chabakauri (2010) ([3]). While, if the investor totally ignores the global interests and bows to local interests, the dynamic mean-CVaR problem can be modelled as a multiperson game, in which the investor at any time instance acts as a Stackelberg leader and makes his/her "best" investment policy by taking into account his/her policies in future time instances. The corresponding subgame Nash equilibrium policy is called time consistent policy (see [4], [5]).

In this paper, we study multiperiod mean-CVaR portfolio selection model with finite states. Instead of adopting martingale method and making complete market assumption (see [8]), we derive the pre-committed policy via linear programming in Section 2. Then, we study the time inconsistent property of the truncated pre-committed policy in Section 3. In Section 4, we show that the time consistent policy takes a piecewise linear form of wealth level, and derive the coefficients via linear programming and integer programming. Finally, we conclude our paper in Section 5.

2   Pre-committed Policy

The capital market under consideration consists of n risky assets with random rates of returns and one riskless asset with deterministic rate of return. The deterministic rate of return of the riskless asset at time period t is $r_t^0 > 0$, and the rates of return of n risky assets at time period t are denoted by a vector $\mathbf{r}_t = [r_t^1, \cdots, r_t^n]'$, where r_t^i is the random return for asset i at time period t. It is assumed in this paper that vectors $\mathbf{r}_t$, $t = 0, 1, \cdots, T-1$, are statistically independent, and $\mathbf{r}_t$ has a discrete distribution with N_t finite states. We use $\mathbf{r}_t(\omega_t)$ to denote the possible realization of $\mathbf{r}_t$ and $p_t(\omega_t)$ to denote its corresponding probability, where ω_t takes value from the scenario set $\{\omega_t^1, \cdots, \omega_t^{N_t}\}$. Then, the sequence $\{\omega_j\}_{j=0}^{t-1}$ denotes a possible path up to time t.

For a given path $\{\omega_j\}_{j=0}^{t-1}$, let $\mathbf{u}_t(\{\omega_j\}_{j=0}^{t-1})$ be the vector of dollar amounts invested in risky assets. Then, the amount invested in riskless asset is $W_t(\{\omega_j\}_{j=0}^{t-1}) - \mathbf{1}'\mathbf{u}_t(\{\omega_j\}_{j=0}^{t-1})$, where $\mathbf{1}$ is the n-dimensional vector of ones and $W_t(\{\omega_j\}_{j=0}^{t-1})$ is the wealth level at time t. An investor of mean-CVaR type who only concerns the global interests, is seeking the pre-committed policy, $\{\mathbf{u}_t^{Pre}(\{\omega_j\}_{j=0}^{t-1})\}_{t=0}^{T-1}$, according to

$$(P_0^T) \quad \begin{cases} \min \nu_C(B - W_T, \alpha), \\ \text{s.t. } \mathbb{E}[W_T] = d, \\ \quad W_{t+1}(\{\omega_j\}_{j=0}^t) = r_t^0 W_t(\{\omega_j\}_{j=0}^{t-1}) + [\mathbf{e}_t(\omega_t)]' \mathbf{u}_t(\{\omega_j\}_{j=0}^{t-1}), \\ \quad \mathbf{u}_t(\{\omega_j\}_{j=0}^{t-1}) \in \mathbb{R}^n, \quad t = 0, 1, \cdots, T-1, \end{cases} \quad (2)$$

where B is a pre-specified benchmark which is used to define the loss, α is the confidential level, d is the preselected expected investment target and $\mathbf{e}_t = \mathbf{r}_t - r_t^0 \mathbf{1}$ is the excess rates of return of risky assets. It is worth to point out that when we choose two constants, B_1 and B_2, as benchmarks, the derived pre-committed policies for two cases are identical and there is a constant difference $B_1 - B_2$ between corresponding optimal CVaRs. Therefore, we just choose the expected investment target d as the benchmark, i.e., $B = d$.

Inspired by Rockafellar and Uryasev's linear programming formulation for CVaR ([9]), we can reformulate the multiperiod mean-CVaR problem (P_0^T) into the following linear programming form,

$$(LP_0^T) \quad \begin{cases} \min z + \dfrac{1}{1 - \alpha} \displaystyle\sum_{\{\omega_j\}_{j=0}^{T-1}} v_T(\{\omega_j\}_{j=0}^{T-1}) p_T(\{\omega_j\}_{j=0}^{T-1}), \\ \text{s.t. } \displaystyle\sum_{\{\omega_j\}_{j=0}^{T-1}} W_T(\{\omega_j\}_{j=0}^{T-1}) p_T(\{\omega_j\}_{j=0}^{T-1}) = d, \\ \quad v_T(\{\omega_j\}_{j=0}^{T-1}) \geq d - W_T(\{\omega_j\}_{j=0}^{T-1}) - z, \\ \quad v_T(\{\omega_j\}_{j=0}^{T-1}) \geq 0, \\ \quad W_{t+1}(\{\omega_j\}_{j=0}^t) = r_t^0 W_t(\{\omega_j\}_{j=0}^{t-1}) + [\mathbf{e}_t(\omega_t)]' \mathbf{u}_t(\{\omega_j\}_{j=0}^{t-1}), \\ \quad \mathbf{u}_t(\{\omega_j\}_{j=0}^{t-1}) \in \mathbb{R}^n, \quad t = 0, 1, \cdots, T-1, \end{cases} \quad (3)$$

where z, $v_T(\{\omega_j\}_{j=0}^{T-1})$ are $1 + \prod_{j=0}^{T-1} N_j$ auxiliary variables, $p_T(\{\omega_j\}_{j=0}^{T-1}) = \prod_{j=0}^{T-1} p_j(\omega_j)$ is the probability of return path $\{\omega_j\}_{j=0}^{T-1}$ and the summation is calculated over all possible return paths. Denote the terminal wealth achieved by the pre-committed policy as W_T^{Pre} and the corresponding optimal objective value of (LP_0^T) as $\nu_C(d - W_T^{Pre}, \alpha)$. By changing d among $(-\infty, +\infty)$, the mean-CVaR pairs $(d, \nu_C(d - W_T^{Pre}, \alpha))$ form the minimum CVaR set. While confining d in $[\prod_{j=0}^{T-1} r_j^0 W_0, +\infty)$, the mean-CVaR pairs form the efficient frontier.

Example 1. There are one riskless asset and two risky assets in the market. The rate of return of riskless asset is $r_0^0 = r_1^0 = 1.01$. The rates of return of risky assets are given as follows

$$\mathbf{r}_0(\omega_0^1) = [0.85, 0.75]', \ p_0(\omega_0^1) = 0.4, \ \mathbf{r}_0(\omega_0^2) = [0.92, 0.91]', \ p_0(\omega_0^2) = 0.1,$$
$$\mathbf{r}_0(\omega_0^3) = [1.06, 1.04]', \ p_0(\omega_0^3) = 0.4, \ \mathbf{r}_0(\omega_0^4) = [1.08, 1.11]', \ p_0(\omega_0^4) = 0.1;$$
$$\mathbf{r}_1(\omega_1^1) = [0.89, 0.85]', \ p_1(\omega_1^1) = 0.25, \ \mathbf{r}_1(\omega_1^2) = [0.92, 0.91]', \ p_1(\omega_1^2) = 0.25,$$
$$\mathbf{r}_1(\omega_1^3) = [1.06, 1.04]', \ p_1(\omega_1^3) = 0.25, \ \mathbf{r}_1(\omega_1^4) = [1.08, 1.11]', \ p_1(\omega_1^4) = 0.25.$$

The market is an incomplete market and arbitrage free.

We consider a two-period mean-CVaR portfolio selection model. Assume that the initial wealth level is $W_0 = 1$ and confidential level is chosen as $\alpha = 0.85$. By changing d among $(0.8, 2.5)$ and solving (LP_0^T), we can derive the minimum CVaR set, which is represented in Figure 1. The upper line is the efficient frontier.

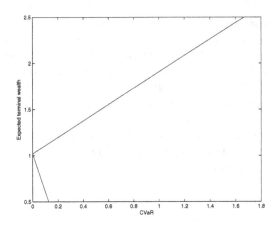

Fig. 1. Minimum CVaR set associated with $W_0 = 1$

Next, fixing $d = 2.2$, we derive the pre-committed policy and the corresponding wealth levels at time 1 and time 2 as follows,

$$\mathbf{u}_0^{Pre} = [16.9372, -18.2401], \; \mathbf{u}_1^{Pre}(\{\omega_0^1\}) = [175.0136, -140.0108]',$$

$$\mathbf{u}_1^{Pre}(\{\omega_0^2\}) = [0,0]', \; \mathbf{u}_1^{Pre}(\{\omega_0^3\}) = [0,0]', \; \mathbf{u}_1^{Pre}(\{\omega_0^4\}) = [40.9123, -36.6057]';$$

$$W_1^{Pre}(\{\omega_0^1\}) = 3.0425, \; W_1^{Pre}(\{\omega_0^2\}) = 1.3097,$$

$$W_1^{Pre}(\{\omega_0^3\}) = 1.3097, \; W_1^{Pre}(\{\omega_0^4\}) = 0.3716,$$

$$W_2^{Pre}(\{\omega_0^1, \omega_1^1\}) = 4.4730, \; W_2^{Pre}(\{\omega_0^1, \omega_1^2\}) = 1.3228,$$

$$W_2^{Pre}(\{\omega_0^1, \omega_1^3\}) = 7.6232, \; W_2^{Pre}(\{\omega_0^1, \omega_1^4\}) = 1.3228,$$

$$W_2^{Pre}(\{\omega_0^2, \omega_1^1\}) = W_2^{Pre}(\{\omega_0^2, \omega_1^2\}) = W_2^{Pre}(\{\omega_0^2, \omega_1^3\}) = W_2^{Pre}(\{\omega_0^2, \omega_1^4\}) = 1.3228,$$

$$W_2^{Pre}(\{\omega_0^3, \omega_1^1\}) = W_2^{Pre}(\{\omega_0^3, \omega_1^2\}) = W_2^{Pre}(\{\omega_0^3, \omega_1^3\}) = W_2^{Pre}(\{\omega_0^3, \omega_1^4\}) = 1.3228,$$

$$W_2^{Pre}(\{\omega_0^4, \omega_1^1\}) = 1.3228, \; W_2^{Pre}(\{\omega_0^4, \omega_1^2\}) = 0.3538,$$

$$W_2^{Pre}(\{\omega_0^4, \omega_1^3\}) = 1.3228, \; W_2^{Pre}(\{\omega_0^4, \omega_1^4\}) = -0.4214.$$

3   Time Inconsistency of Truncated Pre-committed Policy

When comes to time k, a particular return path up to time k, $\{\bar{\omega}_s\}_{s=0}^{k-1}$ is realized. If the investor adopts pre-committed policy up to time k, the investor faces a short term mean-CVaR problem (LP_k^T),

$$
(LP_k^T)\ \begin{cases}
\min\ z + \dfrac{1}{1-\alpha}\ \displaystyle\sum_{\{\omega_j\}_{j=k}^{T-1}} v_T(\{\omega_j\}_{j=k}^{T-1})p_T(\{\omega_j\}_{j=k}^{T-1}),\\[2.5ex]
\text{s.t.}\ \displaystyle\sum_{\{\omega_j\}_{j=k}^{T-1}} W_T(\{\omega_j\}_{j=k}^{T-1})p_T(\{\omega_j\}_{j=k}^{T-1}) = d_k,\\[2.5ex]
v_T(\{\omega_j\}_{j=k}^{T-1}) \geq d - W_T(\{\omega_j\}_{j=k}^{T-1}) - z,\\[1.5ex]
v_T(\{\omega_j\}_{j=k}^{T-1}) \geq 0,\\[1.5ex]
W_{t+1}(\{\omega_j\}_{j=k}^{t}) = r_t^0 W_t(\{\omega_j\}_{j=k}^{t-1}) + [\mathbf{e}_t(\omega_t)]'\mathbf{u}_t(\{\omega_j\}_{j=k}^{t-1}),\\[1.5ex]
\mathbf{u}_t(\{\omega_j\}_{j=k}^{t-1}) \in \mathbb{R}^n,\quad t = k, k+1, \cdots, T-1,
\end{cases}
\tag{4}
$$

where the initial wealth level is the wealth level at time t, i.e., $W_k = W_k^{Pre}(\{\bar{\omega}_s\}_{s=0}^{k-1})$ and the conditional expected investment target is d_k. And the truncated pre-committed policy is denoted as

$$
\{\mathbf{u}_t^{Pre}(\{\omega_s\}_{s=k}^{t-1})\}_{t=k}^{T-1}.
$$

Boda and Filar (2006) ([6]) had shown that multiperiod mean-CVaR model is time inconsistent, which means that the truncated pre-committed policy is NOT optimal for the short term mean-CVaR problem (LP_k^T). But, as multiperiod mean-CVaR model is a multi-objective optimization problem, it is better to consider efficiency instead of optimality. Therefore, we study the time consistency in efficiency property of the pre-committed policy as suggested by Cui et. al. (2012) ([7]), which examines whether the truncated pre-committed policy is still efficient or not for the short term problem (LP_k^T).

To do this, we can derive the efficient frontier of problem (LP_k^T). And then check whether the conditional mean and conditional CVaR pair achieved by the truncated pre-committed policy lies on the efficient frontier or not. The conditional mean and conditional CVaR achieved by the truncated pre-committed policy can be computed easily,

$$
\mathbb{E}(W_T^{Pre}|\ \{\bar{\omega}_j\}_{j=0}^{k-1}) = \sum_{\{\omega_s\}_{s=k}^{T-1}} W_T^{Pre}(\{\bar{\omega}_j\}_{j=0}^{k-1}, \{\omega_s\}_{s=k}^{T-1})p_T(\{\omega_s\}_{s=k}^{T-1}),
\tag{5}
$$

$$
\nu_C(d - W_T^{Pre}, \alpha|\ \{\bar{\omega}_j\}_{j=0}^{k-1})
$$
$$
= \min_z\ z + \frac{1}{1-\alpha}\sum_{\{\omega_s\}_{s=k}^{T-1}} \left(d - W_T^{Pre}(\{\bar{\omega}_j\}_{j=0}^{k-1}, \{\omega_s\}_{s=k}^{T-1}) - z\right)^+ p_T(\{\omega_s\}_{s=k}^{T-1}),
$$
$$
\tag{6}
$$

where $p_T(\{\omega_s\}_{s=k}^{T-1}) = \prod_{s=k}^{T-1} p_s(\omega_s)$ is the conditional probability of return path $\{\omega_s\}_{s=k}^{T-1}$. The following example shows that in general, multiperiod mean-CVaR model is NOT time consistent in efficiency.

Example 2. We continue to study Example 1. At time 1, according to (5) and (6), we have

$$\mathbb{E}(W_2^{Pre}|\ \{\omega_0^1\}) = 3.6854,\ \nu_C(d - W_2^{Pre}, \alpha|\ \{\omega_0^1\}) = 0.8772;$$
$$\mathbb{E}(W_2^{Pre}|\ \{\omega_0^2\}) = 1.3228,\ \nu_C(d - W_2^{Pre}, \alpha|\ \{\omega_0^2\}) = 0.8772$$
$$\mathbb{E}(W_2^{Pre}|\ \{\omega_0^3\}) = 1.3228,\ \nu_C(d - W_2^{Pre}, \alpha|\ \{\omega_0^3\}) = 0.8772;$$
$$\mathbb{E}(W_2^{Pre}|\ \{\omega_0^4\}) = 0.6445,\ \nu_C(d - W_2^{Pre}, \alpha|\ \{\omega_0^4\}) = 2.5937.$$

After deriving the efficient frontiers of (LP_1^2) with different initial wealth levels, we find that the conditional mean-CVaR pair of $\{\omega_0^4\}$ does not lie on the corresponding efficient frontier (see Figure 2). Point 'A' is the conditional mean-CVaR pair achieved by truncated pre-committed optimal policy, which is below the efficient frontier of one-period mean-CVaR problem. Therefore, for the fourth return path $\{\omega_0^4\}$, the truncated pre-committed optimal policy becomes inefficient .

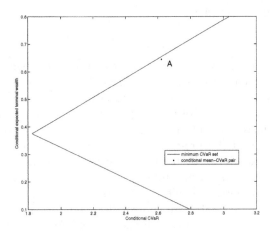

Fig. 2. The efficient frontier and the conditional mean-CVaR pair

It is hard to get the exact conditions under which truncated pre-committed optimal policy is inefficient for short term mean-CVaR problem (LP_k^T). But we can still have some rough results to describe the property of the truncated pre-committed optimal policy.

Proposition 1. *If the pre-committed policy is efficient and the following condition holds,*

$$d - \rho_k^T W_k^{Pre}(\{\bar{\omega}_s\}_{s=0}^{k-1}) \leq \nu(d - W_T^{Pre}, \alpha), \tag{7}$$

where $\rho_k^T = \prod_{j=k}^{T-1} r_j^0$ is the riskless accumulative factor, then the conditional mean of terminal wealth achieved by the truncated pre-committed policy must satisfy

$$\mathbb{E}(W_T^{Pre}|\ \{\bar{\omega}_s\}_{s=0}^{k-1}) \geq \rho_k^T W_k^{Pre}(\{\bar{\omega}_s\}_{s=0}^{k-1}).$$

Proof: We prove the result by contradiction. Assume that the following condition holds

$$\mathbb{E}(W_T^{Pre}|\ \{\bar{\omega}_s\}_{s=0}^{k-1}) < \rho_k^T W_k^{Pre}(\{\bar{\omega}_s\}_{s=0}^{k-1}). \tag{8}$$

Consider a revised portfolio policy $\{\bar{\mathbf{u}}_t(\{\omega_j\}_{j=0}^{t-1})\}_{t=0}^{T-1}$,

$$\bar{\mathbf{u}}_t(\{\omega_j\}_{j=0}^{t-1}) = \begin{cases} \mathbf{u}_t^{Pre}(\{\omega_j\}_{j=0}^{t-1}), & \text{if } \{\omega_s\}_{s=0}^{k-1} \neq \{\bar{\omega}_s\}_{s=0}^{k-1}, \\ \mathbf{0}, & \text{if } \{\omega_s\}_{s=0}^{k-1} = \{\bar{\omega}_s\}_{s=0}^{k-1}, \end{cases}$$

where $\mathbf{0}$ denotes the n-dimensional vector of zeros. This revised policy is different from the pre-committed policy only for the particular return paths following $\{\bar{\omega}_s\}_{s=0}^{k-1}$. Then, the conditional mean of terminal wealth achieved by the revised policy is

$$\mathbb{E}(\overline{W}_T|\ \{\bar{\omega}_s\}_{s=0}^{k-1}) = \rho_k^T W_k^{Pre}(\{\bar{\omega}_s\}_{s=0}^{k-1}).$$

Now let us compare the mean CVaR pairs achieved by pre-committed policy and the revised policy. Due to condition (8) and the smooth property of expectation operator, we have

$$\mathbb{E}(W_T^{Pre})$$
$$=\mathbb{E}(W_T^{Pre}|\{\bar{\omega}_s\}_{s=0}^{k-1})p_T(\{\bar{\omega}_s\}_{s=0}^{k-1}) + \sum_{\{\omega_s\}_{s=0}^{k-1}\neq\{\bar{\omega}_s\}_{s=0}^{k-1}} \mathbb{E}(W_T^{Pre}|\{\omega_s\}_{s=0}^{k-1})p_T(\{\omega_s\}_{s=0}^{k-1})$$
$$<\rho_k^T W_k^{Pre}(\{\bar{\omega}_s\}_{s=0}^{k-1})p_T(\{\bar{\omega}_s\}_{s=0}^{k-1}) + \sum_{\{\omega_s\}_{s=0}^{k-1}\neq\{\bar{\omega}_s\}_{s=0}^{k-1}} \mathbb{E}(W_T^{Pre}|\{\omega_s\}_{s=0}^{k-1})p_T(\{\omega_s\}_{s=0}^{k-1})$$
$$=\mathbb{E}(\overline{W}_T).$$

On the other hand, for given path $\{\bar{\omega}_s\}_{s=0}^{k-1}$, the terminal wealth achieved by the revised policy is a constant

$$\rho_k^T W_k^{Pre}(\{\bar{\omega}_s\}_{s=0}^{k-1}).$$

Condition (7) implies that following the path $\{\bar{\omega}_s\}_{s=0}^{k-1}$, the loss achieved by the revised policy is less than $\nu(d - W_T^{Pre}, \alpha)$, which further implies

$$\nu_C(d - \overline{W}_T, \alpha) \leq \nu_C(d - W_T^{Pre}, \alpha).$$

This finding contradicts to the fact that the corresponding pre-committed policy achieves an efficient mean-CVaR pair. Therefore, the assumption (8) is incorrect and the conclusion holds. □

4   Time Consistent Policy

When the investor totally ignores the global interests and bows to local interests, we reformulate multiperiod mean-CVaR model into a multiperson game and derive the time consistent policy. The multiperson game can be represented as the following nested problem,

$$(NLP_t^T) \begin{cases} \min_{\mathbf{u}_t} \; z + \dfrac{1}{1-\alpha} \sum_{\{\omega_j\}_{j=t}^{T-1}} v_T(\{\omega_j\}_{j=t}^{T-1}) p_T(\{\omega_j\}_{j=t}^{T-1}), \\[2mm] \text{s.t.} \; \sum_{\{\omega_j\}_{j=t}^{T-1}} W_T(\{\omega_j\}_{j=t}^{T-1}) p_T(\{\omega_j\}_{j=t}^{T-1}) = d, \\[2mm] \quad v_T(\{\omega_j\}_{j=t}^{T-1}) \geq d - W_T(\{\omega_j\}_{j=t}^{T-1}) - z, \\[2mm] \quad v_T(\{\omega_j\}_{j=t}^{T-1}) \geq 0, \\[2mm] \quad W_{\tau+1}(\{\omega_j\}_{j=t}^{\tau}) = r_\tau^0 W_\tau(\{\omega_j\}_{j=t}^{\tau-1}) + [\mathbf{e}_\tau(\omega_\tau)]' \mathbf{u}_\tau(\{\omega_j\}_{j=t}^{\tau-1}), \\[2mm] \quad \mathbf{u}_\tau(\{\omega_j\}_{j=t}^{\tau-1}) \text{ solves } (NLP_\tau^T), \quad \tau = t+1, t+2, \cdots, T-1, \\[2mm] \quad W_{t+1}(\{\omega_t\}) = r_t^0 W_t + [\mathbf{e}_t(\omega_t)]' \mathbf{u}_t, \\[2mm] \quad \mathbf{u}_t \in \mathbb{R}^n, \end{cases}$$

with the problem in the final time period given as

$$(NLP_{T-1}^T) \begin{cases} \min_{\mathbf{u}_{T-1}} \; z + \dfrac{1}{1-\alpha} \sum_{\{\omega_{T-1}\}} v_T(\{\omega_{T-1}\}) p_T(\{\omega_{T-1}\}), \\[2mm] \text{s.t.} \; \sum_{\{\omega_{T-1}\}} W_T(\{\omega_{T-1}\}) p_T(\{\omega_{T-1}\}) = d, \\[2mm] \quad v_T(\{\omega_{T-1}\}) \geq d - W_T(\{\omega_{T-1}\}) - z, \\[2mm] \quad v_T(\{\omega_{T-1}\}) \geq 0, \\[2mm] \quad W_T(\{\omega_{T-1}\}) = r_{T-1}^0 W_{T-1} + [\mathbf{e}_{T-1}(\omega_{T-1})]' \mathbf{u}_{T-1}, \\[2mm] \quad \mathbf{u}_{T-1} \in \mathbb{R}^n. \end{cases}$$

The time consistent policy, i.e., subgame Nash equilibrium policy of the multiperson game, can be characterized by the following theorem.

Theorem 1 *The time consistent policy of multiperiod mean-CVaR problem is given as, for $t = 0, 1 \cdots, T-1$,*

$$\mathbf{u}_t^{TC}(W_t) = \left(\mathbf{K}_t^+ 1_{\{(\rho_t^T)^{-1}d \geq W_t\}} + \mathbf{K}_t^- 1_{\{(\rho_t^T)^{-1}d < W_t\}} \right) [(\rho_t^T)^{-1}d - W_t], \quad (9)$$

where $1_{\{\cdot\}}$ is the indicator function, vectors $\mathbf{K}_{T-1}^+$ and $\mathbf{K}_{T-1}^-$ are determined through the following linear programming problems, respectively,

$$(TC_{T-1}^{\pm}) \begin{cases} \min_{\mathbf{K}_{T-1}} \ z + \dfrac{1}{1-\alpha} \sum_{\{\omega_{T-1}\}} v_T(\{\omega_{T-1}\}) p_T(\{\omega_{T-1}\}), \\[2mm] \text{s.t.} \ \sum_{\{\omega_{T-1}\}} (r_{T-1}^0 - [\mathbf{e}_{T-1}(\omega_{T-1})]'\mathbf{K}_{T-1}) p_T(\{\omega_{T-1}\}) = 0, \\[2mm] v_T(\{\omega_{T-1}\}) \geq \pm \left[r_{T-1}^0 - [\mathbf{e}_{T-1}(\omega_{T-1})]'\mathbf{K}_{T-1} \right] - z, \\[2mm] v_T(\{\omega_{T-1}\}) \geq 0, \\[2mm] \mathbf{K}_{T-1} \in \mathbb{R}^n, \end{cases}$$

and vectors $\mathbf{K}_t^+$ and $\mathbf{K}_t^-$ ($t < T-1$) are determined through the following integer programming problems, respectively,

$$(TC_t^{\pm}) \begin{cases} \min_{\mathbf{K}_t} \ z + \dfrac{1}{1-\alpha} \sum_{\{\omega_j\}_{j=t}^{T-1}} v_T(\{\omega_j\}_{j=t}^{T-1}) p_T(\{\omega_j\}_{j=t}^{T-1}), \\[2mm] \text{s.t.} \ \sum_{\{\omega_j\}_{j=t}^{T-1}} Y_{T-1}(\{\omega_j\}_{j=t}^{T-1}) p_T(\{\omega_j\}_{j=t}^{T-1}) = 0, \\[2mm] v_T(\{\omega_j\}_{j=t}^{T-1}) \geq \pm Y_{T-1}(\{\omega_j\}_{j=t}^{T-1}) - z, \\[2mm] v_T(\{\omega_j\}_{j=t}^{T-1}) \geq 0, \\[2mm] Y_j(\{\omega_s\}_{s=t}^{j}) = Y_{j-1}(\{\omega_s\}_{s=t}^{j-1})\{\zeta_{j-1}(\{\omega_s\}_{s=t}^{j-1})(r_j^0 - [\mathbf{e}_j(\omega_j)]'\mathbf{K}_j^+) \\[2mm] \qquad\qquad + [1 - \zeta_{j-1}(\{\omega_s\}_{s=t}^{j-1})](r_j^0 - [\mathbf{e}_j(\omega_j)]'\mathbf{K}_j^-)\}, \\[2mm] Y_{j-1}(\{\omega_s\}_{s=t}^{j-1}) \leq M\zeta_{j-1}(\{\omega_s\}_{s=t}^{j-1}), \\[2mm] -Y_{j-1}(\{\omega_s\}_{s=t}^{j-1}) \leq M[1 - \zeta_{j-1}(\{\omega_s\}_{s=t}^{j-1})], \\[2mm] \zeta_{j-1}(\{\omega_s\}_{s=t}^{j-1}) \in \{0,1\}, \qquad j = t+1, t+2, \cdots, T-1, \\[2mm] Y_t(\omega_t) = r_t^0 - [\mathbf{e}_t(\omega_t)]'\mathbf{K}_t, \\[2mm] \mathbf{K}_t \in \mathbb{R}^n, \end{cases}$$

where M is a large number.

Proof: We prove the main result by backward induction.

First, consider the problem in the final period (NLP_{T-1}^T). When $(\rho_{T-1}^T)^{-1}d > W_{T-1}$, by denoting $\mathbf{u}_{T-1} = \mathbf{K}_{T-1}[(\rho_{T-1}^T)^{-1}d - W_{T-1}]$ and noticing the positive homogeneity of CVaR, (NLP_{T-1}^T) can be reduced into (TC_{T-1}^+). Thus, we have $\mathbf{u}_{T-1}^{TC} = \mathbf{K}_{T-1}^+[(\rho_{T-1}^T)^{-1}d - W_{T-1}]$. Similarly, when $(\rho_{T-1}^T)^{-1}d < W_{T-1}$, by denoting $\mathbf{u}_{T-1} = \mathbf{K}_{T-1}[(\rho_{T-1}^T)^{-1}d - W_{T-1}]$ and noticing the positive homogeneity of CVaR, (NLP_{T-1}^T) can be reduced into (TC_{T-1}^-). Thus, we have $\mathbf{u}_{T-1}^{TC} = \mathbf{K}_{T-1}^-[(\rho_{T-1}^T)^{-1}d - W_{T-1}]$. When $(\rho_{T-1}^T)^{-1}d = W_{T-1}$, we can choose $\mathbf{u}_{T-1}^{TC} = \mathbf{0}$. Therefore, (9) holds for time $T-1$.

Assume that (9) holds from time $t+1$ to time $T-1$. We prove that (9) holds for time t also. At time t, by denoting $\mathbf{u}_t = \mathbf{K}_t[(\rho_t^T)^{-1}d - W_t]$, we have

$$(\rho_{j+1}^T)^{-1}d - W_{j+1}(\{\omega_s\}_{s=t}^j) = [(\rho_j^T)^{-1}d - W_j(\{\omega_s\}_{s=t}^{j-1})]Y_j(\{\omega_s\}_{s=t}^j),$$
$$j = t, t+1, \cdots, T-1,$$

where

$$Y_j(\{\omega_s\}_{s=t}^j) = Y_{j-1}(\{\omega_s\}_{s=t}^{j-1})\Big\{1_{\{Y_{j-1}(\{\omega_s\}_{s=t}^{j-1})\geq 0\}}(r_j^0 - [e_j(\omega_j)]'\mathbf{K}_j^+)$$
$$+ 1_{\{Y_{j-1}(\{\omega_s\}_{s=t}^{j-1})<0\}}(r_j^0 - [e_j(\omega_j)]'\mathbf{K}_j^-)\Big\},$$

and $Y_t(\omega_t) = (r_t^0 - [e_t(\omega_t)]'\mathbf{K}_t)$. Then, when $(\rho_t^T)^{-1}d \geq W_t$, (NLP_t^T) can be reduced into

$$\begin{cases}
\min_{\mathbf{K}_t} z + \dfrac{1}{1-\alpha}\displaystyle\sum_{\{\omega_j\}_{j=t}^{T-1}} v_T(\{\omega_j\}_{j=t}^{T-1})p_T(\{\omega_j\}_{j=t}^{T-1}), \\[2mm]
\text{s.t.} \displaystyle\sum_{\{\omega_j\}_{j=t}^{T-1}} Y_{T-1}(\{\omega_j\}_{j=t}^{T-1})p_T(\{\omega_j\}_{j=t}^{T-1}) = 0, \\[2mm]
v_T(\{\omega_j\}_{j=t}^{T-1}) \geq Y_{T-1}(\{\omega_j\}_{j=t}^{T-1}) - z, \\[1mm]
v_T(\{\omega_j\}_{j=t}^{T-1}) \geq 0, \\[1mm]
Y_j(\{\omega_s\}_{s=t}^j) = Y_{j-1}(\{\omega_s\}_{s=t}^{j-1})\Big\{1_{\{Y_{j-1}(\{\omega_s\}_{s=t}^{j-1})\geq 0\}}(r_j^0 - [e_j(\omega_j)]'\mathbf{K}_j^+) \\[1mm]
\quad\quad + 1_{\{Y_{j-1}(\{\omega_s\}_{s=t}^{j-1})<0\}}(r_j^0 - [e_j(\omega_j)]'\mathbf{K}_j^-)\Big\}, \ j = t+1, \cdots, T-1, \\[1mm]
Y_t(\omega_t) = r_t^0 - [e_t(\omega_t)]'\mathbf{K}_t, \\[1mm]
\mathbf{K}_t \in \mathbb{R}^n,
\end{cases}$$

which can be transfer into integer programming problem (TC_t^+). Thus, we have $\mathbf{u}_t^{TC} = \mathbf{K}_t^+[(\rho_t^T)^{-1}d - W_t]$. Similarly, when $(\rho_t^T)^{-1}d < W_t$, (NLP_t^T) can be reduced into (TC_t^-). Thus, we have $\mathbf{u}_t^{TC} = \mathbf{K}_t^-[(\rho_t^T)^{-1}d - W_t]$. When $(\rho_t^T)^{-1}d = W_t$, we can also choose $\mathbf{u}_t^{TC} = \mathbf{0}$. Therefore, (9) holds for time t.   □

In the following example, we derive the time consistent policy for Example 1 and compare the efficient frontiers achieved by pre-committed policy and time consistent policy.

Example 3. We continue to study Example 1. When fixing $d = 2.2$, according to Theorem 1, we can derive

$$\mathbf{K}_1^+ = [288.5714, -230.8571]', \ \mathbf{K}_1^- = [153.5200, -137.3600]'.$$

Then, at time 0, $(TC_0^{\pm})$ can be rewritten as the following integer programming problems,

$$
\begin{cases}
\min_{\mathbf{K}_0} \; z + \dfrac{1}{1-\alpha} \displaystyle\sum_{\{\omega_0,\omega_1\}} v_2(\{\omega_0,\omega_1\}) p_2(\{\omega_0,\omega_1\}), \\[2ex]
\text{s.t.} \; \displaystyle\sum_{\{\omega_0,\omega_1\}} Y_1(\{\omega_0,\omega_1\}) p_2(\{\omega_0,\omega_1\}) = 0, \\[2ex]
\quad v_2(\{\omega_0,\omega_1\}) \geq \pm Y_1(\{\omega_0,\omega_1\}) - z, \\[1ex]
\quad v_2(\{\omega_0,\omega_1\}) \geq 0, \\[1ex]
\quad Y_1(\{\omega_0,\omega_1\}) = Y_0(\omega_0)\{\zeta(\omega_0)(r_1^0 - [\mathbf{e}_1(\omega_1)]'\mathbf{K}_1^+) \\[0.5ex]
\qquad\qquad\qquad\qquad + [1 - \zeta(\omega_0)](r_1^0 - [\mathbf{e}_1(\omega_1)]'\mathbf{K}_1^-)\}, \\[1ex]
\quad Y_0(\omega_0) \leq M\zeta(\omega_0), \\[1ex]
\quad -Y_0(\omega_0) \leq M(1 - \zeta(\omega_0)), \\[1ex]
\quad \zeta(\omega_0) \in \{0,1\}, \\[1ex]
\quad Y_0(\omega_0) = r_t^0 - [\mathbf{e}_0(\omega_0)]'\mathbf{K}_0, \\[1ex]
\quad \mathbf{K}_0 \in \mathbb{R}^n,
\end{cases}
$$

where M is a large number. We can derive

$$
\mathbf{K}_0^+ = [0,0]', \;\; \mathbf{K}_0^- = [0,0]'.
$$

This means that under the time consistent policy, the investor does not invest the risky assets in the first period.

The following Figure 3 shows the efficient frontiers achieved by pre-committed policy and time consistent policy. We can see that applying pre-committed policy has a better long term investment performance than time consistent policy. The difference between the efficient frontiers represents the degree of time inconsistency.

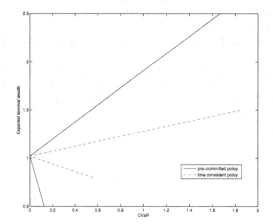

Fig. 3. The efficient frontiers achieved by pre-committed policy and time consistent policy

5   Conclusion

In this paper, we study the multiperiod mean-CVaR model with finite states. As the model suffers time inconsistency, i.e., the truncated global optimal policy (pre-committed policy) is not optimal for the remaining short term problem, we derive both the pre-committed policy and time consistent policy.

We use linear programming to derive the pre-committed policy, and show that the pre-committed policy doesn't satisfy time consistency in efficiency. We also provide a proposition describing the property of the truncated efficient pre-committed policy. Then, we prove that the time consistent policy, subgame Nash equilibrium policy of the multiperson game reformulation, takes a piecewise linear form of the current wealth level and the corresponding coefficients can be derived by a series of integer programming problems and two linear programming problems. The numerical example shows that by adopting the time consistent policy, the investor suffers a loss in long term investment performance.

References

1. Artzner, P., Delbaen, F., Eber, J.M., Heath, D.: Coherent measures of risk. Mathematical Finance 4, 203–228 (1999)
2. Artzner, P., Delbaen, F., Eber, J.M., Heath, D., Ku, H.: Coherent multiperiod risk adjusted values and Bellman's principle. Annals of Operations Research 152, 5–22 (2007)
3. Basak, S., Chabakauri, G.: Dynamic mean-variance asset allocation. Review of Financial Studies 23, 2970–3016 (2010)
4. Björk, T., Murgoci, A.: A general theory of Markovian time inconsistent stochasitc control problem, working paper (2010)
5. Björk, T., Murgoci, A., Zhou, X.Y.: Mean-variance portfolio optimization with state dependent risk aversion. Mathematical Finance 24, 1–24 (2014)
6. Boda, K., Filar, J.A.: Time consistent dynamic risk measures. Mathematical Methods of Operations Reseach 63, 169–186 (2006)
7. Cui, X.Y., Li, D., Wang, S.Y., Zhu, S.S.: Better Than Dynamic Mean-Variance: Time Inconsistency and Free Cash Flow Stream. Mathematical Finance 22, 346–378 (2012)
8. Gao, J.J., Zhou, K., Li, D., Cao, X.R.: Dynamic Mean-LPM and Mean-CVaR Portfolio Optimization in Continuous-time, working paper, arXiv:1402.3464 (2014)
9. Rockafellar, T.R., Uryasev, S.P.: Optimization of Conditional Value-at-Risk. Journal of Risk 2, 21–41 (2000)

Part V
Multiobjective Programming

A Multi Objective Multi Echelon Supply Chain Network Model for a Household Goods Company

Mehmet Alegoz and Zehra Kamisli Ozturk

Anadolu University, Department of Industrial Engineering, Eskisehir, Turkey
{mehmetalegoz,zkamisli}@anadolu.edu.tr

Abstract. In this study, we have developed a multi echelon and multi objective supply chain network model for a company which sells high quality household goods to its customers in different cities. The first objective of the proposed mathematical model tries to minimize the total supply chain network cost while the other objective tries to maximize the service level. We also used a specific transportation cost function which takes care about the distance economy. This specific cost function makes the mathematical model and its results more accurate and realistic. We have concluded the study by solving this real-life problem with the proposed approach. The computational results showed that the proposed approach finds the Pareto optimum solution within a very short time and can be adapted to any company's problem easily.

Keywords: Supply chain network design, multi objective optimization, distance economy.

1 Introduction

A supply chain is a network of suppliers, manufacturing plants, warehouses, and distribution channels organized to acquire raw materials, convert these raw materials to finished products, and distribute these products to customers [1]. The structure of the supply chain may differ from a company to another one. As an example, some companies may use a network including suppliers, production plants, warehouses and markets. Some other companies may not use warehouses and instead, they may prefer to ship from the production plant to the customers via the distribution centers. In fact, these differences come from different objectives of companies. Each company designs their supply chain network according to one or a few objectives. Some of them may focus on total supply chain network cost while designing the network, some others may focus on speed and some others may focus on another performance indicator and this makes the supply chain network different from the others.

The main objective of the supply chain management is to achieve suitable economic results together with the desired consumer satisfaction levels [2]. For this reason a model should also focus on the service level or customer satisfaction level beside the total cost but most of the models in the literature have only one objective [3], [7], [8], which is either minimizing the total cost or maximizing the after tax profit.

© Springer International Publishing Switzerland 2015
H.A. Le Thi et al. (eds.), *Model. Comput. & Optim. in Inf. Syst. & Manage. Sci.*,
Advances in Intelligent Systems and Computing 359, DOI: 10.1007/978-3-319-18161-5_26

For example, Tsiakis, Shah and Pantelides [3] developed a single objective multi echelon supply chain network design model which includes suppliers, plants, warehouses, distribution centers and customers. Their objective was minimizing the total cost by determining the number, capacity and location of warehouses and distribution centers.

There are also some models in the literature which have two or more objectives. For example, Pazhami, Rankumar, Narendran and Ganesh [4] developed a bi-objective supply chain network design model. Their objective was minimizing the total supply chain network cost and maximizing the service level. In order to measure the service level they used Multiple Criteria Decision Making techniques like AHP and TOPSIS. They have found an efficiency score for each warehouse and hybrid facility and as a second objective they have tried to maximize the total efficiency score. They also used different variations of goal programming and compared the results. Wang, Lai and Shi [5] focused on an important but rarely focused issue. They have developed a multi objective green supply chain network design model. One of their objectives was environmental influence and the other was total cost. They tried to find a tradeoff between the cost and environmental influence. Hiremath, Sahu and Tiwari [6] developed a model for outbound logistics network which includes three objectives. Their objectives were minimizing the total cost, maximizing the unit fill rates and maximizing the resource utilization. Production plants, central distribution centers, regional distribution centers and customer zones were the scope of their supply chain network model.

In the rest of the article, in second part, the problem is defined and the methodology to solve this problem is introduced. In third part, the proposed mathematical model is given with all its details. In fourth part, the computational results are given and some analysis about the computational results are made. Finally, in last two parts conclusion and future agenda for interested researchers are given.

2   Problem Definition and Methodology

We have tried to optimize the supply chain network of a household goods company. The company is located in Bilecik, a small city in Turkey. It has enough capacity to produce up to 10000 pallets of products in a month. It has customers in 29 different cities in Turkey. Those cities, their monthly demands (in pallets) and total monthly renting, operating cost of opening a warehouse (ROCW) in each city are summarized in Table 1 on the next page. In fact, the total monthly renting and operating cost of opening a warehouse is the cost which must be accepted by the company if the company wants to open a warehouse in that city. This cost is in Turkish Liras just like all the costs in the model. The demands of the customers can be thought as stable because it changes at most one or two pallets for a city in different months.

The company provides the goods to the customers via the warehouses. According to the company policy, direct shipment from the company to the customers is prohibited. Every month, the company ships enough products from the production plant located in Bilecik to each warehouse which satisfies the monthly demand of all customers assigned to that warehouse. The location of the company is fixed just like the location of the customers. The warehouses can be opened in one or a few of the cities which are given in Table 1. The company does not want to open a warehouse in the city of production plant because of firm policy.

Table 1. Customers, Their Demands and ROCW's

City	Adana	Ankara	Antalya	Balikesir	Bursa	Edirne
Demand	315	493	680	137	367	66
ROCW	28000	44000	27000	19000	26000	21000
City	Eskisehir	G.Antep	Hatay	Mersin	Istanbul	Izmir
Demand	55	51	154	112	2097	432
ROCW	27000	16000	11000	22000	51000	43000
City	Kayseri	Kocaeli	Konya	Malatya	Mugla	Ordu
Demand	158	370	247	69	251	77
ROCW	32000	34000	19000	16000	29000	23000
City	Sakarya	Samsun	Sivas	Tekirdag	Trabzon	S.Urfa
Demand	60	351	94	40	123	48
ROCW	17000	33000	21000	21000	27000	14000
City	Van	Kirikkale	Bartin	Karabuk	Duzce	
Demand	79	34	17	19	23	
ROCW	9000	17000	19000	21000	19000	

In addition to the information given in Table 1, distances between the cities are also needed for the model. Those distances are obtained from the web site of Republic of Turkey, General Directorate of Highways [9].

All in all, the company has the following questions to answer.

- Where should they locate their warehouses?
- How many warehouses should they open?
- Which customer should be assigned to which warehouse?
- What should be the inventory level in each warehouse?

The answers of these questions change according to the objective or objectives of the company. For example, if the company focuses on only cost then they probably decide to open only one warehouse but it brings lots of problems like capacity problems, promised lead time problems etc. From another perspective, if the company focuses on only lead time, they probably decide to open a warehouse in all the cities in order to give the fastest service but this will certainly increase the cost. For this reason, it becomes necessary to create a multi objective supply chain network design model which tries to find a tradeoff between these two objectives.

To find the tradeoff point and answer the questions above, we have proposed a goal programming model which determines the number, location and inventory level of warehouses and assigns the customers to each warehouse. We have written the cost and service level goals as constraints and we tried to minimize the deviation from each goal. In fact, there are lots of definition of service level but in this study we define the service level as closeness to the customers. Because, being close to a customer decrease the lead time and gives the opportunity of fast service. Fast service can even be a reason for a customer to choose the company. Secondly, closeness provides better communication with customers. For all these reasons, our second objective forces the model to open a warehouse as close as possible to the customers who have high monthly demand.

3   The Proposed Mathematical Model

In this part of study, we introduced the mathematical model but before it we briefly explained the concept of distance economy and its importance in transportation cost. We also explained how we will use the distance economy in the model.

3.1   The Concept of Distance Economy

Distance economy (economies of distance) is one of the most important things to take care about while modeling the supply chain network. Most of the models in the literature do not put the distance economy into account but in fact it effects the transportation cost directly. In order to make the model more realistic we also put the distance economy into account.

In fact, transportation cost mainly depends on distance and quantity. Even, it can be thought as the product of distance and quantity. For this reason, if the quantity is fixed, it only changes according to distance. However, this change is not a linear change because of distance economy. According to distance economy rule, the distance based unit cost decreases when the distance increases. Therefore, the cost does not increase linearly. In order to put this rule into account we define a distance economy multiplier. It changes according to distance and determined by using the company data for only some certain distances like in Table 2.

Table 2. Distance Economy Multiplier

Distance (km)	1	200	400	600	800
Multiplier	1,00	0,83	0,81	0,80	0,79

As an example if the distance between two city is 200 km and if we want to transport just one product then the transportation cost is 166 TL (1x0,83x200) but if the distance is 400 km then the cost is 324 TL, not 332 TL (166x2). As it can be seen easily, the transportation cost does not increase linearly. This multiplier makes the model and the cost more realistic. However, the multiplier given in Table 2 is for only some certain distance values, but the distances in our problem can be any number. For example the distance between two cities can be 324 km another one can be 467 km. In these cases, what should be the multiplier? In order to answer these questions we need to find a function for this multiplier which fits best with the data in Table 2.

Our analysis showed that this data fits best to an exponential function $y = x^{-0.035}$ where y is the multiplier and x is the distance. We could not show the curve fitting process due to the space limit but briefly, this function is obtained by minimizing the sum of squares of errors.

3.2   The Proposed Mathematical Model

A multi-objective multi echelon mixed integer supply chain network design model has been developed with the following model parameters, decision variables, constraints and objectives.

Sets
H: Production plant
I: Warehouses ($i = 1 \dots n$)
J: Customers ($j = 1 \dots m$)

Parameters
n_{ij}: Distance between warehouse opened in i^{th} city and customer in j^{th} city
m_{hi}: Distance between production plant and warehouse opened in i^{th} city
q_j: Average monthly demand of jth city
w_i: Average monthly renting and operating cost of warehouse opened in i^{th} city

Decision Variables
s_{ij}: Amount of product transported from warehouse opened in i^{th} city to customer in j^{th} city
a_{hi}: Amount of product transported from production plant to warehouse opened in i^{th} city
p_i: Demand of warehouse opened in i^{th} city

Binary Variables
b_i: If a warehouse is opened in i^{th} city it is 1 otherwise it is 0

Transportation Cost
Since we have found the distance economy multiplier $x^{-0.035}$ where x is the distance between two cities, we can write the transportation cost between the warehouse opened in i^{th} city and customer in j^{th} city as follows.

$$n_{ij}^{-0.035} . n_{ij}. s_{ij} \tag{1}$$

Let $n_{ij}^{-0.035}. n_{ij}$ be n_{ij}^{*} then the equation will be as follows.

$$n_{ij}^{*}. s_{ij} \tag{2}$$

Transportation cost between the production plant and warehouse opened in i^{th} city is written by the same way. Finally, the model can be written as follows.

$$\min z = \ 0.6 \cdot g_1^+ + 0.4 \cdot g_2^+ \tag{3}$$

$$subject \ to$$

$$\sum_{i=1}^{29} m_{hi}^* \cdot a_{hi} + \sum_{i=1}^{29} \sum_{j=1}^{29} n_{ij}^* \cdot s_{ij} + \sum_{i=1}^{29} b_i \cdot w_i + g_1^- - g_1^+ = 2629703 \tag{4}$$

$$\sum_{i=1}^{29} \sum_{j=1}^{29} n_{ij} \cdot s_{ij} + g_2^- - g_2^+ = 0 \tag{5}$$

$$s_{ij} \leq 4000 \cdot b_i \quad \forall i,j \tag{6}$$

$$a_{hi} \leq 10000 \cdot b_i \quad \forall i \tag{7}$$

$$\sum_{i=1}^{29} s_{ij} = q_j \ \forall j \tag{8}$$

$$a_{hi} = p_i \quad \forall i \tag{9}$$

$$p_i \geq \sum_{j=1}^{29} s_{ij} \ \forall i \tag{10}$$

$$p_i \leq 4000 \ \forall i \tag{11}$$

$$\sum_{i=1}^{29} b_i \leq 3 \tag{12}$$

$$s_{ij}, \ a_{hi}, \ p_i \geq 0 \ and \ integer \ and \ b_i \in \{0,1\} \tag{13}$$

The third equation is the objective function which tries to minimize the total positive deviations from the cost goal and service goal. The weighted sum scalarization method is used to scalarize the objectives. We gave 0,6 weight to the first objective (cost goal) and 0,4 weight to the second objective (service goal). These weights are given according to the managers' preferences.

The fourth equation is the cost goal. First part of it, includes the transportation cost between the plant and warehouse opened in i^{th} city. The second part of it includes the transportation cost between the warehouse located in i^{th} city and customer located in j^{th} city. The third part of it is monthly warehouse renting and operating cost of opening a warehouse in i^{th} city. As it can be seen easily, the model adds this cost only if the warehouse is opened in that city. In fact, the fourth equation includes total supply chain network cost (transportation cost, warehouse renting cost and warehouse operating cost). Here the value of 2629703 TL is the ideal cost and this cost was determined as goal.

The fifth equation is the service level equation. As it is mentioned before, we have defined the service level as closeness to the customer. Therefore, the model is forced to satisfy the demand of each customer from a warehouse as close as possible to the customer. The model will probably open warehouses especially near the customers who have high demands in order to minimize the $n_{ij} \cdot s_{ij}$ value. Although opening warehouses close to the customers will decrease the lead time and increase the service level, it also increases the total supply chain network cost. For example, as an ideal solution, if a warehouse in each customer's city is opened, then the fifth equation will be zero as we wish but in this case the transportation cost of first echelon (from the production plant to the warehouse) and warehouse renting and operating costs will be too much. For this reason the fifth equation contradicts the fourth equation and the mathematical model will find a tradeoff point.   Last but not least, as it is seen in fifth equation, the equation is $n_{ij}.s_{ij}$ instead of $n_{ij}^*.s_{ij}$ because this equation is not about cost.

The sixth equation forces the model to open a warehouse in order to make a shipment from a warehouse. If the warehouse is not opened in i^{th} city, then it becomes impossible to make any shipment from i^{th} city but if a warehouse is opened in i^{th} city then it can be made shipment up to 4000 products (the capacity of warehouse).

The seventh equation is just like the previous one. If the warehouse is not opened in i^{th} city then the company cannot send product to that city and similarly if the warehouse is opened in i^{th} city then the company can send up to 10000 products (the capacity of the company) to that city.

The eighth equation ensures that the amount of product which has been sent to j^{th} customer is equal to the demand of that customer.

The ninth equation guarantees that the amount of product which has been sent to i^{th} warehouse from the production plant is equal to the demand of that warehouse.

The tenth equation is a flow logic constraint. It ensures that, the demand of a warehouse must be equal to the amount of product shipped from that warehouse to all its customers.

The eleventh equation limits the maximum capacity of each warehouse. This constraint is added because of firm policy.

The twelfth equation limits the maximum number of warehouses. This constraint is a company policy too just like the previous one.

Finally, the thirteenth equation forces the decision variables to be positive and integer.

4   Computational Results

In order to show its efficiency we have tested the proposed mathematical model by using the firm's data. We used GAMS optimization software and Baron Solver to solve the proposed model. The results showed that the positive deviation from the cost goal (g_1^+) is found as 232656 and the positive deviation from the service goal (g_2^+) is found as 1954618. The deviation from the service goal seems high because our ideal value was zero and it needs to open warehouses in all the cities to be zero. From another perspective, there is a maximum number of warehouses limit. The model cannot open more than 3 warehouses. For this reason the model tried to find the minimum possible deviation as 1954618. The positive deviation from the ideal cost (first objective) is actually the money which must be accepted by the company to give faster service. Because the ideal cost is found by solving the model with single (cost) objective.

The supply chain network created by the model can be seen in Figure 1 on the next page. As it is seen, the model decided to open 3 (maximum number allowed) warehouses. One of them is opened in Adana, a city in south of Turkey. This warehouse is used in order to satisfy the demand of some cities which are in south or east of Turkey. Another warehouse is opened in Eskisehir, nearly in the middle of Turkey. This warehouse satisfies the demand of a few city which are close to it. Finally, the last warehouse is opened in Sakarya, a city in North of the Turkey and it satisfies the demand of some cities which are in north of the Turkey. It also satisfies the demand of Istanbul, which is the most crowded city in Turkey and has the biggest demand (2097 Pallets) Sakarya is so close to Istanbul that a truck can come to Istanbul within at most 3 or 4 hours. In addition, the warehouse renting and operating cost in Sakarya is three times cheaper than Istanbul. For this reason the model results are really logical. Opening a warehouse not in Istanbul but just near the Istanbul is a really good strategy for cost and fast service.

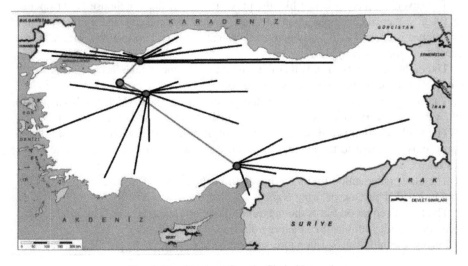

Fig. 1. The Obtained Supply Chain Network

The warehouses and the customers assigned to each warehouse are summarized in Table 3, Table 4 and Table 5 respectively.

Table 3. Warehouse 1 and Assigned Customers

Warehouse Location	Monthly Flow	Cities Assigned to This Warehouse	Product Sent From This Warehouse
Sakarya	3243 pallets	Edirne	66 pallets
		Istanbul	2097 pallets
		Kocaeli	370 pallets
		Ordu	77 pallets
		Sakarya	60 pallets
		Samsun	351 pallets
		Tekirdag	40 pallets
		Trabzon	123 pallets
		Bartin	17 pallets
		Karabuk	19 pallets
		Duzce	23 pallets

When we analyze the warehouse located in Sakarya (by taking care about distances) we see that demands of lots of cities assigned to this warehouse including Istanbul, Kocaeli, Sakarya, Tekirdag, Edirne, Bartin, Karabuk and Duzce can be satisfied within less than six hours. Demands of Ordu and Samsun can be satisfied within less than ten hours. Demand of the farthest city to the warehouse, Trabzon (916 km to the warehouse) can be satisfied within 13 hours. Briefly, it is seen that beside the lowest possible cost our model provides fast service and short lead time.

Table 4. Warehouse 2 and Assigned Customers

Warehouse Location	Monthly Flow	Cities Assigned to This Warehouse	Product Sent From This Warehouse
Eskisehir	2948 pallets	Ankara	493 pallets
		Antalya	680 pallets
		Balikesir	137 pallets
		Bursa	367 pallets
		Eskisehir	55 pallets
		Izmir	432 pallets
		Kayseri	158 pallets
		Konya	247 pallets
		Mugla	251 pallets
		Sivas	94 pallets
		Kirikkale	34 pallets

The same analysis can be done for warehouse 2 and warehouse 3 and it can be again seen that lots of cities' demands can be satisfied within less than six hours, only the demands of a few cities needs more than six hours to be satisfied.

Table 5. Warehouse 3 and Assigned Customers

Warehouse Location	Monthly Flow	Cities Assigned to This Warehouse	Product Sent From This Warehouse
Adana	828 pallets	Adana	315 pallets
		G.Antep	51 pallets
		Hatay	154 pallets
		Mersin	112 pallets
		Malatya	69 pallets
		S.Urfa	48 pallets
		Van	79 pallets

5   Comparing the Model Results with the Existing Company Policy

In the existing situation, the company has only one warehouse which is located in Sakarya. They have opened that warehouse in order to give fast service especially to Istanbul and the cities near the Istanbul. They take care about only the cities which have high demands. From another perspective, since some cities like Van, Gaziantep Malatya etc. are far from Sakarya, it takes more than one day to satisfy the demand of them. In addition, in this case the positive deviation from the cost goal becomes 368876 and the deviation from the service goal becomes 2847515.

The proposed mathematical model showed that opening a warehouse in Sakarya is a good strategy but it is not enough. The model also opened two more warehouses in Eskisehir and Adana respectively. A brief comparison between the supply chain network created by the model and the existing policy is shown in Table 6.

Table 6. Comparison of Model Results and Existing Policy

	Model Results	Existing Policy
Number of Warehouses	3	1
Locations of Warehouses	Sakarya, Eskisehir Adana	Sakarya
Positive Deviation From The Cost Goal	232656	368876
Positive Deviation From The Service Goal	1954618	2847515

Since, we are trying to minimize the sum of positive deviations, we see that the supply chain network obtained from the mathematical model is better than the existing company policy. In other words, as it can be seen easily in Table 6, the supply chain network created by the mathematical model gives faster service with lower cost. It also proves that, opening more warehouses does not always increase the cost.

6   Conclusion

Supply chain networks can be designed by focusing on numerous objectives. Cost, quality, service level, capacity and communication are just a few of the objectives to focus on while designing the supply chain network. Beside the single objective optimization models, there are also some multi objective models which tries to find a tradeoff between the objectives. We have proposed a multi objective model for a household goods company. Since, our problem was a real life problem we tried to make our model as realistic as possible. We used a specific transportation cost function to put the distance economy into account. We also applied some company policies to the model. We prohibited direct shipment from the production plant to the customers, added capacity constraint for warehouses and limited the maximum number of warehouses. All these things made our model a multi objective, multi echelon real life model.

According to the new supply chain network which is designed by the model, the company can satisfy the demands of lots of their customers within less than six hours. In fact, this fast service makes the total supply chain network cost exceed the ideal cost which was 2629703 TL. However, the deviation is only 232656 TL which is acceptable for the company to give the fastest possible service. Because, accepting this cost will increase the customer satisfaction and loyalty. In long term, the company will increase its income and find new customers as a result of this fast service and loyalty. Briefly, it is sometimes beneficial to accept a few excessive cost for long term earnings. Companies should not only focus on cost but also they should focus on other performance indicators like lead time, customer satisfaction and loyalty. Companies must follow a long term strategy and sometimes they should accept some excessive costs for its strategic benefits as our model shows that satisfaction of customers is more important than an excessive 232656 TL cost.

Finally, this model is developed for a company but it can be adapted to any company's problem easily by making only some minor changes in the constraints and/or goals. Because there is not any test problem in the literature, we could not show its adaptability to other problems but by changing some parameters (for example the demands of cities) or some constraints (for example the maximum number and capacities of warehouses), it is possible to adapt the model to any company's problem.

7   Future Agenda

In spite of the fact that this mathematical model is very effective, it may become really difficult (and sometimes impossible) to find an optimum solution when the number of cities increases. For example if we want to solve this model for a company which has customers in 276 cities inside and outside the country, it is almost impossible to find an exact solution in an acceptable time by using any of the optimization software including GAMS. For this reason, it is necessary to adopt a metaheuristic for big problems. That metaheuristic must be suitable both to the problem and to the model. If necessary, the metaheuristic must be modified for the problem and/or model. We currently have been working on developing such an algorithm which includes a metaheuristic (or a few metaheuristics) and gives high quality solutions in a reasonable time for big dimensional problems.

Acknowledgements. This study was supported by the Anadolu University Scientific Research Projects Commission under the grant no 1501F024.

In addition, special thanks to The Scientific and Technological Research Council of Turkey (TUBITAK) BIDEB for their encouragements.

References

1. Santaso, T., Ahmed, S., Goetschalckx, M., Shapiro, A.: A Stochastic Programming Approach for Supply Chain Network Design Under Uncertainty. European Journal of Operational Research 167, 96–115 (2005)
2. Guillén, G., Mele, F.D., Bagajewicz, M.J., Espuña, A., Puigjaner, L.: Multiobjective Supply Chain Design under Uncertainty. Chemical Engineering Science 60, 1535–1553 (2005)
3. Tsiakis, P., Shah, N., Pantelides, C.C.: Design of Multi-echelon Supply Chain Networks under Demand Uncertainty. Ind. Eng. Chem. Res. 40, 3585–3604 (2001)
4. Pazhani, S., Ramkumar, N., Narendran, T.T., Ganesh, K.: A Bi-Objective Network Design Model for Multi-Period, Multi-Product Closed-Loop Supply Chain. Journal of Industrial and Production Engineering 30(4), 264–280 (2013)
5. Wang, F., Lai, X., Shi, N.: A Multi-Objective Optimization for Green Supply Chain Network Design. Decision Support Systems 51, 262–269 (2011)
6. Hiremath, N.C., Sahu, S., Tiwari, M.K.: Multi Objective Outbound Logistics Network Design for a Manufacturing Supply Chain. J. Intell. Manuf. 24, 1071–1084 (2013)
7. Pirkul, H., Jayaraman, V.: A Multi-Commodity, Multi-Plant, Capacitated Facility Location Problem: Formulation and Efficient Heuristic Solution. Computers Ops. Res. 25(10), 869–878 (1998)
8. Pirkul, H., Jayaraman, V.: Production, Transportation, and Distribution Planning in a Multi Commodity Tri-Echelon System. Transportation Science 30, 291–302 (1996)
9. Republic of Turkey General Directorate of Highways, http://www.kgm.gov.tr/

Conic Scalarization Method in Multiobjective Optimization and Relations with Other Scalarization Methods*

Refail Kasimbeyli**, Zehra Kamisli Ozturk, Nergiz Kasimbeyli,
Gulcin Dinc Yalcin, and Banu Icmen

Faculty of Engineering, Department of Industrial Engineering
Aandolu University, Iki Eylul Campus, Eskisehir 26555, Turkey
{rkasimbeyli,zkamisli,nkasimbeyli,gdinc,bicmen}@anadolu.edu.tr

Abstract. The paper presents main features of the conic scalarization method in multiobjective optimization. The conic scalarization method guarantees to generate all proper efficient solutions and does not require any kind of convexity or boundedness conditions. In addition the preference and reference point information of the decision maker is taken into consideration by this method. In this paper, relations with other scalarization methods are investigated and it is shown that some efficient solutions computed by the Pascoletti-Serafini and the Benson's scalarization methods, can be obtained by the conic scalarization method.

Keywords: Conic scalarization method, Weighted sum method, Epsilon constraint method, Benson's method, Weighted Chebyshev method, Pascoletti-Serafini method, Multiobjective optimization, proper efficiency.

1 Introduction

In general, scalarization means the replacement of a multiobjective optimization problem by a suitable scalar optimization problem which is an optimization problem with a real valued objective function.

In this paper we give main features of the conic scalarization method. The conic scalarization method enables to completely characterize the whole set of efficient and properly efficient solutions of multiobjective problems without convexity and boundedness conditions.

In this paper we present theorems which establish relations between the conic scalarization and the Pascoletti-Serafini and the Benson's scalarization methods. It is shown that some efficient solutions computed by the Pascoletti-Serafini and the Benson's scalarization methods, can be obtained by the conic scalarization method.

---

* This study was supported by the Anadolu University Scientific Research Projects Commission under the grant no 1304F062.
** Corresponding author.

319

H.A. Le Thi et al. (eds.), *Model. Comput. & Optim. in Inf. Syst. & Manage. Sci.*,
Advances in Intelligent Systems and Computing 359, DOI: 10.1007/978-3-319-18161-5_27

The rest of the paper is organized as follows. Section 2 gives some preliminaries. Main characteristics of the conic scalarization method are given in Section 3. In section 4, new relations between the conic scalarization method and Pascoletti-Serafini and Benson's scalarization methods are established. Finally, Section 5 draws some conclusions from the paper.

2   Preliminaries

We begin this section with standard definitions from multi-objective optimization.

Let $\mathbb{R}_+^n := \{y = (y_1, ..., y_n) : y_i \geq 0, i = 1, ..., n\}$, and let $\mathbb{Y} \subset \mathbb{R}^n$ be a nonempty set.

Throughout the paper, $\mathbb{R}_+$ denotes the set of nonnegative real numbers. $cl(\mathbb{Y})$, $bd(\mathbb{Y})$, $int(\mathbb{Y})$, and $co(\mathbb{Y})$ denote the *closure*, the *boundary*, the *interior*, and the *convex hull* of a set $\mathbb{Y}$, respectively.

A nonempty subset $\mathbb{C}$ of $\mathbb{R}^n$ is called a *cone* if $y \in \mathbb{C}, \lambda \geq 0 \Rightarrow \lambda y \in \mathbb{C}$. Pointedness of $\mathbb{C}$ means that $\mathbb{C} \cap (-\mathbb{C}) = \{0_{\mathbb{R}^n}\}$.

We will assume that $\mathbb{R}^n$ is partially ordered by a convex pointed cone $\mathbb{C} \subset \mathbb{R}^n$.

Definition 1. *1. An element $y \in \mathbb{Y}$ is called a minimal element of $\mathbb{Y}$ (with respect to the ordering cone $\mathbb{C}$) if $(\{y\} - \mathbb{C}) \cap \mathbb{Y} = \{y\}$.*

2. An element $y \in \mathbb{Y}$ is called a weakly minimal element of $\mathbb{Y}$ if $(\{y\} - int(\mathbb{C})) \cap \mathbb{Y} = \emptyset$.

3. An element $y \in \mathbb{Y}$ is called a properly minimal element of $\mathbb{Y}$ in the sense of Benson [1] if y is a minimal element of $\mathbb{Y}$ and the zero element of $\mathbb{R}^n$ is a minimal element of $cl(cone(\mathbb{Y} + \mathbb{C} - \{y\}))$, where $cone(\mathbb{Y}) := \{\lambda y : \lambda \geq 0, y \in \mathbb{Y}\}$.

4. An element $\overline{y} \in \mathbb{Y}$ is called a properly minimal element of $\mathbb{Y}$ in the sense of Henig [11] if it is a minimal element of $\mathbb{Y}$ with respect to some convex cone $\mathbb{K}$ with $\mathbb{C} \setminus \{0_{\mathbb{R}^n}\} \subset int(\mathbb{K})$.

Henig proved that in the case when the vector space is partially ordered by a closed pointed cone, the two definitions of proper efficiency given in Definition 1, are equivalent (see [11, Theorem 2.1]). Therefore, in the sequel we simply will use the notion of proper efficiency.

Consider a multiobjective optimization problem (in short MOP):

$$\min_{x \in \mathbb{X}}[f_1(x), ..., f_n(x)], \tag{1}$$

where $\mathbb{X}$ is a nonempty set of feasible solutions and $f_i : \mathbb{X} \to \mathbb{R}, i = 1, ..., n$ are real-valued functions. Let $f(x) = (f_1(x), ..., f_n(x))$ for every $x \in \mathbb{X}$ and let $\mathbb{Y} := f(\mathbb{X})$.

Definition 2. *A feasible solution $x \in \mathbb{X}$ is called efficient, weakly efficient or properly efficient solution of multi-objective optimization problem (1) if $y = f(x)$ is a minimal, weakly minimal or properly minimal element of $\mathbb{Y}$, respectively.*

Let $y = (y_1, \ldots, y_n) \in \mathbb{R}^n$. $\|y\|_1 = \sum_{i=1}^n |y_i|$, $\|y\|_2 = (y_1^2 + \cdots + y_n^2)^{1/2}$, and $\|y\|_\infty = \max\{|y_1|, \ldots, |y_n|\}$ denote the l_1, l_2 (Euclidean), and l_∞ norms of y, respectively.

Let $\mathbb{C}$ be a given cone in $\mathbb{R}^n$. Recall that the dual cone $\mathbb{C}^*$ of $\mathbb{C}$ and its quasi-interior $\mathbb{C}^\#$ are defined by

$$\mathbb{C}^* = \{w \in \mathbb{R}^n : w^T y \geq 0 \text{ for all } y \in \mathbb{C}\} \tag{2}$$

and

$$\mathbb{C}^\# = \{w \in \mathbb{R}^n : w^T y > 0 \text{ for all } y \in \mathbb{C} \setminus \{0\}\}, \tag{3}$$

respectively, where w^T denotes the transpose of vector w, and $w^T y = \sum_{i=1}^n w_i y_i$ is the scalar product of vectors $w = (w_1, \ldots w_n)$ and $y = (y_1, \ldots, y_n)$. The elements of these cones define monotone and strongly monotone linear functionals whose level sets (hyperplanes) are used to characterize support points of convex sets.

The following three cones called augmented dual cones of $\mathbb{C}$ were introduced in [14], and it was proven that the elements of these cones define monotone sublinear functionals with conical level sets. Due to this property, these functionals are used to generate efficient solutions of nonconvex multiobjective problems.

$$\mathbb{C}^{a*} = \{(w, \alpha) \in \mathbb{C}^\# \times \mathbb{R}_+ : w^T y - \alpha\|y\| \geq 0 \text{ for all } y \in \mathbb{C}\}, \tag{4}$$

$$\mathbb{C}^{ao} = \{(w, \alpha) \in \mathbb{C}^\# \times \mathbb{R}_+ : w^T y - \alpha\|y\| > 0 \text{ for all } y \in \text{int}(\mathbb{C})\}, \tag{5}$$

and

$$\mathbb{C}^{a\#} = \{(w, \alpha) \in \mathbb{C}^\# \times \mathbb{R}_+ : w^T y - \alpha\|y\| > 0 \text{ for all } y \in \mathbb{C} \setminus \{0\}\}, \tag{6}$$

where $\mathbb{C}$ is assumed to have a nonempty interior in the definition of $\mathbb{C}^{ao}$.

3   Conic Scalarization (CS) Method

The history of development of the CS method goes back to the paper [5], where Gasimov introduced a class of monotonically increasing sublinear functions on partially ordered real normed spaces and showed without convexity and boundedness assumptions that support points of a set obtained by using these functions are properly minimal in the sense of Benson [1]. The question of "can every properly minimal point of a set be calculated in a similar way", was answered only in the case when the objective space is partially ordered by a certain Bishop–Phelps cone. Since then, different theoretical and practical applications by using the suggested class of sublinear functions have been realized [3,6,7,8,9,12,13,14,17,20,22]. The theoretical fundamentals of the conic scalarization method in general form was firstly explained in [14]. The full description of the method is given in [15].

The idea of the CS method is very simple: choose preference parameters which consist of a weight vector $w \in \mathbb{C}^\#$ and a reference point $a \in \mathbb{R}^n$, determine an augmentation parameter $\alpha \in \mathbb{R}_+$ such that $(w, \alpha) \in \mathbb{C}^{a*}$ (or $(w, \alpha) \in \mathbb{C}^{ao}$, or

$(w, \alpha) \in \mathbb{C}^{a\#})$, where for a convenience the l_1−norm is used, and solve the scalar optimization problem:

$$\min_{x \in X} \sum_{i=1}^{n} w_i(f_i(x) - a_i) + \alpha \sum_{i=1}^{n} |f_i(x) - a_i| \qquad (CS(w, \alpha, a))$$

The set of optimal solutions of this scalar problem will be denoted by $Sol(CS(w, \alpha, a))$. Reference point $a = (a_1, \ldots, a_n)$ may be identified by a decision maker in cases when she/he desires to calculate minimal elements that are close to some point. The CS method does not impose any restrictions on the ways for determining reference points. The reference point can be chosen arbitrarily.

The following theorem quoted from [15] explains main properties of solutions obtained by the conic scalarization method in the case when $\mathbb{C} = \mathbb{R}_+^n$. This special case for the cone determining the partial ordering, allows one to explicitly determine augmented dual cones which are used for choosing scalarizing parameters (w, α). For the general case of this theorem see [14, Theorem 5.4].

Theorem 1. *[15, Theorem 6] Let $a \in \mathbb{R}^n$ be a given reference point, and let $\mathbb{C} = \mathbb{R}_+^n$. Assume that $Sol(CS(w, \alpha, a)) \neq \emptyset$ for a given pair $(w, \alpha) \in \mathbb{C}^{a*}$. Then the following hold.*

(i) *If*

$$(w, \alpha) \in \mathbb{C}^{a\circ} = \{((w_1, \ldots, w_n), \alpha) : 0 \leq \alpha \leq w_i, w_i > 0, i = 1, \ldots, n$$
$$\text{and there exists } k \in \{1, \cdots, n\} \text{ such that } w_k > \alpha\},$$

then every element of $Sol(CS(w, \alpha, a))$ is a weakly efficient solution of (1).

(ii) *If $Sol(CS(w, \alpha, a))$ consists of a single element, then it is an efficient solution (1).*

(iii) *If*

$$(w, \alpha) \in \mathbb{C}^{a\#} = \{((w_1, \ldots, w_n), \alpha) : 0 \leq \alpha < w_i, i = 1, \ldots, n\},$$

then every element of $Sol(CS(w, \alpha, a))$ is a properly efficient solution of (1), and conversely, if $\overline{x}$ is a properly efficient solution of (1), then there exists $(w, \alpha) \in \mathbb{C}^{a\#}$ and a reference point $a \in \mathbb{R}^n$ such that $\overline{x}$ is a solution of $Sol(CS(w, \alpha, a))$.

The following theorem gives simple characterization of minimal elements.

Theorem 2. *[15, Theorem 7] Let $\mathbb{Y} \subset \mathbb{R}^n$ be a given nonempty set and let $\mathbb{C} = \mathbb{R}_+^n$. If $\overline{y}$ is a minimal element of $\mathbb{Y}$, then $\overline{y}$ is an optimal solution of the following scalar optimization problem:*

$$\min_{y \in \mathbb{Y}} \{ \sum_{i=1}^{n} (y_i - \overline{y}_i) + \sum_{i=1}^{n} |y_i - \overline{y}_i| \}. \qquad (7)$$

By using assertions of Theorems 1 and 2, we arrive at the following conclusion. By solving the problem $(CS(w, \alpha, a))$ for "all" possible values of the augmentation parameter α between 0 and $\min\{w_1, \ldots, w_n\}$, one can calculate all the efficient solutions corresponding to the decision maker's preferences (the weighting vector $w = (w_1, \ldots, w_n)$ and the reference point a).

The following two remarks illustrate the geometry of the CS method.

Remark 1. It is clear that in the case when $\alpha = 0$ (or, if $f(\mathbb{X}) \subseteq \{a\} \pm \mathbb{C}$) the objective function of the scalar optimization problem $(CS(w, \alpha, a))$ becomes an objective function of the weighted sum scalarization method. The minimization of such an objective function over a feasible set enables to obtain only those efficient solutions x (if the corresponding scalar problem has a solution), for which the minimal vector $f(x)$ is a supporting point of the objective space with respect to some hyperplane

$$H(w) = \{y : w^T y = \beta\},$$

where $\beta = w^T f(x)$. It is obvious that minimal points which are not supporting points of the objective space with respect to some hyperplane, cannot be detected by this way. By augmenting the linear part in $(CS(w, \alpha, a))$ with the norm term (using a positive augmentation parameter α), the hyperplane $H(w)$ becomes a conic surface defined by the cone

$$S(w, \alpha) = \{y \in \mathbb{R}^n : w^T y + \alpha\|y\| \leq 0\}, \tag{8}$$

and therefore the corresponding scalar problem $(CS(w, \alpha, a))$ computes solution x, for which the corresponding vector $f(x)$ is a supporting point of the objective space with respect to this cone. The change of the α, leads to a different supporting cone. The supporting cone corresponding to some weight vector w becomes narrower as α increases, and the smallest cone (which anyway contains the ordering cone) is obtained when α equals its maximum allowable value (for example, $\min\{w_1, \ldots, w_n\}$, if $(w, \alpha) \in \mathbb{C}^{a\#}$). This analysis shows that by changing the α parameter, one can compute different minimal points of the problem corresponding to the same weight vector. And since the method computes supporting points of the decision space with respect to cones (if $\alpha \neq 0$), it becomes clear why this method does not require convexity and boundedness conditions and why it is able to find optimal points which cannot be detected by hyperplanes. Since the cases $\alpha = 0$, or $f(\mathbb{X}) \subseteq \{a\} \pm \mathbb{C}$ leads to the objective function of the weighted sum scalarization method, we can say that the CS method is a generalization of the weighted sum scalarization method.

Remark 2. It follows from the definition of augmented dual cone that $w^T y - \alpha\|y\| \geq 0$ for every $(w, \alpha) \in \mathbb{C}^{a*}$ and all $y \in \mathbb{C}$. Hence

$$\mathbb{C} \subset C(w, \alpha) = \{y \in \mathbb{R}^n : w^T y - \alpha\|y\| \geq 0\}, \tag{9}$$

where $C(w, \alpha)$ is known as the Bishop-Phelps cone corresponding to a pair $(w, \alpha) \in \mathbb{C}^{a*}$. It has been proved that, if $\mathbb{C}$ is a closed convex pointed cone having a weakly compact base, then

$$\mathbb{C} = \cap_{(w,\alpha) \in \mathbb{C}^{a*}} C(w,\alpha),$$

see [14, Theorems 3.8 and 3.9].

On the other hand, since $w^T y - \alpha \|y\| \geq 0$ for every $(w,\alpha) \in \mathbb{C}^{a*}$ and all $y \in \mathbb{C}$, then clearly $w^T y + \alpha \|y\| \leq 0$ for every $y \in -\mathbb{C}$. Thus we conclude that all the cones $S(w,\alpha) = \{y \in \mathbb{R}^n : w^T y + \alpha \|y\| \leq 0\}$ (see (8)) with $(w,\alpha) \in \mathbb{C}^{a*}$, contain the ordering cone $-\mathbb{C}$. Moreover if $(w,\alpha) \in \mathbb{C}^{a\#}$ then we have [14, Lemma 3.6]

$$- \mathbb{C} \setminus \{0\} \in int(S(w,\alpha)) = \{y \in \mathbb{R}^n : w^T y + \alpha \|y\| < 0\}. \tag{10}$$

Due to this property, the CS method guarantees to calculate "all" properly efficient solutions corresponding to the given weights and the given reference point. That is, every solution of the scalar problem $(CS(w,\alpha,a))$, is a properly efficient solutions of the multi-objective optimization problem (1), if $(w,\alpha) \in \mathbb{C}^{a\#}$, see Theorem 1 (iii).

In some cases for a given cone $\mathbb{C}$ and a given norm, there may be available to find a pair $(w,\alpha) \in \mathbb{C}^{a*}$ such that $\mathbb{C} = C(w,\alpha)$. For example if $\mathbb{C} = \mathbb{R}^n_+$ then

$$\mathbb{R}^n_+ = C(w^1, \alpha^1) = \{y \in \mathbb{R}^n : (w^1)^T y - \alpha^1 \|y\|_1 \geq 0\}, \tag{11}$$

where $w^1 = (1,...,1) \in \mathbb{R}^n$, $\alpha^1 = 1$, and the l_1 norm is used in the definition (see [15, Lemma 4]). Similarly, $\mathbb{R}^n_-$ can be represented as a level set $S(w^1, \alpha^1)$ (see (8)) of the function

$$g_{(w^1,\alpha^1)}(y) = y_1 + \ldots + y_n + |y_1| + \ldots + |y_n| \tag{12}$$

in the form:

$$\mathbb{R}^n_- = S(w^1, \alpha^1) = \{(y_1, \ldots, y_n) \in \mathbb{R}^n : y_1 + \ldots + y_n + |y_1| + \ldots + |y_n| \leq 0\}. \tag{13}$$

Hence, it becomes clear that the presented scalarization method enables one to calculate minimal elements which are "supporting" elements of $f(\mathbb{X})$ with respect to the conic surfaces like $S(w,\alpha)$ (see (8)). In practice, one can divide the interval between 0 and $\min\{w_1, \ldots, w_n\}$ into several parts, and for all these values of the augmentation parameter α, the scalar problem $(CS(w,\alpha,a))$ can be solved for the same weights and the same reference point chosen. This will enable decision maker to compute different efficient solutions (if any) with respect to the same set of weights. The scalar problem $(CS(w,\alpha,a))$ is nonsmooth and nonconvex if the original problem is not convex. Such a problem can be solved by using some standard softwares (see, for example [9,12,22]), or special solution algorithms can be applied, see for example [6,8,16].

4   Relations with Other Methods

In this section we present theorems which establish relations between the CS, the Pascoletti-Serafini (PSS) and the Benson's (BS) scalarization methods. It is shown that some efficient solutions computed by the PSS and the BS methods, can be obtained by the CS method.

4.1 Relations between the Conic Scalarization (CS) and the Pascoletti-Serafini Scalarization (PSS) Methods

The method known as the Pascoletti-Serafini scalarization method, is studied in [10,18,21] by Tammer, Weidner, Winkler, Pascoletti and Serafini.

The scalar problem of the PSS method is defined as follows:

$$minimize \quad t \qquad (PSS(a,r))$$

$$\text{s.t.} \qquad a + tr - f(x) \in \mathbb{C}$$

$$x \in \mathbb{X}, t \in \mathbb{R},$$

where $a \in \mathbb{R}^n$ and $r \in \mathbb{C}$ are parameters of $(PSS(a,r))$. The problem $(PSS(a,r))$ can also be written in the form (see [4])

$$minimize \quad t \tag{14}$$

$$\text{s.t.} \qquad a + tr - \mathbb{C} \cap f(X) \neq \emptyset, \qquad t \in \mathbb{R}.$$

This problem can be interpreted in the following form. The ordering cone $\mathbb{C}$ is moved in direction $-r$ along the line $a + tr$ till the set $(a + tr - \mathbb{C}) \cap f(X)$ is reduced to the empty set. The smallest value $\bar{t}$ for which $(a + \bar{t}r - \mathbb{C}) \cap f(X) \neq \emptyset$ is the solution of (14). If the pair $(\bar{t}, \bar{x})$ is a solution of $(PSS(a,r))$ the element $\bar{y} = f(\bar{x})$ with $\bar{y} \in (a + \bar{t}r - \mathbb{C}) \cap f(X)$ will be characterized as a weakly minimal solution of (1).

Theorem 3. *Assume that $\mathbb{C}$ is a closed convex pointed cone with nonempty interior, and that $a \in \mathbb{R}^n$, $r \in int(\mathbb{C})$ and $(\bar{t}, \bar{x})$ is an optimal solution of $(PSS(a,r))$. Then, there exists a weight vector $\bar{w} = (\bar{w}_1, ...\bar{w}_n) \in \mathbb{C}^\#$ and an augmentation parameter $\bar{\alpha} \geq 0$ with $(\bar{w}, \bar{\alpha}) \in C^{a\circ}$ such that*

$$\min_{x \in \mathbb{X}} \bar{w}^T(f(x) - a) + \bar{\alpha}\|f(x) - a\| \leq \bar{t}.$$

Proof. Let $(w, \alpha) \in C^{a\circ}$. By definition of $C^{a\circ}$ (see also (9)), $\mathbb{C} \subset C(w, \alpha)$. Then problem $(PSS(a,r))$ can be written in the following form with possibly a broader set of feasible solutions:

$$minimize \quad t \qquad (PSS_{C(w,\alpha)}(a,r))$$

$$\text{s.t} \qquad a + tr - f(x) \in C(w, \alpha)$$

$$x \in \mathbb{X},$$

By definition of $C(w, \alpha)$, the inclusion $a + tr - f(x) \in C(w, \alpha)$ implies

$$w^T(a + tr - f(x)) - \alpha\|a + tr - f(x)\| \geq 0,$$

or

$$w^T(f(x) - a - tr) + \alpha\|f(x) - a - tr\| \leq 0.$$

Obviously,

$$\alpha(\|f(x) - a\| - \|tr\|) \leq \alpha\|f(x) - a - tr\|.$$

Then if we change the norm term by the left hand side in the above inequality, the set of feasible solutions of $(PSS_{C(w,\alpha)}(a,r))$ will again be extended:

$$w^T(f(x) - a) + \alpha\|f(x) - a\| \leq tw^Tr + |t|\alpha\|r\|. \tag{15}$$

In dependence on the sign of $\bar{t}$ we can consider only positive or only negative range for t in (15). If only negative (or only positive) values of t will be considered then the right hand side of (15) becomes $t(w^Tr - \alpha\|r\|)$ (or $t(w^Tr + \alpha\|r\|)$). Since $r \in int(C)$ and $(w, \alpha) \in C^{a\circ}$ we have $w^Tr - \alpha\|r\| > 0$ (or $w^Tr + \alpha\|r\| > 0$).

Thus, by dividing both sides of (15) with $w^Tr - \alpha\|r\| > 0$ (or $w^Tr + \alpha\|r\| > 0$) and denoting $\bar{w} = w/(w^Tr - \alpha\|r\|)$ and $\bar{\alpha} = \alpha/(w^Tr - \alpha\|r\|)$ (or $\bar{w} = w/(w^Tr + \alpha\|r\|)$ and $\bar{\alpha} = \alpha/(w^Tr + \alpha\|r\|))$, we obtain that the problem $(PSS_{C(w,\alpha)}(a,r))$ can be written (with a possibly broader feasible set) in the form:

$$minimize \quad t \tag{16}$$
$$\text{s.t} \quad \bar{w}^T(f(x) - a) + \bar{\alpha}\|f(x) - a\| \leq t \tag{17}$$
$$x \in \mathbb{X}, \tag{18}$$

This problem is equivalent to the following problem $(CS(\bar{w}, \bar{\alpha}, a))$:

$$\min_{x \in \mathbb{X}}[\bar{w}^T(f(x) - a) + \bar{\alpha}\|f(x) - a\|].$$

Since the set of feasible solutions of problem (16) - (18) is larger than the one of $(PSS(a,r))$, we obtain

$$\min_{x \in \mathbb{X}}[\bar{w}^T(f(x) - a) + \bar{\alpha}\|f(x) - a\|] \leq \bar{t},$$

which completes the proof of theorem.

4.2   Relationship between the Conic Scalarization (CS) and the Benson's (BS) methods.

In this section we explain relationship between the BS and the CS methods. The idea of the BS method is to choose some initial feasible solution $x^0 \in X$ and, if it is not itself efficient, produce a dominating solution that is. To do so, nonnegative deviation variables $l_i = f_i(x^0) - f_i(x)$ are introduced, and their sum is maximized. This results in an x dominating x^0, if one exists, and the objective ensures that it is efficient, pushing x as far from x^0 as possible. The corresponding scalar problem for given x^0 is:

$$\max \sum_{i=1}^{n} l_i \quad (BS(x^0))$$
$$\text{s.t.}$$
$$f_i(x^0) - l_i - f_i(x) = 0, \quad i = 1, \ldots, n$$
$$l \geq 0, x \in \mathbb{X}.$$

Theorem 4. *Let $\bar{x}$ be an efficient solution to (1). Suppose that $\bar{x}$ is an optimal solution of Benson scalar problem $(BS(x^0))$ for a feasible solution $x^0 \in X$. Then $\bar{x}$ is an optimal solution of the conic scalar problem $CS(w^1, \alpha^1, f(x^0))$, where $w^1 = (1, \ldots, 1) \in \mathbb{R}^n$, $\alpha^1 = 1$, and the l_1 norm is used:*

$$min_{x \in X} \sum_{i=1}^{n}(f_i(x) - f_i(x^0)) + \sum_{i=1}^{n}|f_i(x) - f_i(x^0)|.$$

Proof. Since $f(x^0) \in f(X)$, $f(\bar{x}) \in f(x^0) - R_+^n$ and (see (13))

$$-R_+^n = \{y : (w^1)^T y + \alpha^1 \|y\|_1 \leq 0\},$$

we have:

$$\sum_{i=1}^{n}(f_i(x) - f_i(x^0)) + \sum_{i=1}^{n}|f_i(x) - f_i(x^0)| = 0$$

for all $x \in X_0 = \{x \in X : f(x) \in f(x^0) - R_+^n\}$, and in particular for $x = \bar{x}$. Obviously,

$$\sum_{i=1}^{n}(f_i(x) - f_i(x^0)) + \sum_{i=1}^{n}|f_i(x) - f_i(x^0)| > 0$$

for all $x \in X \setminus X_0$ which completes the proof.

Theorem 5. *Let $\bar{x}$ be an optimal solution of Benson scalar problem $(BS(x^0))$ for a feasible solution $x^0 \in X$. Assume that $\bar{x}$ is a properly efficient solution to (1). Then there exists $\bar{\alpha} \in [0, 1)$ such that $\bar{x}$ is an optimal solution of the conic scalar problem $CS(w^1, \bar{\alpha}, f(\bar{x}))$, where $w^1 = (1, \ldots, 1) \in \mathbb{R}^n$ with the l_1 norm:*

$$min_{x \in X} \sum_{i=1}^{n}(f_i(x) - f_i(\bar{x})) + \bar{\alpha} \sum_{i=1}^{n}|f_i(x) - f_i(\bar{x})|. \tag{19}$$

Proof. We have:

$$-R_+^n = \{y : (w^1)^T y + \alpha^1 \|y\|_1 \leq 0\},$$

and clearly

$$-\mathbb{R}_+^n \setminus \{0\} \in int(\{y \in \mathbb{R}^n : (w^1)^T y + \alpha\|y\| \leq 0\}),$$

for every $\alpha \in [0, 1)$ (see (10)), where

$$int(\{y \in \mathbb{R}^n : (w^1)^T y + \alpha\|y\| \leq 0\}) = \{y \in \mathbb{R}^n : (w^1)^T y + \alpha\|y\| < 0\}).$$

Since $\bar{x}$ is a properly efficient solution to (1), there exists $\bar{\alpha} \in [0, 1)$ such that

$$\{f(\bar{x})\} + \{y \in \mathbb{R}^n : (w^1)^T y + \bar{\alpha}\|y\| \leq 0\} \cap f(X) = \{f(\bar{x})\}.$$

This leads

$$\{y \in \mathbb{R}^n : (w^1)^T(y - f(\bar{x})) + \bar{\alpha}\|y - f(\bar{x})\| \leq 0\} \cap f(X) = \{f(\bar{x})\}.$$

The last relation means that

$$(w^1)^T(f(x) - f(\bar{x})) + \bar{\alpha}\|f(x) - f(\bar{x})\| \geq 0$$

for every $x \in X$. which proves the theorem.

5 Conclusion

In this paper, the conic scalarization method is analyzed and main properties of solutions obtained by this method are explained. Additionally simple characterization of minimal elements is given. It has been emphasized that the conic scalarization method guarantee to generate the proper efficient solutions while it does not require any kind of convexity and/or boundedness assumptions. In addition the preference and reference point information of decision maker is taken into consideration by this method.

The paper also discussed relations between the conic scalarization method and Pascoletti-Serafini and Benson's scalarization methods. It has been shown that some solutions obtained by the Pascoletti-Serafini and Benson's scalarization methods, can also be obtained by the conic scalarization method.

Acknowledgments. The authors thank the Anadolu University Scientific Research Projects Commission, who supported this project under the Grant No 1304F062.

References

1. Benson, H.P.: An improved definition of proper efficiency for vector maximization with respect to cones. J. Math. Anal. Appl. 71, 232–241 (1979)
2. Ehrgott, E.: Multicriteria Optimization. Springer, Heidelberg (2005)
3. Ehrgott, M., Waters, C., Kasimbeyli, R., Ustun, O.: Multiobjective Programming and Multiattribute Utility Functions in Portfolio Optimization. INFOR: Information Systems and Operational Research 47, 31–42 (2009)
4. Eichfelder, G.: Adaptive Scalarization Methods in Multiobjective Optimization. Springer, Heidelberg (2008)
5. Gasimov, R.N.: Characterization of the Benson proper efficiency and scalarization in nonconvex vector optimization. In: Koksalan, M., Zionts, S. (eds.) Multiple Criteria Decision Making in the New Millennium. Lecture Notes in Econom. and Math. Systems, vol. 507, pp. 189–198 (2001)
6. Gasimov, R.N.: Augmented Lagrangian duality and nondifferentiable optimization methods in nonconvex programming. J. Global Optimization 24, 187–203 (2002)
7. Gasimov, R.N., Ozturk, G.: Separation via polyhedral conic functions. Optim. Methods Softw. 21, 527–540 (2006)
8. Gasimov, R.N., Rubinov, A.M.: On augmented Lagrangians for optimization problems with a single constraint. Journal of Global Optimization 28(5), 153–173 (2004)
9. Gasimov, R.N., Sipahioglu, A., Sarac, T.: A multi-objective programming approach to 1.5-dimensional assortment problem. European J. Oper. Res. 179, 64–79 (2007)
10. Gerth, C., Weidner, P.: Nonconvex separation theorems and some applications in vector optimization. J. Optimization Theory Appl. 67(2), 297–320 (1990)
11. Henig, M.I.: Proper efficiency with respect to cones. J. Optim. Theory Appl. 36, 387–407 (1982)
12. Ismayilova, N.A., Sagir, M., Gasimov, R.N.: A multiobjective faculty-course-time slot assignment problem with preferences. Math. Comput. Modelling. 46, 1017–1029 (2007)

13. Kasimbeyli, R.: Radial Epiderivatives and Set-Valued Optimization. Optimization 58, 521–534 (2009)
14. Kasimbeyli, R.: A Nonlinear Cone Separation Theorem and Scalarization in Nonconvex Vector Optimization. SIAM J. on Optimization 20, 1591–1619 (2010)
15. Kasimbeyli, R.: A conic scalarization method in multi-objective optimization. Journal of Global Optimization 56(2), 279–297 (2013)
16. Kasimbeyli, R., Ustun, O., Rubinov, A.M.: The Modified Subgradient Algorithm Based on Feasible Values. Optimization 58(5), 535–560 (2009)
17. Ozdemir, M.S., Gasimov, R.N.: The analytic hierarchy process and multiobjective 0-1 faculty course assignment. European J. Oper. Res. 157, 398–408 (2004)
18. Pascoletti, A., Serafini, P.: Scalarizing Vector Optimization Problems. Journal of Optimization Theory and Applications 42, 499–524 (1984)
19. Petschke, M.: On a theorem of Arrow, Barankin and Blackwell. SIAM J. Control Optim. 28, 395–401 (1990)
20. Rubinov, A.M., Gasimov, R.N.: Scalarization and nonlinear scalar duality for vector optimization with preferences that are not necessarily a pre-order relation. Journal of Global Optimization 29, 455–477 (2004)
21. Tammer, C., Winkler, K.: A new scalarization approach and applications in multicriteria d.c. optimization. J. Nonlinear Convex Anal. 4(3), 365–380 (2003)
22. Ustun, O., Kasimbeyli, R.: Combined forecasts in portfolio optimization:a generalized approach. Computers & Operations Research 39, 805–819 (2012)

Part VI
Numerical Optimization

A Hybrid Direction Algorithm with Long Step Rule for Linear Programming: Numerical Experiments

Mohand Bentobache[1,2] and Mohand Ouamer Bibi[1]

[1] LAMOS, Laboratory of Modelling and Optimization of Systems,
University of Bejaia, 06000, Bejaia, Algeria
[2] LMPA, Laboratory of Pure and Applied Mathematics,
University of Laghouat, 03000, Laghouat, Algeria
mbentobache@yahoo.com, mobibi.dz@gmail.com

Abstract. In a previous work [M.O. Bibi and M. Bentobache, *A hybrid direction algorithm for solving linear programs*, International Journal of Computer Mathematics, Vol. 92, N°1, pp. 201–216, 2014], a new search direction for the adaptive method, called hybrid direction, was suggested. For testing optimality, the optimality estimate was defined and used. However, a suboptimality criterion was not given and the updating formula, when we change the support, was not derived. In this paper, we overcome all the difficulties encountered in previous works. Indeed, by using the suboptimality estimate of the current solution, we derive a more general updating formula for the suboptimality estimate when we change the feasible solution and when we change the support too. Furthermore, we present a long step rule procedure to change the support in our method. Finally, the adaptive method with hybrid direction and long step rule is described and computational experiments showing the efficiency of our algorithm are presented.

Keywords: linear programming, adaptive method, hybrid direction, suboptimality estimate, long step rule, computational experiments.

1 Introduction

In [11], the authors developed the support method which is a generalization of the simplex method [10] for solving Linear Programming (LP) problems. The principle of this method is to start by a support feasible solution comprising a basis and a feasible solution and to go through interior or extreme points to achieve an optimal one. Later, they have developed the adaptive method to solve, particularly, linear optimal control problems [12]. This method is generalized to solve general linear and convex quadratic problems [6,9,13,14].

In [1,3,4,7,8], we suggested a new search direction for the adaptive method. This direction is called a hybrid direction because it takes for some solution components extreme values in order to bring them to their bounds and it takes for others the reduced gradient values. For testing optimality, the optimality

© Springer International Publishing Switzerland 2015
H.A. Le Thi et al. (eds.), *Model. Comput. & Optim. in Inf. Syst. & Manage. Sci.*,
Advances in Intelligent Systems and Computing 359, DOI: 10.1007/978-3-319-18161-5_28

estimate is defined and used. However, a suboptimality criterion was not given and the updating formula, when we change the support, was not derived.

In this work, we overcome all the difficulties encountered in previous works. Indeed, by using the suboptimality estimate of the current solution, we derive a more general updating formula for the suboptimality estimate when we change the feasible solution and when we change the support too. Hence, the updating formula given in [11] is a special case of our formula. In the previous work [1], the long step rule is not described. In this paper, we describe the long step rule for our method and we suggest a procedure for updating appropriately the parameter η.

The paper is organized as follows: in Section 2, we give some definitions. In Section 3, we present theoretic aspects of the suggested algorithm and we give some numerical examples for illustration purpose. In Section 4, we give some computational experiments which show the efficiency of the suggested method. Finally, Section 5 concludes the paper and provides some perspectives.

2   Problem Statement and Definitions

Consider the linear programming problem with bounded variables presented in the following standard form:

$$\max z = c^T x,$$
$$\text{subject to } Ax = b, \; l \leq x \leq u, \tag{1}$$

where c and x are n-vectors; b an m-vector; A an $(m \times n)$-matrix with $rank A = m < n$; l and u are finite-valued n-vectors. The symbol $(^T)$ is the transposition operation. We define the following sets of indices:

$$I = \{1, 2, \ldots, m\}, \quad J = \{1, 2, \ldots, n\}, \quad J = J_B \cup J_N, \quad J_B \cap J_N = \emptyset, \quad |J_B| = m.$$

- If v is an arbitrary n-vector and M an arbitrary $(m \times n)$-matrix, then we can write

$$v = v(J) = (v_j, j \in J) \text{ and } M = M(I, J) = (m_{ij}, i \in I, \; j \in J).$$

Moreover, we can partition v and M as follows:

$$v^T = (v_B^T, v_N^T), \text{ where } v_B = v(J_B) = (v_j, j \in J_B), \; v_N = v(J_N) = (v_j, j \in J_N);$$

$$M = (M_B, M_N), \text{ where } M_B = M(I, J_B), \; M_N = M(I, J_N).$$

- A vector x verifying the constraints of problem (1) is called a *feasible solution*.
- A feasible solution x^0 is called *optimal* if

$$z(x^0) = c^T x^0 = \max \; c^T x,$$

where x is taken from the set of all feasible solutions of the problem (1).
- A feasible solution x^ϵ is said to be ϵ-*optimal* or *suboptimal* if

$$z(x^0) - z(x^\epsilon) = c^T x^0 - c^T x^\epsilon \leq \epsilon,$$

where x^0 is an optimal solution for the problem (1) and ϵ is a positive number chosen beforehand.

- We consider the index subset $J_B \subset J$ such that $|J_B| = |I| = m$. Then the set J_B is called a *support* if $\det(A_B) \neq 0$.
- The pair $\{x, J_B\}$ comprising a feasible solution x and a support J_B will be called a *support feasible solution* (SFS).
- An SFS is called *nondegenerate* if $l_j < x_j < u_j$, $j \in J_B$.
- We define the m-vector of multipliers π and the n-vector of reduced costs Δ as follows:

$$\pi^T = c_B^T A_B^{-1}, \quad \Delta^T = \pi^T A - c^T = (\Delta_B^T, \Delta_N^T),$$

where $\Delta_B^T = c_B^T A_B^{-1} A_B - c_B^T = 0$, $\Delta_N^T = c_B^T A_B^{-1} A_N - c_N^T$.

For an SFS $\{x, J_B\}$, we make the following partition: $J_N = J_N^{++} \cup J_N^{--} \cup J_N^0$, where

$$J_N^{++} = \{j \in J_N : \Delta_j > 0\}, J_N^{--} = \{j \in J_N : \Delta_j < 0\} \text{ and } J_N^0 = \{j \in J_N : \Delta_j = 0\}. \tag{2}$$

- The quantity $\beta(x, J_B)$ defined by

$$\beta = \beta(x, J_B) = \sum_{j \in J_N^{++}} \Delta_j (x_j - l_j) + \sum_{J_N^{--}} \Delta_j (x_j - u_j) \tag{3}$$

is called the *suboptimality estimate* [11]. Thus, we have the following results [11]:

Theorem 1. *(Sufficient condition for suboptimality). Let $\{x, J_B\}$ be an SFS for the problem (1) and ϵ an arbitrary positive number. If $\beta(x, J_B) \leq \epsilon$, then the feasible solution x is ϵ-optimal.*

Corollary 1. *Let $\{x, J_B\}$ be an SFS for the problem (1). The condition $\beta(x, J_B) = 0$ is sufficient for the optimality of the feasible solution x.*

3   An Iteration of the Adaptive Method with Hybrid Direction (AMHD)

Let $\{x, J_B\}$ be an SFS for the problem (1) and $\eta \in [0, 1]$. Let $x^+ \in \mathbb{R}_+^n$ and $x^- \in \mathbb{R}_-^n$ be two vectors defined as follows:

$$x^+ = \eta(x - l) \text{ and } x^- = \eta(x - u).$$

We introduce the following set of indices:

$$J_N^+ = \{j \in J_N : \Delta_j > x_j^+\}, \quad J_N^- = \{j \in J_N : \Delta_j < x_j^-\},$$

$$J_N^{P+} = \{j \in J_N : 0 < \Delta_j \leq x_j^+\}, \quad J_N^{P-} = \{j \in J_N : x_j^- \leq \Delta_j < 0\}, \tag{4}$$

$$J_N^P = \{j \in J_N : x_j^- \leq \Delta_j \leq x_j^+\} = J_N^{P+} \cup J_N^{P-} \cup J_N^0.$$

Thus,

$$J_N^{++} = J_N^+ \cup J_N^{P+}, \quad J_N^{--} = J_N^- \cup J_N^{P-}, \quad J_N = J_N^+ \cup J_N^- \cup J_N^P.$$

Let us define the nonnegative quantities $\gamma = \gamma(\eta, x, J_B)$ and μ as follows:

$$\gamma = \begin{cases} \frac{1}{\eta}\left[\sum_{j \in J_N^+} \Delta_j x_j^+ + \sum_{j \in J_N^-} \Delta_j x_j^- + \sum_{j \in J_N^{P+} \cup J_N^{P-}} \Delta_j^2\right], & \text{if } \eta > 0; \\ \beta(x, J_B), & \text{if } \eta = 0, \end{cases} \tag{5}$$

$$\mu = \begin{cases} \frac{1}{\eta}\left[\sum_{j \in J_N^{P+}} \Delta_j(x_j^+ - \Delta_j) + \sum_{j \in J_N^{P-}} \Delta_j(x_j^- - \Delta_j)\right], & \text{if } \eta > 0; \\ 0, & \text{if } \eta = 0. \end{cases} \tag{6}$$

The quantity $\gamma(\eta, x, J_B)$ is called the optimality estimate [1] and we recall that the SFS $\{x, J_B\}$ is optimal if $\gamma(\eta, x, J_B) = 0$.

Remark 1. When $\eta \longrightarrow 0$, we get $J_N^{P+} = J_N^{P-} = \emptyset$. Then $\lim_{\eta \longrightarrow 0} \mu = 0$.

Lemma 1. *For all $\eta \geq 0$, the optimality estimate can be written as follows:*

$$\gamma = \beta - \mu \leq \beta. \tag{7}$$

Proof. For $\eta = 0$, we have $\mu = 0$ and $\gamma = \beta$, so $\gamma = \beta - \mu$. For $\eta > 0$,

$$\begin{aligned} \beta = \beta(x, J_B) &= \sum_{j \in J_N^{++}} \Delta_j(x_j - l_j) + \sum_{j \in J_N^{--}} \Delta_j(x_j - u_j) \\ &= \frac{1}{\eta} \sum_{j \in J_N^+ \cup J_N^{P+}} \Delta_j x_j^+ + \frac{1}{\eta} \sum_{j \in J_N^- \cup J_N^{P-}} \Delta_j x_j^- \\ &= \frac{1}{\eta} \sum_{j \in J_N^+} \Delta_j x_j^+ + \frac{1}{\eta} \sum_{j \in J_N^-} \Delta_j x_j^- + \frac{1}{\eta} \sum_{j \in J_N^{P+}} \Delta_j x_j^+ + \frac{1}{\eta} \sum_{j \in J_N^{P-}} \Delta_j x_j^- \\ &= \gamma - \frac{1}{\eta} \sum_{j \in J_N^{P+} \cup J_N^{P-}} \Delta_j^2 + \frac{1}{\eta} \sum_{j \in J_N^{P+}} \Delta_j x_j^+ + \frac{1}{\eta} \sum_{j \in J_N^{P-}} \Delta_j x_j^- \\ &= \gamma + \frac{1}{\eta} \sum_{j \in J_N^{P+}} \Delta_j(x_j^+ - \Delta_j) + \frac{1}{\eta} \sum_{j \in J_N^{P-}} \Delta_j(x_j^- - \Delta_j) \\ &= \gamma + \mu. \end{aligned}$$

Since $\mu \geq 0$, we get $\gamma = \beta - \mu \leq \beta$.

3.1   Change of the Feasible Solution

Let $\{x, J_B\}$ be an SFS for the problem (1) and $\eta \in [0, 1]$. When $\gamma(\eta, x, J_B) > 0$, we define the feasible direction d as follows:

$$
\begin{aligned}
&d_j = l_j - x_j, \text{ if } j \in J_N^+;\ d_j = u_j - x_j, \text{ if } j \in J_N^-;\\
&d_j = \frac{-\Delta_j}{\eta}, \text{ if } j \in J_N^P \text{ and } \eta > 0;\\
&d_j = 0, \text{ if } j \in J_N^P = J_N^0 \text{ and } \eta = 0;\\
&d_B = -A_B^{-1} A_N d_N.
\end{aligned}
\tag{8}
$$

This direction, with respect to the standard direction of the adaptive method is called a hybrid direction. Contrarily to the direction used in the adaptive method, which takes only extreme or zero values, the hybrid direction takes extreme values for components with relatively big values of the reduced cost vector and it takes for others the reduced gradient values.

In order to improve the objective function while remaining in the feasible region, we compute the step length θ^0 along the direction d as follows: $\theta^0 = \min\{\theta_{j_1}, 1\}$, where

$$
\theta_{j_1} = \min\{\theta_j, j \in J_B\} \text{ and } \theta_j = \begin{cases} (u_j - x_j)/d_j, & \text{if } d_j > 0;\\ (l_j - x_j)/d_j, & \text{if } d_j < 0;\\ \infty, & \text{if } d_j = 0. \end{cases}
\tag{9}
$$

Then the new feasible solution is $\bar{x} = x + \theta^0 d$. Since $Ad = 0$, the vector d is a feasible direction. Furthermore, in [1], we proved that d is an ascent direction: the objective function increment is given by

$$
\bar{z} - z = \theta^0 \gamma(\eta, x, J_B) \geq 0,\ \theta^0 \in [0, 1].
\tag{10}
$$

By replacing the expression of $\gamma(\eta, x, J_B)$ in (10), we get

$$
\bar{z} - z = \theta^0 (\beta - \mu).
\tag{11}
$$

Lemma 2. *For all $\eta \geq 0$, the quantity $\bar{\beta} = \beta(\bar{x}, J_B)$ can be computed as follows:*

$$
\bar{\beta} = (1 - \theta^0)\beta + \theta^0 \mu \leq \beta.
\tag{12}
$$

Proof. For $\eta > 0$, we have

$$\bar{\beta} = \sum_{j \in J_N^{++}} \Delta_j(\bar{x}_j - l_j) + \sum_{j \in J_N^{--}} \Delta_j(\bar{x}_j - u_j)$$

$$= \sum_{j \in J_N^{++}} \Delta_j(x_j - l_j) + \sum_{j \in J_N^{--}} \Delta_j(x_j - u_j) + \theta^0 \sum_{j \in J_N^{++}} \Delta_j d_j + \theta^0 \sum_{j \in J_N^{--}} \Delta_j d_j$$

$$= \beta + \theta^0 \sum_{j \in J_N^+ \cup J_N^{P+}} \Delta_j d_j + \theta^0 \sum_{j \in J_N^- \cup J_N^{P-}} \Delta_j d_j$$

$$= \beta + \theta^0 \sum_{j \in J_N^+} \Delta_j d_j + \theta^0 \sum_{j \in J_N^-} \Delta_j d_j + \theta^0 \sum_{j \in J_N^{P+}} \Delta_j d_j + \theta^0 \sum_{j \in J_N^{P-}} \Delta_j d_j$$

$$= \beta - \theta^0 \sum_{j \in J_N^+} \Delta_j(x_j - l_j) - \theta^0 \sum_{j \in J_N^-} \Delta_j(x_j - u_j) - \frac{\theta^0}{\eta} \sum_{j \in J_N^{P+} \cup J_N^{P-}} \Delta_j^2$$

$$= \beta - \frac{\theta^0}{\eta} \sum_{j \in J_N^+} \Delta_j x_j^+ - \frac{\theta^0}{\eta} \sum_{j \in J_N^-} \Delta_j x_j^- - \frac{\theta^0}{\eta} \sum_{j \in J_N^{P+} \cup J_N^{P-}} \Delta_j^2$$

$$= \beta - \theta^0 \gamma \le \beta.$$

Since for $\eta \ge 0$, $\gamma = \beta - \mu$, then we find

$$\bar{\beta} = (1 - \theta^0)\beta + \theta^0 \mu.$$

For $\eta = 0$, we have $\mu = 0$. Hence, we find the classical updating formula of $\beta(x, J_B)$ [11]: $\bar{\beta} = \beta(\bar{x}, J_B) = (1 - \theta^0)\beta(x, J_B)$.

Theorem 2. *(Sufficient conditions for optimality and suboptimality of $\bar{x}$)*
If $\theta^0 = 1$ and $\mu = 0$, then the feasible solution $\bar{x}$ is optimal.
If $\theta^0 = 1$ and $\mu \le \epsilon$, then the feasible solution $\bar{x}$ is ϵ-optimal.

Proof. We assume that $\theta^0 = 1$ and $\mu = 0$. Following Lemma 2, we have $\bar{\beta} = 0$. By using Corollary 1, we deduce the optimality of $\bar{x}$. If $\theta^0 = 1$ and $\mu \le \epsilon$, then Lemma 2 yields $\bar{\beta} = \mu \le \epsilon$. By using Theorem 1, we deduce the suboptimality of $\bar{x}$.

If $\theta^0 = \theta_{j_1} < 1$ and $\bar{\beta} > \epsilon$, then we change the support J_B.

3.2   Change of the Support

Short Step Rule. We define the n-vector κ and the real number α_0 as follows:

$$\kappa = x + d \text{ and } \alpha_0 = \kappa_{j_1} - \bar{x}_{j_1},$$

where j_1 is the leaving index computed in (9). So the dual direction is

$$t_{j_1} = -sign(\alpha_0); \ t_j = 0, \ j \ne j_1, \ j \in J_B; \ t_N^T = t_B^T A_B^{-1} A_N. \tag{13}$$

Remark 2. We have $\alpha_0 = \kappa_{j_1} - \bar{x}_{j_1} = x_{j_1} + d_{j_1} - x_{j_1} - \theta^0 d_{j_1} = (1 - \theta^0)d_{j_1}$. Since $0 \leq \theta^0 < 1$, then $t_{j_1} = -sign(\alpha_0) = -sign(d_{j_1})$.

Remark 3. The dual direction t and the primal direction d are orthogonal. Indeed,

$$t^T d = t_N^T d_N + t_B^T d_B = (t_B^T A_B^{-1} A_N)d_N + t_B^T(-A_B^{-1} A_N d_N) = 0.$$

Let us define the following sets:

$$J_N^{0^+} = \{j \in J_N^0 : t_j > 0\} \text{ and } J_N^{0^-} = \{j \in J_N^0 : t_j < 0\}, \tag{14}$$

and the quantity:

$$\alpha = -|\alpha_0| + \sum_{j \in J_N^{0^+} \cup J_N^{P^+}} t_j(\kappa_j - l_j) + \sum_{j \in J_N^{0^-} \cup J_N^{P^-}} t_j(\kappa_j - u_j). \tag{15}$$

The new reduced cost vector and the new support are computed as follows:

$$\bar{\Delta} = \Delta + \sigma^0 t \text{ and } \bar{J}_B = (J_B \setminus \{j_1\}) \cup \{j_0\},$$

where

$$\sigma^0 = \sigma_{j_0} = \min_{j \in J_N} \{\sigma_j\}, \text{ with } \sigma_j = \begin{cases} \frac{-\Delta_j}{t_j}, & \text{if } \Delta_j t_j < 0; \\ 0, & \text{if } j \in J_N^{0^-} \text{ and } \kappa_j \neq u_j; \\ 0, & \text{if } j \in J_N^{0^+} \text{ and } \kappa_j \neq l_j; \\ \infty, & \text{otherwise.} \end{cases} \tag{16}$$

Remark 4. If $\sigma^0 = \infty$, then the problem (1) is infeasible.

We assume that $\sigma^0 < \infty$. The suboptimality estimate corresponding to the new feasible solution and the new support is given by

$$\bar{\bar{\beta}} = \beta(\bar{x}, \bar{J}_B) = \sum_{j \in \bar{J}_N, \bar{\Delta}_j > 0} \bar{\Delta}_j(\bar{x}_j - l_j) + \sum_{j \in \bar{J}_N, \bar{\Delta}_j < 0} \bar{\Delta}_j(\bar{x}_j - u_j).$$

Lemma 3. *The suboptimality estimate $\beta(\bar{x}, \bar{J}_B)$ can be written as follows:*

$$\beta(\bar{x}, \bar{J}_B) = \beta(\bar{x}, J_B) + \sigma^0 \alpha. \tag{17}$$

Remark 5. If $\eta = 0$ and $J_N^P = J_N^0 = \emptyset$, then $\alpha = -|\alpha_0|$. So we find the classical updating formula of $\beta(\bar{x}, J_B)$ [11]: $\beta(\bar{x}, \bar{J}_B) = \beta(\bar{x}, J_B) - \sigma^0 |\alpha_0|$.

If $\alpha \leq 0$, then $\bar{\bar{\beta}} \leq \bar{\beta}$. However, when $\alpha > 0$ we update appropriately the value of the parameter η by applying the following procedure:

Procedure 1 *(Procedure of updating the value of η)*

(1) *Compute* $\bar{x}^+ = \eta(\bar{x} - l)$, $\bar{J}_N^{P^+} = \{j \in J_N : 0 < \Delta_j \leq \bar{x}_j^+\}$;
(2) *compute* $\bar{x}^- = \eta(\bar{x} - u)$, $\bar{J}_N^{P^-} = \{j \in J_N : \bar{x}_j^- \leq \Delta_j < 0\}$;
(3) *if* $\bar{J}_N^{P^+} \cup \bar{J}_N^{P^-} = \emptyset$, *then set* $\bar{\eta} = \eta$;
  else compute $\eta_0 = \min_{j \in \bar{J}_N^{P^+}} \frac{\Delta_j}{\bar{x}_j - l_j}$, $\eta_1 = \min_{j \in \bar{J}_N^{P^-}} \frac{\Delta_j}{\bar{x}_j - u_j}$;
(4) *if* $\bar{J}_N^{P^+} \neq \emptyset$ *and* $\bar{J}_N^{P^-} = \emptyset$, *then set* $\bar{\eta} = \eta_0$;
(5) *if* $\bar{J}_N^{P^+} = \emptyset$ *and* $\bar{J}_N^{P^-} \neq \emptyset$, *then set* $\bar{\eta} = \eta_1$;
(6) *if* $\bar{J}_N^{P^+} \neq \emptyset$ *and* $\bar{J}_N^{P^-} \neq \emptyset$, *then set* $\bar{\eta} = \min\{\eta_0, \eta_1\}$.

Proposition 1. *After applying Procedure 1, we get a new value $\bar{\eta}$ which satisfies $\bar{\eta} \leq \eta$.*

Proof. We prove that if $\bar{J}_N^{P^+} \cup \bar{J}_N^{P^-} \neq \emptyset$, then $\bar{\eta} \leq \eta$. Indeed, three cases can occur:
Case 1: if $\bar{J}_N^{P^+} \neq \emptyset$ and $\bar{J}_N^{P^-} = \emptyset$, then

$$\exists j \in \bar{J}_N^{P^+} : 0 < \Delta_j \leq \eta(\bar{x}_j - l_j) \Rightarrow \eta \geq \frac{\Delta_j}{\bar{x}_j - l_j} \geq \min_{j \in \bar{J}_N^{P^+}} \frac{\Delta_j}{\bar{x}_j - l_j} = \eta_0 = \bar{\eta}.$$

Case 2: if $\bar{J}_N^{P^+} = \emptyset$ and $\bar{J}_N^{P^-} \neq \emptyset$, then

$$\exists j \in \bar{J}_N^{P^-} : \eta(\bar{x}_j - u_j) \leq \Delta_j < 0 \Rightarrow \eta \geq \frac{\Delta_j}{\bar{x}_j - u_j} \geq \min_{j \in \bar{J}_N^{P^-}} \frac{\Delta_j}{\bar{x}_j - u_j} = \eta_1 = \bar{\eta}.$$

Case 3: if $\bar{J}_N^{P^+} \neq \emptyset$ and $\bar{J}_N^{P^-} \neq \emptyset$, then

$$\exists j \in \bar{J}_N^{P^+} : 0 < \Delta_j \leq \eta(\bar{x}_j - l_j) \Rightarrow \eta \geq \frac{\Delta_j}{\bar{x}_j - l_j} \geq \min_{j \in \bar{J}_N^{P^+}} \frac{\Delta_j}{\bar{x}_j - l_j} = \eta_0 \geq \bar{\eta},$$

and

$$\exists j \in \bar{J}_N^{P^-} : \eta(\bar{x}_j - u_j) \leq \Delta_j < 0 \Rightarrow \eta \geq \frac{\Delta_j}{\bar{x}_j - u_j} \geq \min_{j \in \bar{J}_N^{P^-}} \frac{\Delta_j}{\bar{x}_j - u_j} = \eta_1 \geq \bar{\eta}.$$

Therefore, in all the cases we have proved that $\bar{\eta} \leq \eta$.

Long Step Rule. In this paper, thanks to the updating formula of the suboptimality estimate, we can modify the procedure presented in [12], called long step rule, in order to compute the dual step length in our method. This procedure is described in the following steps:

Procedure 2 *(Long step rule procedure for AMHD)*

(1) *Compute* $\sigma_j, j \in J_N$, $\sigma^0 = \min_{j \in J_N} \sigma_j$ *and* j_0 *with (16)*;
(2) *if* $\sigma^0 = \infty$, *then the problem (1) is infeasible*;

(3) *sort the indices* $\{i \in J_N : \sigma_i \neq \infty\}$ *by the increasing order of the numbers* σ_i:

$$\sigma_{i_1} \leq \sigma_{i_2} \leq \cdots \leq \sigma_{i_p}; \ i_k \in J_N, \ \sigma_{i_k} \neq \infty, \ k = 1, \ldots, p;$$

(4) *if* $p = 1$, *then* $j_0 = i_1$; *go to step (8)*;
(5) *for all* $i_k, k = 1, \ldots, p$, *compute* $\Delta V_{i_k} = |t_{i_k}|(u_{i_k} - l_{i_k})$;
(6) *compute* V_{i_k}, $k = 0, \ldots, p$, *where*

$$\begin{cases} V_{i_0} = V_0 = \alpha, \\ V_{i_k} = V_0 + \sum_{s=1}^{k} \Delta V_{i_s} = V_{i_{k-1}} + \Delta V_{i_k}, \ k = 1, \ldots, p; \end{cases}$$

(7) *choose the index* $j_0 = i_q$ *such that* $V_{i_{q-1}} < 0$ *and* $V_{i_q} \geq 0$;
(8) *let* j_1 *be the index computed by (9). Set* $\bar{J}_B = (J_B \setminus \{j_1\}) \cup \{j_0\}$, $\sigma^0 = \sigma_{j_0}$
  and $\bar{\Delta} = \Delta + \sigma^0 t$;
(9) *compute* $\beta(\bar{x}, \bar{J}_B) = \beta(\bar{x}, J_B) + \sum_{k=1}^{q}(\sigma_{i_k} - \sigma_{i_{k-1}})V_{i_{k-1}}$, *with* $\sigma_{i_0} = 0$.

3.3 Scheme of the Algorithm

Let $\{x, J_B\}$ be an initial SFS for the problem (1), ϵ be a nonnegative number and $\eta \in [0, 1]$. The scheme of the adaptive method with hybrid direction and long step rule (AMHDLS) is described in the following steps:

Algorithm 1 *(AMHDLS for solving linear programs with bounded variables)*

(1) *compute* $\pi^T = c_B^T A_B^{-1}$, $\Delta_N^T = \pi^T A_N - c_N^T$;
(2) *compute the suboptimality estimate* β *with (3)*;
(3) *if* $\beta = 0$, *then the algorithm stops with the optimal SFS* $\{x, J_B\}$;
(4) *if* $\beta \leq \epsilon$, *then the algorithm stops with the ϵ-optimal SFS* $\{x, J_B\}$;
(5) *compute the vectors* $x^+ = \eta(x - l)$ *and* $x^- = \eta(x - u)$;
(6) *compute the sets* J_N^+, J_N^-, J_N^{P+} *and* J_N^{P-} *with (4)*;
(7) *compute* μ *with (6)*;
(8) *compute the primal search direction* d *with (8)*;
(9) *compute the primal step length* θ^0 *with (9)*;
(10) *compute* $\bar{x} = x + \theta^0 d$ *and* $\bar{z} = z + \theta^0(\beta - \mu)$;
(11) *if* $\theta^0 = 1$, *then*
   (11.1) *if* $\mu = 0$, *then* $\bar{x}$ *is optimal. Stop*;
   (11.2) *if* $\mu \leq \epsilon$, *then* $\bar{x}$ *is ϵ-optimal. Stop*;
   (11.3) *else, compute the value* $\bar{\eta}$ *with Procedure 1; set* $\eta = \bar{\eta}$, $x = \bar{x}$, $z = \bar{z}$,
      $\beta = \mu$ *and go to step (5)*;
(12) *if* $\theta^0 = \theta_{j_1} < 1$, *then compute* $\bar{\beta} = (1 - \theta^0)\beta + \theta^0 \mu$; *if* $\bar{\beta} \leq \epsilon$, *then the*
   algorithm stops with the ϵ-optimal SFS $\{\bar{x}, J_B\}$;
(13) *compute* $\kappa = x + d$ *and* $\alpha_0 = \kappa_{j_1} - \bar{x}_{j_1}$;
(14) *compute the dual direction* t *with (13)*;
(15) *compute* J_N^0 *with (2)*, J_N^{0+} *and* J_N^{0-} *with (14)*;
(16) *compute* α *with (15); if* $\alpha > 0$ *then*
   (16.1) *if* $J_N^0 \neq \emptyset$, *then choose an index* $j_0 \in J_N^0$ *with* $t_{j_0} \neq 0$, *set* $\bar{\Delta} = \Delta$,
      $\bar{J}_B = (J_B \setminus \{j_1\}) \cup \{j_0\}$, $\bar{\beta} = \bar{\beta}$ *and go to step (18)*;

(16.2) *if $J_N^0 = \emptyset$, then compute the value $\bar{\eta}$ with Procedure 1; set $\eta = \bar{\eta}$,*
   $x = \bar{x}$, $z = \bar{z}$, $\beta = \bar{\beta}$ and go to step (5);
(17) *compute the dual step length σ^0, the new reduced costs vector $\bar{\Delta}$, the new support $\bar{J}_B$ and the new suboptimality estimate $\bar{\bar{\beta}}$ with the long step rule (Procedure 2);*
(18) *set $x = \bar{x}$, $J_B = \bar{J}_B$, $J_N = \bar{J}_N$, $z = \bar{z}$, $\Delta = \bar{\Delta}$, $\beta = \bar{\bar{\beta}}$, go to step (3).*

4   Experimental Results

In order to compare the Adaptive Method with Hybrid Direction and Long Step Rule (AMHDLS) with the Adaptive Method with Hybrid Direction and Short Step Rule suggested in [1] (AMHDSS), and the Primal Simplex Algorithm with Dantzig's rule (PSA), we have developed an implementation under the MATLAB programming language. The implementation details of the three algorithms and the randomly generated test LP problems are similar to those presented in [1]. The test problems have constraints matrix A of size $n \times n$, where $n \in \{100, 200, 300, 400, 500, 600, 700, 800, 900, 1000\}$ and each size contains ten generated test problems.

We have solved the randomly generated set of test problems with the three considered algorithms on a computer with Intel(R) Core(TM) i5 CPU M560 @ 2.67GHz machine with 4 GB of RAM, working under the Windows 7 operating system. We have initialized AMHDLS, AMHDSS and PSA with the same pair $\{x, J_B\}$, where $x = (l; b - Al)$ and $J_B = \{p+1, p+2, \ldots, p+m\}$. Moreover, we have set $\eta = 1$ in AMHDLS and AMHDSS.

Numerical results are reported in Table 1, where "CPU" and "Niters" represent respectively the mean CPU time in seconds and the average number of iterations of the ten problems generated for each problem size. We plot the CPU time and the number of iterations for the three algorithms. The graphs are shown in Figure 1.

Table 1. CPU time and number of iterations for AMHDLS, AMHDSS and PSA

	AMHDLS		AMHDSS		PSA	
Size	CPU	Niters	CPU	Niters	CPU	Niters
100×100	0.06	61.80	0.03	68.20	0.05	66.90
200×200	0.34	257.60	0.23	341.10	0.55	588.40
300×300	0.93	620.90	0.95	987.10	5.69	3094.50
400×400	2.53	991.10	2.46	1695.40	20.13	5991.40
500×500	5.57	1444.90	5.80	2606.50	65.27	10355.40
600×600	12.75	1917.90	13.85	3679.60	168.76	14492.70
700×700	24.37	2417.80	26.27	4708.30	378.29	20424.90
800×800	46.18	2965.30	53.30	6055.80	738.57	25756.90
900×900	74.58	3532.60	84.09	7146.30	1381.86	34083.70
1000×1000	122.60	4105.10	134.39	8375.10	2347.92	41118.70
Mean	**28.99**	**1831.50**	**32.14**	**3566.34**	**510.71**	**15597.35**

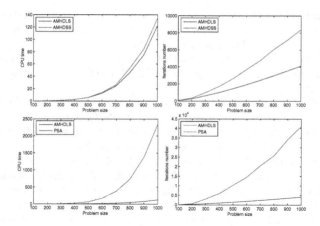

Fig. 1. Graphs of CPU time and iterations number of the three algorithms

We remark that AMHDLS outperforms AMHDSS and PSA particularly in solving LP problems with higher dimensions. So we expect a good performance of our method in solving practical problems. However, a robust implementation and further numerical experiments are needed to prove its efficiency on solving real benchmarks LP problems.

5 Conclusion

In addition to [1,3,4,7,8] where we use the optimality estimate to test the optimality of the current feasible solution, in this work the conditions used to characterize the optimality are based on the suboptimality estimate. A general formula for updating the suboptimality estimate is derived. A long step rule is used for changing the support in our method and a procedure which updates, when necessary, the parameter η is suggested. Hence, an algorithm called the adaptive method with hybrid direction and long step rule is described. Finally, numerical experiments have shown the efficiency of our method in solving randomly generated test problems. In future work, we will apply some crash procedure like that presented in [2] in order to initialize AMHDLS with a good initial SFS, then we will test its performance on solving practical LP test problems.

References

1. Bibi, M.O., Bentobache, M.: A hybrid direction algorithm for solving linear programs. International Journal of Computer Mathematics 92(1), 201–216 (2015)
2. Bentobache, M., Bibi, M.O.: A Two-phase Support Method for Solving Linear Programs: Numerical Experiments, Mathematical Problems in Engineering 2012, Article ID 482193, 28 pages, doi:10.1155/2012/482193
3. Bentobache, M.: On mathematical methods of linear and quadratic programming, Ph.D. diss., University of Bejaia (2013)

4. Bentobache, M., Bibi, M.O.: Adaptive method with hybrid direction: theory and numerical experiments. In: Proceedings of Optimization 2011, Universidade Nova de Lisboa, Portugal, July 24-27, p. 112 (2011)
5. Bentobache, M., Bibi, M.O.: The adaptive method with hybrid direction for solving linear programs. In: Proceedings of COSI 2013, CDTA, Algiers, Algeria, Juin 09-11, pp. 277–288 (2013)
6. Bibi, M.O.: Support method for solving a linear-quadratic problem with polyhedral constraints on control. Optimization 37, 139–147 (1996)
7. Bibi, M.O., Bentobache, M.: The adaptive method with hybrid direction for solving linear programming problems with bounded variables. In: Proceedings of COSI 2011, University of Guelma, Algeria, April 24-27, pp. 80–91 (2011)
8. Bibi, M.O., Bentobache, M.: A hybrid direction algorithm for solving linear programs. In: Proceedings of DIMACOS 2011, University of Mohammedia, Morocco, May 5-8, pp. 28–30 (2011)
9. Brahmi, B., Bibi, M.O.: Dual support method for solving convex quadratic programs. Optimization 59, 851–872 (2010)
10. Dantzig, G.B.: Linear Programming and Extensions. Princeton University Press, Princeton (1963)
11. Gabasov, R., Kirillova, F.M.: Methods of linear programming, Vol. 1, 2 and 3, Edition of the Minsk University (1977, 1978 and 1980) (in Russian)
12. Gabasov, R., Kirillova, F.M., Prischepova, S.V.: Optimal Feedback Control. Springer, London (1995)
13. Kostina, E.A., Kostyukova, O.I.: An algorithm for solving quadratic programming problems with linear equality and inequality constraints. Computational Mathematics and Mathematical Physics 41, 960–973 (2001)
14. Kostina, E.: The long step rule in the bounded-variable dual simplex method: Numerical experiments. Mathematical Methods of Operations Research 55, 413–429 (2002)

A Monotone Newton-Like Method for the Computation of Fixed Points

Lotfi Mouhadjer and Boubakeur Benahmed

[1] Département de Mathématiques, EPST de Tlemcen,
BP 165 Bel Horizon Tlemcen, Algérie
lotfi.mouhadjer@gmail.com
[2] Département de Mathématiques et Informatique, ENP d'Oran,
BP 1523 El m'naouer, Algérie
benahmed_b@yahoo.fr

Abstract. In this paper we introduce a monotone Newton-like method for the computations of fixed points of a class of nonlinear operators in ordered Banach spaces. We use a Lakshmikantham's fixed point theorem [4, Theorem 1.2] and the classical Banach fixed point theorem to prove the convergence of this method. We prove also that under a suitable condition, the rate of convergence of the proposed method is superlinear. As an application we consider a class of nonlinear matrix equations.

1 Introduction

Let E be a Banach space, $\mathcal{L}(E)$ denote the space of bounded linear operators on E. Let D be a non empty closed subset of E and $T : D \to D$ a contractive operator. The classical Banach fixed point theorem assure that T has a unique fixed point $w^* = Tw^*$ in D and assure the convergence to w^* of the sequence $(w_n)_{n\geq 0}$ defined by

$$w_0 \in D \quad , \quad w_{n+1} = Tw_n \tag{1.1}$$

The convergence of $(w_n)_{n\geq 0}$ may be quite slow however if the constant of contraction of T is close to 1; more rapid convergence can be achieved, for example by application of Newton method to the nonlinear equation

$$x - Tx = 0$$

This procedure lead to consider the sequense $(x_n)_{n\geq 0}$ defined by (assume that T is Fréchet derivable on D)

$$x_0 \in D \quad , \quad x_{n+1} = x_n - [I - T'(x_n)]^{-1}(x_n - Tx_n) \tag{1.2}$$

The calculation of the approximation x_{n+1} by scheme (1.2) require the knowledge of $[I - T'(u)]^{-1}$ for every $u \in D$, or equivalently solve at every iteration the equation

$$[I - T'(x_n)]y_n = x_n - Tx_n$$

where y_n is the unknown.

© Springer International Publishing Switzerland 2015

H.A. Le Thi et al. (eds.), *Model. Comput. & Optim. in Inf. Syst. & Manage. Sci.*,
Advances in Intelligent Systems and Computing 359, DOI: 10.1007/978-3-319-18161-5_29

To avoid this situation we propose to replace $[I - T'(u)]^{-1}$ by an approximated operator. Since $\|T'(u)\|_{\mathcal{L}(E)} \leq q < 1$ for every $u \in D$, then we have

$$[I - T'(u)]^{-1} = \sum_{k=0}^{\infty} T'^{(k)}(u) \qquad (1.3)$$

where $T'^{(0)}(.) = I$, and $T'^{(k)}(.) = T'(.) \circ T'^{(k-1)}(.)$ for $k \geq 1$

Consider now for any nondecreasing integer sequence $(a_n)_{n \geq 0}$ the following operator

$$S_{a_n}(.) = \sum_{k=0}^{a_n} T'^{(k)}(.) \qquad (1.4)$$

The idea is to replace the convergent series (1.3) by its partial sum $S_{a_n}(u)$. This idea lead to a Newton-like Method for computing fixed point of T, defined by the sequence $(x_n)_{\geq 0}$ such that $x_0 \in D$ and

$$x_{n+1} = x_n - S_{a_n}(x_n)(x_n - Tx_n) \qquad (1.5)$$

We remark that if $a_n = 0$ then $x_{n+1} = Tx_n$. This means that the basic fixed point method (1.1) is particular case of the proposed Newton-like method (1.5). In this paper, using a Lakshmikantham's fixed point theorem and the classical Banach fixed point theorem, we prove that in ordered Banach space setting and under some hypotheses on T, the sequence $(x_n)_{n \geq 0}$ defined by relation (1.5) is monotone and convergent to fixed point of T with superlinear rate.

2   Preliminaries

Thoughout this paper $E := (E, \|.\|)$ Stands for real Banach space. We denote the zero element of E by 0 as in $\mathbb{R}$. Let $K \neq \{0\}$ be always a closed nonempty subset of E. K is called cone if $ax + by \in K$ for all $x, y \in K$ and nonnegative real numbers a, b where $K \cap (-K) = \{0\}$. For a given cone K one can define a partial ordering (denoted by $\preceq$) with respect to K by : $x \preceq y$ (or $y \succeq x$) if and only if $y - x \in K$. The notation $x \prec y$ means that $x \preceq y$ and $x \neq y$
The cone K is called:

1. *Normal* if there exists a positive constant δ such that for every $(x, y) \in E \times E$, we have:

$$0 \preceq x \preceq y \Rightarrow \|x\| \leq \delta \|y\| \qquad (2.1)$$

   The least positive number satisfying (2.1) is called the normal constant of K. If $\delta = 1$; then the norm $\|.\|$ is called a monotonic norm.

2. *Regular* if any increasing sequence and bounded from above is convergent. That is if $\{x_n\}_{n \geq 1}$ is a sequence such that $x_1 \preceq x_2 \preceq ... \preceq y$ for some $y \in E$ then there is $x \in E$ such that $\lim_{n \to \infty} \|x_n - x\| = 0$

It is well known that any regular cone is normal and the converse is true if the Banach space E is reflexive. See [1,2,3] for more details. If u, v are two elements of an ordered Banach space, such that $u \preceq v$, then the order interval $[u, v]$ is defined as

$$[u, v] = \{z \in E : u \preceq z \preceq v\}$$

The following result is proved in [1]

Proposition 1. *Let (E, K) be an ordered Banach space by a cone K The following statements are equivalent*

1. *The cone K is normal*
2. *Any decreasing sequence and bounded from below is convergent*
3. *Any order interval $[u, v]$ is bounded*
4. $x_n \preceq y_n \preceq z_n$, $n \geq 1$, $\|x_n - l\| \to 0$, $\|z_n - l\| \to 0$, then

$$\|y_n - l\| \to 0$$

An operator $S : D \subset E \to E$ defined on an ordered Banach space (E, K) is called:

1. *Positive if $S(D \cap K) \subset K$.*
2. *Increasing if for every $(x, y) \in D \times D$, such that $x \preceq y$, we have $S(x) \preceq S(y)$*

3   Main Result

Let (E, K) be an ordered Banach space by a regular cone K. Let $T : E \to E$ be a nonlinear operator. Asume the following hypotheses :

(H1) There exists x_0 , $y_0 \in E$ such that $x_0 \prec y_0$ and $x_0 \preceq Tx_0$, $Ty_0 \preceq y_0$.
(H2) The Fréchet derivative $T'(u)$ of T exists for every $u \in [x_0, y_0]$ and the mapping $u \mapsto T'(u)h$ is increasing for every $h \in K$, $u \in [x_0, y_0]$.
(H3) $T'(u)$ is positive operator for each $u \in [x_0, y_0]$ and there exist $0 < q < 1$ such that $\|T'(u)\|_{\mathcal{L}(E)} \leq q$ for every $u \in [x_0, y_0]$.
(H4) The mapping $u \in [x_0, y_0] \mapsto T'(u)$ is η- lipschitzian

It is clear that if hypothesis $(H3)$ is satisfied, then hypothesis (iii) of theorem [4, Theorem 1.2] is fulfilled. This is a direct consequence of Neumann series and positivity of the operator $T'(u)$. Moreover in this case the uniqueness of fixed points is assured.
We define the operator $S_{a_n} : [x_0, y_0] \to \mathcal{L}(E)$ by

$$S_{a_n}(u) = \sum_{k=0}^{a_n} T'^{(k)}(u) \tag{3.1}$$

where $(a_n)_{n \geq 0}$ is any nondecreasing integer sequence, and

$$T'^{(0)}(u) = I \quad , \quad T'^{(k+1)}(u) = T'(u) \circ T'^{(k)}(u)$$

It is clear that if the hypothesis $(H3)$ is satisfied, then $S_{a_n}(u)$ is positive operator for every $u \in [x_0, y_0]$ and

$$S_{a_n}(u)h \preceq [I - T'(u)]^{-1}h \quad \text{for every} \quad h \in K$$

Now we present the main result

Theorem 1. *Let (E, K) be an ordered Banach space by a regular cone K. Let $T : E \to E$ be a nonlinear operator satisfying the hypotheses $(H1) - (H3)$. Consider the Newton-like method defined by the sequence $(x_n)_{n \geq 0}$ such that*

$$x_{n+1} = x_n - S_{a_n}(x_n)(x_n - Tx_n) \tag{3.2}$$

Then $(x_n)_{n \geq 0}$ is an increasing sequence wich converge to the unique fixed point $u^* \in [x_0, y_0]$ of T. Moreover if the hypothesis $(H4)$ is satisfied, then the rate of convergence of $(x_n)_{n \geq 0}$ is

1. *Linear if $(a_n)_{n \geq 0}$ is a constant sequence.*
2. *Superlinear if $a_n \to +\infty$ and $\eta < 1$.*

Remark 1. If we choose $a_n = 0$, then $S_{a_n} = S_0 = I$, and consequently $x_{n+1} = Tx_n$. So the basic fixed point method is a particular case of a monotone Newton-like method defined by relation (3.2).

To prove Theorem 1 we need the following important two Lemmas

Lemma 1. *[4] If the hypothesis $(H2)$ is satisfied then*

$$Tx - Ty \preceq T'(x)(x - y) \quad and \tag{3.3}$$

$$Ty - Tx \preceq T'(y)(y - x) \tag{3.4}$$

whenever $x_0 \preceq x \prec y \preceq y_0$

Lemma 2. *Consider the sequence $(x_n)_{n \geq 0}$ defined in Theorem 1. Then for all $n \in \mathbb{N}$ we have*

$$x_n \preceq Tx_n$$

Proof. We use the induction principle. By hypothesis $(H1)$ we have $x_0 \preceq Tx_0$. Now assume that for some k, $x_k \preceq Tx_k$ and we prove that $x_{k+1} \preceq Tx_{k+1}$. To show this we prove at first that

$$x_{k+1} \preceq Tx_k + T'(x_k)(x_{k+1} - x_k) \tag{3.5}$$

In fact by definition of the sequence $(x_n)_{n \geq 0}$ we have

$$x_{k+1} - x_k = S_{a_k}(x_k)(Tx_k - x_k)$$

Let $z_k = Tx_k - x_k$, $R_{a_n} := \sum_{k=a_n+1}^{\infty} T'^{(k)}(x_k)$. Then

$$S_{a_k}(x_k) + R_{a_k} = \sum_{k=0}^{\infty} T'^{(k)}(x_k) = [I - T'(x_k)]^{-1}$$

So

$$S_{a_k}(x_k) = [I - T'(x_k)]^{-1} - R_{a_k}$$

Hence the relation (3.5) became

$$x_{k+1} - x_k = [I - T'(x_k)]^{-1} z_k - R_{a_k} z_k$$

So

$$[I - T'(x_k)](x_{k+1} - x_k) = z_k - [I - T'(x_k)] \circ R_{a_k} z_k$$

Thus

$$x_{k+1} = Tx_k + T'(x_k)(x_{k+1} - x_k) - r_k$$

where $r_k = [I - T'(x_k)] \circ R_{a_k} z_k$. Next we show that $r_k \succeq 0$. In fact

$$r_k = R_{a_k} z_k - T'(x_k) \circ R_{a_k} z_k$$

So

$$r_k = \sum_{j=a_k+1}^{\infty} T'^{(j)}(x_k) z_k - \sum_{j=a_k+1}^{\infty} T'^{(j+1)}(x_k) z_k$$

Hence

$$r_k = T'^{(a_k+1)}(x_k) z_k$$

Since $T'^{(a_k+1)}(x_k)$ is positive linear operator and $z_k \succeq 0$ we have $r_k \succeq 0$. So the relation (3.5) is true. Now using Lemma 1 we obtain

$$x_{k+1} - Tx_{k+1} \preceq Tx_k - Tx_{k+1} + T'(x_k)(x_{k+1} - x_k)$$
$$\preceq T'(x_k)(x_k - x_{k+1}) - T'(x_k)(x_{k+1} - x_k) = 0$$

Consequently $x_n \preceq Tx_n$ for all $n \in \mathbb{N}$.

Remark 2. The sequence $(x_n)_{n \geq 0}$ defined in Theorem 1 is increasing. In fact, since $S_{a_n}(.)$ is positive linear operator, by Lemma 2 we have for every $n \in \mathbb{N}$

$$x_{n+1} - x_n = S_{a_n}(x_n)(Tx_n - x_n) \succeq 0$$

Proof of Theorem 1.
Part one : Existence and uniqueness of fixed point of T
First we show that T has a unique fixed point in $[x_0, y_0]$. Since $T'(u)$ is positive for every $u \in [x_0, y_0]$, the relation (3.3) implies that

$$Tx \preceq Ty \quad whenever \quad x_0 \preceq x \prec y \preceq y_0$$

Then for every $z \in [x_0, y_0]$ we have

$$x_0 \preceq Tx_0 \preceq Tz \preceq Ty_0 \preceq y_0$$

So $Tz \in [x_0, y_0]$ or equivalently $T([x_0, y_0]) \subset [x_0, y_0]$. Next let x, y be in $[x_0, y_0]$, then $x + t(y - x) \in [x_0, y_0]$ for all $t \in [0, 1]$. This is due to convexity of $[x_0, y_0]$. By the mean value Theorem we get

$$\|T(y) - T(x)\| \leq \sup_{0 \leq t \leq 1} \|T'(x + t(y - x))\|_{\mathcal{L}(E)} \|y - x\| \leq q \|y - x\|$$

Hence T is contraction mapping on $[x_0, y_0]$. So by the Banach fixed point theorem, T has a unique fixed point $u^* \in [x_0, y_0]$. Moreover the sequence $(w_n)_{n \geq 0}$ such that

$$w_0 = x_0 \quad w_{n+1} = Tw_n$$

converge to u^*.

Part two : Convergence of $(x_n)_{n \geq 0}$ **to** u^*

Let $(u_n)_{n \geq 0}$, $(w_n)_{n \geq 0}$ be two sequences defined by : $u_0 = w_0 = x_0$ and for every $n \in \mathbb{N}$

$$u_{n+1} = u_n - [I - T'(u_n)]^{-1}(u_n - Tu_n) \quad \text{and} \quad w_{n+1} = Tw_n$$

It is easy to show that for every $n \in \mathbb{N}$

$$u_{n+1} = Tu_n - T'(u_n)(u_{n+1} - u_n)$$

Then by the Lakshmikantham Theorem [4, Theorem 1.2], we conclude that the sequence $(u_n)_{n \geq 0}$ converge to the unique fixed point u^* of T. Now we will show that for every $n \in \mathbb{N}$

$$w_n \preceq x_n \preceq u_n \tag{3.6}$$

For this purpose we use the induction. For $n = 0$, relation (3.6) is true. Next assume that for $k \in \mathbb{N}$ $w_k \preceq x_k \preceq u_k$. Then we have

$$x_{k+1} - w_{k+1} = x_k - S_{a_k}(x_k - Tx_k) - Tw_k$$

$$= Tx_k + \sum_{j=1}^{a_k} T'^{(j)}(x_k)(Tx_k - x_k) - Tw_n$$

Hence by lemma 1 we get

$$x_{k+1} - w_{k+1} \succeq T'(w_k)(x_k - w_k) + \sum_{j=1}^{a_k} T'^{(j)}(x_k)(Tx_k - x_k) \tag{3.7}$$

Since $T'(w_k)$ is positive, and $x_k \preceq Tx_k$ by Lemma 2, the second side of (3.7) is positive. So $x_{k+1} \succeq w_{k+1}$. Now we show that $x_{k+1} \preceq u_{k+1}$. In fact we have

$$u_{k+1} - x_{k+1} = u_k - [I - T'(u_k)]^{-1}(u_k - Tu_k) - x_k + S_{a_k}(x_k - Tx_k)$$

$$= (u_k - x_k) + \sum_{j=0}^{+\infty} T'^{(j)}(u_k)(Tu_k - u_k) - \sum_{j=0}^{a_k} T'^{(j)}(x_k)(Tx_k - x_k)$$

$$\succeq (u_k - x_k) + \sum_{j=0}^{+\infty} T'^{(j)}(x_k)(Tu_k - u_k) - \sum_{j=0}^{a_k} T'^{(j)}(x_k)(Tx_k - x_k)$$

$$= (u_k - x_k) + \sum_{j=a_k+1}^{+\infty} T'^{(j)}(x_k)(Tu_k - Tx_k + u_k - x_k)$$

Now since $x_k \preceq u_k$ by Lemma 1 we obtain

$$Tu_k - Tx_k \succeq T'(x_k)(u_k - x_k)$$

Hence

$$T'^{(j)}(x_k)(Tu_k - Tx_k + u_k - x_k) \succeq T'^{(j+1)}(x_k)(u_k - x_k) - T'^{(j)}(x_k)(u_k - x_k)$$

So

$$\sum_{j=a_k+1}^{+\infty} T'^{(j)}(Tu_k - Tx_k + u_k - x_k) \succeq \sum_{j=a_k+1}^{+\infty} [T'^{(j+1)}(x_k)(u_k - x_k) - T'^{(j)}(x_k)(u_k - x_k)]$$

$$= T'^{(a_k+1)}(x_k)(u_k - x_k) - (u_k - x_k)$$

Consequently

$$u_{k+1} - x_{k+1} \succeq T'^{(a_k+1)}(x_k)(u_k - x_k) \succeq 0$$

So by the induction principle we deduce that for every $n \in \mathbb{N}$, the relation (3.6) is true. Now since $\lim_{n \to +\infty} u_n = \lim_{n \to +\infty} w_n = u^*$; $w_n \preceq x_n \preceq u_n$ and the cone K is normal, then the proposition 1 implies that $\lim_{n \to +\infty} x_n = u^*$.

Part three: The rate of convergence

Assume that hypothesis $(H1) - (H4)$ are satisfied. Let

$$e_n := e_n(a_n) = u^* - x_n$$

be the error of approximation of the fixed point u^* by the sequence $(x_n)_{n \geq 0}$. It is clear that $e_n \succeq o$. Now for every $n \in \mathbb{N}$ we have

$$e_{n+1} = u^* - x_{n+1} = Tu^* - [x_n - S_{a_n}(x_n)(x_n - Tx_n)]$$

$$= Tu^* - [Tx_n + \sum_{k=1}^{a_n} T'^{(k)}(x_n)(Tx_n - x_n)]$$

$$= Tu^* - Tx_n - \sum_{k=1}^{a_n} T'^{(k)}(x_n)(Tx_n - Tu^* + u^* - x_n)$$

Thus, by lemma 1 we get

$$e_{n+1} \preceq T'(u^*)(u^* - x_n) + \sum_{k=1}^{a_n} T'^{(k)}(x_n)(Tx_n - Tu^*) - \sum_{k=1}^{a_n} T'^{(k)}(x_n)(u^* - x_n)$$

$$\preceq T'(u^*)e_n + \sum_{k=1}^{a_n} T'^{(k)}(x_n) \circ T'(u^*)e_n - \sum_{k=1}^{a_n} T'^{(k)}(x_n)e_n$$

$$\preceq T'(u^*)e_n + \sum_{k=1}^{a_n} T'^{(k+1)}(u^*)e_n - \sum_{k=1}^{a_n} T'^{(k)}(x_n)e_n$$

$$= \sum_{k=1}^{a_n+1} T'^{(k)}(u^*)e_n - \sum_{k=1}^{a_n} T'^{(k)}(x_n)e_n$$

$$= \sum_{k=1}^{a_n} [T'^{(k)}(u^*) - T'^{(k)}(x_n)]e_n + T'^{(a_n+1)}(u^*)e_n$$

Now, hypothesis (H4) implies that

$$\left\|T'^{(k)}(u^*) - T'^{(k)}(x_n)\right\|_{\mathcal{L}(E)} \leq \eta^k \|e_n\|$$

Since the cone K is normal with constant of normality δ, since the hypothesis (H3) is satisfied, we get the following estimation

$$\|e_{n+1}\| \leq \delta[(\sum_{k=1}^{a_n} \eta^k) \|e_n\| + q^{a_n+1}] \|e_n\|$$

Hence

$$\frac{\|e_{n+1}\|}{\|e_n\|} = \alpha_n \|e_n\| + \beta_n \tag{3.8}$$

Where

$$\alpha_n = \delta(\sum_{k=1}^{a_n} \eta^k) \qquad \beta_n = \delta q^{a_n+1}$$

then we have the following two cases :

1. *First case: If there exists a constant a such that $a_n = a$ for every $n \in \mathbb{N}$ then*

$$lim_{n \to +\infty} \alpha_n \|e_n\| = 0 \qquad and \quad \beta_n := \beta = \delta q^{a+1} < 1$$

So there exists an integer N such that for every $n \geq N$ we have

$$\frac{\|e_{n+1}\|}{\|e_n\|} \leq \beta \tag{3.9}$$

The relation (3.9) means that the rate of convergence of $(x_n)_{n \geq 0}$ is linear

2. *Second case: If $a_n \to +\infty$ and $\eta < 1$, then*

$$\alpha_n \leq \frac{\delta}{1 - \eta} \qquad and \quad \beta_n \to 0$$

Consequently we have

$$lim_{n \to +\infty} \frac{\|e_{n+1}\|}{\|e_n\|} = 0 \tag{3.10}$$

The relation (3.10) means that the rate of convergence of $(x_n)_{n \geq 0}$ is super-linear.

4   Application to a Nonlinear Matrix Equation

In this section we apply the monotone Newton like method (3.2) to the following nonlinear matrix equation

$$X + A^* X^{-1} A + B^* X^{-1} B = I \tag{4.1}$$

where A, B are two nonsingular complex matrices, and the positive definite solution is required.

The problem of solving the matrix equation (4.1), is related to problems of solving system of linear equations $Px = f$, arising in solving discretized elliptic partial differential equations [5]. It is proved in [5] that equation (4.1) has positive definite solution, and two iterative algorithms for finding positive definite solution of equation (4.1) are proposed.

For a complex matrix A, we use A^*, $\lambda_{\max}(A^*A)$, $\lambda_{\min}(A^*A)$, $\|A\| = \lambda_{\max}^{\frac{1}{2}}(A^*A)$ to denote respectively the conjugated transpose of A, the maximal eigenvalue and the minimal eigenvalue of A^*A, the spectral norm of the matrix A.

By a simple manipulations we can prove that equation (4.1) is equivalent to the following fixed point equation

$$Y = TY := I + Y(A^*YA + B^*YB) \tag{4.2}$$

where $Y = X^{-1}$.

Let $E = H^n$ be the Banach space of $n \times n$ complex hermitian matrices normed by the spectrale norme, and ordered by the cone $K = H_+^n$ of positive semi definite matrices. Consider the operator $T : H^n \to H^n$ defined by (4.2).

Lemma 3. *If*

$$\lambda_{\max}(A^*A) + \lambda_{\max}(B^*B) < \frac{1}{4} \tag{4.3}$$

Then there exists X_0 , Y_0 in H_+^n such that

$$X_0 \prec Y_0 \quad and \quad X_0 \preceq TX_0 \quad , \quad TY_0 \preceq Y_0$$

where $X_0 = s_1 I$, $Y_0 = s_2 I$ and s_2, s_1, $(1 < s_1 \leq s_2 < 2)$ are the smallest positive real solutions of the equations
$(\lambda_{\max}(A^*A) + \lambda_{\max}(B^*B)).x^2 - x + 1 = 0$ and $(\lambda_{\min}(A^*A) + \lambda_{\min}(B^*B)).x^2 - x + 1 = 0$ *respectively.*

Proof. Let $t \in \mathbb{R}$, by a simple calculus we obtain

$$T(tI) - tI = (1 - t)I + t^2(A^*A + B^*B).$$

Hence

$$T(tI) - tI \succeq \left[1 - t + (\lambda_{\min}(A^*A) + \lambda_{\min}(B^*B)).t^2\right].I$$

and

$$T(tI) - tI \preceq \left[1 - t + (\lambda_{\max}(A^*A) + \lambda_{\max}(B^*B)).t^2\right].I$$

Putting

$$\lambda_{\min} = \lambda_{\min}(A^*A) + \lambda_{\min}(B^*B)$$
$$\lambda_{\max} = \lambda_{\max}(A^*A) + \lambda_{\max}(B^*B)$$
$$p(t) = 1 - t + \lambda_{\min}t^2,$$
$$q(t) = 1 - t + \lambda_{\max}t^2.$$

Let φ be a numerical functions defined on $]0, \frac{1}{4}[$ as follow :

$$\varphi(t) = \frac{1 - \sqrt{1 - 4t}}{2t}$$

It is easy to show that φ is increasing on $]0, \frac{1}{4}[$, and that for every $t \in]0, \frac{1}{4}[$, we have

$$1 < \varphi(t) < 2$$

On the other hand, the smallest solutions of equations $p(t) = 0$ and $q(t) = 0$, are respectively:

$$s_1 = \varphi(\lambda_{\min}) \ , \ s_2 = \varphi(\lambda_{\max}) \tag{4.4}$$

It follows that

$$1 < s_1 \leq s_2 < 2$$

Let be $X_0 = s_1 I$, $Y_0 = s_2 I$, then we have $X_0 \prec Y_0$ and

$$TX_0 - X_0 = (1 - s_1)I + (A^*A + B^*B)s_1^2 \succeq p(s_1)I = 0$$

and

$$TY_0 - Y_0 = (1 - s_2)I + (A^*A + B^*B)s_2^2 \preceq q(s_2)I = 0$$

Theorem 2. *If*

$$\lambda_{\max}(A^*A) + \lambda_{\max}(B^*B) < \frac{1}{4} \tag{4.5}$$

Then the equation (4.2) has a unique solution Y_+ in $[s_1 I, s_2 I]$ where s_1 , s_2 are defined in Lemma 3. Thus the equation (4.1) has a unique solution X_+ in $\left[\frac{1}{s_2}I, \frac{1}{s_1}I\right]$. Morover the monotone Newton-like method, defined by the sequence $(x_n)_{n \geq 0}$ where

$$x_0 = s_1 I \quad and \quad x_{n+1} = x_n - S_{a_n}(x_n)(x_n - Tx_n) \tag{4.6}$$

and T is the matricial operator defined by (4.2), converge to Y_+ with superlinear rate of convergence. Hence we can compute X_+ since $X_+ = Y_+^{-1}$.

Proof. By lemma 3 there exists X_0 , Y_0 in H_+^n such that

$$X_0 \prec Y_0 \quad and \quad X_0 \preceq TX_0 \ , \quad TY_0 \preceq Y_0$$

Hence hypothesis $(H1)$ is satisfied. In the other hand, T is Fréchet derivable on $[X_0, Y_0]$ and for every $h \in H^n$, $Z \in [X_0, Y_0]$ we have

$$T'(Z)h = ZF(h) + hF(Z)$$

where

$$F(X) = A^*XA + B^*XB \ , \quad X \in H^n$$

Let $Z_1, Z_2 \in H^n$, such that $Z_1 \preceq Z_2$ and $h \in K$, then we get

$$T'(Z_1)h - T'(Z_2)h = (Z_1 - Z_2)F(h) + hF(Z_1 - Z_2) \preceq 0 \tag{4.7}$$

Hence the mapping $Z \mapsto T'(Z)h$ is increasing for every $h \in K$. Then hypothesis $(H2)$ also satisfied.

Now let $Z \in [X_0, Y_0]$ and $h \in H^n$, then

$$\|T'(Z)h\| \leq 2 \|Z\| \|F\| \|h\|$$

Since the spectral norm is monotone we have $\|Z\| \leq \|Y_0\| = s_2 < 2$ and $\|F\| \leq \lambda_{\max}(A^*A) + \lambda_{\max}(B^*B) < \frac{1}{4}$. So we conclude that

$$\|T'(Z)\| \leq q := 4\left[\lambda_{\max}(A^*A) + \lambda_{\max}(B^*B)\right] < 1$$

Hence the hypothesis $(H3)$ is satisfied. Now, the relation (4.7) implies that for every $Z_1, Z_2 \in H^n$

$$\|T'(Z_1) - T'(Z_2)\| \leq 2 \|F\| \|Z_1 - Z_2\| < \frac{1}{2} \|Z_1 - Z_2\|$$

Hence $(H4)$ is satisfied with $\eta = \frac{1}{2}$. Consequently, Theorem 1 is applicable.

4.1 Numerical Examples

In this subsection we give a numerical example to illustrate Theorem 2. All numerical experiments are run in MATLAB version R2010a. We recall that if X_+ and Y_+ are solutions of equations (4.1) and (4.2) respectively, then $X_+ = Y_+^{-1}$. We denote the residual error by

$$er(X) = \left\|X + A^*X^{-1}A + B^*X^{-1}B - I\right\| \qquad (4.8)$$

Consider the equation (4.1) with:

$$A = \frac{1}{452}\begin{pmatrix} 30 & 22 & 23 & 35 & 40 & 52 \\ 22 & 17 & 19 & 66 & 30 & 10 \\ 23 & 19 & 11 & 13 & 25 & 21 \\ 35 & 66 & 13 & 19 & 17 & 6 \\ 40 & 30 & 25 & 7 & 20 & 15 \\ 52 & 10 & 21 & 6 & 15 & 9 \end{pmatrix}, \quad B = \frac{1}{452}\begin{pmatrix} 11 & 12 & 15 & 17 & 20 & 45 \\ 12 & 7 & 19 & 21 & 51 & 13 \\ 15 & 19 & 65 & 44 & 23 & 18 \\ 17 & 21 & 44 & 31 & 32 & 33 \\ 20 & 51 & 23 & 32 & 13 & 41 \\ 45 & 19 & 18 & 33 & 41 & 24 \end{pmatrix}$$

We have

$$\lambda_{\max}(A^*A) + \lambda_{\max}(B^*B) = 0.2483$$

and

$$s_1 = 1.0046, \quad s_2 = 1.6991 \qquad (s_1 \text{ and } s_2 \text{ are defined by (4.4)})$$

Let $u_0 = s_1 I$. We fixe the number of iterations $n = 6$. We obtain the following solution:

$$X_+ := X_+(a_n) = \begin{pmatrix} 0.9186 & -0.0667 & -0.0594 & -0.0649 & -0.0682 & -0.0571 \\ -0.0667 & 0.9244 & -0.0564 & -0.0601 & -0.0591 & -0.0560 \\ -0.0594 & -0.0564 & 0.9268 & -0.0700 & -0.0654 & -0.0592 \\ -0.0649 & -0.0633 & -0.0700 & 0.9133 & -0.0744 & -0.0645 \\ -0.0682 & -0.0591 & -0.0654 & -0.0744 & 0.9205 & -0.0646 \\ -0.0571 & -0.0560 & -0.0592 & -0.0645 & -0.0646 & 0.9303 \end{pmatrix}$$

With different choices of the sequence $(a_n)_{n \geq 0}$, we obtain the following results

Table 1. Residual error according to choice of a_n

a_n	$er(X_+)$
5	1.4×10^{-3}
10	4.86×10^{-5}
15	2.57×10^{-6}
n^2	2.64×10^{-8}
$n^2 + 5n$	5.58×10^{-12}
n^3	3.42×10^{-16}

Now taking the matrices A and B as in example 4.2 of [5]. We choose $a_n = 3^n$, we get the following results

Table 2. Residual errors with $a_n = 3^n$

n	$er(X_+)$
2	3.50×10^{-3}
3	6.77×10^{-5}
4	3.73×10^{-8}
5	1.14×10^{-14}
6	1.27×10^{-16}

References

1. Guo, D., Cho, Y.J., Zhu, J.: Partial ordering methods in nonlinear problems. Nova Science Publishers (2004)
2. Guo, D., Lakshmikantham, V.: Nonlinear Problems in Abstract Cones. Academic Press, New York (1988)
3. Krasnoselskii, M.: Positive Solutions of Operator Equations. Noordhoff, Groningen (1964)
4. Lakshmikantham, V., Carl, S., Heikkilä, S.: Fixed point theorems in ordered Banach spaces via quasilinearization. Nonlinear Analysis: Theory, Methods & Applications, 3448–3458 (2009)
5. Long, J.H.: On the Hermitian positive definite solution of the nonlinear matrix equation $X + A^*X^{-1}A + B^*X^{-1}B = I$. Bull. Braz. Math. Soc., New Series. Sociedade Brasileira de Matematica 39(3), 371–386 (2008)

A Numerical Implementation of an Interior Point Methods for Linear Programming Based on a New Kernel Function

Mousaab Bouafia[1], Djamel Benterki[2], and Adnan Yassine[3]

[1] LabCAV, Laboratoire de Contrôle Avancé, Université de Guelma, Algérie
mousaab84@yahoo.fr
[2] LMFN, Laboratoire de Mathématiques Fondamentales et Numériques,
Sétif-1, Algérie
dj_benterki@yahoo.fr
[3] LMAH, Laboratoire de Mathématiques Appliquées du Havre,
Université du Havre, France
adnan.yassine@univ-lehavre.fr

Abstract. In this paper, we define a new barrier function and propose a new primal-dual interior point methods based on this function for linear optimization. The proposed kernel function which yields a low algorithm complexity bound for both large and small-update interior point methods. This purpose is confirmed by numerical experiments showing the efficiency of our algorithm which are presented in the last of this paper.

Keywords: Linear Optimization, Kernel function, Interior Point methods, Complexity Bound.

1 Introduction

We consider the standard linear optimization

$$(P) \ \min\{c^t x : Ax = b, x \geq 0\},$$

where $A \in \mathbb{R}^{m \times n}$, $rank(A) = m$, $b \in \mathbb{R}^m$, and $c \in \mathbb{R}^n$, and its dual problem

$$(D) \ \max\{b^t y : A^t y + s = c, s \geq 0\}.$$

In 1984, Karmarkar [13] proposed a new polynomial-time method for solving linear programs. This method and its variants that were developed subsequently are now called interior point methods **IPMs**. For a survey, we refer to [17]. The primal-dual interior point algorithm which is the most efficient for a computational point of view [1]. It is known, before the reference [9], that the iteration complexity of an algorithm is in an appropriate measure for its efficiency. At present, the best known theoretical iteration bound for small-update **IPMs** is better than the one for large-update **IPMs**. However, in practice, large-update **IPMs** are

© Springer International Publishing Switzerland 2015
H.A. Le Thi et al. (eds.), *Model. Comput. & Optim. in Inf. Syst. & Manage. Sci.*,
Advances in Intelligent Systems and Computing 359, DOI: 10.1007/978-3-319-18161-5_30

much more efficient than small-update **IPMs** [16,17]. Many researchers are proposed and analyzed various primal-dual interior point methods for linear optimization **LO** based on the logarithmic barrier function [1,11,12]. Peng et al [14,15] introduced self-regular barrier functions for primal-dual **IPMs** for **LO** and obtained the best complexity result so far $\mathbf{O}\left(\sqrt{n}\log n \log \frac{n}{\epsilon}\right)$ for large-update primal-dual **IPMs** with some specific self-regular barrier functions. Recently, Bai et al $[2-7]$, Ghami et al [9,10] and G. M. Cho [7] proposed new primal-dual **IPMs** for **LO** problems based on various kernel functions to improve the iteration bound for large-update methods from $\mathbf{O}\left(n\log \frac{n}{\epsilon}\right)$ to $\mathbf{O}\left(\sqrt{n}\log n \log \frac{n}{\epsilon}\right)$. For its part, EL Ghami et al [8] used a new kernel function with a trigonometric barrier term and proposed a new primal-dual **IPMs** and proved that the iteration bound of large-update methods is $\mathbf{O}\left(n^{\frac{3}{4}}\log \frac{n}{\epsilon}\right)$. Motivated by their work, we define a new kernel function and propose a new primal-dual **IPMs** based on this kernel function for **LO**. We show that the iteration bounds are $\mathbf{O}\left(\sqrt{n}\left(\log n\right)^2 \log \frac{n}{\epsilon}\right)$ for large-update methods and $\mathbf{O}\left(\sqrt{n}\log \frac{n}{\epsilon}\right)$ for small-update methods.

Without loss of generality, we assume that (P) and (D) satisfy the interior point condition **IPC**, i.e., there exist (x^0, y^0, s^0) such that

$$Ax^0 = b, x^0 > 0, A^t y^0 + s^0 = c, s^0 > 0. \tag{1}$$

It is well known that finding an optimal solution of (P) and (D) is equivalent to solve the following system:

$$\begin{aligned} Ax &= b, x \geq 0, \\ A^t y + s &= c, s \geq 0, \\ xs &= 0. \end{aligned} \tag{2}$$

The paper is organized as follows. In Section **2**, we recall how a given kernel function defines a primal-dual corresponding **IPMs**, and we present the generic form of this algorithm. In Section **3**, we define a new kernel function and give its properties which are essential for the complexity analysis. In Section **4,** we derive decrease of the barrier function during an inner iteration result for both large-update and small-update methods. In Section **5,** we present some numerical results. Finally, concluding remarks are given in Section **6**.

We use the following notations throughout the paper. $\mathbb{R}^n_+$ and $\mathbb{R}^n_{++}$ denote the set of n-dimensional nonnegative vectors and positive vectors respectively. For $x, s \in \mathbb{R}^n$, $x_{\min}$ and xs denote the smallest component of the vector x and the vector componentwise product of the vector x and s, respectively. We denotes by $X = diag(x)$ the $n \times n$ diagonal matrix with the components of the vector $x \in \mathbb{R}^n$ are the diagonal entries. e denotes the n-dimensional vector of ones. For $f(x), g(x) : \mathbb{R}^n_{++} \rightarrow \mathbb{R}^n_{++}$, $f(x) = \mathbf{O}(g(x))$ if $f(x) \leq C_1 g(x)$ for some positive constant C_1 and $f(x) = \Theta(g(x))$ if $C_2 g(x) \leq f(x) \leq C_3 g(x)$ for some positive constant C_2 and C_3 and finally, $\|\|$ denotes the 2-norm of a vector.

2   The Prototype Algorithm

The basic idea of primal-dual **IPMs** is to replace the equation of complementarity condition for (P) and (D) define in (2), by the parameterized equation $xs = \mu e$, with $\mu > 0$. Thus we consider the system

$$
\begin{aligned}
Ax &= b, x \geq 0, \\
A^t y + s &= c, s \geq 0, \\
xs &= \mu e.
\end{aligned}
\tag{3}
$$

If the **IPC** is satisfied, then there exists a solution, for each $\mu > 0$, and this solution is unique. It is denoted as $(x(\mu), y(\mu), s(\mu))$, and we call $x(\mu)$ the μ-center of (P) and $(y(\mu), s(\mu))$ the μ-center of (D). If $\mu \to 0$, then the limit of the central path exists, and since the limit points satisfy the complementarity condition, the limit yields optimal solutions for (P) and (D). From a theoretical point of view, the **IPC** can be assumed without loss of generality. In fact, we may, and will, assume that $x^0 = s^0 = e$. In practice, this can be realized by embedding the given problems (P) and (D) into a homogeneous self-dual problem which has two additional variables and two additional constraints. For this and the other properties mentioned above, see [16].

2.1   The Search Directions

Without loss of generality, we assume that $(x(\mu), y(\mu), s(\mu))$ is known for some positive μ. For example, due to the above assumption, we assume that for $\mu = 1$, $x(1) = s(1) = e$. We then decrease μ to $\mu = (1 - \theta)\mu$ for some fixed $\theta \in]0, 1[$, and we solve the following Newton system:

$$
\begin{aligned}
A\Delta x &= 0, \\
A^t \Delta y + \Delta s &= 0, \\
s\Delta x + x\Delta s &= \mu e - xs.
\end{aligned}
\tag{4}
$$

This system uniquely defines a search direction $(\Delta x, \Delta y, \Delta s)$. If necessary, we repeat the procedure until we find iterates that are close to $(x(\mu), y(\mu), s(\mu))$. Then μ is again reduced by the factor $1 - \theta$, and we apply Newton's method targeting the new μ-centers, and so on. This process is repeated until μ is small enough, i.e., until $n\mu \leq \epsilon$; at this stage, we have found an ϵ-optimal solution of problems (P) and (D). The result of a Newton step with stepsize α is defined as:

$$x^+ = x + \alpha \Delta x, \; y^+ = y + \alpha \Delta y, \; s^+ = s + \alpha \Delta s, \tag{5}$$

where the stepsize α satisfies $(0 < \alpha \leq 1)$.

Now, we introduce the scaled vector v and the scaled search directions d_x and d_s as follows:

$$v = \sqrt{\frac{xs}{\mu}}, \; d_x = \frac{v\Delta x}{x}, \; d_s = \frac{v\Delta s}{s}. \tag{6}$$

The system (4) can be rewritten as follows:

$$\begin{aligned} \overline{A}d_x &= 0, \\ \overline{A}^t \Delta y + d_s &= 0, \\ d_x + d_s &= v^{-1} - v, \end{aligned} \tag{7}$$

where $\overline{A} = \frac{1}{\mu}AV^{-1}X$, $V = diag(v)$. Note that the right-hand side of the third equation in (7) equals to the negative gradient of the logarithmic barrier function $\Phi(v)$, i.e.,

$$d_x + d_s = -\nabla \Phi(v), \tag{8}$$

where the barrier function $\Phi(v) : \mathbb{R}^n_{++} \to \mathbb{R}_+$ is defined as follows:

$$\Phi(v) = \Phi(x, s; \mu) = \sum_{i=1}^n \psi(v_i), \tag{9}$$

where

$$\psi(v_i) = \frac{v_i^2 - 1}{2} - \log v_i. \tag{10}$$

$\psi(t)$ is called the kernel function of the logarithmic barrier function $\Phi(v)$. In this paper, we replace $\psi(t)$ by a new kernel function $\psi_M(t)$, which will be defined in Section **2**. It is clear from the above description that the closeness of (x, s) to $(x(\mu), s(\mu))$ is measured by the value of $\Phi(v)$, with $\tau > 0$ as a threshold value. If $\Phi(v) \leq \tau$, then we start a new outer iteration by performing a μ–update; otherwise, we enter an inner iteration by computing the search directions at the current iterates with respect to the current value of μ and apply (5) to get new iterates. If necessary, we repeat the procedure until we find iterates that are in the neighborhood of $(x(\mu), s(\mu))$. Then μ is again reduced by the factor $1 - \theta$ with $0 < \theta < 1$, and we apply Newton's method targeting the new μ–centers, and so on. This process is repeated until μ is small enough, i.e., until $n\mu < \epsilon$. The choice of the stepsize α, $(0 < \alpha \leq 1)$ is another crucial issue in the analysis of the algorithm.

Prototype Algorithm for LO

Begin algorithm

a proximity the function $\Phi_M(v)$;

A threshold parameter $\tau > 0$;

an accuracy parameter $\epsilon > 0$;

a fixed barrier update parameter $\theta, 0 < \theta < 1$;

begin

$x = e$; $s = e$; $\mu = 1$; $v = e$.

while $n\mu \geq \epsilon$ **do**

begin (outer iteration)

  $\mu = (1 - \theta)\mu$;

  while $\Phi_M(x, s; \mu) > \tau$ **do**

  begin (inner iteration)

    Solve the system (7) via (6) to obtain $(\Delta x, \Delta y, \Delta s)$;

    choose a suitable a stepsize α and take

  $x = x + \alpha \Delta x$;

  $y = y + \alpha \Delta y$;

  $s = s + \alpha \Delta s$;

  $v = \sqrt{\frac{xs}{\mu}}$;

  end (inner iteration)

end (outer iteration)

End algorithm.

Fig 1. Algorithm

3   The New Kernel Function and Its Properties

In this section, we define a new kernel function and give its properties which are essential to our complexity analysis.

We call $\psi(t) : \mathbb{R}_{++} \to \mathbb{R}_+$ a kernel function if ψ is twice differentiable and satisfies the following conditions:

$$
\begin{aligned}
&\psi'(1) = \psi(1) = 0, \\
&\psi''(t) > 0, \\
&\lim_{t \to 0^+} \psi(t) = \lim_{t \to +\infty} \psi(t) = +\infty.
\end{aligned}
\tag{11}
$$

Now, we define a new function $\psi_M(t)$ as follows:

$$
\psi_M(t) = \frac{p}{2}t^2 + \exp(p(\frac{1}{t} - 1)) - (1 + \frac{p}{2}), \quad p > 0.
\tag{12}
$$

For convenience of reference, we gives the first three derivatives with respect to t as follows:

$$
\begin{aligned}
\psi'_M(t) &= pt - \frac{p}{t^2}\exp(p(\frac{1}{t} - 1)), \\
\psi''_M(t) &= p + \left(\frac{2p}{t^3} + \frac{p^2}{t^4}\right)\exp(p(\frac{1}{t} - 1)), \\
\psi'''_M(t) &= -\left(\frac{6p}{t^4} + \frac{6p^2}{t^5} + \frac{p^3}{t^6}\right)\exp(p(\frac{1}{t} - 1)).
\end{aligned}
\tag{13}
$$

Obviously, $\psi_M(t)$ is a kernel function and

$$\psi_M''(t) > p. \tag{14}$$

In this paper, we replace the barrier function $\Phi(v)$ in (8) with the function barrier $\Phi_M(v)$ as follows:

$$d_x + d_s = -\nabla\Phi_M(v), \tag{15}$$

where

$$\Phi_M(v) = \sum_{i=1}^{n} \psi_M(v_i), \tag{16}$$

$\psi_M(t)$ is defined in (12). Hence, the new search direction $(\Delta x, \Delta y, \Delta s)$ is obtained by solving the following modified Newton system:

$$\begin{aligned} A\Delta x &= 0, \\ A^t\Delta y + \Delta s &= 0, \\ s\Delta x + x\Delta s &= -\mu v\nabla\Phi_M(v). \end{aligned} \tag{17}$$

We use $\Phi_M(v)$ as the proximity function to measure the distance between the current iterate and the $\mu-$center for given $\mu > 0$. We also define the norm-based proximity measure, $\delta(v) : \mathbb{R}_{++}^n \to \mathbb{R}_+$, as follows:

$$\delta(v) = \tfrac{1}{2}\|\nabla\Phi_M(v)\| = \tfrac{1}{2}\|d_x + d_s\|. \tag{18}$$

Lemma 1. *For $\psi_M(t)$, we have the following results.*

(i) $\psi_M(t)$ is exponentially convex for all $t > 0$; that is,

$$\psi_M\left(\sqrt{t_1 t_2}\right) \leq \frac{1}{2}\left(\psi_M(t_1) + \psi_M(t_2)\right).$$

(ii) $\psi_M''(t)$ is monotonically decreasing for all $t > 0$.
(iii) $t\psi_M''(t) - \psi_M'(t) > 0$ for all $t > 0$.
(iv) $\psi_M''(t)\psi_M'(\beta t) - \beta\psi_M'(t)\psi_M''(\beta t) > 0, \ \ t > 1, \ \beta > 1.$

Lemma 2. *For $\psi_M(t)$, we have*

$$\frac{p}{2}(t-1)^2 \leq \psi_M(t) \leq \frac{1}{2p}[\psi_M'(t)]^2, \quad t > 0 \tag{19}$$

$$\psi_M(t) \leq \frac{p^2 + 3p}{2}(t-1)^2, \quad t > 1 \tag{20}$$

Let $\sigma : [0, +\infty[\to [1, +\infty[$ be the inverse function of $\psi_M(t)$ for $t \geq 1$ and $\rho : [0, +\infty[\to]0, 1]$ be the inverse function of $\frac{-1}{2}\psi_M'(t)$ for all $t \in]0, 1]$. Then we have the following lemma.

Lemma 3. *For $\psi_M(t)$, we have*

$$1 + \sqrt{\frac{2}{p^2 + 3p}}\, s \leq \sigma(s) \leq 1 + \sqrt{\frac{2}{p}}\, s, \, s \geq 0 \tag{21}$$

$$\rho(z) > \frac{1}{\sqrt{\frac{2}{p}z + 1}}, \quad z \geq 0. \tag{22}$$

Let $\bar{p} : [0, +\infty[\to \,]0, 1]$ be the inverse function of

$$\varphi_M(t) = \frac{p}{t^2}\exp(p(\frac{1}{t} - 1)), \, p > 0, \text{ for all } t \in \,]0, 1].$$

Then, we have the following lemma.

Lemma 4. *For $\varphi_M(t)$, we have*

$$\bar{p}(z) > \frac{1}{1 + \log\left(\frac{z}{p}\right)^{\frac{1}{p}}}, \quad z \geq 0 \tag{23}$$

$$\rho(z) \geq \bar{p}(p + 2z), \quad z \geq 0. \tag{24}$$

Lemma 5. *Let $0 \leq \theta < 1$, $v_+ = \frac{v}{\sqrt{1-\theta}}$, If $\Phi_M(v) \leq \tau$ then, we have*

$$\Phi_M(v_+) \leq \frac{\left(\theta np + 2\tau + 2\sqrt{2\tau np}\right)}{2(1 - \theta)}.$$

Denote

$$(\Phi_M)_0 = \frac{\left(\theta np + 2\tau + 2\sqrt{2\tau np}\right)}{2(1 - \theta)} = L(n, \theta, \tau), \tag{25}$$

then, $(\Phi_M)_0$ is an upper bound for $\Phi_M(v_+)$ during the process of the algorithm.

4 Decrease of the Barrier Function during an Inner Iteration

In this section, we compute a default, stepsize α and the resulting decrease of the barrier function. After a damped step we have

$$x^+ = x + \alpha\Delta x; \quad y^+ = y + \alpha\Delta y; \quad s^+ = s + \alpha\Delta s; \quad \text{Using (6), we have}$$

$$x^+ = x\left(e + \alpha\frac{\Delta x}{x}\right) = x\left(e + \alpha\frac{d_x}{v}\right) = \frac{x}{v}(v + \alpha d_x),$$

$$s^+ = s\left(e + \alpha\frac{\Delta s}{s}\right) = s\left(e + \alpha\frac{d_s}{v}\right) = \frac{s}{v}(v + \alpha d_s),$$

So, we have

$$v_+ = \sqrt{\frac{x^+ s^+}{\mu}} = \sqrt{(v + \alpha d_x)(v + \alpha d_s)}.$$

Define, for $\alpha > 0$,

$$f(\alpha) = \Phi_M(v_+) - \Phi_M(v).$$

Then $f(\alpha)$ is the difference of proximities between a new iterate and a current iterate for fixed μ. By Lemma 1 (i), we have

$$\Phi_M(v_+) = \Phi_M\left(\sqrt{(v + \alpha d_x)(v + \alpha d_s)}\right) \le \frac{1}{2}\left(\Phi_M(v + \alpha d_x) + \Phi_M(v + \alpha d_s)\right).$$

Therefore, we have $f(\alpha) \le f_1(\alpha)$, where

$$f_1(\alpha) = \frac{1}{2}\left(\Phi_M(v + \alpha d_x) + \Phi_M(v + \alpha d_s)\right) - \Phi_M(v). \tag{26}$$

Obviously, $f(0) = f_1(0) = 0$. Taking the first two derivatives of $f_1(\alpha)$ with respect to α, we have

$$f_1'(\alpha) = \frac{1}{2}\sum_{i=1}^{n}\left(\psi_M'(v_i + \alpha d_{xi})d_{xi} + \psi_M'(v_i + \alpha d_{si})d_{si}\right),$$

$$f_1''(\alpha) = \frac{1}{2}\sum_{i=1}^{n}\left(\psi_M''(v_i + \alpha d_{xi})d_{xi}^2 + \psi_M''(v_i + \alpha d_{si})d_{si}^2\right),$$

Using (15) and (18), we have

$$f_1'(0) = \frac{1}{2}\nabla\Phi_M(v)^t(d_x + d_s) = -\frac{1}{2}\nabla\Phi_M(v)^t \nabla\Phi_M(v) = -2\delta(v)^2. \tag{27}$$

Lemma 6. *Let $\delta(v)$ be as defined in (18). Then, we have*

$$\delta(v) \ge \sqrt{\frac{p}{2}\Phi_M(v)}. \tag{a}$$

From Lemmas 4.1–4.4 in [3], we have the following Lemma **7**.

Lemma 7. *The largest stepsize $\overline{\alpha}$ satisfying*

$$\overline{\alpha} \ge \frac{1}{\psi_B''(\rho(2\delta))}.$$

and, we have

$$f_1(\alpha) \le 0, \text{ for } \alpha \le \overline{\alpha}$$

Lemma 8. *Let ρ and $\overline{\alpha}$ be as defined in Lemma 7. If*

$$\Phi_M = \Phi_M\left(v\right) \geq \tau \geq 1,$$

then we have

$$\overline{\alpha} \geq \frac{1}{p + \left[\left(2+p\right)\left(p+4\right)\sqrt{\frac{p}{2}\Phi_M}\right]\left[1 + \log\left(1 + \frac{4}{p}\sqrt{\frac{p}{2}\Phi_M}\right)^{\frac{1}{p}}\right]^2}.$$

Denoting

$$\widetilde{\widetilde{\alpha}} = \frac{1}{p + \left[\left(2+p\right)\left(p+4\right)\sqrt{\frac{p}{2}\Phi_M}\right]\left[1 + \log\left(1 + \frac{4}{p}\sqrt{\frac{p}{2}\Phi_M}\right)^{\frac{1}{p}}\right]^2}. \tag{28}$$

We have that $\widetilde{\widetilde{\alpha}}$ is the default stepsize and that $\widetilde{\widetilde{\alpha}} \leq \overline{\alpha}$.

Lemma 9. *Let $\widetilde{\widetilde{\alpha}}$ be the default stepsize as defined in (28) and let*

$$\left(\Phi_M\right)_0 \geq \Phi_M\left(v\right) \geq 1.$$

Then

$$f\left(\widetilde{\widetilde{\alpha}}\right) \leq \frac{-\sqrt{\frac{p}{2}}\left[\left(\Phi_M\right)_0\right]^{\frac{1}{2}}}{\left[\sqrt{\frac{2}{p}}p + \left(2+p\right)\left(p+4\right)\right]\left[1 + \log\left(1 + \frac{4}{p}\sqrt{\frac{p}{2}\left(\Phi_M\right)_0}\right)^{\frac{1}{p}}\right]^2} \tag{29}$$

After the update of μ to $\left(1-\theta\right)\mu$, we have

$$\Phi_M\left(v_+\right) \leq \left(\Phi_M\right)_0 = \frac{\left(\theta n p + 2\tau + 2\sqrt{2\tau n p}\right)}{2\left(1-\theta\right)} = L\left(n, \theta, \tau\right).$$

Lemma 10. *Let K be the total number of inner iterations in the outer iteration. Then we have*

$$K \leq \frac{4\sqrt{p} + 2\sqrt{2}\left(2+p\right)\left(p+4\right)}{\sqrt{p}}\left[1 + \log\left(1 + \frac{4}{p}\sqrt{\frac{p}{2}\left(\Phi_M\right)_0}\right)^{\frac{1}{p}}\right]^2\left[\left(\Phi_M\right)_0\right]^{\frac{1}{2}}.$$

Theorem 1. *Let an* **LO** *problem be given, let $\left(\Phi_M\right)_0$ be as defined in (25) and let $\tau \geq 1$. Then, the total number of iterations to have an approximate solution with $n\mu < \epsilon$ is bounded by*

$$\frac{4\sqrt{p} + 2\sqrt{2}\left(2+p\right)\left(p+4\right)}{\sqrt{p}}\left[1 + \log\left(1 + \frac{4}{p}\sqrt{\frac{p}{2}\left(\Phi_M\right)_0}\right)^{\frac{1}{p}}\right]^2\left[\left(\Phi_M\right)_0\right]^{\frac{1}{2}}\frac{\log\frac{n}{\epsilon}}{\theta}.$$

For large-update methods with $\tau = \mathbf{O}(n)$ and $\theta = \mathbf{\Theta}(1)$, we distinguish the two cases:

The first case: if $p \in [1, +\infty[$, we get for large-update methods $(\Phi_M)_0 = \mathbf{O}(pn)$ and $\mathbf{O}\left(\sqrt{n}\,(p\log n)^2 \log \frac{n}{\epsilon}\right)$ iterations.

The second case: if $p \in]0, 1[$, we get for large-update methods $(\Phi_M)_0 = \mathbf{O}(n)$ and $\mathbf{O}\left(\sqrt{\frac{n}{p^5}}\left(\log \frac{n}{p}\right)^2 \log \frac{n}{\epsilon}\right)$ iterations.

In case of a small-update methods, we have $\tau = \mathbf{O}(1)$ and $\theta = \mathbf{\Theta}\left(\frac{1}{\sqrt{n}}\right)$. Substitution of these values into theorem 1 does not give the best possible bound.

$$\Phi_M(v_+) \leq n\psi_M\left(\frac{1}{\sqrt{1-\theta}}\sigma\left(\frac{\Phi_M(v)}{n}\right)\right) \leq \frac{(p^2+3p)}{2(1-\theta)}\left(\theta\sqrt{n}+\sqrt{\frac{2}{p}}\tau\right)^2 = (\Phi_M)_0,$$

Using this upper bound for $(\mathbf{\Phi}_M)_0$, we get the following iteration bound:

$$\frac{4\sqrt{p}+2\sqrt{2}\,(2+p)\,(p+4)}{\sqrt{p}}\left[1+\log\left(1+\frac{4}{p}\sqrt{\frac{p}{2}(\Phi_M)_0}\right)^{\frac{1}{p}}\right]^2 [(\Phi_M)_0]^{\frac{1}{2}} \frac{\log \frac{n}{\epsilon}}{\theta}.$$

The first case: if $p \in [1, +\infty[$, we get for small-update methods $(\Phi_M)_0 = \mathbf{O}(p^2)$ and $\mathbf{O}\left(\sqrt{np^5}\log \frac{n}{\epsilon}\right)$ iterations.

The second case: if $p \in]0, 1[$, we get for small-update methods $(\Phi_M)_0 = \mathbf{O}(1)$ and $\mathbf{O}\left(\sqrt{\frac{n}{p^5}}\log \frac{n}{\epsilon}\right)$ iterations.

5   Numerical Tests

In this section, we present some numerical results. We consider the following example: $n = 2m$,

$$A(i,j) = \begin{cases} 0 \; if \; i \neq j \; or \; j \neq i+m \\ 1 \; if \; i = j \; or \; j = i+m \end{cases}$$
$$c(i) = -1, \, c(i+m) = 0 \text{ and } b(i) = 2, \text{ for } i = 1, ..., m.$$

Point of departure (Installation):

$$x^0(i) = x^0(i+m) = 1, s^0(i) = 1, s^0(i+m) = 2 \text{ and } y^0(i) = -2, \text{ for } i = 1, ..., m.$$

We take $\mu^0 = 1$, $\tau = 1$ and $\epsilon = 10^{-4}$. Furthermore, the parameters m, p and θ are taken as follows: $m \in \{5, 15, 25, 35, 50\}$, $p \in \{0.5, 1, 4\}$, $\theta \in \{0.1, 0.5, 0.9\}$.

In the table of results, (ex (m, n)) represents the size of the example, (Inn Itr) represents the total number of inner iterations and (Out Itr) represents the total number of outer iterations. We summarize these numerical study in the tables 1, 2, 3.

$p = 0.5$	$\theta = 0.1$		$\theta = 0.5$		$\theta = 0.9$	
$ex\,(m,n)$	Inn Itr	*Out* Itr	Inn Itr	*Out* Itr	Inn Itr	*Out* Itr
$(5,10)$	3243	117	7244	19	43395	7
$(15,30)$	5297	125	14893	20	99326	7
$(25,50)$	6960	129	22501	21	149611	7
$(35,70)$	8474	132	29042	21	197663	7
$(50,100)$	10559	135	40447	22	311676	8

table 1

$p = 1$	$\theta = 0.1$		$\theta = 0.5$		$\theta = 0.9$	
$ex\,(m,n)$	Inn Itr	*Out* Itr	Inn Itr	*Out* Itr	Inn Itr	*Out* Itr
$(5,10)$	1438	115	2538	19	8641	7
$(15,30)$	2408	124	4673	20	16268	7
$(25,50)$	3145	128	6583	21	22241	7
$(35,70)$	3802	131	8035	21	27491	7
$(50,100)$	4671	134	10501	22	40309	8

table 2

$p = 4$	$\theta = 0.1$		$\theta = 0.5$		$\theta = 0.9$	
$ex\,(m,n)$	Inn Itr	*Out* Itr	Inn Itr	*Out* Itr	Inn Itr	*Out* Itr
$(5,10)$	2028	112	3194	18	8893	7
$(15,30)$	3545	122	5962	20	15325	7
$(25,50)$	4677	127	8073	21	19900	7
$(35,70)$	5619	130	9568	21	23652	7
$(50,100)$	6837	133	12028	22	33136	8

table 3

Comments

These numerical tests for different dimensions confirm our purpose and consolidates our theoretical results.

6 Concluding Remarks

In this paper, we have analyzed large-update and small-update versions of the primal-dual interior point algorithm described in Fig 1 that are based on the new function (12). The proposed function is not logarithmic and not self-regular. We proved that the iteration bound of a large-update interior point method based on the kernel function considered in this paper is $\mathbf{O}\left(\sqrt{n}\,(\log n)^2 \log \frac{n}{\epsilon}\right)$ and for small-update methods, we obtain the best know iteration bound, namely $\mathbf{O}\left(\sqrt{n}\log \frac{n}{\epsilon}\right)$, just take $p = \Theta\,(1)$. These results are an important contribution to improve the computational complexity of the problem studied.

References

1. Andersen, E.D., Gondzio, J., Meszaros, C., Xu, X.: Implementation of interior point methods for large scale linear programming. In: Terlaky, T. (ed.) Interior Point Methods of Mathematical Programming, pp. 189–252. Kluwer Academic Publisher, The Netherlands (1996)
2. Bai, Y.Q., El Ghami, M., Roos, C.: A new efficient large-update primal-dual interior point method based on a finite barrier. SIAM Journal on Optimization 13, 766–782 (2003)
3. Bai, Y.Q., El Ghami, M., Roos, C.: A comparative study of kernel functions for primal-dual interior point algorithms in linear optimization. SIAM Journal on Optimization 15, 101–128 (2004)
4. Bai, Y.Q., Roos, C.: A primal-dual interior point method based on a new kernel function with linear growth rate. In: Proceedings of the 9th Australian Optimization Day, Perth, Australia (2002)
5. Bai, Y.Q., Roos, C.: A polynomial-time algorithm for linear optimization based on a new simple kernel function. Optimization Methods and Software 18, 631–646 (2003)
6. Bai, Y.Q., Guo, J., Roos, C.: A new kernel function yielding the best known iteration bounds for primal-dual interior point algorithms. Acta Mathematica Sinica 49, 259–270 (2007)
7. Cho, G.M.: An interior point algorithm for linear optimization based on a new barrier function. Applied Mathematics and Computation 218, 386–395 (2011)
8. El Ghami, M., Guennoun, Z.A., Bouali, S., Steihaug, T.: Interior point methods for linear optimization based on a kernel function with a trigonometric barrier term. Journal of Computational and Applied Mathematics 236, 3613–3623 (2012)
9. El Ghami, M., Ivanov, I., Melissen, J.B.M., Roos, C., Steihaug, T.: A polynomial-time algorithm for linear optimization based on a new class of kernel functions. Journal of Computational and applied Mathematics 224, 500–513 (2009)
10. El Ghami, M., Roos, C.: Generic primal-dual interior point methods based on a new kernel function. RAIRO-Operations Research 42, 199–213 (2008)
11. Gonzaga, C.C.: Path following methods for linear programming. SIAM Review 34, 167–227 (1992)
12. den Hertog, D.: Interior point approach to linear, quadratic and convex programming. Mathematical Application, vol. 277. Kluwer Academic Publishers, Dordrecht (1994)
13. Karmarkar, N.K.: A new polynomial-time algorithm for linear programming. In: Proceedings of the 16th Annual ACM Symposium on Theory of Computing, vol. 4, pp. 373–395 (1984)
14. Peng, J., Roos, C., Terlaky, T.: Self-regular functions and new search directions for linear and semidefinite optimization. Mathematical Programming 93, 129–171 (2002)
15. Peng, J., Roos, C., Terlaky, T.: Self-Regularity: A New Paradigm for Primal-Dual interior point Algorithms. Princeton University Press, Princeton (2002)
16. Roos, C., Terlaky, T., Vial, J.P.: Theory and algorithms for linear optimization. In: An Interior Point Approach. John Wiley & Sons, Chichester (1997)
17. Ye, Y.: Interior point algorithms. Theory and Analysis. John-Wiley & Sons, Chichester (1997)

A Smoothing Method for Sparse Optimization over Polyhedral Sets

Tangi Migot and Mounir Haddou

IRMAR-INSA de Rennes, 20 avenue des Buttes de Coësmes,
CS 14315, 35043 Rennes Cedex, France
{tangi.migot,mounir.haddou}@insa-rennes.fr

Abstract. We investigate a method to solve NP-hard problem of minimizing $\ell 0$-norm of a vector over a polyhedral set P. A simple approximation is to find a solution of the problem of minimizing $\ell 1$-norm. We are concerned about finding improved results. Using a family of smooth concave functions $\theta_r(.)$ depending on a parameter r we show that $\min_{x \in P} \sum_{i=1}^{n} \theta_r(x_i)$ is equivalent to $\min_{x \in P} ||x||_0$ for r sufficiently small and $\min_{x \in P} ||x||_1$ for r sufficiently large. This gives us an algorithm based on a homotopy-like method. We show convergence results, error estimates and numerical simulations.

Keywords: smoothing functions, $\ell 0$-minimization, sparsity, concave minimization.

1   Introduction

Consider a polyhedron P such that $P = \{x \in \mathbb{R}^n | \ b \in \mathbb{R}^m, \ Ax \le b\} \cap \mathbb{R}_+^n$ which is non-empty and does not contain just a single element. The system $Ax \le b$ is said underdetermined, $m < n$, so it has an infinite number of solutions. One should note that the hypothesis of considering polyhedron in the nonnegative orthant is not a loss of generality, it just simplifies the presentation as the absolute value disappear in the norm expressions. We are interested in finding the sparsest solution over this polyhedron, which is equivalent to minimize the $\ell 0$-norm

$$\forall x \in \mathbb{R}^n, \ ||x||_0 = \sum_{i=1}^{n} s(|x_i|), \text{ where } s(t) = \begin{cases} 0 & si \ t = 0 \\ 1 & sinon \end{cases}. \quad (1)$$

We will study in this document the following NP-hard problem

$$(P_0) \quad \min \ ||x||_0 \ . \quad (2)$$
$$x \in P$$

This problem has several applications, the most notable are in machine learning and compressed sensing [3,4]. This problem being difficult to solve, a more simple approach is often used, which consists in solving the convex problem in $\ell 1$-norm

$$(P_1) \quad \min \ ||x||_1 \ . \quad (3)$$
$$x \in P$$

© Springer International Publishing Switzerland 2015   369
H.A. Le Thi et al. (eds.), *Model. Comput. & Optim. in Inf. Syst. & Manage. Sci.*,
Advances in Intelligent Systems and Computing 359, DOI: 10.1007/978-3-319-18161-5_31

This can be seen as a convexification of (P_0). We are interested in finding a method to improve the solution we get by solving (P_1). In this way several relaxation of $||.||_0$ have been tried, for instance in [5] they consider a concave minimization problem reformulating $\ell 0$-norm with ℓp-norm, in [8] and [9] they also consider a concave minimization problem using concave function to relax $\ell 0$-norm such that $(t+r)^p$, $-(t+r)^{-p}$, $\log(t+r)$ or $1-e^{-rt}$ with $r > 0$ and $p \in \mathbb{N}$. The purpose of this paper is to provide a theoretical context of a relaxation method giving concave minimization problem in a more general way, meaning using a general family of concave reformulation functions. This family has already been used in the different context of complementarity [6]. These functions are non-decreasing continuous smooth concave functions such

$$\theta : \mathbb{R} \to]-\infty, 1[\text{ with } \theta(t) < 0 \text{ if } t < 0, \ \theta(0) = 0 \text{ and } \lim_{t \to +\infty} \theta(t) = 1 . \quad (4)$$

We introduce a nonnegative parameter r, so

$$\theta_r(t) = \theta\left(\frac{t}{r}\right) \text{ with } \theta_r(0) = 0 \ \forall r > 0 \text{ and } \lim_{r \to 0} \theta_r(t) = 1 \ \forall t > 0 . \quad (5)$$

Examples of this family are $\theta_r^1(x) = \frac{x}{x+r}$ or $\theta_r^2(x) = 1 - e^{-\frac{x}{r}}$. In this paper we will consider the following concave optimization problem

$$(P_r) \quad \min_{x \in P} \sum_{i=1}^{n} \theta_r(x_i) . \quad (6)$$

This document will be organised as follow : section 2 presents convergence results, then, section 3 shows a sufficient convergence condition, finally, section 4 gives the algorithm with, in section 5, numerical results.

2   Convergence

In this section, we will show the link between problems (P_r), (P_0) and (P_1). Theorem 1 gives convergence of (P_r) to (P_0) for r sufficiently small.

Theorem 1 (Convergence to $\ell 0$-norm). *Given $S^*_{||.||_0}$ the set of solutions of (P_0) and S^*_r the set of solutions of (P_r). Every limit point of any sequence $\{x_r\}_r$, such that $x_r \in S^*_r$, is an optimal point of (P_0).*

Proof. Suppose $\bar{x} = \lim_{r \to 0} x_r$ the limits of a subsequence of $\{x_r\}_r$ and $x^* \in S^*_{||.||_0}$, we have for r sufficiently small

$$\sum_{i=1}^{n} \theta_r(\bar{x}_i) \leq \sum_{i=1}^{n} \theta_r(x_i^*) . \quad (7)$$

With the definition of $\theta_r(.)$'s function, for $r > 0$ and $t \in \mathbb{R}^n$

$$\sum_{i=1}^{n} \lim_{r \to 0} \theta_r(t_i) = ||t||_0 . \quad (8)$$

Replacing into (7) we get

$$||\bar{x}||_0 \leq ||x^*||_0 \; , \tag{9}$$

thanks to the definition of $\bar{x}$

$$||\bar{x}||_0 = ||x^*||_0 \; . \tag{10}$$

$\square$

Remark 1. Now we can wonder whether there exist for each $\bar{x} \in S^*_{||\cdot||_0}$ a sequence of $\{x^r\}_r$ which converges to this point. We don't give formal proof here, but with the hypothesis that our polyhedron contains no half-line, we have that $\bar{x}$ is then necessary one of the extreme points of P. Moreover there is a finite number of extreme points so we can build an infinite subsequence which goes by $\bar{x}$.

The next theorem shows for r sufficiently large that solutions of (P_r) are the same than solutions of (P_1). We will denote $\lim\limits_{r \to +\infty} S^*_r$ the set of limit points of subsequence of a sequence $\{x_r\}_r$, with x_r optimal solution of (P_r).

Theorem 2 (Convergence to $\ell 1$-norm). *Given $S^*_{||\cdot||_1}$ the set of solutions of (P_1) and S^*_r the set of solutions of (P_r). Passing to the limit when r is going to $+\infty$ we have*

$$\lim_{r \to +\infty} S^*_r \subset S^*_{||\cdot||_1} \; . \tag{11}$$

Proof. As $r > 0$, we can use a scaling technique for $S^{*(2)}_r = \arg\min\limits_{x \in P} \sum_{i=1}^n r\theta_r(x_i)$

$$\min_{x \in P} \sum_{i=1}^n \theta_r(x_i) \iff \min_{x \in P} \sum_{i=1}^n r\theta_r(x_i)$$
$$S^*_r = S^{*(2)}_r \; .$$

Given $x^r \in S^{*(2)}_r$ and $\bar{x} \in S^*_{||\cdot||_1}$. We use the first order Taylor's theorem for $\theta(t)$ in 0 :

$$\theta(t) = t\theta'(0) + g(t), \text{ where } \lim_{t \to 0} \frac{g(t)}{t} = 0 \; . \tag{12}$$

Functions θ are concave, so we have $\theta'(0) > 0$. By definition of $\bar{x}$ and then using (12) we have

$$\sum_{i=1}^{n} r\theta_r(x_i^r) \leq \sum_{i=1}^{n} r\theta_r(\bar{x}_i)$$

$$\sum_{i=1}^{n} x_i^r \theta'(0) + rg\left(\frac{x_i^r}{r}\right) \leq \sum_{i=1}^{n} x_i^r \theta'(0) + rg\left(\frac{\bar{x}_i}{r}\right)$$

$$\sum_{i=1}^{n} x_i^r - \sum_{i=1}^{n} \bar{x}_i \leq \frac{r}{\theta'(0)} \sum_{i=1}^{n} g\left(\frac{\bar{x}_i}{r}\right) - \frac{r}{\theta'(0)} \sum_{i=1}^{n} g\left(\frac{x_i^r}{r}\right)$$

$$\leq \frac{1}{\theta'(0)} \left| \sum_{i=1}^{n} \frac{g\left(\frac{\bar{x}_i}{r}\right)}{\frac{\bar{x}_i}{r}} \bar{x}_i \right| + \frac{1}{\theta'(0)} \left| \sum_{i=1}^{n} \frac{g\left(\frac{x_i^r}{r}\right)}{\frac{x_i^r}{r}} x_i^r \right|$$

$$\leq \frac{1}{\theta'(0)} \left(\sum_{i=1}^{n} \left| \frac{g\left(\frac{\bar{x}_i}{r}\right)}{\frac{\bar{x}_i}{r}} \right| \right) \left(\sum_{i=1}^{n} \bar{x}_i \right)$$

$$+ \frac{1}{\theta'(0)} \left(\sum_{i=1}^{n} \left| \frac{g\left(\frac{x_i^r}{r}\right)}{\frac{x_i^r}{r}} \right| \right) \left(\sum_{i=1}^{n} x_i^r \right)$$

$$\sum_{i=1}^{n} x_i^r \leq \left(\sum_{i=1}^{n} \bar{x}_i \right) \frac{1 + \frac{1}{\theta'(0)} \left(\sum_{i=1}^{n} \left| \frac{g\left(\frac{\bar{x}_i}{r}\right)}{\frac{\bar{x}_i}{r}} \right| \right)}{1 - \frac{1}{\theta'(0)} \left(\sum_{i=1}^{n} \left| \frac{g\left(\frac{x_i^r}{r}\right)}{\frac{x_i^r}{r}} \right| \right)} .$$

We have

$$\lim_{r \to +\infty} \frac{\bar{x}}{r} = 0 \ , \tag{13}$$

and now,

$$\sum_{i=1}^{n} \theta_r(x_i^r) \leq \sum_{i=1}^{n} \theta_r(\bar{x}_i) \tag{14}$$

$$\lim_{r \to +\infty} \sum_{i=1}^{n} \theta_r(x_i^r) \leq \lim_{r \to +\infty} \sum_{i=1}^{n} \theta_r(\bar{x}_i) \tag{15}$$

$$\leq 0 \tag{16}$$

$$\lim_{r \to +\infty} \sum_{i=1}^{n} \theta_r(x_i^r) = 0 \ . \tag{17}$$

As for $r > 0 : \theta_r(x_i) \in [0, 1[$ and $\theta_r^{-1}(0) = 0$, we have

$$\lim_{r \to +\infty} \frac{x_i^r}{r} = 0 \ \forall i \ . \tag{18}$$

Using (13) and (18) it becomes

$$\lim_{r \to +\infty} \frac{\bar{x}_i}{r} = 0 \implies \lim_{r \to +\infty} \frac{g(\frac{\bar{x}_i}{r})}{\frac{\bar{x}_i}{r}} = 0 \tag{19}$$

$$\lim_{r \to +\infty} \frac{x_i^r}{r} = 0 \implies \lim_{r \to +\infty} \frac{g(\frac{x_i^r}{r})}{\frac{x_i^r}{r}} = 0 \ . \tag{20}$$

And then going to the limit

$$\lim_{r \to +\infty} \sum_{i=1}^{n} x_i^r \leq \sum_{i=1}^{n} \bar{x}_i \ . \tag{21}$$

So, with (21) and by definition of $\bar{x}$ we have the equality and then the result.   □

Finally, the next theorem gives a monotonicity result, which gives a relation between the three problems (P_r), (P_0) and (P_1).

Theorem 3 (Monotonicity of solutions). *Given $x \in P$, we set $y = \frac{x}{||x||_\infty + \epsilon}$ where $\epsilon > 0$, so $y \in [0, 1[^n$. We define a function $\Psi_r(t)$ as*

$$\Psi_r(t) = \frac{\theta_r(t)}{\theta_r(1)} \ , \tag{22}$$

where $\theta_r(t)$ is the smooth function described in the introduction, which we will consider here as convex in r. For r and $\bar{r}$ such that $0 < \bar{r} < r < +\infty$, we have the following inequality

$$||y||_1 \leq \sum_{i=1}^{n} \Psi_r(y_i) \leq \sum_{i=1}^{n} \Psi_{\bar{r}}(y_i) \leq ||y||_0 \ . \tag{23}$$

Proof. We start with first inequality

$$||y||_1 \leq \sum_{i=1}^{n} \Psi_r(y_i) \ , \tag{24}$$

$$\sum_{i=1}^{n} \Psi_r(y_i) - ||y||_1 = \sum_{i=1}^{n} (\frac{\theta_r(y_i)}{\theta_r(1)} - y_i) \tag{25}$$

$$\geq 0 \ . \tag{26}$$

Because we have by subadditivity of θ (concave and $\theta(0) = 0$)

$$\theta_r(y_i) \geq y_i \, \theta_r(1) \ . \tag{27}$$

We continue with the second inequality showing that $\Psi_r(y)$ functions are non-increasing in r,

$$\sum_{i=1}^{n} \Psi_r(y_i) \leq \sum_{i=1}^{n} \Psi_{\bar{r}}(y_i) \ . \tag{28}$$

$\Psi_r(y)$ function is non-increasing in r if its derivative to r is negative,

$$\Psi_r'(y) = \frac{\theta_r'(y)\theta_r(1) - \theta_r'(1)\theta_r(y)}{\theta_r^2(1)} . \tag{29}$$

Using convexity of $\theta_r(y)$ in r, we have

$$\theta_r(y) < 0 \tag{30}$$

$$\frac{\theta_r'(y)}{\theta_r'(1)} > 1 . \tag{31}$$

Moreover as $\theta_r(y)$ is non-decreasing in y

$$\frac{\theta_r(y)}{\theta_r(1)} < 1 . \tag{32}$$

So, in (29) the derivative in r is negative and then we have (28). Finally, we move to the last inequality,

$$\sum_{i=1}^{n} \Psi_{\bar{r}}(y_i) \leq ||y||_0 . \tag{33}$$

The following expression is positive, because $\theta_r(y)$ is non-decreasing in y and $y \in [0, 1[$

$$||y||_0 - \sum_{i=1}^{n} \Psi_{\bar{r}}(y_i) = \sum_{i=1:\ y_i \neq 0}^{n} 1 - \frac{\theta_{\bar{r}}(y_i)}{\theta_{\bar{r}}(1)} \geq 0 . \tag{34}$$

Associating the three inequalities we have the theorem. □

All this results leads us to the general behaviour of the method. First, we start from one solution of (P_1) then by decreasing parameter r the solution of (P_r) becomes closer of the desired solution, a solution of (P_0).

3 Error Estimate

In this section we focus on what happened when r becomes sufficiently small.

Lemma 1. *Consider θ functions where $\theta \geq \theta^1$, with $\theta_r^1(t) = \frac{t}{t+r}$ for $t, r \in \mathbb{R}$. Let $k = ||x^*||_0 < n$ be the optimal value of problem (P_0). Set $I(x, r) = \{i | x_i \geq kr\}$. Then, we have*

$$x^r \in \arg\min_{x \in P} \sum_{i=1}^{n} \theta_r(x_i) \Rightarrow card(I(x^r, r)) \leq k . \tag{35}$$

Proof. We use a proof by contradiction. Consider that $card(I(x^r, r)) \geq k + 1$ and we have $x^r \in \arg\min_{x \in P} \sum_{i=1}^{n} \theta_r(x_i)$, then

$$\sum_{i=1}^{n} \theta_r(x_i^r) \geq (k+1)\theta_r(kr) \geq (k+1)\theta_r^1(kr) = (k+1)\frac{kr}{kr+r} = k , \tag{36}$$

which is in contradiction to the definition of x^r. □

This lemma gives us a theoretical stopping criterion for the decrease of r, as for $r < \bar{r} = \frac{\min_{x_i^r \neq 0} x_i^r}{k}$, x^r becomes an optimal solution. In the following lemma we look at the consequences for the evaluation of θ function.

Lemma 2. *Consider θ functions where $\theta \geq \theta^1$, with $\theta_r^1(t) = \frac{t}{t+r}$ for $t, r \in \mathbb{R}$. Let $k = \|x^*\|_0 < n$ be the optimal value of problem (P_0) and $\bar{r} = \frac{\min_{x_i^r \neq 0} x_i^r}{k}$. Then, we have*

$$r \leq \bar{r} \iff \theta_r(\min_{x_i^r \neq 0} x_i^r) \geq \frac{k}{k+1} . \tag{37}$$

Proof. We just have to develop expression of θ^1,

$$\theta_r^1(\min_{x_i^r \neq 0} x_i^r) = \frac{\min_{x_i^r \neq 0} x_i^r}{\min_{x_i^r \neq 0} x_i^r + r} \geq \frac{k}{k+1} \tag{38}$$

$$\iff \min_{x_i^r \neq 0} x_i^r (k+1) \geq k(\min_{x_i^r \neq 0} x_i^r + r) \tag{39}$$

$$\iff \min_{x_i^r \neq 0} x_i^r \geq kr \tag{40}$$

$$\iff \bar{r} = \frac{\min_{x_i^r \neq 0} x_i^r}{k} \geq r , \tag{41}$$

so we have the results. $\square$

Both previous lemma lead us to the following theorem, which is a sufficient condition for our method.

Theorem 4 (Sufficient condition). *Consider θ functions where $\theta \geq \theta^1$, with $\theta_r^1(t) = \frac{t}{t+r}$ for $t, r \in \mathbb{R}$. Let $k = \|x^*\|_0 < n$ be the optimal value of problem (P_0) and $x^r \in S_r$. Then, $\theta_r(\min_{x_i^r \neq 0} x_i^r) \geq \frac{k}{k+1}$ is a sufficient condition to have x^r solution of $S_{\|\cdot\|_0}^*$, i.e*

$$\theta_r(\min_{x_i^r \neq 0} x_i^r) \geq \frac{k}{k+1} \implies x^r \in S_{\|\cdot\|_0}^* . \tag{42}$$

Proof. Lemma [2] gives with $\bar{r}$ defines as in the lemma

$$\theta_r(\min_{x_i^r \neq 0} x_i^r) \geq \frac{k}{k+1} \iff r \leq \bar{r} , \tag{43}$$

then by lemma [1] and using $x_r \in S_r$ we have

$$x^r \in \arg\min_{x \in C} \sum_{i=1}^{n} \theta_r(x_i) \implies card(I(x_r, r)) \leq k . \tag{44}$$

Using $r \leq \bar{r}$ and the fact that k is the optimal value of problem in $\ell 0$-norm, we have the results. $\square$

4   Algorithm

The previous study allows us to build a generic algorithm

$$[\text{Algorithm 1}] \quad \begin{cases} \{r^k\}_{k\in\mathbb{N}}, \ r^0 > 0 \text{ and } \lim\limits_{k\to+\infty} r^k = 0 \\ \text{find } x^k : \ x^k = \inf\limits_{x\in P} \sum_{i=1}^{n} \theta_{r^k}(x_i) \end{cases} . \tag{45}$$

Now, as we can see several questions remain about initialization, choice of the sequence $\{r^k\}$ and the method used to solve the concave minimization problem. In the previous section, we have shown a sufficient condition to converge, which will help us building a stopping criterion. We make a few remarks about these questions, note that interesting related remarks can be found in [8].

Remark 2 (On the behaviour of θ functions). These concave functions are acting, for r sufficiently small, as step function. So we have the following behaviour

$$\theta_r(t) \simeq \begin{cases} 1 \text{ if } t >> r \\ 0 \text{ if } t << r \end{cases} \tag{46}$$

which gives us a strategy to update r. Let x^k be our current iterate and r^k the corresponding parameter. We divide our iterate into two sets those with indices in $I = \{i \mid x_i^k >= r^k\}$ and the others with indices in $\bar{I} = \{i \mid x_i^k < r^k\}$. We can see I as the non-zero components and $\bar{I}$ the zero components of x^k. So we will choose r^{k+1} around $\max\limits_{i\in\bar{I}} x_i^k$ to ask whether or not it belongs to zeros and we repeat this operation until r is sufficiently small to consider $\bar{I}$ the set of real zeros. Also this is a global behaviour, to be sure to have decrease of r we can choose a fixed parameter of decrease.

Remark 3 (Initialization). It is the main purpose of our method to start with the solution x^0 of the problem (P_1) which is a convex problem. So, we need to find the r^0 which corresponds to this x^0. A natural way of doing this is to find the parameter which minimizes the following problem

$$\min_{r>0} \| \sum_{i=1}^{N} \theta_r(x_i^0) - \|x^0\|_1 \|_2^2 . \tag{47}$$

A simpler idea is to be inspired from last remark and put r^0 as a value which is just beyond the top value of x_i^0.

Remark 4 (Stopping criterion). It has been shown, in the previous section, a sufficient condition to have convergence, which uses the quantity $\frac{k}{k+1}$, which depends on the solution we are looking for. Numerically, we can make more iterations but being sure to satisfy this criterion using the fact that $\|x^0\|_0 \geq k$, which gives us the following criterion

$$\theta_r(\min_{x_i^r \neq 0} x_i^r) \geq \frac{\|x^0\|_0}{\|x^0\|_0 + 1} \geq \frac{k}{k+1} . \tag{48}$$

5 Numerical Simulations

Thanks to the previous sections we have keys for an algorithm, we will show now some numerical results. These simulations have been done using Python programming language with the free software package for convex optimization CVXOPT, [1].

In the same way as in [5] and [9] we will use SLA (successive linearisation algorithm) algorithm to solve our concave minimization problem at each iteration in r. This algorithm is a finitely timestep Franck & Wolf algorithm, [7].

Proposition 1 (SLA algorithm for concave minimization). *Given ϵ sufficiently small and r^k. We know x^k and we find x^{k+1} as a solution of the linear problem*

$$\min_{x \in P} x^t \nabla \theta_{r^k}(x^k) \ , \tag{49}$$

with x^0 a solution of the problem (P_1). We stop when

$$x^{k+1} \in P \ and \ (x^{k+1} - x^k)^t \nabla \theta_{r^k}(x^k) \leq \epsilon \ . \tag{50}$$

This algorithm generates a finite sequence with strictly decreasing objective function values.

Proof. see [[7], Theorem 4.2].

We note that this algorithm didn't provide necessarily a global optimum, as it ends in a local solution. For the precision in our simulations we choose $\epsilon = 1e-8$.

We make simulations on various polyhedron $P = \{x \in \mathbb{R}^n | \ b \in \mathbb{R}^m, \ Ax \leq b\} \cap \mathbb{R}^n_+$ with $m < n$. In the same way as in [5] we choose $n = (500, 750, 1000)$ and in each case $m = (40\%, 60\%, 80\%)$. For each n and m we choose randomly one hundred problems. We take a random matrix A of size $m \times n$ and a default sparse solution x_{init} (sparsity : 5% of non-zeros). We get b by calculating the product $b = Ax_{init}$. At the end we will compare the sparsity of our solution using $\theta^1_r(t) = \frac{t}{t+r}$ ($\#\theta^1$), the sparse default solution ($\#\ell 0$) and the initial iterate ($\#\ell 1$). We get the initial iterate as a solution of problem (P_1). The item $\#$ indicates the number of non-zeros components in a vector.

We can see the results in table (1). First we must say that in many cases the $\ell 1$-norm minimization solution solve the problem in $\ell 0$-norm, which is not completely surprising according to [3]. This results confirm the interest of this method, as in every case it manages to find an equivalent solution to the default sparse solution. Also it improves in every case the solution we get by $\ell 1$-norm minimization problem, which was our principle aim.

Table 1. Numerical results for $\min_{x \in P} ||x||_0$ with $P = \{x \in \mathbb{R}^n_+ |\ b \in \mathbb{R}^m,\ Ax \leq b\}$, dimensions of the problem are first 2 columns. Then, we compare a default sparse solution with sparsity 5% of non-zeros, the initial iterate solution of $\min_{x \in P} ||x||_1$ and the solution by θ-algorithm with function θ^1. The item # indicates the number of non-zeros.

n	m	$\#\ell 0 = \#\theta^1$	$\#\ell 1 = \#\ell 0$	$\#\theta^1 < \#\ell 1$	$\#\theta^1 = \#\ell 1$
1000	800	100	97	3	97
1000	600	100	89	11	89
1000	400	100	52	48	52
750	600	100	98	2	98
750	450	100	82	18	82
750	300	100	48	52	48
500	400	100	81	19	81
500	300	100	82	18	82
500	200	100	38	62	38

6   Conclusion and Outlook

We have shown a general method to solve NP-hard problem of minimizing $\ell 0$-norm. This method requires to find a sequence of solution from concave minimization problem, which we solved with a successive linearization algorithm. We gave convergence results, a sufficient convergence condition and keys to implement the method. To confirm that we improved the solution obtained from the $\ell 1$-norm problem we gave some numerical results.

Further studies can investigate the case where the $\ell 1$-norm solve the $\ell 0$-norm problem, to find an improved stopping condition. We can also study a very similar problem [4,2] which is the minimizing $\ell 0$-norm problem with noise.

References

1. Andersen, M.S., Dahl, J., Vandenberghe, L.: A python package for convex optimization, version 1.1.6. (2013), cvxopt.org
2. Babaie-Zadeh, M., Jutten, C.: On the stable recovery of the sparsest overcomplete representations in presence of noise. IEEE Transactions on Signal Processing 58(10), 5396–5399 (2010)
3. Donoho, D.L.: Neighborly polytopes and sparse solutions of underdetermined linear equations. Tech. rep., Department of Statistics, Stanford University (2004)
4. Donoho, D.L., Elad, M., Temlyakov, V.N.: Stable recovery of sparse overcomplete representations in the presence of noise. IEEE Transactions on Information Theory 52(1), 6–18 (2006)
5. Fung, G.M., Mangasarian, O.L.: Equivalence of minimal $\ell 0$ and ℓp norm solutions of linear equalities, inequalities and linear programs for sufficiently small p. Journal of Optimization Theory and Applications 151(1), 1–10 (2011)
6. Haddou, M., Maheux, P.: Smoothing methods for nonlinear complementarity problems. Journal of Optimization Theory and Applications (2014)

7. Mangasarian, O.L.: Machine learning via polyhedral concave minimization. In: Applied Mathematics and Parallel Computing, pp. 175–188. Springer (1996)
8. Mohimani, H., Babaie-Zadeh, M., Jutten, C.: A fast approach for overcomplete sparse decomposition based on smoothed norm. IEEE Transactions on Signal Processing 57(1), 289–301 (2009)
9. Rinaldi, F.: Mathematical Programming Methods for minimizing the zero-norm over polyhedral sets. Ph.D. thesis, Sapienza, University of Rome (2009)

Bilevel Quadratic Fractional/Quadratic Problem

Nacéra Maachou[1] and Mustapha Moulaï[2]

[1] USTHB, Faculty of Mathematics, BP 32 El-Alia, 16111 Algiers, Algeria
nacera_maachou@yahoo.fr
[2] USTHB University, LaROMaD Laboratory, BP 32 El-Alia, 16111 Algiers, Algeria
mmoulai@usthb.dz

Abstract. The bilevel programming problems are useful tools for solving the hierarchy decision problems. The purpose of this paper is to find the optimality conditions and a solution procedure to solve a bilevel quadratic fractional-quadratic programming problem in which the leader's objective is quadratic fractional and the follower's objective is quadratic. The variables associated with both the level problems are related by linear constraints. The proposed method is based on Karush-Kuhn-Tucker conditions and a related bilevel linear fractional-quadratic problem is constructed in which the leader's objective is linear fractional and the follower's objective is quadratic in order to obtain an optimal solution of a bilevel quadratic fractional-quadratic programming problem. The main idea behind our method is to scan the basic feasible solutions of the related bilevel linear fractional- quadratic programming problem in a systematic manner till an optimal solution of the problem is obtained.

Keywords: linear fractional programming, Bilevel Programming, quadratic fractional programming, Karush-Kuhn-Tucker conditions.

1 Introduction

Bilevel problems can be formulated as

$$(P) \max_{(x,y) \in S} F(x,y) \tag{1}$$

$$\text{where } y \in arg \max_{y \in S(x)} f(x,y).$$

where

$$x \in \mathbb{R}^{n_1} \ \ and \ \ y \in \mathbb{R}^{n_2} \tag{2}$$

are variables controlled by the first level and the second level decision maker, respectively;

$$F, f : \mathbb{R}^n \longmapsto \mathbb{R}, \ n = n_1 + n_2; \ S \subset \mathbb{R} \tag{3}$$

defines the common constraint region and

$$S(x) = \left\{ y \in \mathbb{R}^{n_2} : (x,y) \in S \right\}. \tag{4}$$

© Springer International Publishing Switzerland 2015
H.A. Le Thi et al. (eds.), *Model. Comput. & Optim. in Inf. Syst. & Manage. Sci.*,
Advances in Intelligent Systems and Computing 359, DOI: 10.1007/978-3-319-18161-5_32

Let S_1 be the projection of S onto $\mathbb{R}^{n1}$. For each $x \in S_1$, the second level decision maker solves problem (P_1):

$$(P_1) \begin{cases} \max \ f(x,y) \\ \text{s.t} \\ \quad y \in S(x). \end{cases} \tag{5}$$

The feasible region of the first level decision maker, called inducible region IR, is implicitly defined by the second level optimization problem:

$$IR = \{(x,y^*) : x \in S_1, y^* \in M(x)\}$$

where $M(x)$ denotes the set of optimal solutions to (P_1). We assume that S is not empty and that for all decision taken by the first level decision maker, the second level decision maker has some room to respond, i.e $M(x) \neq \emptyset$.

Bilevel optimization problems have seen a rapid development and broad interests both from the practical and the theoretical point of view [17].

A large number of possible applications have been reported: principal-agency problems in economy [15,16], the coordination of multidivisional firms [1], neural network training [9] and many others. Considering some of these applications, it seems to be imaginable that one level objective function is nonlinear.

On the other hand, fractional programming and quadratic programming when there exists only one level of decision have received remarkable attention in the literature [8,7]. It is worth mentioning that objective functions which are ratios frequently appear, for instance, when an efficiency measure of a system is to be optimized or when approaching a stochastic programming problem.

Quadratic problems arise directly in such applications as least-squares regression with bounds or linear constraints, portfolio optimization, or robust data fitting. They also arise as subproblem in optimization algorithms for stochastic nonlinear programming [14]. Fractional bilevel problems have been considered in [3,4,5]. Quadratic bilevel problems have been addressed in [10,12,17].

Calvete and Galé [3] studied optimality conditions for the linear fractional/ quadratic bi-level programming problem based on Karush Kuhn Tucker conditions and duality theory. For a largest class of fractional/quadratic bi-level problem, we are interested to bilevel quadratic fractional/quadratic programming (BQFQP) problem, where the first level objective function is fractional quadratic [6] and the second level objective function is quadratic. The variables associated with both the level problems are related by linear constraints. For solving the problem, we construct the related bilevel linear fractional/quadratic problem and use Kuhn -Tucker optimality conditions with duality conditions [1], by applying some propositions, we are able to determinate an optimal solution for (BQFQP). The contents of the paper is as follows: In Section 2 the BQFQP problem is formulated. Section 3 provides the main theoretical results on optimality conditions. A more detailed description of the algorithm for the whole problem is given in Section 4. Section 5 concludes the paper.

2   Notations and Definitions

Let S the set of the feasible solutions $(x, y) : x \in \mathbb{R}^{n_1}, y \in \mathbb{R}^{n_2}$ satisfying the constraints $\begin{cases} Ax + By \leq r \\ x \geq 0, y \geq 0 \end{cases}$

where A is $m \times n_1$ matrix, B is $m \times n_2$ matrix and r a vector of $\mathbb{R}^m$. Let C, H, P vectors of $\mathbb{R}^{n_1}$, D, M, R vectors of $\mathbb{R}^{n_2}$; E, F, L, G, Q are real symmetric matrices and α, β two elements of $\mathbb{R}$.

The Bilevel quadratic fractional/quadratic programming problem $(BQFQP)$ [13], intended to be studied can be mathematically stated as:

$$(BQFQP) \begin{cases} \max_{x} F(x, y) = \dfrac{C^T x + x^T E x + D^T y + y^T F y + \alpha}{H^T x + x^T L x + M y + y^T G y + \beta} \\ \text{where } y \text{ solves} \\ \qquad \begin{cases} \max f(x, y) = Px + Ry + (x, y)^T Q(x, y) \\ s.t \quad (x, y) \in S \\ \qquad (x, y) \geq 0 \end{cases} \end{cases} \quad (6)$$

Definition 1. *The set of feasible solutions (x, y) of (BQFQP) problem is the set S defined by :*

$$S = \{(x, y) : Ax + By \leq r, x, y \geq 0\} .$$

Definition 2. *The projection of S onto the leader's decision space is*

$$S(x) = \{x \geq 0 : \exists y \in \mathbb{R}^{n_2} : (x, y) \in S\}.$$

In order to ensure that the problem (6) is well posed we make an assumption that S is non empty and bounded and

$C^T x + x^T E x + D^T y + y^T F y + \alpha \geq 0,$
$H^T x + x^T L x + My + x^T Gy + \beta > 0$ for all $(x, y) \in S$

In the problem (6), let
$\mathcal{Q} = \begin{pmatrix} Q_3 & Q_2^T \\ Q_2 & Q_1 \end{pmatrix}$ where Q_1, Q_2 and Q_3 are matrices of conformal dimensions.
Then $f(x, y)$ is transformed into

$$f(x, y) = P^T x + x^T Q_2 x + y^T Q_1 y + (R + 2Q_2 x)^T y$$

Because x is fixed prior to the maximization of f, the follower's problem is equivalent to

$$(P_x) \begin{cases} \max_{y} f(x, y) = y^T Q_1 y + (R + 2Q_2 x)^T y \\ s.t \\ \qquad By \leq r - Ax \\ \qquad (x, y) \geq 0 \end{cases} \quad (7)$$

For all $x \in S_1$ (S_1 the projection of S onto $\mathbb{R}^{n1}$), we assume that Q_1 is positively definite so as there will be a unique optimal solution to the second level problem. That is to say, $M(x)$ is a singleton for all S_1, where $M(x)$ is the set of optimal solutions to (P_x). For each $x \in S_1$, $S(x)$ is also nonempty compact polyhedron. Finally, the inducible region, or the feasible region of the first level decision maker, called inducible region IR, is implicitly defined by the second level optimization problem:

$$IR = \{(x, y^*) : x \in S_1, y^* \in M(x)\}$$

Therefore $(BQFQP)$ is equivalent to:

$$(BQFQP') \begin{cases} \max\limits_{x} F(x,y) = \dfrac{C^T x + x^T E x + D^T y + y^T F y + \alpha}{H^T x + x^T L x + M y + y^T G y + \beta} \\ \text{where } y \text{ solves} \\ \quad \begin{cases} \max\limits_{y} f(x,y) = y^T Q_1 y + (R + 2Q_2 x)^T y \\ s.t \qquad By \leq r - Ax \\ \qquad (x,y) \geq 0 \end{cases} \end{cases} \qquad (8)$$

Definition 3. . A point (x^*, y^*) is said to be optimal to $(BQFQP')$ if $(x^*, y^*) \in IR$ and $F(x^*, y^*) \geq F(x, y)$ for all $(x, y) \in IR$

A $(BQFQP')$ may be written:

$$(BQFQP') \begin{cases} \max\limits_{x} F(x,y) = \dfrac{C^T x + x^T E x + D^T y + y^T F y + \alpha}{H^T x + x^T L x + M y + y^T G y + \beta} \\ s.t \\ \qquad (x,y) \in IR \end{cases} \qquad (9)$$

To find an optimal solution of problem $(BQFQP')$, a related bilevel linear fractional/quadratic programming $(BLFQP)$ problem is constructed.

$$(BLFQP) \begin{cases} \max\ g(x,y) = \dfrac{a^T x + b^T y + \alpha}{c^T x + d^T y + \beta} \\ s.t \\ \qquad (x,y) \in IR \end{cases} \qquad (10)$$

where

$$a = \max\limits_{x \in IR} (C + x^T E) \text{ and } b = \max\limits_{x \in IR} (D + y^T F).$$

$$c = \min\limits_{x \in IR} (H + x^T L) \text{ and } d = \min\limits_{x \in IR} (M + y^T G).$$

3   Some Results

Proposition 1.

$$g(x, y) \geq F(x, y), \quad for \ all \ (x, y) \in IR. \tag{11}$$

Proof. By definition of a, b, c and d, we have for all $(x, y) \in IR$:

$$a \geq C + x^T E \quad \text{and} \quad c \leq H + x^T L$$
$$b \geq D + y^T F \qquad \qquad d \leq M + y^T G \qquad .$$

As $(x, y) \geq 0$,

$$a^T x \geq (C^T + x^T E) x \quad \text{and} \quad b^T y \geq (D^T + y^T F) y \ , \quad \text{for all } (x, y) \in IR$$

Then

$$a^T x + b^T y \geq (C^T + x^T E) x + (D^T + y^T F) y, \quad \forall (x, y) \in IR.$$

In the same way: $c^T x + d^T y \leq (H_j + x^T L_j) x + (M_j + y^T G_j) y$, for all $(x, y) \in IR$.

Clearly it follows that $g(x, y) \geq F(x, y)$, for all $(x, y) \in IR$.

Recall that the bilevel linear fractional/quadratic programming problem $(BLFQP)$ can be solved by applying the optimality condition based on duality and Karush-Kuhn-Tucker conditions [8].

Consider the problem $(BLFQP)$:

$$(BLFQP) \begin{cases} \max \ g(x, y) = \dfrac{a^T x + b^T y + \alpha}{c^T x + d^T y + \beta} \\ \text{s.t} \\ \qquad (x, y) \in IR \end{cases} \tag{12}$$

By applying Karush-Kuhn-Tucker necessary and sufficient conditions to (P_x), there exists $\omega \in \mathbb{R}^m$, such that (x, y, ω) satisfies:

$$Ax + By \leq r \tag{13}$$
$$\omega^T(Ax + By - r) = 0 \tag{14}$$
$$2Q_2 x + 2Q_1 y + B^T \omega = -R^T \tag{15}$$
$$\omega \geq 0 \tag{16}$$

Similarly, if (x, y, ω) satisfies $(13) - (16)$ then $(x, y) \in IR$

Thus, the given $(BLFQP)$ problem becomes a linear fractional programming (LFP) problem given by:

$$(LFP) \begin{cases} \max\limits_{(x,y,\omega)} \ g(x, y) = \dfrac{a^T x + b^T y + \alpha}{c^T x + d^T y + \beta} \\ \text{s.t} \qquad Ax + By \leq r \\ \qquad \quad 2Q_2 x + 2Q_1 y + B^T \omega = -R^T \\ \qquad \quad x, y, \omega \geq 0 \end{cases} \tag{17}$$

with the condition

$$\omega^T(Ax + By - r) = 0 \tag{18}$$

Theorem. (x, y) is an optimal solution to the $(BLFQP)$ problem if and only of there exists $\omega^* \in \mathbb{R}^m$ such that (x^*, y^*, ω^*) is an optimal solution to the following one level nonlinear programming problem:

$$(NLP) \begin{cases} \max\limits_{(x,y,\omega)} & g(x,y) = \dfrac{a^T x + b^T y + \alpha}{c^T x + d^T y + \beta} \\ \text{s.t} & (13) - (16) \end{cases} \tag{19}$$

Using duality theory, we conclude the existence of $\omega^* \in \mathbb{R}^m$ such that (x^*, y^*, ω^*) is optimal solution of the (NLP)[8]. Hence, from Theorem [8] we get that (x^*, y^*) is an optimal solution to the $(BLFQP)$ problem [8].

The nonlinear programming (NLP) can be rewrite as:

$$(LFP) \begin{cases} \max\limits_{(x,y,\omega)} & g(x,y) = \dfrac{a^T x + b^T y + \alpha}{c^T x + d^T y + \beta} \\ \text{s.t} & Ax + By \le r \\ & 2Q_2 x + 2Q_1 y + B^T \omega = -R^T \\ & x, y, \omega \ge 0 \end{cases} \tag{20}$$

with the condition

$$\omega^T(Ax + By - r) = 0 \tag{21}$$

So, for determine the optimal solution of $(BLFQP)$, we solve the linear fractional problem (LFP) with the condition (21).

As the objective function is linear fractional, therefore, it is both pseudo-concave and pseudoconvexe and thus, its optimal solution will be basic feasible solution. We focus to find a basic feasible solution which satisfies the condition

$$\omega^T(Ax + By - r) = 0 \tag{22}$$

Recall that, for obtaining the basic feasible solutions, the linear fractional program (LFP) is transformed into a linear program (LP) by Charnes and Cooper's method [2]. Murty's method [11] is applied to a linear program (LP) to determine all basic feasible solutions. This method is based by generating an objective coefficient vector in each iteration, such that the optimization of this objective function subject to the specified constraints by linear programming techniques leads to a new basic feasible solution, until all basic feasible solutions are obtained.

3.1   Notations

$\Delta_i = $ Set of the ith best feasible solution of $(BLFQP)$,
g_i the value corresponding at Δ_i

Clearly

Δ_1 =Set of optimal solution (first best feasible solution) of $(BLFQP)$,

g_1 the optimal value corresponding at Δ_1 and

Δ_2 = Set of the 2nd best feasible solution of $(BLFQP)$, g_2 the value corresponding at Δ_2.

It follows that $g_1 > g_2$.

Obviously for $i = k$ we have $g_k > g_{k+1}$.

Introduce the notations

$\quad T^k = \Delta_1 \cup \Delta_2 \cup ... \cup \Delta_k,$

$\quad \max \{F(x,y) \mid (x,y) \in IR\} = F(x^\star, y^\star) = F_1$

Proposition 2. *If for some $k \geq 1$, $g_k \leq \max \{F(x,y) \mid (x,y) \in T^k\} = F(\hat{x}, \hat{y})$, then $(\hat{x}, \hat{y})$ is the optimal solution of $(BQFQP')$.*

Proof. We have

$$\forall (x,y) \in T^k \quad F(\hat{x}, \hat{y}) \geq F(x,y) \tag{23}$$

On the other hand from Proposition 1 and our hypothesis it follows that

$$\forall (x,y) \in IR/T^k \quad F(x,y) \leq g(x,y) \leq g_{k+1} < g_k \leq F(\hat{x}, \hat{y}). \tag{24}$$

Conditions (23) and (24) implies that $(\hat{x}, \hat{y})$ is the optimal solution of $(BQFQP')$ problem.

Next proposition shows that when the hypothesis of Proposition 2 is not satisfied, we have information about the maximum value of the initial problem $(BQFQP')$.

Proposition 3. *If $g_k > \max \{F(x,y) \mid (x,y) \in T^k\}$, then $g_k > F_1 \geq \max \{F(x,y) \mid (x,y) \in T^{k+1}\}$.*

Proof. The second inequality is obvious. For the first part, we note as in proposition 2 that

$$for \ all \ (x,y) \in IR/T^k \ , \ F(x,y) \leq g(x,y) \leq g_{k+1} < g_k \tag{25}$$

Therefore

$$\max \{F(x,y) : (x,y) \in IR/T^k\} < g_k \tag{26}$$

By hypothesis,

$$\max \{F(x,y) : (x,y) \in T^k\} < g_k \tag{27}$$

Then conditions (26) and (27) imply that $F_1 < g_k$.

3.2 Algorithm

Step 0 Solving the bilevel linear-quadratic problems (using Karush-Kuhn-Tucker necessary and sufficient conditions[1]):

$$\begin{cases} a = \max_{(x,y)\in IR} (C + x^T E), b = \max_{(x,y)\in IR} (D + y^T F), \\ c = \min_{(x,y)\in IR} (H + x^T L), d = \min_{(x,y)\in IR} (M + y^T G). \end{cases} \tag{28}$$

and construct the related bilevel linear fractional/quadratic programming problem ($BLFQP$), go to step 1.

Step 1 Solve The linear fractional programming problem (LFP) .

$$(LFP) \begin{cases} \max_{(x,y,\omega)} \ g(x,y) = \dfrac{a^T x + b^T y + \alpha}{c^T x + d^T y + \beta} \\ \text{s.t} \quad Ax + By \le r \\ \qquad 2Q_2 x + 2Q_1 y + B^T\omega = -R^T \\ \qquad x, y, \omega \ge 0 \end{cases} \tag{29}$$

Let (x^*, y^*, ω^*) be its optimal solution. Set $l = 1$ and go to step 2.

Step 2 . If the solution obtained is such that: $\omega^*(Ax^* + By^* - r) = 0$, then $(x^*, y^*) = (x^l, y^l)$ is an optimal solution of ($BLFQP$) problem, let $\Delta_l = \{(x^l, y^l)\}$, compute $g(x^l, y^l) = g_l$. Go to step 4.
Otherwise, go to step 3.

Step 3 Find the next best basic feasible solution to the (LFP) problem using murty's method [11] and go to step 2.

Step 4 Test
  - if $g_l \le \max\{F(x,y) \mid (x,y) \in T^l\} = F(\hat{x}, \hat{y})$. Then $(\hat{x}, \hat{y})$ is the optimal solution of problem ($BQFQP$).
  - If $g_l > \max\{F(x,y) \mid (x,y) \in T^l\}$, $l = l + 1$ and go to step 3.

As seen previously, the algorithm converges in a finite number of steps since the number of basic feasible solutions in S is finite and once a point is tested for optimality of ($BQFQP$), it is deleted from the set and hence does not reappear in the process.

4 Conclusion

In this paper, we have studied the bilevel quadratic fractional/quadratic programming problem. Our procedure is based on related bilevel linear fractional /quadratic programming problem and Karush-Kuhn-Tucker conditions. These conditions have been used in literature for solving a large number of nonlinear programming problems. Bilevel linear fractional/quadratic programming problem related to the main problem is formulated and its basic feasible solutions are scanned in a systematic manner till an optimal solution of the problem is obtained. Our new procedure can be applied to bilevel linear/quadratic programming problems since they are special cases of this mathematical program. we hope that this article motivates the researchers to develop better solution procedures for this problem.

Acknowledgments. The authors are grateful to anonymous referees for their substantive comments and suggestions that improved the quality and presentation of the paper.

References

1. Bard, J.F.: Coordination of mutivisional organization through two levels of management. OMEGA 11, 457–468 (1983)
2. Charnes, A., Cooper, W.W.: Programming with linear fractional functions. Naval Research Logistics Quaterly 9, 181–186 (1962)
3. Calvete, H.I., Galé, C.: The bilevel linear/ linear fractional programming problem. European J. of Operational Research 114(1), 188–197 (1999)
4. Calvete, H.I., Galé, C.: Bilevel fractional programming. In: Flouads, C.A., Pardalos, P.M. (eds.) Encyclopedia of Optimization, pp. 135–137. Kluwer Academic Publishers, Dordrecht (2001)
5. Calvete, H.I., Galé, C.: Solving linear fractional bilevel programs. Operations Research Letters 32(2), 143–151 (2004b)
6. Gotoh, J.Y., Konno, H.: Maximization of the Ratio of Two Convex Quadratic Functions over a Polytope. Computational Optimization and Applications 20(1), 43–60 (2001)
7. Gupta, R., Puri, M.C.: Extreme point quadratic fractional programming problem. Optimization 30(3), 205–214 (1994)
8. Calvete, H.I., Galé, C.: Optimality conditions for the linear fractional/quadratic bilevel problem. Monografias del Seminario Matemático Garcia de Galdeano 31, 285–294 (2004)
9. Mangasarian, O.L.: Misclassification minimization. Journal of Global Optimization 5, 309–323 (1994)
10. Migdalas, A., Pardalos, P.M., Värbrand, P.: Multilevel optimization: Algorithm and applications. Kluwer Academic Publishers, Dordrecht (1998)
11. Murt, K.G., Chung, S.J.: Extreme Point Enumeration. College of engineering, University of Michigan. Technical Report 92–21 (1992)
12. Muu, L.D., Quy, N.V.: A global optimization method for solving convex quadratic bilevel programming problems. J. of Global Optimization 26, 199–219 (2003)
13. Mishra, S., Ghosh, A.: Interactive fuzzy programming approach to Bi-level quadratic fractional programming problems. Ann. Oper. Res. 143, 251–263 (2006)
14. Ozaltin, O.Y., Prokopyev, O.A., Schaefer, A.J.: Two-Stage Quadratic Integer Programs with Stochastic Right-Hand Sides. Optimization Online Stochastic Programming, 1–53 (October 2009)
15. Rees, R.: The theory of Principal and Agent-Part I. Bulletin of Economic Research 37(1), 3–26 (1985)
16. Rees, R.: The theory of Principal and Agent–Part II. Bulletin of Economic Research 37(2), 75–97 (1985)
17. Vicente, L.N., Calamai, P.H.: Bilevel and multilevel programming: A bibliography review. Journal of Global Optimization 5, 291–306 (1994)

Derivative-Free Optimization for Population Dynamic Models

Ute Schaarschmidt[1,*], Trond Steihaug[1], and Sam Subbey[2]

[1] Department of Informatics, University of Bergen, Bergen, Norway
{ute.schaarschmidt,trond.steihaug}@ii.uib.no
[2] Institute of Marine Research, Bergen, Norway
samuel.subbey@imr.no

Abstract. Quantifying populations in changing environments involves fitting highly non-linear and non-convex population dynamic models to distorted observations. Derivatives of the objective function with respect to parameters might be expensive to obtain, unreliable or unavailable.

The aim of this paper is to illustrate the use of derivative-free optimization for estimating parameters in continuous population dynamic models described by ordinary differential equations. A set of non-linear least squares problems is used to compare several solvers in terms of accuracy, computational costs and robustness. We also investigate criteria for a good optimization method which are specific to the type of objective function considered here. We see larger variations in the performances of the derivative-free methods when applied for parameter estimation in population dynamic models than observed for standard noisy benchmark problems.

1 Introduction

In this paper, we consider the problem of finding a parameter vector $\boldsymbol{\theta} \in \Theta \subset \mathbb{R}^m$, which minimizes the least squares error (1) between a solution $\mathbf{x}(t; \boldsymbol{\theta})$ of an ordinary differential equation (ODE) (2)–(3) and a set of r data points $\mathbf{d}(t_j)$, $j = 1, \ldots, r$.

$$\min_{\boldsymbol{\theta} \in \Theta} f(\boldsymbol{\theta}) = \sum_{j=1}^{r} \|\mathbf{x}(t_j; \boldsymbol{\theta}) - \mathbf{d}(t_j)\|_2^2 \qquad (1)$$

$$\text{s.t.} \quad \frac{d}{dt} \mathbf{x}(t; \boldsymbol{\theta}) = \mathbf{g}(\mathbf{x}(t; \boldsymbol{\theta}), \boldsymbol{\theta}) \qquad (2)$$

$$\mathbf{x}(t_0) = \mathbf{d}(t_0) . \qquad (3)$$

This might be a challenging task, when the objective function is highly non-linear, non-convex and derivatives of the objective function f with respect to parameters $\boldsymbol{\theta}$ are expensive to obtain, unavailable or can only be calculated with modest accuracy. Even if a unique solution of the differential equation

---

* Corresponding author.

© Springer International Publishing Switzerland 2015
H.A. Le Thi et al. (eds.), *Model. Comput. & Optim. in Inf. Syst. & Manage. Sci.*,
Advances in Intelligent Systems and Computing 359, DOI: 10.1007/978-3-319-18161-5_33

exists, its numerical approximation might not be differentiable. In addition, the ODE solver might fail to return a proper value for $\mathbf{x}(t; \boldsymbol{\theta})$, a case of hidden constraints.

Derivative-free optimization methods utilize the objective function, but not its gradients. In general, two main classes of derivative-free methods may be distinguished. Direct search methods explore the parameter space by generating trial points according to pre-defined geometric patterns. The second class combines trust region methods with local models obtained by interpolating sample points. For an introduction to derivative-free methods, see [4]. A recent review of algorithms and software implementations may be found in [13].

The literature on derivative-free optimization includes examples of objective functions, which are defined by (partial) differential equations. In [8], a set of methods has been employed for minimizing costs for groundwater supply. Derivative-free solvers have also been used for parameter estimation in energy density functionals in computational nuclear physics [9]. In both cases, simulation of the model involves numerical solution of a partial differential equation.

The following criteria for a good derivative-free method have been established (see e.g., [12]): the ability to improve the objective function using few function evaluations, the accuracy of the solution, when a higher computational budget is available, and the robustness to the initial iterate for the parameter vector. Benchmarking of derivative-free solvers often relies on the process described in [12] and involves the use of data and performance profiles. Results of benchmark processes for derivative-free solvers (such as in [13]) depend on the set of problems and the set of solvers used for comparison, and can in general not be extrapolated.

In [12], perturbation of objective function values is introduced to simulate numerical noise. Problems of the form (1)–(3) and defined by differential equations exemplify objective functions subject to numerical noise.

We employ several derivative-free solvers and compute data and performance profiles. Similarities and dissimilarities of the results with performances for the set of noisy benchmark problems are investigated. Such a comparison is of considerable interest, [12].

This paper illustrates how the benchmark procedure can be applied to nonlinear least squares problems defined by a set of differential equations describing population dynamics. We address challenges specific to this class of problems. The set of differential equations investigated here is able to represent several classes of dynamic behaviour. This gives us the opportunity to examine what effect the non-linearity of the solution of the differential equation has on the quality of the parameter estimates.

The set of non-linear least squares problems defined by population dynamic systems is described in Chap. 2. For the purpose of this paper, the number of derivative-free methods is limited, but we choose a representative for each of the main classes classified in [4]. Main properties of the derivative-free algorithms employed are sketched in Chap. 3. Benchmarking of the set of solvers for problems defined by differential equations is based on the established criteria, which are described in Chap. 4. This chapter also includes details about the numerical experiments. Results and discussions of challenges specific to objective functions defined by differential equations are given in Chap. 5. Some concluding remarks can be found in Chap. 6.

2 Test Problems Defined by Dynamic Systems

The systems of differential equations defining the non-linear problems (1)–(3) investigated here, describe changes of numbers of individuals $x_i(t)$ in age-classes $i = 0, 1, \ldots, n$. Dropping the time-dependency, the (simulation) model in dimensionless form is given by (4), with $\epsilon > 0$.[1] The value of parameter $\gamma \geq 1$ influences the curvature and the asymptotic behaviour of the solution of the differential equation, [14]. The limit of the differential equation as $\gamma \to \infty$ is implemented using the exponential function and we refer to it as the case $\gamma \to \infty$. For details about the dynamic system, which is a parametrised version of a model introduced in [16], we refer the reader to [14].

$$
\frac{d}{dt}x_i = \begin{cases} -x_0 + \frac{1}{\epsilon}\left[-x_0\left(1 + \frac{1}{\gamma}\sum_{l=1}^{n}\theta_{n+l}x_l\right)^{\gamma} + \sum_{l=1}^{n}\theta_l x_l\right] , i = 0 \\ x_{i-1} - \theta_{2n+i}x_i \qquad\qquad\qquad\qquad\qquad\qquad , i = 1, \ldots, n \end{cases} \tag{4}
$$

Data is generated by the differential algebraic system (5)–(6) assuming the vector of parameters $\tilde{\boldsymbol{\theta}} = (\tilde{\theta}_1 \ldots \tilde{\theta}_m)^t$ to be given. Each parameter θ_k, $k = 1, \ldots, m$ determining the simulation model has an equivalent $\tilde{\theta}_j$ associated with the data generating model. The solution $\mathbf{x}(t; \tilde{\boldsymbol{\theta}})$ of the differential equation converges for $\epsilon \to 0$ to the solution $\bar{\mathbf{d}}(t; \tilde{\boldsymbol{\theta}})$ of the differential algebraic system, [14].

$$
\bar{d}_0(t) = \left(\sum_{l=1}^{n} \tilde{\theta}_l \bar{d}_l(t)\right)\left(1 + \frac{1}{\gamma}\sum_{l=1}^{n} \tilde{\theta}_{n+l}\bar{d}_l(t)\right)^{-\gamma} , \tag{5}
$$

$$
\frac{d}{dt}\bar{d}_i(t) = \bar{d}_{i-1}(t) - \tilde{\theta}_{2n+i}\bar{d}_i(t) , \quad i = 1, \ldots, n . \tag{6}
$$

Distorted data points $\mathbf{d}(t_j)$ are defined by (7). We assume observational errors $\mathbf{u}_j$ to be additive and normally distributed with zero mean and standard deviation $\sigma > 0$. To avoid zero solutions of the differential equation, data points are bounded from below by $\mathbf{d}_{low} = 10^{-1}$. Assuming no observational error and for $\epsilon > 0$ but sufficiently small, the residual $f(\tilde{\boldsymbol{\theta}})$ of the non-linear least squares problem (1)–(3) at $\tilde{\boldsymbol{\theta}}$ is an upper limit for the solution of the optimization problem and close to zero. Distorted data points correspond to non-zero residual problems.

$$
\mathbf{d}(t_j) = \min\left\{\bar{\mathbf{d}}(t_j) + \mathbf{u}_j, \mathbf{d}_{low}\right\}, \text{ with } \mathbf{u}_j \sim \mathcal{N}(0, \sigma) \text{ and } j = 1, \ldots, r . \tag{7}
$$

We denote by P a set of non-linear least squares problems (1)–(3) defined by population dynamic systems (4) and (5)–(6). A specific element p of the

---

[1] We use bold symbols to represent vectors, e.g., $(\theta_1 \ldots \theta_m)^t = \boldsymbol{\theta}$.

set P is defined by a curvature γ of models (4) and (5)–(6), an initial condition $\mathbf{d}(t_0) = \mathbf{x}(t_0)$, a level of distortion $\sigma \geq 0$ and a starting point $\boldsymbol{\theta}^{(0)}$. The variations of model and data points aim at representing a broad spectrum of problems.

3 Optimization Methods

FMINSEARCH. The Nelder-Mead algorithm is one of the most popular derivative-free methods and has for example be described in [4]. A simplex of $m + 1$ sample points is iteratively improved by reflections, extractions or contractions. In each iteration, the vertex with highest objective function value is replaced by a point with a lower function value. The solver can therefore adapt to the shape of the objective function. Here, we use a Matlab version of the Nelder-Mead algorithm, which is described in [10].

NOMAD. Directional direct search methods explore the parameter space by generating sample points on a mesh. Each iteration may consist of search and poll steps. The search step allows to freely explore the mesh. Convergence is ensured by the poll step, in which the neighbourhood of the current iterate is explored. Mesh adaptive direct search (MADS), which was introduced in [2], allows the set of poll directions to become asymptotically dense in the unit sphere. The algorithm is globally convergent to a first order critical point. The version of MADS employed here is called NOMAD and available from [1]. In addition to a user guide [3], a description of the algorithm can be found in [11]. Bound constraints of the form $\mathbf{a} \leq \boldsymbol{\theta} \leq \mathbf{b}$ are obtained by box-projections.

SID-PSM. A generalized pattern search method, which uses simplex gradient and Hessian information for ordering poll steps has been introduced in [5]. Simplex derivatives are approximations of the gradients and can be considered as coefficients of linear multivariate polynomial interpolation models. SID-PSM [5,6] is a pattern search method guided by simplex derivatives, which uses quadratic minimum Frobenius norm models to define the search directions. The efficiency of this ordering for smooth and noisy problems has been shown in [6].

Here, we use version 1.2 of SID-PSM and choose the default option to use the negative simplex gradient as direction of potential decrease. The poll vectors are ordered according to the increasing amplitude of their angles to the direction of potential descent. Bound constraints are obtained by box projections.

ORBIT. ORBIT is a trust region interpolation-based algorithm introduced in [17], which employs radial basis function models with linear polynomial tails. Global convergence to first order critical points has been shown in [18,19]. The same authors illustrate that the method is able to achieve high improvements of the objective function from the initial iterate.

While several options are available, we use cubic radial basis functions based on maximal $3m$ interpolation points, as recommended in [18]. We employ a version from June, 2014, which handles bound constraints by box-projection.

4   Description of the Numerical Experiments

4.1   Comparison of Optimization Methods

Two ways of aggregating information about performance of derivative-free solvers for sets of test problems are data and performance profiles introduced in [12] and [7], respectively. Regarding the term performance profile, we follow the definition given in [7,12].

As outlined in [12], the number $t_{p,s}$ of function evaluations required to solve a problem $p \in P$ by solver $s \in S$ is an appropriate measure for the computational costs of derivative-free optimization. Data profiles $d_s(\alpha)$ measure the fraction of problems, which an algorithm s solves with a computational budget corresponding to α simplex gradient evaluations (8). Here, $|P|$ and size$P$ denote the cardinality of a set P, m the number of unknown parameters of problem p and $\alpha > 0$ a tolerance.

$$d_s(\alpha) = \frac{1}{|P|}\text{size}\left\{p \in P : \frac{t_{p,s}}{m+1} \leq \alpha\right\} \tag{8}$$

Performance profiles have been introduced in [7] as cumulative distribution function of a performance metric- the ratio between the computational costs of a specific algorithm s and the minimum effort which allows any solver to achieve convergence. For derivative-free optimization, performance profiles may base on measurements of computational costs in terms of numbers of function evaluations, [12]. The performance profile $\rho_s(\alpha)$ is defined by (9). Data and performance profiles can both be interpreted as fraction of problems, which are solved with a limited computational budget. The difference is that data profiles use an absolute value of computational costs, while performance profiles employ a budget proportional to the minimum cost which would allow any solver to solve a specific problem.

$$\rho_s(\alpha) = \frac{1}{|P|}\text{size}\left\{p \in P : \frac{t_{p,s}}{\min\{t_{p,s} : s \in S\}} \leq \alpha\right\} \tag{9}$$

The convergence of derivative-free algorithms may be tested by comparing the decrease of the objective function value to the maximal possible reduction (10), as suggested in [12]. Here, $\tau > 0$ denotes some tolerance and f_{L} is the minimum value of the function value. If f_{L} is unknown, it may be approximated by the smallest objective function value obtained by any solver using the maximum number of function evaluations.

$$f(\boldsymbol{\theta}^{(0)}) - f(\boldsymbol{\theta}) \geq (1 - \tau)(f(\boldsymbol{\theta}^{(0)}) - f_{\mathrm{L}}) \tag{10}$$

4.2   Implementation

This section describes details about the implementation of the test problems. Data points arise from numerical solution of the differential algebraic system (5)–(6) with $n = 2$ and for $\tilde{\theta}_1 = 6$, $\tilde{\theta}_2 = 8$, $\tilde{\theta}_3 = \tilde{\theta}_4 = 0.05$, $\tilde{\theta}_5 = 2$, $\tilde{\theta}_6 = 1.5$. The set

Table 1. Characteristics of the test problems.

Initial condition	$d_1(t_0) = 25$, $d_2(t_0) = 60$ and $d_1(t_0) = 0.5$, $d_2(t_0) = 0.6$
Curvature of model	$\gamma = 1$, $\gamma = 2$ and $\gamma \to \infty$
Level of distortion	$\sigma \in \{0, 1, 3, 5, 7, 9\}$
Starting point	$\boldsymbol{\theta}^{(0)} \in \{1.2\tilde{\boldsymbol{\theta}}, 0.8\tilde{\boldsymbol{\theta}}, 1.5\tilde{\boldsymbol{\theta}}, 0.5(\mathbf{b} - \mathbf{a})\}$

of data points $\mathbf{d}(t_j)$, $j = 1, \ldots, r$ consists of $r = 10$ equally distributed points in time $t_j \in [0, 4]$.

Assuming no observational error and for ϵ sufficiently small, a close upper bound for the minimum of the objective function is given by $f(\tilde{\boldsymbol{\theta}})$. We assume ϵ to be known and fixed with value $\epsilon = 0.004$. Numerical experiments show that the residual $|\bar{d}_i(t; \tilde{\boldsymbol{\theta}}) - x_i(t; \tilde{\boldsymbol{\theta}})|$ is of order 10^{-1} for all $i = 0, 1, 2$, $t \in [0, 4]$ and for all undistorted data sets. As described in Chap. 2, a specific problem $p \in P$ is defined by a unique combination of characteristics, which are summarized in Table 1.

The solution of differential equation (4) may explode for some values of parameter $\boldsymbol{\theta}$. Then, the ODE solver might return an error message. We explicitly allow cases of hidden constraints, as they are known to be a common problem when estimating parameters in population dynamic models. We choose parameter space $0 = \mathbf{a} \leq \boldsymbol{\theta} \leq \mathbf{b}$, with $\mathbf{b} = (100, 100, 2, 2, 5, 5)^t$. Whenever the ODE solver returns an error message, the value 'NaN' is assigned to the objective function. For the numerical experiments considered here, all optimization methods investigated continue after meeting hidden constraints.

The system of differential equations representing the models is singularly perturbed and stiff. Thus, it is solved by numerical differentiation formulas (NDFs) as described in [15]. The ODE solver approximates the solution of the differential equations at specific points in time t_j, $j = 1, \ldots, r$ by interpolation. The resulting error is of the same order as the local error with an upper bound equal to 10^{-7}. On average, about 100 steps are taken while solving the differential equation and the global error of the solution of the differential equation is roughly of order 10^{-5}. The global error of the solution of the ODE is a limit for the precision τ, which we can expect in the convergence test (10). The components of the parameter vector may differ significantly in their magnitude. Therefore, all parameters are linearly transformed to values in interval $[0, 1]$.

All test problems have $m = 6$ unknown parameters. We employed a maximum number of 700 function evaluations corresponding to 100 simplex gradient evaluations. The choice of values for termination parameters ensures that no solver terminates before the maximum number of function evaluations is reached. We use the default options for algorithm parameter values, if not stated otherwise in Chap. 3. Minimum values of the objective functions are not available and we compare the solutions with the lowest objective function value f_L obtained by any solver.

5   Results

In this chapter, we compare the set of solvers described in Chap. 3 for the non-linear least squares problems defined by ODEs and outlined in Chap. 2. Data and performance profiles have been computed for $\tau \in \{10^{-1}, 10^{-3}, 10^{-5}\}$. For the sake of brevity, we present a subset of the results in Fig. 1.

For $\tau = 10^{-1}$, the data profile in Fig. 1(a) displays the number of problems, for which the objective function is reduced by 90% compared with the best reduction. For example, for a budget of 30 simplex gradient evaluations, ORBIT, NOMAD, SID-PSM and FMINSEARCH solve 53%, 72%, 79% and 83% of the problems, respectively. Performance profiles for $\tau = 10^{-1}$ are presented in Fig. 1(b). FMINSEARCH solves 67% of the problems with at most twice as many function evaluations as the fastest solver. SID-PSM is fastest to solve 18% of the problems. This type of information is readily available in performance profiles and not in data profiles.

Overall, FMINSEARCH solves the highest number of problems, followed by SID-PSM and NOMAD. The Nelder-Mead algorithm also solves the highest percentage of problems fastest. For a computational budget of 7 simplex gradient evaluations, ORBIT solves the highest fraction of problems. Data profiles for tolerances $\tau = 10^{-3}$ and $\tau = 10^{-5}$ are displayed in Fig. 1(c) and (d). FMINSEARCH solves the highest fraction of problems, namely 80% and 60%, respectively. SID-PSM performs second best. Further examples, for which Nelder-Mead found a more accurate solution than ORBIT, while the trust region interpolation-based method improved the objective function well with few function evaluations, can be found in [17,18,19].

A solution of the non-linear least squares problem has to satisfy hidden constraints and bound constraints. In contrast to the other solvers, the simplicial direct search method solves unbounded problems. The numerical experiments indicate that box constraints are only active under assumptions $\gamma \to \infty$, $\theta^{(0)} = 0.5b$ and $d_1(t_0) = 0.5$, $d_2(t_0) = 0.6$. These assumptions are valid for about 4% of the problems. Then, FMINSEARCH returns an infeasible solution and none of the solvers is able to reduce the objective function value to 10^2 within 100 simplex gradient evaluations. Numerical experiments show that hidden constraints may be active for the subset of optimization problems defined by $\gamma \to \infty$ and $\theta^{(0)} = 0.5b$. This corresponds to about 8% of the problems. These numbers indicate the influence of constraints on the results.

5.1   Comparison with Noisy Benchmark Problems

The benchmark procedure proposed in [12] investigates a set of noisy benchmark problems. The objective functions are perturbations of non-linear least squares functions from the CUTEr collection, see (11). The deterministic noise function $\phi : \mathbb{R}^n \to [-1, 1]$ is defined in terms of the cubic Chebyshev polynomial. For details, we refer the reader to [12].

$$f(\boldsymbol{\theta}) = (1 + \epsilon_f \phi(\boldsymbol{\theta})) \sum_{k=1}^{m} f_k(\boldsymbol{\theta})^2 \qquad (11)$$

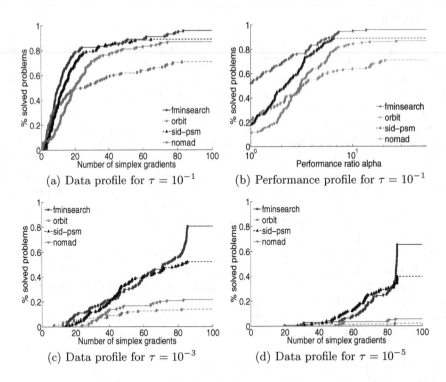

(a) Data profile for $\tau = 10^{-1}$ (b) Performance profile for $\tau = 10^{-1}$

(c) Data profile for $\tau = 10^{-3}$ (d) Data profile for $\tau = 10^{-5}$

Fig. 1. Percentage of problems solved and relative performance by a solver for the set of non-linear least squares problems

The objective functions considered in this paper are noisy due to the numerical solution of the differential equations defining the parameter estimation problems. In this section, we investigate whether derivative-free solvers perform similarly for the set of noisy benchmark problems and the problems defined by population dynamic models. Data and performance profiles considered rely on comparing a solution found by a specific solver with the best solution obtained by any algorithm. Thus, results of the benchmark process depend on the set of solvers S and we repeat the comparison for the set of solvers used in this paper. We employ the relative noise level $\epsilon_f = 10^{-5}$, as this corresponds the level of noise of the non-linear least squares problems (see Chap. 4). Data and performance profiles have been computed for $\tau \in \{10^{-1}, 10^{-3}, 10^{-5}\}$. A representative subset of results is presented in Fig. 2.

As illustrated in Fig. 2(a), for accuracy $\tau = 10^{-1}$, the Nelder-Mead algorithm solves the highest percentage of problems employing up to 100 simplex gradient evaluations. However, SID-PSM solves the highest number of problems for computational budgets smaller or equal to 10 simplex gradient evaluations. Numerical results for $\tau = \{10^{-3}, 10^{-5}\}$ indicate that the simplex derivative based search method performs best for all computational budgets and solves the highest number of problems in total (see e.g., Fig. 2(b)). Summarizing, we observe

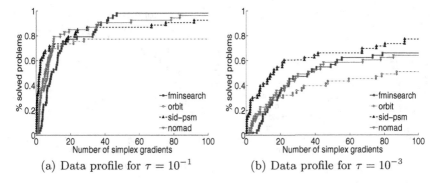

(a) Data profile for $\tau = 10^{-1}$ (b) Data profile for $\tau = 10^{-3}$

Fig. 2. Percentage of problems solved and the relative performance by a solver for the set of noisy benchmark problems

that the Nelder-Mead algorithm achieves better results for the parameter estimation problems than for the noisy benchmark problems, while SID-PSM performs better for the latter type of objective function.

When comparing Fig. 2(a),(b) and Fig. 1(a),(c) one may observe higher variations in performances for the problems defined by differential equations than for the standard test problems. For example, for accuracy $\tau = 10^{-3}$, the algorithms solve 14%-80% of the problems defined by dynamic systems and 51%-77% of the noisy benchmark problems. All solvers except for ORBIT solve 92%-98% of the standard benchmark problems and 87%-96% of the non-linear least squares problems with accuracy $\tau = 10^{-1}$. The results indicate that the problems defined by dynamic systems are more challenging than the noisy benchmark problems.

5.2 Non-Linearity of the Differential Equation

The structure of the differential equation depends on parameter γ. For $\gamma \to \infty$, the solution of the differential equation explodes for a large set of parameters and the objective function is highly sensitive to the values of θ_3 and θ_4. We classify the set of non-linear least squares problems by the value of γ and compare the corresponding data profiles.

As illustrated in Fig. 3, all derivative-free methods solve more problems under assumption $\gamma = 1$ than for the case $\gamma \to \infty$. The solver which is most sensitive to the high non-linearity of the differential equation is ORBIT. It converges for 94% of the problems defined by $\gamma = 1$ and for 34% for the case $\gamma \to \infty$. We hypothesize that the approximation of the objective function by radial basis functions might be sensitive to the high non-linearity of the objective function for this particular set of problems.

5.3 Robustness to the Starting Point

When knowledge about the parameters is limited, it is important to know how robust a solver is to the initial iterate. In the same manner as in the previous part,

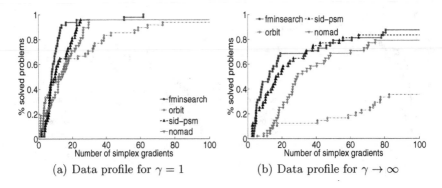

(a) Data profile for $\gamma = 1$ (b) Data profile for $\gamma \to \infty$

Fig. 3. Percentage of problems solved for the set of non-linear least squares problems, classified by the structure of the population dynamic model

we investigate the sensitivity of solvers to the initial iterate $\boldsymbol{\theta}^{(0)}$ by classifying the set of parameter estimation problems. Data and performance profiles for the subsets of problems defined by $\boldsymbol{\theta}^{(0)} \in \{1.2\tilde{\boldsymbol{\theta}}, 0.8\tilde{\boldsymbol{\theta}}, 1.5\tilde{\boldsymbol{\theta}}, 0.5\mathbf{b}\}$ are compared. For the sake of brevity, Fig. 4 presents results for $\boldsymbol{\theta}^{(0)} = 0.8\tilde{\boldsymbol{\theta}}$ and $\boldsymbol{\theta}^{(0)} = 0.5\mathbf{b}$. The latter case corresponds to the highest distance between starting point and $\tilde{\boldsymbol{\theta}}$.

Numerical results indicate that FMINSEARCH is the best solver overall for the cases $\boldsymbol{\theta}^{(0)}/\tilde{\boldsymbol{\theta}} = 0.8, 1.2, 1.5$. The performance profiles presented in Fig. 4(a) illustrate that FMINSEARCH solves almost 90% of the problems defined by $\boldsymbol{\theta}^{(0)}/\tilde{\boldsymbol{\theta}} = 0.8$ fastest. However, the Nelder-Mead algorithm solves less problems, when the distance between initial iterate and $\tilde{\boldsymbol{\theta}}$ increases. For $\boldsymbol{\theta}^{(0)} = 0.5\mathbf{b}$, FMINSEARCH solves the least number of problems fastest and in total (see Fig. 4(b)). We conclude that SID-PSM, NOMAD and ORBIT are more robust to the starting point than FMINSEARCH.

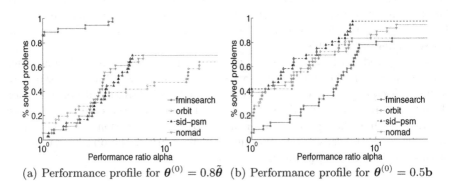

(a) Performance profile for $\boldsymbol{\theta}^{(0)} = 0.8\tilde{\boldsymbol{\theta}}$ (b) Performance profile for $\boldsymbol{\theta}^{(0)} = 0.5\mathbf{b}$

Fig. 4. Relative performance for the set of non-linear least squares problems, classified by the starting point

6   Conclusion

The optimization problems investigated here indicate that derivative-free optimization methods allow to solve parameter estimation problems which are defined by stiff differential equations. We have compared several solvers based on standard criteria such as accuracy, computational costs and robustness to the starting point. In addition, we have described how a set of test problems may be split into subsets in order to examine questions specific for objective functions defined by differential equations. For example, the sensitivity of performances to the non-linearity of the solution of the differential equation has been investigated. A possible extension of the work presented in this paper will be to examine the challenge of hidden constraints more closely.

We caution that the conclusions drawn here are only valid for the specific set of solvers and optimization problems. Which solver would achieve best results for a specific problem, depends on several factors. While the Nelder-Mead algorithm FMINSEARCH solves the highest fraction of the parameter estimation problems, it is less robust to the starting point. ORBIT is effective when the computational budget is small. The limited testing indicates that trust-region interpolation-based methods may be more sensitive to the non-linearity of the solution of the differential equation.

Our observations are not completely in accordance with results obtained when solving the set of standard benchmark problems. In particular, we see larger variations in the performances of the derivative-free methods for the parameter estimation problem defined by population dynamic models than for the set of noisy benchmark problems. This indicates that benchmarking of derivative-free methods for population dynamic models should be based on problems defined by differential equations.

References

1. Abramson, M.A., Audet, C., Couture, G., Dennis Jr., J.E., Le Digabel, S., Tribes, C.: The NOMAD Project, https://www.gerad.ca/nomad/
2. Audet, C., Dennis Jr., J.E.: Mesh Adaptive Direct Search Algorithms for Constrained Optimization. SIAM J. Optim. 17(1), 188–217 (2006)
3. Audet, C., Le Digabel, S., Tribes, C.: NOMAD user guide. Tech. Rep. G-2009-37, Les cahiers du GERAD (2009)
4. Conn, A.R., Scheinberg, K., Vicente, L.N.: Introduction to Derivative-Free Optimization. MPS-SIAM Series on Optimization. SIAM, Philadelphia (2009)
5. Custódio, A.L., Vicente, L.N.: Using sampling and simplex derivatives in pattern search methods. SIAM J. Optim. 18, 537–555 (2007)
6. Custódio, A.L., Rocha, H., Vicente, L.N.: Incorporating minimum Frobenius norm models in direct search. Comput. Optim. Appl. 46, 265–278 (2010)
7. Dolan, E.D., Moré, J.J.: Benchmarking optimization software with performance profiles. Math. Program. 91, 201–213 (2002)

8. Fowler, K.R., Reese, J.P., Kees, C.E., Dennis Jr., J.E., Kelley, C.T., Miller, C.T., Audet, C., Booker, A.J., Couture, G., Darwin, R.W., Farthing, M.W., Finkel, D.E., Gablonsky, J.M., Gray, G., Kolda, T.G.: A comparison of derivative-free optimization methods for groundwater supply and hydraulic capture community problems. Adv. Water Resour. 31, 743–757 (2008)

9. Kortelainen, M., McDonnell, J., Nazarewicz, W., Reinhard, P.G., Sarich, J., Schunck, N., Stoitsov, M.V., Wild, S.M.: Nuclear energy density optimization: Large deformations. Phys. Rev. C 85(2), 024304 (2012)

10. Lagarias, J.C., Reeds, J.A., Wright, M.H., Wright, P.E.: Convergence properties of the Nelder-Mead simplex method in low dimensions. SIAM J. Optim. 9(1), 112–147 (1998)

11. Le Digabel, S.: Algorithm 909: NOMAD: Nonlinear optimization with the MADS algorithm. ACM Transactions on Mathematical Software 37(4), 44:1–44:15 (2011)

12. Moré, J.J., Wild, S.M.: Benchmarking derivative-free optimization algorithms. SIAM J. Optim. 20, 172–191 (2009)

13. Rios, L.M., Sahinidis, N.V.: Derivative-free optimization: a review of algorithms and comparison of software implementations. J. Glob. Optim. 56, 1247–1293 (2013)

14. Schaarschmidt, U., Subbey, S., Steihaug, T.: Application of slow-fast population dynamic models. Technical Report 409, Institute of Informatics, University of Bergen (2014)

15. Shampine, L.F., Reichelt, M.W.: The MATLAB ODE Suite. SIAM J. Sci. Comput. 18(1), 1–22 (1997)

16. Touzeau, S., Gouzé, J.L.: On the stock-recruitment relationships in fish population models. Environ. Model. Assess. 3, 87–93 (1998)

17. Wild, S.M., Regis, R.G., Shoemaker, C.A.: ORBIT: Optimization by radial basis function interpolation in trust-regions. SIAM J. Sci. Comput. 30(6), 3197–3219 (2008)

18. Wild, S.M., Shoemaker, C.A.: Global convergence of radial basis function trust region derivative-free algorithms. SIAM J. Optim. 21(3), 761–781 (2011)

19. Wild, S.M., Shoemaker, C.A.: Global convergence of radial basis function trust-region algorithms for derivative-free optimization. SIAM Review 55(2), 349–371 (2013)

New Underestimator for Univariate Global Optimization

Mohand Ouanes[1], Hoai An Le Thi[2], and Ahmed Zidna[2]

[1] Laromad, Département de Mathématiques,
Faculté des Sciences, Université de Tizi-Ouzou, Algérie
`ouanes_mohand@yahoo.fr`
[2] Laboratoire d'Informatique Théorique et Appliquée,
Université de Lorraine, Ile du Saulcy, 57045 Metz, France
`{hoai-an.le-thi,ahmed.zidna}@univ-lorraine.fr`

Abstract. The aim of this paper is to propose a new underestimator for solving univariate global optimization problems, which is better than the underestimator used in the classical αBB method [1], and the quadratic underestimator developed in [4]. We can propose an efficient algorithm based on Branch and Bound techniques and an efficient w-subdivision for branching. A convex/concave test is added to accelerate the convergence of the algorithm.

Keywords: Global optimization, αBB method, quadratic underestimator, Branch and Bound.

1 Introduction

Univariate global optimization problems attract attention of researchers not only because they arise in many real-life applications, but also because the methods proposed to solve them can be extended to the multivariate case. Often problems in multidimensional case can be reduced to one dimensional case. One class of deterministic approaches, which called lower bounding method, emerged from the natural strategy to find a global minimum for sure. The efficiency of a method is in the construction of tight underestimator and to discard a big regions which do not contain the global minimum as quickly as possible. In this paper, we consider the following problem

$$(P) \begin{cases} \min f(x), \\ x \in [x^0, x^1] \subset \mathbf{R}, \end{cases}$$

where $f(x)$ is a nonconvex and C^2-continuous function on the real interval $[x^0, x^1]$.

Several methods have been studied in the literature for univariate global optimization problems (see [3] and references therein). We can cite among them the classical αBB method developed in [1] and the method proposed in [4]. The latter consists in the construction of an explicit quadratic underestimator. A generalization to the multivariate case is proposed in [5].

© Springer International Publishing Switzerland 2015   403
H.A. Le Thi et al. (eds.), *Model. Comput. & Optim. in Inf. Syst. & Manage. Sci.*,
Advances in Intelligent Systems and Computing 359, DOI: 10.1007/978-3-319-18161-5_34

The main contributions of our theoretical paper are: a construction of a new underestimator which is better than the two underestimators proposed in [1,4]) and a convex/concave test which accelerate the convergence of the proposed branch and bound algorithm.

The structure of the paper is as follows : In section 2, the two underestimators developed in [1] and in [4] are given. In section 3, a new underestimator of the objective function is proposed and the convex/concave test is given. In section 4, the algorithm is described and its convergence is shown.

2   Background

In what follows, we give two underestimators developed by the authors respectively in [1] and [4].

2.1   Underestimator in αBB Method [1]

The underestimator in αBB method on the interval $[x^0, x^1]$ is

$$LB_\alpha(x) = f(x) - \frac{K_\alpha}{2}(x - x^0)(x^1 - x),$$

where $K_\alpha \geq \max\{0, -f''(x)\}$, for all $x \in [x^0, x^1]$.
This underestimator satisfies the following properties :

1. It is convex (i.e. $LB_\alpha''(x) = f''(x) + K_\alpha \geq 0$ because $K_\alpha \geq \max\{0, -f''(x)\}$, $\forall x \in [x^0, x^1]$).
2. It coincides with the function $f(x)$ at the endpoints of the interval $[x^0, x^1]$.
3. It is a underestimator of the objective function $f(x)$.

For more details see [1].

2.2   Quadratic Underestimator [4]

The quadratic underestimator developed in [4] on the interval $[x^0, x^1]$ is

$$LB_{LO}(x) = f(x^0)\frac{x^1 - x}{x^1 - x^0} + f(x^1)\frac{x - x^0}{x^1 - x^0} - \frac{K}{2}(x - x^0)(x^1 - x),$$

where $K \geq |f''(x)|$, for all $x \in [x^0, x^1]$.
This quadratic underestimator satisfies the following properties:

1. It is convex and quadratic (i.e. $LB_{LO}''(x) = K \geq 0$).
2. It coincides with the function $f(x)$ at the endpoints of the interval $[x^0, x^1]$.
3. It is a underestimator of the objective function $f(x)$.

For more details see [4].

3   A New Underestimator

Let $L_h f(x)$ be the linear interpolant of $f(x)$ on the interval $[x^0, x^1]$, given by

$$L_h f(x) = f(x^0) \frac{x^1 - x}{x^1 - x^0} + f(x^1) \frac{x - x^0}{x^1 - x^0}, \tag{1}$$

and assume that $K_q \geq \max\{0, f''(x)\}, \forall x \in [x^0, x^1]$.

We propose a new underestimator on the interval $[x^0, x^1]$ as follows

$$LB(x) = \frac{K_q f(x) + K_\alpha L_h f(x)}{K_\alpha + K_q} - \frac{K_\alpha K_q}{2(K_\alpha + K_q)}(x - x^0)(x^1 - x). \tag{2}$$

Proposition 1. *The following properties hold:*

(i) $LB(x)$ *coincides with* $f(x)$ *at the endpoints of the interval* $[x^0, x^1]$.
(ii) $LB(x)$ *is convex on the interval* $[x^0, x^1]$.
(iii) $LB(x) \leq f(x), \forall x \in [x^0, x^1]$.

Proof. **(i)** Obvious from the construction of $LB(x)$.

(ii) Since $K_\alpha \geq 0$, $K_q \geq 0$ and $(K_\alpha + f''(x)) \geq 0$, and for all x in $[x^0, x^1]$, we have

$$LB''(x) = \frac{K_q(K_\alpha + f''(x))}{K_\alpha + K_q} \geq 0.$$

Then $LB(x)$ is convex on the interval $[x^0, x^1]$.
(iii) One has for all x in $[x^0, x^1]$, $K_\alpha \geq 0$, $K_q \geq 0$, and $(K_q - f''(x)) \geq 0$, then

$$(LB(x) - f(x))'' = \frac{K_\alpha(K_q - f''(x))}{K_\alpha + K_q} \geq 0.$$

which implies that $(LB(x) - f(x))$ is convex, moreover $(LB(x) - f(x))$ vanishes at the endpoints of the interval $[x^0, x^1]$, hence $LB(x) - f(x) \leq 0, \forall x \in [x^0, x^1]$.

In the next proposition, we will show that the new underestimator is better than the classical underestimator developed in αBB method [1] and than the quadratic underestimator developed in [4].

Proposition 2. *For all x in $[x^0, x^1]$, The following inequalities hold*

1. $LB(x) \geq LB_\alpha(x)$,
2. $LB(x) \geq LB_{LO}(x)$.

Proof. 1. We have

$$(LB(x) - LB_\alpha(x))'' = -\frac{K_\alpha(K_\alpha + f''(x))}{K_\alpha + K_q} \leq 0, \forall x \in [x^0, x^1],$$

since $K_\alpha \geq 0$, $K_q \geq 0$, and for all x in $[x^0, x^1]$, $(K_\alpha + f''(x)) \geq 0$, then $(LB(x) - LB_\alpha(x))$ is a concave function on $[x^0, x^1]$. Moreover it vanishes at the endpoints of this interval, hence $LB(x) \geq LB_\alpha(x), \forall x \in [x^0, x^1]$.

2. We have

$$(LB(x) - LB_{LO}(x))'' = \frac{K_q f''(x) + K_\alpha K_q - K K_\alpha - K K_q}{K_\alpha + K_q}.$$

For the values of K, we have two cases:

case 1: $K = K_\alpha \geq K_q$. By a simple calculation, one obtains

$$(LB(x) - LB_{LO}(x))'' = (f''(x) - K_q)\frac{K_q}{K_\alpha + K_q} \leq 0.$$

Since $(f''(x) - K_q) \leq 0, \forall x \in [x^0, x^1]$, $K_q \geq 0, K_\alpha \geq 0$ then $(LB(x) - LB_{LO}(x))$ is concave on $[x^0, x^1]$, moreover it vanishes at the endpoints of the interval. Hence $(LB(x) \geq LB_{LO}(x), \forall x \in [x^0, x^1]$.

case 2: $K = K_q$. We have

$$(LB(x) - LB_{LO}(x))'' = (f''(x) - K_q)\frac{K_q}{K_\alpha + K_q} \leq 0.$$

Since $(f''(x) - K_q) \leq 0$, $K_q \geq 0$, and $K_\alpha \geq 0$ then $(LB(x) - LB_{LO}(x))$ is a concave function which vanishes at the endpoints of the interval. Hence $LB(x) \geq LB_{LO}(x), \forall x \in [x^0, x^1]$.

Example 1. $f(x) = \sin x, x \in [0, 2\pi]$.
We propose to compare $LB_\alpha(x)$ and $LB_{LO}(x)$ with $LB(x)$.
For x in $[0, 2\pi]$, we have

$$LB_\alpha(x) = \sin x - \frac{1}{2}x(2\pi - x); LB_\alpha''(x) = -\sin x + 1 \geq 0, \forall x \in [0, 2\pi],$$

$$LB_{LO}(x) = -\frac{1}{2}x(2\pi - x); LB_{LO}''(x) = 1 \geq 0,$$

$$LB(x) = \frac{1}{2}(\sin x - \frac{1}{2}x(2\pi - x)) = \frac{1}{2}LB_\alpha(x); LB''(x) = \frac{1}{2}(-\sin x + 1) \geq 0, \forall x \in [0, 2\pi].$$

All this lower bound functions are convex because their second derivatives are positive, moreover they vanishes at the endpoints of $[0, 2\pi]$, then

$$LB_\alpha(x) \leq 0, LB_{LO}(x) \leq 0, LB(x) \leq 0. \forall x \in [0, 2\pi]$$

Now we compare $LB_\alpha(x)$ and $LB_{LO}(x)$ with $LB(x)$.

i) One has $LB(x) = \frac{1}{2}(\sin x - \frac{1}{2}x(2\pi - x)) = \frac{1}{2}LB_\alpha(x)$. As $LB_\alpha(x) \leq 0$, we have $LB(x) \geq LB_\alpha(x)$ which means that $LB(x)$ is better than $LB_\alpha(x)$.
ii) For all x in $[0, 2\pi]$, one has $(LB(x) - LB_{LO}(x))'' = \frac{1}{2}(-\sin x + 1) - 1 = \frac{1}{2}(-\sin x - 1) \leq 0$, then $(LB(x) - LB_{LO}(x))$ is a concave function which vanishes at 0 and 2π, consequently $(LB(x) - LB_{LO}(x))$ is nonnegative on $[0, 2\pi]$. Hence $LB(x) \geq LB_{LO}(x), \forall x \in [0, 2\pi]$ which means that $LB(x)$ is better than $LB_{LO}(x)$.

3.1   Convex/Concave Test

In order to accelerate the convergence of the proposed BB algorithm, we give the following description of the Convex/concave test.

At iteration k, for all x in $[a_k, b_k]$, we compute K_α^k and K_q^k by using interval analysis method.

One has

$$K_\alpha^k \geq \max\{0, -f''(x)\}, \text{ and } K_q^k \geq \max\{0, f''(x)\}, \forall x \in [a_k, b_k].$$

- If $K_\alpha^k = 0$ then $-f''(x) \leq 0, \forall x \in [a_k, b_k]$, which implies that $f(x)$ is convex on the interval $[a_k, b_k]$, hence any local search leads to a global minimum on this interval.
- If $K_q^k = 0$ then $f''(x) \leq 0, \forall x \in [a_k, b_k]$, which implies that $f(x)$ is concave on the interval $[a_k, b_k]$, hence its global minimum is reached at the endpoints of $[a_k, b_k]$.
- If $K_\alpha^k = K_q^k = 0$ then $f(x)$ is affine on $[a_k, b_k]$ and its global minimum is reached at one of the endpoints of this interval.

Remark 1. If the convex/concave test is satisfied for all subintervals, then the algorithm stops because the global solution for each subinterval is obtained either at the endpoints of the intervals or by applying a local search.

Example 2. $f(x) = \sin x, x \in [0, \pi]$.
We have $K_\alpha = 1, K = 1$, and $K_q = 0$.

Since $K_q = 0$ then f is concave on $[0, \pi]$, hence its global minimum is reached at the endpoints of this interval (i.e. at 0 and π). This example shows the usefulness of the convex/concave test, It allows us to find the optimal solutions at the first iteration and the algorithm stops, but the two other methods presented in [1] and [4] are not able to find the optimal solution at the first iteration.

4   Algorithm and Its Convergence

We now present our efficient branch and bound algorithm.

4.1   Algorithm

Initialization step: – Let ε be a given tolerance number, let $[x^0, x^1]$ the initial interval.
- Compute K_α^0 and K_q^0 such that $K_\alpha^0 \geq \max\{0, -f''(x)\}$, and $K_q^0 \geq \max\{0, f''(x)\}, \forall x \in [x^0, x^1]$ by using interval analysis method.
- Convex/concave test:
   If $K_\alpha^0 = 0$ then f is convex, any local search gives an optimal solution and the algorithm stops
   If $K_q^0 = 0$ then f is concave, the optimal solution is reached at the endpoints of $[x^0, x^1]$ and the algorithm stops.

- Set $k := 0; T^0 = [a_0, b_0] = [x^0, x^1]; M := T^0$.
- solve the convex problem

$$\min \left\{ LB_0(x) : x \in T^0 \right\},$$

to obtain an optimal solution x_0^*.

- Set $UB_0 := \min \{f(x_0), f(x_1), f(x_0^*)\} = f(\overline{x}^0)$.
- Set $LB_0 = LB(T^0) := LB^0(x_0^*)$.

Iteration step While $UB_k - LB_k \geq \varepsilon$ do

1. Let $T^k = [a_k, b_k] \in M$ be the interval such that $LB_k = LB(T^k)$
2. Bisect T^k into two intervals into $T_1^k = [a_k, x_k^*]; T_2^k = [x_k^*, b_k]$
   Set $T_1^k := [a_k^1, b_k^1]$ and $T_2^k := [a_k^2, b_k^2]$
3. For $i = 1, 2$ do
   (a) Convex/concave test : Compute K_α^{ki} and K_q^{ki} on T_i^k.
       If $K_\alpha^{ki} = 0$, f is convex, any local search gives an optimal solution
       x_{ki}^* on T_i^k, then update $LB(T_i^k) = UB(T_i^k) = f(x_i^k)$ and goto c).
       If $K_q^{ki} = 0$, f is concave on T_i^k, then update $LB(T_i^k) = UB(T_i^k) = \min\{f(a_k^i), f(b_k^i)\}$ and goto c).
   (b) Set $T_i^k = [a_k^i, b_k^i]$ and Compute $LB^{ki}(x)$.
       Set x_{ki}^* the solution of the convex problem.

$$\min \left\{ LB^{ki}(x), x \in T_i^k \right\}.$$

   (c) To fit into M the intervals $T_i^k : M \leftarrow M \bigcup \{T_i^k : UB_k - LB(T_i^k) \geq \varepsilon, i = 1, 2\} \setminus \{T^k\}$.
   (d) Update $UB_k := \min\{UB_k, f(a_k^i), f(b_k^i), f(x_{ki}^*)\} := f(\overline{x}^k)$.
4. Update $LB_k := \min\{LB(T) : T \in M\}$
5. Delete from M all intervals T such that $LB(T) > UB_k - \varepsilon$.
6. Set $k := k + 1$.

End while

Result step : $\overline{x}^k$ is an $\varepsilon-$ optimal solution to (P)

4.2   Convergence

In what follows, we establish the convergence of our branch and bound algorithm.

Proposition 3. *The sequence $\{\overline{x}^k\}$ generated by the algorithm converges to an optimal solution of problem (P).*

Proof. If the algorithm stops at iteration k which may be obtained by the convex/concave test or by the rule $UB_k - LB_k < \varepsilon$ then the solution is exact or $\varepsilon-$optimal.

If the algorithm is infinite then it generates an infinite sequence $\{T^k\}$ of intervals whose lengths h_k decrease to zero, then the whole sequence $\{T^k\}$ shrinks to a singleton.

We must show that
$$\lim_{k \to \infty} (UB_k - LB_k) = 0.$$

We have $0 \leq UB_k - LB_k = f(\overline{x}^k) - LB(x_k^*) =$

$$f(\overline{x}^k) - \frac{K_q f(x_k^*) + K_\alpha L_h f(x_k^*)}{K_\alpha + K_q} + \frac{K_\alpha K_q}{2(K_\alpha + K_q)}(x_k^* - a_k)(b_k - x_k^*) =$$

$$\frac{K_q(f(\overline{x}^k) - f(x_k^*)) + K_\alpha(f(\overline{x}^k) - L_h f(x_k^*)) + \frac{K_\alpha K_q}{2}(x_k^* - a_k)(b_k - x_k^*)}{K_\alpha + K_q}$$

The terms of the numerator of the above expression can be bounded as follows

- The first term can be bounded by using the mean value theorem

$$K_q(f(\overline{x}^k) - f(x_k^*)) = K_q f'(\xi_1^k)(\overline{f}^k - f_k^*) \leq K_q C_1(b_k - a_k),$$

  where $C_1 \geq |f'(\xi_1^k)| \geq 0$, and ξ_1^k is a real number between $\overline{x}^k$ and x_k^*.
- For the second term, we have

$$K_\alpha(f(\overline{x}^k) - L_h f(x_k^*)) \leq K_\alpha |f(\overline{x}^k) - L_h f(x_k^*)| \leq K_\alpha C_2(b_k - a_k)^2,$$

  where C_2 is a real positif number [2].
- For the third term, we have

$$\frac{K_\alpha K_q}{2}(x_k^* - a_k)(b_k - x_k^*) \leq \frac{K_\alpha K_q}{2}(b_k - a_k)^2.$$

Then, it results that

$$0 \leq \lim_{k \to \infty} UB_k - LB_k \leq \lim_{k \to \infty} (b_k - a_k) \frac{K_q C_1 + K_\alpha C_2(b_k - a_k) + \frac{K_\alpha K_q}{2}(b_k - a_k)}{K_\alpha + K_q} = 0,$$

i.e. $\{T^k\} = \{[a_k, b_k]\}$ shrinks to a singleton. Hence the sequence $\{\overline{x}^k\}$ converges to an optimal solution of problem (P).

5 Conclusion

We proposed in this theoretical paper a new underestimator in branch and bound algorithm for solving univariate global optimization problems. We showed that this new underestimator is better than the classical underestimator developed in αBB method and the quadratic underestimator developed in [4]. The convergence of our algorithm is shown. Tow simple examples are presented to illustrate the superiority of the new underestimator and the usefulness of the convex/concave test. A future work is to extend this method to the multidimensional case. A work in this direction is currently in progress.

References

1. Androulakis, I.P., Marinas, C.D., Floudas, C.A.: αBB: A global optimization method for general constrained nonconvex problems. J. Glob. Optim. 7, 337–363 (1995)
2. de Boor, C.: A practical method Guide to Splines. Applied Mathematical Sciences. Springer (1978)
3. Floudas, C.A., Gounaris, C.E.: A review of recent advances in global optimization. Journal of Global Optimization 45, 3–38 (2009)
4. Le Thi, H.A., Ouanes, M.: Convex quadratic underestimation and Branch and Bound for univariate global optimization with one nonconvex constraint. RAIRO Oper. Res. 40, 285–302 (2006)
5. Ouanes, M., Le Thi, H.A., Nguyen, T.P., Zidna, A.: New quadratic lower bound for multivariate functions in global optimization. Mathematics and Computers in Simulation 109, 197–211 (2015)

Semismooth Reformulation and Nonsmooth Newton's Method for Solving Nonlinear Semidefinite Programming

Boubakeur Benahmed[1] and Amina Alloun[2]

[1] Département de Mathématiques et Informatique,
Ecole Nationale Polytechnique d'Oran (ENPO),
BP 1523 El m'naour, Algérie
[2] Ecole Préparatoire en Sciences Economiques,
Commerciales et Sciences de Gestion d'Oran (EPSECG), Algérie
{amina_alloun,benahmed_b}@yahoo.fr

Abstract. In this paper, our interest is to solve canonical nonlinear semidefinite programming (NLSDP). First, we give a reformulation of the (KKT) system associated to the NLSDP as a nonsmooth equation by using the Fischer-Burmeister (FB) function. The nonsmooth equation is then solved by the nonsmooth Newton's method using formulas of the generalized Jacobian of the FB function given by L. Zhang et al. in [14]. Under mild conditions, we prove that the convergence is locally quadratic.

Keywords: Nonlinear semidefinite programming, Optimality conditions, Fischer-Burmeister function, Nonsmooth Newton's method, Generalized Jacobian of semismooth function.

1 Introduction

Let S^n denote the space of $n \times n$ symmetric matrices endowed with the inner product $\langle X, S \rangle_{tr} := tr(XS)$, (where $tr(X)$ denotes the trace of the matrix X). We use $\|X\|_F$ (resp. $\|\lambda\|_2$) to denote the F-norm in S^n (resp. 2-norm in $\mathbb{R}^m$). For $X \in S^n$, in the following the inequality $X \succeq 0$ ($X \succ 0$) means that X is positive semidefinite (positive definite) matrix. The cone of symmetric positive semidefinite matrices is denoted S^n_+.

In this paper, we consider the canonical nonlinear semidefinite programming ($NLSDP$) problem of the form

$$\begin{cases} \min f(X) \\ g(X) = 0 \\ \quad X \succeq 0 \end{cases} \tag{1}$$

where $f : S^n \longrightarrow \mathbb{R}$, $g : S^n \longrightarrow \mathbb{R}^m$ are twice continuously differentiable functions.

© Springer International Publishing Switzerland 2015   411
H.A. Le Thi et al. (eds.), *Model. Comput. & Optim. in Inf. Syst. & Manage. Sci.*,
Advances in Intelligent Systems and Computing 359, DOI: 10.1007/978-3-319-18161-5_35

If f and g are linear (affine) functions, the $NLSDP$ (1) is reduced to a canonical linear $LSDP$, which have been extensively studied during the last decades.

Nonlinear semidefinite programming ($NLSDP$) is an extension of the linear semidefinite programming ($LSDP$) and arises in various application fields such as system control and finantial engineering ([4]).

There are several numerical approaches for solving $NLSDP$, among them we can cite: the program package LOQO based on primal-dual method ([12],[13]), the sequential quadratic programming method ([5], [4]) and the interior point method ([3], [11]). An other approach, consists to reformulate the KKT system associated to the $NLSDP$ via a semidefinite cone (SDC) complementarity function and then solve the reformulated system by the nonsmooth Newton method.

Let the Lagrangian function associated to the $NLSDP$ (1) be

$$L(X, \lambda, S) = f(X) + \sum_{i=1}^{m} \lambda_i g_i(X) - \langle X, S \rangle \tag{2}$$

Under mild assumptions (convexity and regularity), the $NLSDP$ has a solution if and only if the following optimality conditions (KKT) hold

$$\begin{cases} \nabla_X L(X, \lambda, S) = Df(X) + \sum_{i=1}^{m} \lambda_i Dg_i(X) - S = 0 \\ g(X) = 0 \\ X \succeq 0, \ S \succeq 0, \ \langle X, S \rangle = 0 \end{cases} \tag{3}$$

where $\lambda = (\lambda_1, \lambda_2, ..., \lambda_m)^T \in \mathbb{R}^m$ is the Lagrange multiplier, Df and Dg_i are the Fréchet-derivatives of f and g_i respectively.

Definition 1. *A function $\phi : S^n \times S^n \longrightarrow S^n$ is called a semidefinite cone (SDC) complementarity function if*

$$\phi(X, S) = 0 \Longleftrightarrow X \succeq 0, \ S \succeq 0, \ \langle X, S \rangle_{tr} = 0$$

In [7], the authors used the SDC complementarity function named the natural residual function defined by

$$\phi_{NR}(X, S) = P_{S_+^n}(X - S) - X, ...\text{for all } X, S \in S^n,$$

where $P_{S_+^n}$ denote the orthogonal projection onto S_+^n.

In this paper, we use the Fischer-Burmeister (F-B) function defined by

$$\phi_{FB}(X, S) = X + S - \sqrt{X^2 + S^2}, ...\text{for all } X, S \in S^n \tag{4}$$

Then, with an (SDC) complementarity function ϕ the (KKT) system can be reformulated as the following nonsmooth equation

$$\Phi(X, \lambda, S) = \begin{pmatrix} \psi(X, \lambda, S) \\ \phi(X, S) \end{pmatrix} = 0 \tag{5}$$

With $\Phi : S^n \times \mathbb{R}^m \times S^n \longrightarrow S^n \times \mathbb{R}^m \times S^n$ and

$$\psi(X, \lambda, S) = \begin{pmatrix} Df(X) + \sum_{i=1}^{m} \lambda_i Dg_i(X) - S \\ g(X) \end{pmatrix} \qquad (6)$$

Note that the nonsmoothness of Φ is due to ϕ since the function ψ is smooth.

Then one way to solve $\Phi(X, \lambda, S) = 0$ is the nonsmooth Newton's method.

This paper is organized as follows: In section 2, we give the nonsmooth Newton's method for solving NLSDP and prove the convergence quadratic of the proposed algorithm. Section 3, is devoted to the calculus of the Clarke's generalized Jacobian of the F-B function.

2 Nonsmooth Newton's Method for Solving NLSDP

We use the nonsmooth Newton's method to solve the nonsmooth equation (5):

$$\begin{pmatrix} X^{k+1} \\ \lambda^{k+1} \\ S^{k+1} \end{pmatrix} = \begin{pmatrix} X^k \\ \lambda^k \\ S^k \end{pmatrix} - V_k^{-1} \Phi(X^k, \lambda^k, S^k) \qquad (7)$$

where V_k is the Clarke generalized Jacobian of Φ at the point (X^k, λ^k, S^k).

Here we need the genaralized Jacobian of ϕ since the other components are differentiable.

In this paper, we use the particular choice of $\phi = \phi_{FB}$, since it is proved that ϕ_{FB} is globally Lipschitz continuous, continuously differentiable around any $(X, S) \in S^n \times S^n$ if $[X \ S]$ is of full row rank, and strongly semismooth everywhere in $S^n \times S^n$ (see [10]).

Let $Y^k = (X^k, \lambda^k, S^k)$, $\Delta Y^k = (\Delta X^k, \Delta \lambda^k, \Delta S^k)$, where $\Delta X^k = X^{k+1} - X^k$ and let $Y^* = (X^*, \lambda^*, S^*)$ a KKT point of problem (1). Then (7) is equivalent to solve

$$V_k . \Delta Y^k = -\Phi(X^k, \lambda^k, S^k)$$

That is

$$\begin{cases} \nabla^2_X L(Y^k)(\Delta X^k) + \sum_{i=1}^{m} \Delta \lambda_i^k Dg_i(X^k) - \Delta S^k = -\nabla_X L(Y^k) \\ \langle Dg_i(X^k), \Delta X^k \rangle = -g_i(X^k), ...i = 1, ..., m \\ \mathcal{U}_k(\Delta X^k) + \mathcal{V}_k(\Delta \lambda^k) = -\phi_{FB}(X^k, S^k) \end{cases} \qquad (8)$$

where $(\mathcal{U}_k, \mathcal{V}_k) \in \partial \phi_{FB}(X^k, S^k)$.

Note that for solving the system (8), we get ΔX^k from the second equation then we replace it in the third equation for obtaining $\Delta \lambda^k$ and finally ΔS^k is calculated from the first equation.

Algorithm 1. *Step 0: Choose* $Y^k = (X^k, \lambda^k, S^k) \in S^n \times \mathbb{R}^m \times S^n$, $\varepsilon \succ 0$, *and set* $k := 0$.

Step 1: If $\|\Phi(Y^k)\| \leq \varepsilon$, *STOP. (here* $\|(X^k, \lambda^k, S^k)\| = (\|X^k\|_F^2 + \|\lambda^k\|_2^2 + \|S^k\|_F^2)^{1/2})$.

Step 2: Choose $(\mathcal{U}_k, \mathcal{V}_k) \in \partial\phi_{FB}(X^k, S^k)$ *and find* $\Delta Y^k = (\Delta X^k, \Delta\lambda^k, \Delta S^k)$ *solution of (8).*

Step 3: Set $Y^{k+1} = Y^k + \Delta Y^k$, $k := k + 1$, *and go to step 1.*

Theorem 2. *Let* $Y^* = (X^*, \lambda^*, S^*)$ *be a KKT point of problem (1) such that any* $V \in \partial\Phi(Y^*)$ *is nonsingular. Then, the algorithm (1) converges locally and quadratically to* Y^*.

Proof. First, since ϕ_{FB} is strongly semismooth then Φ is also strongly semismooth (see [10]). Now, thanks to [Theorem 2.1; [9]] and the assumption that any $V \in \partial\Phi(Y^*)$ is nonsingular then the algorithm converges locally and quadratically to Y^*.

Remark 1. To apply the nonsmooth Newton's method, the major difficulties are how to calculate a generalized Jacobien of $\mathcal{G}$ and how to satisfy the local convergence conditions. So, the next section is devoted to the calculus of function ϕ_{FB}.

3   Generalized Jacobian of the Symmetric Matrix-valued Fischer-Burmeister Function

In ([14]) the authors investigate the differential properties of the matrix-valued F-B function, including the formulas of the directional derivative, the B-subdifferential and the generalized Jacobian.

For any $m \times n$ matrix A and index sets $I \subseteq \{1, 2, ..., m\}$ and $J \subseteq \{1, 2, ..., \}$, A_{IJ} denotes the submatrix of A with rows and columns specified by I, J, respectively. Particularly, A_{ij} is the entry of A at (i, j) position. We use " $\circ$ " to denote the Hadamard product between matrices i.e for any matrices A and B, $C = A \circ B$ with $C_{ij} = A_{ij}B_{ij}$. For all $X, H \in S^n$, we define $\mathcal{L}_X(H)$ by

$$\mathcal{L}_X(H) = XH + HX.$$

For any $X \in S^n$, we use $\lambda_1(X) \geq \lambda_2(X) \geq ..... \geq \lambda_n(X)$ to denote the real eigenvalues of X being arranged in the non-increasing order. Let $\Lambda(X) = diag(\lambda_1(X), \lambda_2(X), ..., \lambda_n(X)) \in S^n$ the diagonal matrix whose $i - th$ diagonal entry is $\lambda_i(X), i = 1, ..., n$.

Denote by O^n the set of all $n \times n$ orthogonal matrices in $\mathbb{R}^{n \times n}$ and $O^n(X)$ a subset of O^n by

$$O^n(X) = \{P \in O^n / X = P\Lambda(X)P^T\}$$

Let $f^{[1]}(\Lambda(X)) \in S^n$ be the first divided difference matrix whose (i,j) entry is given by

$$f^{[1]}(\Lambda(X))_{ij} = \begin{cases} \frac{f(\lambda_i(X)) - f(\lambda_j(X))}{\lambda_i(X) - \lambda_j(X)} & \text{if } i \neq j = 1, 2, .., n \\ f'(\lambda_i(X)) & \text{if } i = j = 1, 2, ..., n \end{cases}$$

In the following, f is defined by

$$f(t) = \begin{cases} \sqrt{t}, & \text{if } t \geq 0, \\ \sqrt{-t}, & \text{if } t \leq 0, \end{cases}$$

Let $G : S^n \times S^n \longrightarrow S^n$ be the function defined by

$$G(X,Y) := X^2 + Y^2 \qquad \forall X, Y \in S^n$$

and $\sigma_1 \geq \sigma_2 \geq ... \geq \sigma_n$ be eigenvalues of G. Assume that $G(X,Y)$ have the spectral decomposition

$$G(X,Y) = PDP^T = P\,diag(\sigma_1, \sigma_2, ..., \sigma_n)P^T \tag{9}$$

where $P \in O^n(G)$.

Note that D can be writen as

$$D = \begin{bmatrix} D_\alpha & 0 \\ 0 & 0_\beta \end{bmatrix}$$

with

$$\alpha := \{i : \sigma_i > 0\} \text{ and } \beta := \{i : \sigma_i = 0\} \tag{10}$$

Assume that $Z = [X\ Y] \in \mathbb{R}^{n \times 2n}$ admit the following singular value decomposition

$$Z = P[\Sigma(Z)\ 0]Q^T = P[\Sigma(Z)\ 0][Q_1 Q_2]^T = P[\Sigma(Z)\ 0]Q_1{}^T \tag{11}$$
$$= P[\Sigma(Z)\ 0][Q_\alpha Q_\beta]^T$$

where $Q \in O^n$, $Q \in O^{2n}$, $Q_1, Q_2 \in \mathbb{R}^{2n \times n}$, $Q_\alpha \in \mathbb{R}^{2n \times |\alpha|}$, $Q_\beta \in \mathbb{R}^{2n \times |\beta|}$ and $Q = [Q_1 Q_2]$, $Q_1 = [Q_\alpha Q_\beta]$ and $\Sigma(Z) = diag(\sqrt{\sigma_1}, \sqrt{\sigma_2}, ..., \sqrt{\sigma_n})$.

3.1 Formulas for B-subdifferential and Generalized Jacobian of ϕ_{FB}

In this subsection, we characterise the $B-$subdifferential and the generalized Jacobian of ϕ_{FB}.

Let $A \in S^n_+$ be given and have the eigenvalue decomposition (9). Let

$$\Pi_\beta(A) = \left\{ \Theta \in \mathbb{R}^{|\beta| \times |\beta|} : \frac{\sqrt{\lambda_{i+|\alpha|}(A^m)}}{\sqrt{\lambda_{i+|\alpha|}(A^m)} + \sqrt{\lambda_{j+|\alpha|}(A^m)}} \to \Theta_{ij}, \right. \tag{12}$$
$$\left. A^m \longrightarrow A \text{ with } A^m \succ 0 \right\}$$

where α and β are the corresponding subsets given by (10).

Theorem 3. *([14])For any given $X, Y \in S^n$, let $G = X^2 + Y^2$ have the spectral decomposition as in 9 and $Z = [X\ Y] \in \mathbb{R}^{n \times 2n}$ admit the singular value decomposition 9*
Let be $L_Z(H) = L_Z(H_X, H_Y) = \mathcal{L}_X(H_X) + \mathcal{L}_Y(H_Y)$ and

$$K(H_X, H_Y) = \left\{ P_\beta^T [(H_X)^2 + (H_Y)^2] P_\beta - P_\beta^T L_Z(H) P_\alpha D_\alpha^{-1} P_\alpha^T L_Z(H) P_\beta \right\}^{\frac{1}{2}}$$

Then $W \in \partial_B \phi_{FB}(X, Y)$ (resp. $\partial \phi_{FB}(X, Y)$)if and only if there exists $S \in \partial_B K(0, 0)$ (resp. $\partial K(0, 0)$) such that for any $(H_X, H_Y) \in S^n \times S^n$

$$W(H_X, H_Y) = H_X + H_Y - P \begin{bmatrix} f^{[1]}(D)_{\alpha\alpha} \circ P_\alpha^T L_Z(H) P_\alpha & f^{[1]}(D)_{\alpha\beta} \circ P_\alpha^T L_Z(H) P_\beta \\ f^{[1]}(D)_{\beta\alpha} \circ P_\beta^T L_Z(H) P_\alpha & S(H_X, H_Y) \end{bmatrix} P^T$$

For $S \in \partial_B K(0, 0)$, there exist $(U, V) \in O^{|\beta|} \times O^{n+|\beta|}$ and $\Theta \in \Pi_\beta(G)$ such that for any $(H_X, H_Y) \in S^n \times S^n$

$$S(H_X, H_Y) = U(\Theta \circ \tilde{Q}_\beta^T \begin{bmatrix} H_X \\ H_Y \end{bmatrix} \tilde{P}_\beta + (1_{|\beta|} 1_{|\beta|}^T - \Theta) \tilde{P}_\beta^T [H_X\ H_Y] \tilde{Q}_\beta) U^T$$

where $\tilde{P} = [P_\alpha\ P_\beta U]$ and $\tilde{Q} = [Q_\alpha\ Q_{\bar\alpha} V]$

Conclusion 4. *We proposed to solve nonlinear semidefinite programming problem by solving a nonsmooth reformulation to the associated KKT system by the nonsmooth Newton's method. We proved that the algorithm converges locally and quadratically. For linear semidefinite programming, equivalent conditions are given for the nonsingularity of all elements of the Clarke's generalized Jacobian of Φ (see, [6]), we want to extend them to the nonlinear case. The question of implementation of the proposed algorithm and numerical examples are under considerations.*

References

1. Bi, S., Pan, S., Chen, J.-S.: Nonsingularity Conditions for the Fischer-Burmeister System of Nonlinear SDPs. SIAM J. Optim. 21(4), 1392–1417 (2011)
2. Clarke, F.H.: Optimization and nonsmooth optimization. Wiley, New York (1983)
3. Fares, B., Apkarian, P., Noll, D.: An augmented Lagrangian method for a class of LMI-constrained problems in robust control theory. Internat. J. Control. 74, 348–360 (2001)
4. Fares, B., Noll, D., Apkarian, P.: Robust control via sequential semidefinite programming. SIAM J. Control Optim. 40, 1791–1820 (2002)
5. Freund, R.W., Fleurian, J., Christoph, V.: A sequential semidefinite programming method and an application in passive reduced-order modeling (2005)
6. Han, L., Bi, S., Pan, S.: Nonsingularity of F-B system and Constraint Nondegeneracy in Semidefinite Programming. Numer. Algor. 62, 79–113 (2013)
7. Li, C., Sun, W.: A nonsmooth Newton-type method for nonlinear semidefinite programming. Journal of Nanjing Normal University. Natural Science Edition 31(2) (January 2008)

8. Li, C., Sun, W., de Sampaio, R.J.B.: An equivalency condition of nonsingularity in nonlinear semidefinite programming. Jrl. Syst. Sci. & Complexity 22, 1–8 (2009)

9. Qi, L., Sun, D.: A survey of Some Nonsmooth Equations and Smoothing Newton Methods. School of Mathematics. The University of New South Wales Sydney 2052, Australia (1998)

10. Sun, D.F., Sun, J.: Strong Semismoothness of the Fischer-Burmeister SDC and SOC Complementarity Functions. Math. Program. 103, 575–581 (2005)

11. Tuan, H.D., Apkarian, P., Nakashina, Y.: A new Lagrangian dual global optimization algorithm for solving bilinear matrix inequalities. Internat. J. Robust Nonlinear Control 10, 561–578 (2000)

12. Vanderbei, R.J.: LOQO usre's manual version 3. 10. Report SOR 97-08. Princeton University, Princeton (1997)

13. Vanderbei, R.J., Benson, H.Y., Shanno, D.F.: Interior point method for nonconvexe nonlinear programming: filter method and merit functions. Compt. Optim. Appl. 23, 257–272 (2002)

14. Zhang, L., Zhang, N., Pang, L.: Differential Properties of the Symmetric Matrix-Valued Fischer-Burmeister Function. J. Optim Theory Appl. 153, 436–460 (2012)

The Nonlinear Separation Theorem and a Representation Theorem for Bishop–Phelps Cones*

Refail Kasimbeyli** and Nergiz Kasimbeyli

Faculty of Engineering, Department of Industrial Engineering
Aandolu University, Iki Eylul Campus, Eskisehir 26555, Turkey
{rkasimbeyli,nkasimbeyli}@anadolu.edu.tr

Abstract. The paper presents a theorem for representation a given cone as a Bishop–Phelps cone in normed spaces and studies interior and separation properties for Bishop–Phelps cones. The representation theorem is formulated in the form of a necessary and sufficient condition and provides relationship between the parameters determining the Bishop–Phelps cone. The necessity is given in reflexive Banach spaces. The representation theorem is used to establish the theorem on interior of the Bishop–Phelps cone representing a given cone, and the nonlinear separation theorem. It is shown that every Bishop–Phelps cone in finite dimensional space satisfies the separation property for the nonlinear separation theorem. The theorems on the representation, on the interior and on the separation property studied in this paper are comprehensively illustrated on examples in finite and infinite dimensional spaces.

Keywords: Nonlinear Separation Theorem, Bishop–Phelps Cone, Representation Theorem, Augmented Dual Cone.

1 Introduction

In this paper we present a theorem for representation a given cone as a Bishop-Phelps cone (BP cone for short) in normed spaces.

This cone was introduced by Bishop and Phelps in 1962 (see [1,13]). Since then BP cones played a crucial role in many investigations on characterization of supporting elements of certain subsets of normed linear spaces.

Most remarkable characteristics for BP cones were given in [12,4,5]. Petschke has shown that every nontrivial convex cone C with a closed and bounded base in a real normed space is representable as a BP cone [12, Theorem 3.2]. Another basic result which follows from this theorem says that [12, Theorem 3.4], every nontrivial convex cone in a finite dimensional space is representable as a BP cone if and only if it is closed and pointed (see also [4, Theorem 4.4] and [5, Proposition 2.18, Proposition 2.19]).

---

* This study was supported by the Anadolu University Scientific Research Projects Commission under the grants no 1404F227 and 1306F245.
** Corresponding author.

© Springer International Publishing Switzerland 2015   419
H.A. Le Thi et al. (eds.), *Model. Comput. & Optim. in Inf. Syst. & Manage. Sci.,*
Advances in Intelligent Systems and Computing 359, DOI: 10.1007/978-3-319-18161-5_36

In this paper, we present a necessary and sufficient condition for representation of a given cone as a BP cone, where we do not impose any condition on the existence of a base or on a base. Generally, not every cone in infinite dimensional spaces has a base, or some cones may have a base which is unbounded.

The representation theorem presented in this paper uses the given norm of the normed space and there is no need to prove the existence of an additional equivalent norm.

This theorem guarantees not only the availability for a representation of some class of cones as a BP cone, but also provides relationship between the parameters (the linear functional, the norm, and the scalar coefficient of the norm) determining this BP cone, and explicitly defines the BP cone which equals the given cone. This is of great importance, because it provides an analytical expression for the given cone and thus provides a convenient mathematical tool in investigations. There are many existence and characterization theorems for optimal solutions in the literature where the objective space is assumed to be partially ordered by a BP cone (see e.g. [2,3,6]).

In this paper we also prove theorems on interior of the BP cone and by using the representation theorem we present a detailed discussion on the relationship between the augmented dual cones and the representation of the interior of BP cones.

Finally, by using the representation theorem we show that every BP cone and its conic neighborhood satisfy the nonlinear separation property in finite dimensional spaces. This property was suggested by R. Kasimbeyli in [7] where he proved that two cones satisfying the separation property, can be separated by some BP cone. Such a BP cone is defined by some element from the augmented dual cone. Note that, the augmented dual cones, BP cones and the nonlinear separation theorem are used to develop optimality conditions and solution approaches for a certain class of nonconvex optimization problems both in single objective optimization theory and in vector optimization (see e.g, [6,8,9,10,11,2]).

The theorems on the representation, on the interior and on the separation property studied in this paper are comprehensively illustrated on examples in finite and infinite dimensional spaces.

The paper is organized as follows. Section 2 gives some preliminaries. The general nonlinear separation property and separation theorems are given in Section 3. In this section a sufficient condition for separation property is also presented. The representation theorem, the theorems on the interior of BP cones and the relationship between the nonlinear separation property and the BP cones are given in Section 4. Section 5 presents illustrative examples and detailed discussions of the representation and characterization theorems in finite and infinite dimensional spaces. Finally, Section 6 draws some conclusions from the paper.

2   Preliminaries

In this section, we recall some concepts of cones, separability and proper efficiency. Throughout the paper, we will assume always, unless stated specifically otherwise, that:

(i) Y is a reflexive Banach space with dual space Y^*, and $C \subsetneqq Y$ is a cone which contains nonzero elements;

(ii) $\mathsf{cl}(S)$, $\mathsf{bd}(S)$, $\mathsf{int}(S)$, and $\mathsf{co}(S)$ denote the *closure* (in the norm topology), the *boundary*, the *interior*, and the *convex hull* of a set S, respectively;

(iii) R_+ and R_{++} denote the sets of nonnegative and positive real numbers, respectively;

The unit sphere and unit ball of Y are denoted by

$$U = \{y \in Y : \|y\| = 1\} \tag{1}$$

and

$$B = \{y \in Y : \|y\| \le 1\},$$

respectively.

A nonempty subset C of Y is called a *cone* if

$$y \in C, \lambda \ge 0 \Rightarrow \lambda y \in C.$$

Pointedness of C means that

$$C \cap (-C) = \{0_Y\}.$$

$$\mathsf{cone}(S) = \{\lambda s : \lambda \ge 0 \text{ and } s \in S\}$$

denotes the cone *generated* by a set S.

$C_U = C \cap U = \{y \in C : \|y\| = 1\}$ denotes the *base norm* of the cone C. The term *base norm* is justified by the obvious assertion that $C = \mathsf{cone}(C_U)$, and is firstly used in [14].

Recall that the dual cone C^* of C and its quasi-interior $C^\#$ are defined by

$$C^* = \{y^* \in Y^* : y^*(y) \ge 0 \text{ for all } y \in C\} \tag{2}$$

and

$$C^\# = \{y^* \in Y^* : y^*(y) > 0 \text{ for all } y \in C \setminus \{0\}\}, \tag{3}$$

respectively.

The following three cones called augmented dual cones of C were introduced in [7].

$$C^{a*} = \{(y^*, \alpha) \in C^\# \times R_+ : y^*(y) - \alpha\|y\| \ge 0 \text{ for all } y \in C\}, \tag{4}$$

$$C^{a\circ} = \{(y^*, \alpha) \in C^\# \times R_+ : y^*(y) - \alpha\|y\| > 0 \text{ for all } y \in \mathsf{int}(C)\}, \tag{5}$$

and

$$C^{a\#} = \{(y^*, \alpha) \in C^\# \times R_+ : y^*(y) - \alpha\|y\| > 0 \text{ for all } y \in C \setminus \{0\}\}, \tag{6}$$

where C is assumed to have a nonempty interior in the definition of $C^{a\circ}$.

3   The Nonlinear Separation Theorem

In this section, we recall the nonlinear separation theorem given by R. Kasimbeyli in [7]. This theorem enables to separate two cones (which are not necessarily convex, having only the vertex in common) by a level set of some monotonically increasing (with respect to the ordering cone) sublinear function. In this section, we present this theorem without proof.

Definition 1. *Let C and K be closed cones of a normed space $(Y, \|\cdot\|)$ with base norms C_U and K_U, respectively. Let $K_U^\partial = K_U \cap bd(K)$, and let $\widetilde{C}$ and $\widetilde{K}^\partial$ be the closures of the sets $co(C_U)$ and $co(K_U^\partial \cup \{0_Y\})$, respectively. The cones C and K are said to have the separation property with respect to the norm $\|\cdot\|$ if*

$$\widetilde{C} \cap \widetilde{K}^\partial = \emptyset. \tag{7}$$

Definition 2. *Let C and K be nonempty cones of a normed space $(Y, \|\cdot\|)$ with $int(K) \neq \emptyset$. A cone K is called a conic neighborhood of C if $(C \setminus \{0_Y\}) \subset int(K)$. For a positive real number ε, a cone $C_\varepsilon = cone(C_U + \varepsilon B)$ is called an ε-conic neighborhood of C.*

The following two theorems proved in [7, Theorems 4.3 and 4.4] concern the existence of a pair $(y^*, \alpha) \in C^{a\#}$ for which the corresponding sublevel set $S(y^*, \alpha)$ of the strongly monotonically increasing sublinear function $g(y) = y^*(y) + \alpha\|y\|$ separates the given cones C and K, where $S(y^*, \alpha)$ is defined as

$$S(y^*, \alpha) = \{y \in Y : y^*(y) + \alpha\|y\| \leq 0.\} \tag{8}$$

Theorem 1. *Let C and K be closed cones in a reflexive Banach space $(Y, \|\cdot\|)$. Assume that the cones $-C$ and K satisfy the separation property defined in Definition 1,*

$$-\widetilde{C} \cap \widetilde{K}^\partial = \emptyset. \tag{9}$$

Then, $C^{a\#} \neq \emptyset$, and there exists a pair $(y^*, \alpha) \in C^{a\#}$ such that the corresponding sublevel set $S(y^*, \alpha)$ of the strongly monotonically increasing sublinear function $g(y) = y^*(y) + \alpha\|y\|$ separates the cones $-C$ and $bd(K)$ in the following sense:

$$y^*(y) + \alpha\|y\| < 0 \leq y^*(z) + \alpha\|z\| \tag{10}$$

for all $y \in -C \setminus \{0_Y\}$, and $z \in bd(K)$. In this case the cone $-C$ is pointed.

Conversely, if there exists a pair $(y^*, \alpha) \in C^{a\#}$ such that the corresponding sublevel set $S(y^*, \alpha)$ of the strongly monotonically increasing sublinear function $g(y) = y^*(y) + \alpha\|y\|$ separates the cones $-C$ and $bd(K)$ in the sense of (10) and if either the cone C is closed and convex or $(Y, \|\cdot\|)$ is a finite dimensional space, then the cones $-C$ and K satisfy the separation property (9).

Remark 1. It follows from Theorem 1 that two cones satisfying the separation property (9) can be separated by a BP cone defined for some pair $(y^*, \alpha) \in C^{a\#}$, and conversely, if there exists a pair $(y^*, \alpha) \in C^{a\#}$ such that the corresponding BP cone separates the given cones, then these cones satisfy the separation property (9).

Theorem 2. *Let C be a closed cone of a reflexive Banach space $(Y, \|\cdot\|_Y)$, and let C_ε be its ε-conic neighborhood for a real number $\varepsilon \in (0,1)$. Suppose that C and C_ε satisfy the separation property given in Definition 1. Then, there exists a pair $(y^*, \alpha) \in C^{a\#}$ such that*

$$- C \setminus \{0_Y\} \subset \text{int}(S(y^*, \alpha)) \subset -C_\varepsilon, \tag{11}$$

where $\text{int}(S(y^*, \alpha))$ *can be defined as*

$$\text{int}(S(y^*, \alpha)) = \{y \in Y : y^*(y) + \alpha\|y\| < 0\}. \tag{12}$$

Remark 2. It follows from definition of the augmented dual cone that every nontrivial cone $C \subset Y$ is a subset of the BP cone

$$C(y^*, \alpha) = \{y \in Y : \alpha\|y\| \le y^*(y)\}$$

if $(y^*, \alpha) \in C^{a*}$. Theorem 2 strengthens this assertion by saying that for a cone C satisfying conditions of this theorem, there exists a BP cone which contains the given cone being contained in the ε-conic neighborhood of C for a real number $\varepsilon \in (0,1)$. In other words, under the conditions of Theorem 2, there exists a BP cone which is as close to the given cone as possible.

The following theorem is presented in [9, Lemma 3] and gives a general sufficient condition for the separation property in R^n.

Theorem 3. *Let C be a closed convex cone in R^n. Assume that there exist a pair $(y^*, \alpha) \in R^n \times R_{++}$ such that,*

$$cl(co(C_U)) = \{y \in B : y^*(y) \ge \alpha\}. \tag{13}$$

Then for an arbitrary closed cone $K \subset R^n$ with $C \cap K = \{0\}$, the cones C and K satisfy the separation property given in Definition 1.

4   Main Results

In this section, we show that the condition (13) of Theorem 3 is necessary and sufficient for the representation of a given cone as a BP cone in reflexive Banach spaces. Moreover, this BP cone is defined for the same norm (which is the given norm of the normed space) and the same pair $(y^*, \alpha) \in Y^* \times R_{++}$ used in condition (13). Thus, the theorem presented in this paper guarantees not only the availability of a representation of some class of cones as a BP cone, but also gives its exact expression by explaining properties of parameters determining this BP cone.

The following definition for BP cones is used in this paper:

Definition 3. *Let $(Y, \|\cdot\|)$ be a real normed space. For some positive number $\alpha > 0$ and some continuous linear functional y^* from the dual space Y^* the cone*

$$C(y^*, \alpha) = \{y \in Y : \alpha\|y\| \le y^*(y)\} \tag{14}$$

is called Bishop-Phelps cone. In this definition, the triple $(y^*, \alpha, \|\cdot\|)$ will be referred to as parameters determining the given BP cone.

In the original definition of Bishop and Phelps it is required that $\|y^*\|_* = 1$ and $\alpha \in (0,1]$.

Some authors (see for example, [2,4,5]) do not use the constant α and the assumption $\|y^*\|_* = 1$. This paper follows Definition 3. It easily follows from the definition that every BP cone is closed and pointed [5,12].

We first present a sufficient condition for characterizing interior of every BP cone. Then, the representation theorem will be presented. We begin with the following lemma characterizing the quasi-interior of the augmented dual cone.

Lemma 1. *Let $C \in Y$ be a nonempty cone. If $(y^*, \alpha) \in C^{a\#}$ then $\|y^*\|_* > \alpha$.*

Proof. Let $(y^*, \alpha) \in C^{a\#}$ and let $y \in C \setminus \{0\}$. Then

$$0 < y^*(y) - \alpha\|y\| \leq \|y^*\|_*\|y\| - \alpha\|y\| = \|y\|(\|y^*\|_* - \alpha),$$

which completes the proof.

The following theorem characterizes interior of BP cones. Note that this theorem is given in [7] in a slightly different setting, therefore we present this theorem without the proof for which we refer reader to [7, Lemma 3.6].

Theorem 4. *Let $C(y^*, \alpha) = \{y : y^*(y) \geq \alpha\|y\|\}$ be a given BP cone for some pair $(y^*, \alpha) \in C^{a*}$. If $(y^*, \alpha) \in C^{a\#}$ then*

$$int(C(y^*, \alpha)) = \{y : y^*(y) > \alpha\|y\|\} \neq \emptyset. \tag{15}$$

Remark 3. A sufficient condition on the characterization of interior of BP cones was also presented in [5, Theorem 2.5(b)], which is equivalent to that of Theorem 4. Below we present examples which demonstrate that the condition of Theorem 4 is not necessary in general (see, Section 5).

Now we present the representation theorem.

Theorem 5. *Let C be a nonempty closed convex cone of a real normed space $(Y, \|\cdot\|$. Assume that*

$$cl(co(C_U)) = \{y \in B : y^*(y) \geq \alpha\} \tag{16}$$

for some $(y^*, \alpha) \in Y^* \times R_{++}$. Then C is representable as a Bishop–Phelps cone with the same norm and the same pair (y^*, α) defining the condition (16). Conversely, if $C = \{y \in Y : y^*(y) - \alpha\|y\| \geq 0\}$ is a Bishop–Phelps cone of a reflexive Banach space $(Y, \|\cdot\|)$, then C satisfies condition (16).

Proof. **Necessity.** Let $y^* \in Y^*$ and let $\alpha > 0$ be a real number, and let $C = \{y \in Y : y^*(y) - \alpha\|y\| \geq 0\}$ be a given Bishop–Phelps cone in $(Y, \|\cdot\|)$. Show that C satisfies condition (16) with the same $y^* \in Y^*$, $\alpha > 0$ and the same norm.

Let

$$\tilde{C} = cl(co(C_U)). \tag{17}$$

It is clear that the base norm of C can be represented as

$$C_U = \{y \in U : y^*(y) - \alpha\|y\| \geq 0\} = \{y \in U : y^*(y) - \alpha \geq 0\}. \quad (18)$$

As $\alpha > 0$, in particular, it follows from the definition that C is convex and pointed.

We define the following set

$$D = \{y \in B : y^*(y) \geq \alpha\}. \quad (19)$$

First we show that

$$\mathrm{co}(C_U) = D. \quad (20)$$

Let $y \in \mathrm{co}(C_U)$. Then, by definition of convex hull, there exists a set of nonnegative numbers β_i, $i \in I$ such that, y can be represented as

$$y = \sum_{i \in I} \beta_i y_i, \quad \text{where } y_i \in C_U \text{ and } \sum_{i \in I} \beta_i = 1.$$

Clearly, $y \in B$. On the other hand

$$y^*(y) = \sum_{i \in I} \beta_i y^*(y_i) \geq \alpha.$$

Then, from (19) we have $y \in D$; that is, $\mathrm{co}(C_U) \subset D$.

Now, let $y \in D$. We will show that $y \in \mathrm{co}(C_U)$.

If $\|y\| = 1$ then $y \in U$ and the inclusion $y \in C_U \subset \mathrm{co}(C_U)$ follows from (18).

Consider the case $\|y\| < 1$, that is $y \in \mathrm{int}(B)$. Denote $\nu = y^*(y)$. Clearly $\nu \geq \alpha$. Take any non-zero vector $b \in Y$ satisfying $y^*(b) = 0$. Consider

$$y_\lambda = y + \lambda b, \quad \lambda \in (-\infty, \infty).$$

We have

$$y^*(y_\lambda) = y^*(y) + \lambda y^*(b) = \nu \geq \alpha. \quad (21)$$

As $b \neq \mathbf{0}$, we have $\|y_\lambda\| \to \infty$ if $|\lambda| \to \infty$ which means that $y \notin B$ for sufficiently large values of λ. On the other hand, since $y \in \mathrm{int}(B)$, the inclusion $y_\lambda \in \mathrm{int}(B)$ holds for sufficiently small in absolute value numbers $\lambda > 0$ and $\lambda < 0$. Then, since $\|y_\lambda\|$ is a weakly upper semicontinuous function of λ, and B is weakly compact, there exist numbers $\lambda_1 > 0$ and $\lambda_2 < 0$ such that the corresponding points $y_1 \doteq y_{\lambda_1}$ and $y_2 \doteq y_{\lambda_2}$ belong to the boundary of B (as maximum values of $\|y_\lambda\|$ w.r.t. $\lambda > 0$ and $\lambda < 0$ respectively). That is,

$$y_i \in U, \quad i = 1, 2.$$

These inclusions together with (21) and (18) imply that $y_i \in C_U$, $i = 1, 2$.

Finally, denoting $\lambda' = \lambda_1/(\lambda_1 - \lambda_2)$, it is not difficult to check that,

$$\lambda' \in (0,1) \text{ and } y = (1 - \lambda')y_1 + \lambda'y_2.$$

Therefore, $y \in \text{co}(C_U)$, which means that $D \subset \text{co}(C_U)$.

Thus, we have shown that the relation (20) is true. From this relation, we have

$$\widetilde{C} = \{y \in B : y^*(y) \geq \alpha\},$$

and the proof of Necessity is completed.

Sufficiency. Now let C be a nonempty closed convex cone of Y, and suppose that condition (16) is satisfied for some $(y^*, \alpha) \in Y^* \times R$ with $\alpha > 0$. Show that C is representable as a Bishop–Phelps cone, that is show that $C = C(y^*, \alpha)$.

Let $y \in C \setminus \{0\}_Y$. Then there exists a positive real number β such that $\beta y \in C_U$, and hence $\beta y \in cl(co(C_U))$. Then by condition (16) we have:

$$y^*(\beta y) \geq \alpha.$$

Then, since $\beta y \in C_U$, we have $\alpha = \alpha\|\beta y\|$, and $y^*(\beta y) \geq \alpha\|\beta y\|$. Thus, $y^*(y) \geq \alpha\|y\|$, which means that $C \subset C(y^*, \alpha)$.

Now let $y \in C(y^*, \alpha)$. Then for every $y \in C(y^*, \alpha)$ there exists a scalar $\beta > 0$ such that $\beta y \in U \cap C(y^*, \alpha)$ and therefore

$$y^*(\beta y) \geq \alpha\|\beta y\| = \alpha,$$

which implies by condition (16) that $\beta y \in cl(co(C_U))$. Since C is a closed and convex cone, we obtain $y \in C$, which establishes the inclusion $C(y^*, \alpha) \subset C$, and the proof of the theorem is completed.   □

The next theorem establishes an additional property for parameters of the BP cone representing the given cone.

Theorem 6. *Let $C \subset Y$ be a given cone which is representable as a BP cone. If $C(y^*, \alpha)$ is a BP cone representing the given cone C, then $(y^*, \alpha) \in C^{a*} \setminus C^{a\#}$.*

Proof. Assume that $C = C(y^*, \alpha)$ for some pair $(y^*, \alpha) \in C^{a*}$. Then $C = C(y^*, \alpha) = \{y \in Y : y^*(y) - \alpha\|y\| \geq 0,\}$ and clearly $(y^*, \alpha) \in C^{a*}$ by the definition of C^{a*}. Obviously, the set $\{y \in Y : y^*(y) - \alpha\|y\| = 0\}$ represents the boundary of C, and since C is assumed to contain nonzero elements, there exists some $y \in C \setminus \{0\}$ in the boundary with $y^*(y) - \alpha\|y\| = 0$ which completes the proof.   □

Theorem 7. *Let $C \subset Y$ be a given cone and let $(y^*, \alpha) \in C^{a*}$ with $\alpha > 0$, be the pair for which the representation property (16) is satisfied. Let $C(y^*, \alpha)$ be the BP cone representing the given cone C. Then $(y^*, \beta) \in C^{a\#}$ for every $\beta \in (0, \alpha)$, and*

$$(C \setminus \{0\}) \subset int(C(y^*, \beta)) = \{y : y^*(y) > \beta\|y\|\} \neq \emptyset.$$

Proof. Let $C = C(y^*, \alpha)$, where $(y^*, \alpha) \in C^{a*}$ with $\alpha > 0$. Then by Lemma [7, Lemma 3.2 (ii)] $(y^*, \beta) \in C^{a\#}$ for every $\beta \in (0, \alpha)$. Now let $y \in C(y^*, \alpha)$, and $\beta \in (0, \alpha)$ be arbitrary elements. Then

$$y^*(y) - \beta\|y\| > y^*(y) - \alpha\|y\| \geq 0,$$

which means by Theorem 4 that $y \in int(C(y^*, \beta))$ and the proof is completed.

□

The following theorem establishes that every BP cone in R^n, satisfies the separation property together with its ε conic neighborhood.

Theorem 8. *Let C be a BP cone in R^n. Then for every $\varepsilon \in (0, 1)$, cones C and $bd(C_\varepsilon)$ satisfy the separation property given in Definition 1.*

Proof. Since $C \cap bd(C_\varepsilon) = \{0\}$, the proof follows from theorems 5 and 3 and the definition of the ε conic neighborhood of a cone (see Definition 2).

5   Illustrative Examples

In this section we present illustrative examples for the representation and separation theorems, and for the theorem on interior of BP cones in both finite and infinite dimensional spaces.

5.1   Example 1

Let $C = R_+^n$. Due to Kasimbeyli [7, Theprem 5.9], this cone satisfies the separation property (7) with respect to l_1 norm for arbitrary n with $y^* = (1, \ldots, 1)$, and $\alpha = 1$. Then by Theorem 8, it satisfies the representation property, and its BP representation is given by

$$C(y^*, \alpha)_{l_1} = \{(y_1, \ldots, y_n) : \sum_{i=1}^{n} y_i - \sum_{i=1}^{n} |y_i| \geq 0\}.$$

It is evident that $int(R_+^n) = \{(y_1, \ldots, y_n) : y_i > 0, i = 1, \ldots, n\} \neq \emptyset$. On the other hand, $(y^*, \alpha) \in C^{a*} \setminus C^{a\#}$ (see, Theorem 6) and thus the interior of the BP cone $C(y^*, \alpha)_{l_1}$ cannot be represented by

$$\{(y_1, \ldots, y_n) : \sum_{i=1}^{n} y_i - \sum_{i=1}^{n} |y_i| > 0\},$$

which is empty set.

The nonnegative orthant R_+^n has interesting and different interpretations for different values of n and different norms. Therefore we consider each case separately.

In the case $n = 1$, all three norms l_1, l_2, l_∞ have the same formulation and therefore the BP representation of R_+ for all the three norms is given by (see also [4, Example 2.6 (a)])

$$C(1, 1) = \{y \in R : y - |y| \geq 0\}.$$

In the case $n = 2$, the l_2 and the l_∞ norms representations of R_+^2 are respectively:

$$C((1, 1), 1)_{l_2} = \{(y_1, y_2) : y_1 + y_2 - \sqrt{y_1^2 + y_2^2} \geq 0\},$$

and

$$C((1, 1), 1)_{l_\infty} = \{(y_1, y_2) : y_1 + y_2 - \max(|y_1|, |y_2|) \geq 0\}.$$

It is remarkable that, the condition (16) of the representation theorem is not satisfied for R_+^n with $n \geq 3$ in the cases of l_2 and l_∞ norms. Hence the nonnegative orthant of R^n with $n \geq 3$ can not be represented as a BP cone in the cases of l_2 and l_∞ norms. For example, consider the vector $y = (-1, 2, 2) \in R^3$. Let $y_3^* = (1, 1, 1)$ and let $\alpha = 1$. Then, the relation

$$y_3^*(y) \geq \alpha \|y\|$$

is satisfied for both l_2 and l_∞ norms, but $y \notin R_+^3$.

Remark 4. The nonnegative orthant of R^n is also considered in [4, Example 2.6 (c)], where the BP cone representation with l_1 norm and the interpretation on the interior are not correct.

5.2   Example 2

Let Y be the Banach space l^1, and let C be the nonnegative orthant of l^1. Then $(l^1)^* = l^\infty$ and taking $y^*(y) = \sum_{i=1}^{\infty} y_i$ and $\alpha = 1$ it can easily be shown that the representation condition (16) is satisfied and the BP cone

$$C(y^*, \alpha) = \{y \in l^1 : \sum_{i=1}^{\infty} y_i \geq \sum_{i=1}^{\infty} |y_i|\}$$

is a cone representing the nonnegative orthant of l^1. The augmented dual cone C^{a*} and its quasi interior $C^{a\#}$ can easily be calculated for the nonnegative orthant C of l^1.

By definition of the augmented dual cone, we have

$$\mathbb{C}^{a*} = \left\{ ((w_1, w_2, \ldots), \alpha) \in \mathbb{C}^\# \times \mathbb{R}_+ : \sum_{i=1}^{\infty} w_i y_i \geq \alpha \sum_{i=1}^{\infty} y_i \right.$$
$$\left. \text{for all } (y_1, y_2, \ldots) \in \mathbb{C} \right\}$$

or

$$\mathbb{C}^{a*} = \{((w_1, w_2, \ldots), \alpha) : w_i > 0, i = 1, 2, \ldots, 0 \leq \alpha \leq \inf\{w_1, w_2, \ldots\}\}$$

and

$$\mathbb{C}^{a\#} = \{((w_1, w_2, \ldots), \alpha) : 0 \leq \alpha < \inf\{w_1, w_2, \ldots\}\}.$$

It is evident that $(y^*, \alpha) \in \mathbb{C}^{a*} \setminus \mathbb{C}^{a\#}$, where $y^*(y) = \sum_{i=1}^{\infty} y_i$ and $\alpha = 1$.

Note that similar interpretation on the interior presented in subsection 5.1 for R_+^2 with l_1 norm is also valid for the nonnegative orthant of l^1.

In this case we have:

$$\{(y_1, y_2, \ldots) : \sum_{i=1}^{\infty} y_i - \sum_{i=1}^{\infty} |y_i| > 0\} = \emptyset,$$

thus, the relation (15) for the interior of a BP cone is not satisfied. Again, by [7, Lemma 3.2 (ii)]), we have that the pair (y^*, β) belongs to $C^{a\#}$ for every $\beta \in (0, 1)$. Then by Theorem 4, the interior of BP cone $C(y^*, \beta)$ with $\beta \in (0, 1)$, can be represented in the following form:

$$int(C(y^*, \beta)) = \{(y_1, y_2, \ldots) : \sum_{i=1}^{\infty} y_i - \beta \sum_{i=1}^{\infty} |y_i| > 0\} \neq \emptyset$$

and $C \setminus \{0_{l^1}\} \subset int(C(y^*, \beta))$ for every $\beta \in (0, 1)$.

6   Conclusions

In this paper, we present a representation theorem which establishes that every cone of a real normed space satisfying condition (16) is representable as a Bishop–Phelps cone and conversely, every BP cone of a reflexive Banach space, representing given cone C satisfies this condition. This theorem is formulated without any conditions neither on the existence of a base, nor on a base itself. The condition (16) uses the given norm of the normed space (the presented theorem does not need to construct another norm for representation) and gives an explicit formulation of how a given cone can be expressed in the form of a BP cone. Note that such a representation theorem appears in the literature firstly. Earlier a representation theorem was given by Petschke, who showed that every nontrivial convex cone C with a closed and bounded base in a real normed space is representable as a BP cone.

The paper studies two important properties of BP cones in relation with the representation theorem. One of them is the interior of BP cones, the other one is the separation property used in the nonlinear separation theorem for not necessarily convex cones. The paper presents characterization theorems on interior of BP cones. It has been shown that every BP cone satisfies the separation property together with its ε conic neighborhood in R^n. This property is very important in both theoretical investigations and practical applications in nonconvex analysis (see e.g. [3,6,7,8,9,10,11,2]).

References

1. Bishop, E., Phelps, R.R.: The support functionals of a convex set. In: Klee, V. (ed.) Convexity. Proceedings of Symposia in Pure Mathematics, vol. VII, pp. 27–35. American Mathematical Society, Providence (1962)
2. Eichfelder, E., Kasimbeyli, R.: Properly optimal elements in vector optimization with variable ordering structures. Journal of Global Optimization 60, 689–712 (2014)
3. Gasimov, R.N.: Characterization of the Benson proper efficiency and scalarization in nonconvex vector optimization. In: Koksalan, M., Zionts, S. (eds.) Multiple Criteria Decision Making in the New Millennium. Lecture Notes in Econom. and Math. Systems, vol. 507, pp. 189–198 (2001)
4. Ha, T.X.D., Jahn, J.: Properties of Bishop-Phelps Cones. Preprint no. 343. University of Erlangen-Nuremberg (2011)
5. Jahn, J.: Bishop-Phelps cones in optimization. International Journal of Optimization: Theory, Methods and Applications 1, 123–139 (2009)
6. Kasimbeyli, N.: Existence and characterization theorems in nonconvex vector optimization. Journal of Global Optimization (to appear), doi:10.1007/s10898-014-0234-7.
7. Kasimbeyli, R.: A Nonlinear Cone Separation Theorem and Scalarization in Nonconvex Vector Optimization. SIAM J. on Optimization 20, 1591–1619 (2010)
8. Kasimbeyli, R.: Radial epiderivatives and set-valued optimization. Optimization 58, 521–534 (2009)
9. Kasimbeyli, R.: A conic scalarization method in multi-objective optimization. Journal of Global Optimization 56, 279–297 (2013)
10. Kasimbeyli, R., Mammadov, M.: On weak subdifferentials, directional derivatives and radial epiderivatives for nonconvex functions. SIAM J. Optim. 20, 841–855 (2009)
11. Kasimbeyli, R., Mammadov, M.: Optimality conditions in nonconvex optimization via weak subdifferentials. Nonlinear Analysis: Theory, Methods and Applications 74, 2534–2547 (2011)
12. Petschke, M.: On a theorem of Arrow, Barankin and Blackwell. SIAM J. Control Optim. 28, 395–401 (1990)
13. Phelps, R.R.: Support cones in Banach spaces and their applications. Adv. Math. 13, 1–19 (1974)
14. Salz, W.: Eine topologische Eigenschaft der effizienten Punkte konvexer Mengen. Operations Research. Verfahren XXIII, 197–202 (1976)

Part VII
Spline Approximation
and Optimization

Convergence of the Simplicial Rational Bernstein Form

Jihad Titi[1], Tareq Hamadneh[1], and Jürgen Garloff[2]

[1] University of Konstanz, Konstanz, Germany
jihadtiti@yahoo.com, tareq.hamadneh@uni-konstanz.de
[2] University of Applied Sciences/HTWG Konstanz
Konstanz, Germany
garloff@htwg-konstanz.de

Abstract. Bernstein polynomials on a simplex V are considered. The expansion of a given polynomial p into these polynomials provides bounds for the range of p over V. Bounds for the range of a rational function over V can easily obtained from the Bernstein expansions of the numerator and denominator polynomials of this function. In this paper it is shown that these bounds converge monotonically and linearly to the range of the rational function if the degree of the Bernstein expansion is elevated. If V is subdivided then the convergence is quadratic with respect to the maximum of the diameters of the subsimplices.

Keywords: Bernstein polynomial, simplex, range bounds, rational function, degree elevation, subdivision.

1 Introduction

During the last decade, polynomial minimization over simplices has attracted the interest of many researchers, see [1,2], [4], [6,7,8,9,10,11,12]. Special attention was paid to the use of the expansion of the given polynomial into Bernstein polynomials over a simplex, the so-called *simplicial Bernstein expansion*, [2], [4], [8], [10,11,12]. In [10,11,12], R. Leroy gave results on degree elevation and subdivision of the underlying simplex of this expansion. In [13] the Bernstein form of a polynomial over a box, the so-called *tensorial Bernstein form*, was used to find an enclosure of the range of a given (multivariate) rational function over a box. Convergence properties of this *tensorial rational Bernstein form* were investigated in [5]. In this paper we present convergence properties of the corresponding *simplicial rational Bernstein form* based on Leroy's results.

The organization of our paper is as follows: In the next section we briefly recall the simplical polynomial Bernstein form and its basic properties, e.g., their range enclosing property. We also provide additional properties like sharpness, monotonicity, and convergence of the bounds which will be used later on. In Section 3 we present our main results, viz. convergence properties of the simplicial rational Bernstein form.

With the exception of the range enclosing property of the polynomial and rational forms, we state the results in each case only for the maximum of the quantities under consideration because the respective statements for the minimum are analogous.

© Springer International Publishing Switzerland 2015
H.A. Le Thi et al. (eds.), *Model. Comput. & Optim. in Inf. Syst. & Manage. Sci.*,
Advances in Intelligent Systems and Computing 359, DOI: 10.1007/978-3-319-18161-5_37

2   Bernstein Expansion over a Simplex

In this section we present firstly the most important and fundamental properties of
the Bernstein expansion over a simplex we will employ throughout the paper.

We follow the notation and definitions that have been used in [11], [12]. Firstly, we
recall the definition of a simplex.

Definition 1. *Let $v_0, \ldots, v_n$ be $n+1$ points of $\mathbb{R}^n$. The ordered list $V = [v_0, \ldots, v_n]$ is
called simplex of vertices $v_0, \ldots, v_n$. The realization $|V|$ of the simplex V is the set of
$\mathbb{R}^n$ defined as the convex hull of the points $v_0, \ldots, v_n$. The diameter of V is the length
of the largest edge of $|V|$.*

Throughout our paper we will assume that the points $v_0, \ldots, v_n$ are affinely independent in which case the simplex V is non-degenerate. We will often consider the sim-
plex $\Delta := [0, e_1, \ldots, e_n]$, called the *standard simplex*, where 0 is the zero vector in $\mathbb{R}^n$
and e_i is the ith vector of the canonical basis of $\mathbb{R}^n$, $i = 1, \ldots, n$. This is no restriction
since any simplex V in $\mathbb{R}^n$ can be mapped affinely upon Δ.

Recall that any vector $x \in \mathbb{R}^n$ can be written as an affine combination of the ver-
tices $v_0, \ldots, v_n$ with weights $\lambda_0, \ldots, \lambda_n$ called barycentric coordinates. If $x = (x_1, \ldots, x_n)$
$\in \Delta$, then $\lambda = (\lambda_0, \ldots, \lambda_n) = (1 - \sum_{i=1}^n x_i, x_1, \ldots, x_n)$.

For every multi-index $\alpha = (\alpha_0, \ldots, \alpha_n) \in \mathbb{N}^{n+1}$ and $\lambda = (\lambda_0, \ldots, \lambda_n) \in \mathbb{R}^{n+1}$ we write
$|\alpha| := \alpha_0 + \ldots + \alpha_n$ and $\lambda^\alpha := \prod_{i=0}^n \lambda_i^{\alpha_i}$. The relation $\leq$ on $\mathbb{N}^{n+1}$ is understood entrywise.
For $\alpha, \beta \in \mathbb{N}^{n+1}$ with $\beta \leq \alpha$ we define

$$\binom{\alpha}{\beta} := \prod_{i=0}^n \binom{\alpha_i}{\beta_i}.$$

If k is any natural number such that $|\alpha| = k$, we use the notation $\binom{k}{\alpha} := \frac{k!}{\alpha_0! \cdots \alpha_n!}$.

Definition 2. *[3, Section 8] Let k be a natural number. The Bernstein polynomials of
degree k with respect to V are the polynomials $(B_\alpha^k)_{|\alpha|=k}$, where*

$$B_\alpha^k := \binom{k}{\alpha} \lambda^\alpha. \tag{1}$$

The Bernstein polynomials of degree k with respect to V take nonnegative values
on V and sum up to 1: $1 = \sum_{|\alpha|=k} B_\alpha^k$.

Let p be a polynomial of degree l,

$$p(x) = \sum_{|\beta| \leq l} a_\beta x^\beta.$$

Since the Bernstein polynomials of degree k form a basis of the vector space $\mathbb{R}_k[X]$
of polynomials of degree at most k, see, e.g, [10, Proposition 1.6], p can be uniquely
expressed as $(l \leq k)$

$$p(x) = \sum_{|\alpha|=k} b_\alpha(p, k, V) B_\alpha^k. \tag{2}$$

The numbers $b_\alpha(p, k, V)$ are called the *Bernstein coefficients of p of degree k with respect to V*. If $V = \Delta$, we obtain by multinomial expansion the following representation of the Bernstein coefficients in terms of the coefficients of p ($|\alpha| = k$)

$$b_\alpha(p, k, \Delta) = \sum_{\beta \leq \alpha} \frac{\binom{\alpha}{\beta}}{\binom{k}{\beta}} a_\beta. \tag{3}$$

It is easy to see from (3) that the Bernstein coefficients are linear with respect to p.

Definition 3. *Let $V = [v_0, \ldots, v_n]$ be a non-degenerate simplex of $\mathbb{R}^n$ and $p \in \mathbb{R}_l[X]$.*

- *The grid points of degree k associated to V are the points ($|\alpha| = k$)*

$$v_\alpha(k, V) := \frac{1}{k}(\alpha_0\, v_0 + \ldots + \alpha_n\, v_n) \in \mathbb{R}^n. \tag{4}$$

- *The control points associated to p of degree k with respect to V are the points $(v_\alpha(k, V), b_\alpha(p, k, V)) \in \mathbb{R}^{n+1}$.*
  The set of control points of p forms its control net of degree k.
- *The discrete graph of p of degree k with respect to V is formed by the collection $(v_\alpha(k, V), p(v_\alpha(k, V)))_{|\alpha|=k}$.*

In the sequel, $(e_0, \ldots, e_n)$ denotes the standard basis of $\mathbb{R}^{n+1}$.

Proposition 1. *[11, Proposition 2.7] For $p \in \mathbb{R}_l[X]$ the following properties hold.*

(i) *Interpolation at the vertices:*

$$b_{ke_i} = p(v_i), \quad 0 \leq i \leq n; \tag{5}$$

(ii) *convex hull property: the graph of p over V is contained in the convex hull of its associated control points;*

(iii) *range enclosing property:*

$$\min_{|\alpha|=k} b_\alpha(p, k, V) \leq p(x) \leq \max_{|\alpha|=k} b_\alpha(p, k, V), x \in V. \tag{6}$$

The following proposition gives necessary and sufficient conditions when equality holds in (6).

Proposition 2. *Let $p \in \mathbb{R}_l[X]$. Then*

$$\max_{x \in \Delta} p(x) = \max_{|\alpha|=k} b_\alpha(p, k, \Delta) \tag{7}$$

if and only if

$$\max_{|\alpha|=k} b_\alpha(p, k, \Delta) = b_{\alpha^*}(p, k, \Delta) \text{ for some } \alpha^* = k e_{i_0}, \; i_0 \in \{0, \ldots, n\}. \tag{8}$$

A similar statement holds for the minimum.

Proof. Without loss of generality we consider the standard simplex (of dimension n). If the maximum is attained at an index $k e_{i_0}$ then the statement holds trivially by Proposition 1 (i). Conversely, suppose first that the $b_\alpha(p, k, \Delta)$ are equal for all $|\alpha| = k$. Then the statement is trivial. Otherwise not all $b_\alpha(p, k, \Delta)$ are equal. Suppose that the maximum of $p(x)$ occurs at $x^* \in \Delta$ with barycentric coordinates λ^*. If x^* is in the interior of Δ then $0 < B_\alpha^k(\lambda^*) < 1$ for all $|\alpha| = k$ and hence

$$p(x^*) = \sum_{|\alpha|=k} b_\alpha(p, k, \Delta) B_\alpha^k(\lambda^*)$$

$$< \max_{|\alpha|=k} b_\alpha(p, k, \Delta) \sum_{|\alpha|=k} B_\alpha^k(\lambda^*) = \max_{|\alpha|=k} b_\alpha(p, k, \Delta),$$

a contraction. If x^* is lying on the boundary of Δ then it is contained in a subsimplex of dimension $n-1$, Δ' say. The Bernstein coefficients of p over Δ' coincide with the respective coefficients contained in the part of the array $(b_\alpha(p, k, \Delta), |\alpha| = k)$ between the (extreme) coefficients associated with the vertices of Δ' according to (5). Now we can apply the arguments used above to Δ'. Continuing in this way of examining all possible cases, we decrease the dimension of the simplices to be investigated step by step and arrive finally at the situation in which x^* is a vertex of Δ which completes the proof. $\square$

Recall that the barycentric coordinates can be written in terms of the components of the variable $x \in \mathbb{R}^n$. By multiplying both sides of (2) with $1 = \sum_{i=0}^n \lambda_i = (1 - \sum_{i=1}^n x_i) + \sum_{i=1}^n x_i$ and rearranging the result we obtain, see also [10, Proposition 1.12],

$$p(x) = \sum_{|\beta|=k+1} b_\beta(p, k+1, V) B_\beta^{k+1}, \tag{9}$$

where

$$b_\beta(p, k+1, V) = b_{\alpha+e_l}(p, k+1, V) = \frac{1}{k+1} \sum_{i=0, i \neq l}^n \alpha_i b_{\alpha+e_l-e_i}(p, k, V)$$

$$+ \frac{\alpha_l + 1}{k+1} b_\alpha(p, k, V). \tag{10}$$

It is easy to see from (10) that the coefficients $b_\beta(p, k+1, V)$ are convex combinations of the coefficients $b_\alpha(p, k, V)$. Hence the upper bounds decrease monotonically.

Proposition 3. *Let $p \in \mathbb{R}_l [X]$. Then it holds that*

$$\max_{|\beta|=k+1} b_\beta(p, k+1, V) \leq \max_{|\alpha|=k} b_\alpha(p, k, V). \tag{11}$$

In order to relate the control net and the discrete graph of a given polynomial, R. Leroy [11], [12] suggested to use the so-called *second differences* which are given in the following definition.

Definition 4. *Let* $V = [v_0, \ldots, v_n]$ *be a non-degenerate simplex of* $\mathbb{R}^n$. *For* $|\gamma| = k - 2$ *and* $0 \leq i < j \leq n$, *define the* second differences *of* p *of degree* k *with respect to* V *as*

$$\nabla^2 b_{\gamma,i,j}(p, k, V) := b_{\gamma + e_i + e_{j-1}} + b_{\gamma + e_{i-1} + e_j} - b_{\gamma + e_{i-1} + e_{j-1}} - b_{\gamma + e_i + e_j},$$

with the convention $e_{-1} := e_n$. *The second differences constitute the collection*

$$\nabla^2 b(p, k, V) := (\nabla^2 b_{\gamma,i,j}(p, k, V))_{|\gamma| = k-2, \, 0 \leq i < j \leq n}. \tag{12}$$

The maximum of the second differences, i.e.,

$$\|\nabla^2 b(p, k, V)\|_\infty = \max_{|\gamma| = k-2, \, 0 \leq i < j \leq n} |\nabla^2 b_{\gamma,i,j}(p, k, V)|, \tag{13}$$

will play an important role in the subsequent convergence analysis.

The following theorem gives the convergence of the control net to the discrete graph of a given polynomial with respect to degree elevation. Since any simplex can be mapped upon the standard simplex by an affine transformation, we present the following statements only for Δ.

Theorem 1. *[12, Theorem 4.2] Let* $p \in \mathbb{R}_l[X]$ *and* $l < k$. *Then*

$$\max_{|\alpha|=k} |p(v_\alpha(k, \Delta)) - b_\alpha(p, k, \Delta)| \leq \frac{n(n+2)l(l-1)}{24(k-1)} \|\nabla^2 b(p, l, \Delta)\|_\infty. \tag{14}$$

The following corollary is an immediate consequence of Theorem 1.

Corollary 1. *Let* $p \in \mathbb{R}_l[X]$. *If* $l < k$, *then*

$$\max_{|\alpha|=k} b_\alpha(p, k, \Delta) - \max_{x \in \Delta} p(x) \leq \frac{T_1}{k-1}, \tag{15}$$

where

$$T_1 := \frac{n(n+2)l(l-1)}{24} \|\nabla^2 b(p, l, \Delta)\|_\infty. \tag{16}$$

Proof. Assume that the maximum of $(b_\alpha(p, k, \Delta))$ such that $|\alpha| = k$ is attained at $b_{\alpha^*}(p, k, \Delta)$. Then we have

$$\max_{|\alpha|=k} b_\alpha(p, k, \Delta) - \max_{x \in \Delta} p(x) \leq b_{\alpha^*}(p, k, \Delta) - p(v_{\alpha^*})$$

$$= |b_{\alpha^*}(p, k, \Delta) - p(v_{\alpha^*})| \leq \frac{T_1}{k-1}.$$

The second inequality follows since v_{α^*} is a grid point in Δ, while the last inequality follows by using Theorem 1. $\square$

Definition 5. *Let* $V = [v_0, \ldots, v_n]$ *be a non-degenerate simplex of* $\mathbb{R}^n$ *and* $v' \in \mathbb{R}^n$. *The simplices* V^i *generated by subdivision with respect to the point* v' *are defined as*

$$V^i := [v_0, \ldots, v_{i-1}, v', v_{i+1}, \ldots, v_n], \; 0 \leq i \leq n.$$

Assume that Δ has been subdivided with respect to a point $\acute{v}$ in Δ, $\Delta = V^1 \cup \ldots \cup V^\sigma$ say. By using [12, Algorithm 4.6 (de Casteljau)] it is easy to see that the Bernstein coefficients of p over any V^i are contained in the interval $[\min_{|\alpha|=k} b_\alpha(p,k,\Delta),$ $\max_{|\alpha|=k} b_\alpha(p,k,\Delta)], 1 \le i \le \sigma$, see [12, Remark 4.7]. The following theorem gives the convergence of the control net to the discrete graph with respect to subdivision.

Theorem 2. *[12, Theorem 4.9] Let $\Delta = V^1 \cup \ldots \cup V^\sigma$ be a subdivision of the standard simplex Δ and h be an upper bound on the diameters of the V^i's. Then, for each $i \in \{1, \ldots, \sigma\}$ and $|\alpha| = k$, we have*

$$|p(\mathbf{v}_\alpha(k, V^i)) - b_\alpha(p, k, V^i)| \le h^2 k \frac{n^2(n+1)(n+2)^2(n+3)}{576} \|\nabla^2 b(p, k, \Delta)\|_\infty.$$

The following corollary can similarly be proven as Corollary 1.

Corollary 2. *Let $p \in \mathbb{R}_l[X]$, $\Delta = V^1 \cup \ldots \cup V^\sigma$ be a subdivision of the standard simplex Δ and h be an upper bound on the diameters of the V^i's. Then*

$$\max_{\substack{|\alpha|=k, \\ i=1,\ldots,\sigma}} b_\alpha(p, k, V^i) - \max_{x \in \Delta} p(x) \le h^2 T_2, \tag{17}$$

where

$$T_2 := k \frac{n^2(n+1)(n+2)^2(n+3)}{576} \|\nabla^2 b(p, k, \Delta)\|_\infty. \tag{18}$$

3 The Simplicial Rational Bernstein Form

In this section we present our results in the case of rational functions. Throughout this section we assume that p and q are polynomials of degree less than or equal to l with Bernstein coefficients $b_\alpha(p, k, \Delta)$ and $b_\alpha(q, k, \Delta)$, $|\alpha| = k$, respectively, over the standard simplex Δ, where $l \le k$. We also assume that all Bernstein coefficients $b_\alpha(q, k, \Delta)$ have the same sign and are non-zero (this implies that $q(x) \ne 0$, for all $x \in \Delta$) and without loss of generality we may assume that all of them are positive. For the negative case replace q by $-q$. The following theorem provides bounds for the range of $f := \frac{p}{q}$ over Δ. We use the notation

$$b_\alpha(f, k, V) := \frac{b_\alpha(p, k, V)}{b_\alpha(q, k, V)} \quad \text{for all } \alpha, \ |\alpha| = k.$$

Theorem 3. *[13, Theorem 3.1, Remark 6] The range of f over Δ can be bounded by*

$$\min_{|\alpha|=k} b_\alpha(f, k, \Delta) \le f(x) \le \max_{|\alpha|=k} b_\alpha(f, k, \Delta), \ x \in \Delta. \tag{19}$$

We now extend results from the polynomial to the rational case.

Theorem 4. *The equality holds in the right inequality in (19) if and only if*

$$\max_{|\alpha|=k} b_\alpha(f, k, \Delta) = b_{\alpha^*}(f, k, \Delta) \text{ for some } \alpha^* = k\mathbf{e}_{i_0} \ i_0 \in \{0, \ldots, n\}. \tag{20}$$

A similar statement holds for the equality in the left inequality.

Proof. Put $M := \max_{x \in \Delta} f(x)$. Suppose that the equality holds in the right inequality of (19). Then there exists $x^* \in \Delta$ and α^* with $|\alpha^*| = k$ such that $f(x^*) = M = \max_{|\alpha|=k} b_\alpha(f, k, \Delta) = b_{\alpha^*}(f, k, \Delta)$. Hence $s(x) := p(x) - Mq(x) \leq 0$ for all $x \in \Delta$, $b_\alpha(s, k, \Delta) \leq 0$ for all α with $|\alpha| = k$ (by linearity of the Bernstein coefficients), and $s(x^*) = 0$. By Propositions 1 (iii) and 2 there exists $i_0 \in \{0, \ldots, n\}$ such that $\alpha^* = k e_{i_0}$ and $s(x^*) = b_{\alpha^*}(s, k, \Delta) = 0$. By linearity, $b_{\alpha^*}(p, k, \Delta) = M b_{\alpha^*}(q, k, \Delta)$. Hence the first implication follows. The converse holds by Proposition 1 (i). □

Theorem 5. *The upper bounds decrease monotonically*

$$\max_{|\beta|=k+1} b_\beta(f, k+1, \Delta) \leq \max_{|\alpha|=k} b_\alpha(f, k, \Delta). \tag{21}$$

Proof. Put

$$M^{(k)} := \max_{|\alpha|=k} b_\alpha(f, k, \Delta), \text{ and } s(x) := p(x) - M^{(k)} q(x).$$

Then by the linearity of the Bernstein coefficients, we have for all β with $|\beta| = k+1$

$$b_\beta(s, k+1, \Delta) \leq \max_{|\beta|=k+1} (b_\beta(p, k+1, \Delta) - M^{(k)} b_\beta(q, k+1, \Delta))$$

$$\leq \max_{|\alpha|=k}(b_\alpha(p, k, \Delta) - M^{(k)} b_\alpha(q, k, \Delta)) \leq 0.$$

The second inequality follows by application of Proposition 3 to the polynomial s and the last inequality is a consequence of the definition of $M^{(k)}$. This implies

$$b_\beta(p, k+1, \Delta) \leq M^{(k)} b_\beta(q, k+1, \Delta)$$

from which the result follows. □

Now we turn to the convergence of the bounds for the range of rational functions provided by the Bernstein coefficients under degree elevation and subdivision.

Theorem 6. *For $l < k$ it holds that*

$$\max_{|\alpha|=k} b_\alpha(f, k, \Delta) - \max_{x \in \Delta} f(x) \leq \frac{A_1}{k-1}, \tag{22}$$

where

$$A_1 := \frac{n(n+2)l(l-1)}{24 B_1} (\|\nabla^2 b(p, l, \Delta)\|_\infty + \max_{|\alpha|=l} |b_\alpha(f, l, \Delta)| \, \|\nabla^2 b(q, l, \Delta)\|_\infty) \tag{23}$$

and $B_1 := \min_{|\alpha|=l} b_\alpha(q, l, \Delta)$.

Proof. The proof follows by using Corollary 1 and arguments similar to that given in the proof of the following theorem. □

Remark 1. For any $0 < \epsilon$ we can guarantee that

$$\max_{|\alpha|=k} b_\alpha(f, k, \Delta) - \max_{x \in \Delta} f(x) < \epsilon$$

if we choose $\frac{A_1}{\epsilon} + 1 < k$.

The last theorem shows that the convergence of the bounds to the range is only linear (in k) if we elevate the degree. Instead, if we subdivide Δ we obtain quadratic convergence with respect to the maximum diameter of the subsimplices.

Theorem 7. *Let $\Delta = V^1 \cup \ldots \cup V^\sigma$ be a subdivision of the standard simplex Δ and h be an upper bound on the diameters of the V^i's. Then we have*

$$\max_{\substack{|\alpha|=k, \\ i=1,\ldots,\sigma}} b_\alpha(f, k, V^i) - \max_{x \in \Delta} f(x) \le h^2 A_2, \tag{24}$$

where

$$A_2 := k \frac{n^2(n+1)(n+2)^2(n+3)}{576 B_1}(\|\nabla^2 b(p, k, \Delta)\|_\infty + B_2\|\nabla^2 b(q, k, \Delta)\|_\infty), \tag{25}$$

B_1 is given in Theorem 6, and $B_2 := \frac{\max_{|\alpha|=k} |b_\alpha(p,k,\Delta)|}{B_1}$.

Proof. Suppose that $\max_{\substack{|\alpha|=k, \\ i=1,\ldots,\sigma}} b_\alpha(f, k, V^i)$ is attained at $b_{\alpha^*}(f, k, V^{i_0})$ with $|\alpha^*| = k$, $i_0 \in \{1, \ldots, \sigma\}$. Then

$$\max_{\substack{|\alpha|=k, \\ i=1,\ldots,\sigma}} b_\alpha(f, k, V^i) - \max_{x \in \Delta} f(x) \le b_{\alpha^*}(f, k, V^{i_0}) - \max_{x \in V^{i_0}} f(x)$$

$$\le \left| \frac{[p(v_{\alpha^*}(k, V^{i_0})) - b_{\alpha^*}(p, k, V^{i_0})] - b_{\alpha^*}(f, k, V^{i_0})[q(v_{\alpha^*}(k, V^{i_0})) - b_{\alpha^*}(q, k, V^{i_0})]}{q(v_{\alpha^*}(k, V^{i_0}))} \right|$$

$$\le \frac{|p(v_{\alpha^*}(k, V^{i_0})) - b_{\alpha^*}(p, k, V^{i_0})| + |b_{\alpha^*}(f, k, V^{i_0})||q(v_{\alpha^*}(k, V^{i_0})) - b_{\alpha^*}(q, k, V^{i_0})|}{|q(v_{\alpha^*}(k, V^{i_0}))|}$$

$$\le h^2 A_2,$$

where the second inequality follows since $v_{\alpha^*}(k, V^{i_0})$ is a grid point in V^{i_0}, the third follows by using the triangle inequality, and the fourth is a consequence of Theorem 2 and the fact that the Bernstein coefficients of a polynomial over V^{i_0} are contained in the interval spanned by the Bernstein coefficients over Δ. $\square$

We conclude the paper with a lower bound on the number of subdivision steps needed in order to obtain a tolerance $\epsilon > 0$ between the maximum of the Bernstein coefficients of the given rational function over the subsimplices and its maximum over Δ. Before we present our result we need the following definition.

Definition 6. *[12, Definition 5.5] Let V be a non-degenerate simplex of $\mathbb{R}^n$, $S(V)$ be a subdivision of the simplex V, i.e., $S(V) = (V^1, \ldots, V^\sigma)$ with $V = V^1 \cup \ldots \cup V^\sigma$.*

- *By $m(S(V))$ the largest diameter of the subsimplices V^i is denoted.*
- *The subdivision scheme S is said to have a shrinking factor C, $0 \le C \le 1$, if for every simplex V,*

$$m(S(V)) \le Cm(V). \tag{26}$$

Theorem 8. *Let S be a subdivision scheme with shrinking factor C, $0 < C < 1$. Then*

$$\max_{\substack{|\alpha|=k, \\ i=1,\dots,\sigma}} b_\alpha(f,k,V^i) - \max_{x \in \Delta} f(x) < \epsilon, \tag{27}$$

if $\frac{\ln \frac{\epsilon}{2A_2}}{2\ln C} < N$, *where A_2 is given in Theorem 7.*

Proof. For any $0 < \epsilon$ take N such that $2C^{2N}A_2 < \epsilon$. By using Definition 6 and the fact that $m(\Delta) = \sqrt{2}$ we may choose $h = \sqrt{2}C^N$. Hence by using Theorem 7 and $0 < C < 1$ the result follows.   □

References

1. Basu, S., Leroy, R., Roy, M.-F.: A Bound on the Minimum of a Real Positive Polynomial over the Standard Simplex, arxiv: 0902.3304v1 (February 19, 2009), www.arxiv.org
2. Boudaoud, F., Caruso, F., Roy, M.-F.: Certificates of Positivity in the Bernstein Basis. Discrete Comput. Geom. 39, 639–655 (2008)
3. Farouki, R.T.: The Bernstein Polynomial Basis: A Centennial Retrospective. Comput. Aided Geom. Design 29, 379–419 (2012)
4. Garloff, J.: Convergent Bounds for the Range of Multivariate Polynomials. In: Nickel, K. (ed.) Interval Mathematics 1985. LNCS, vol. 212, pp. 37–56. Springer, Heidelberg (1986)
5. Garloff, J., Hamadneh, T.: Convergence and Inclusion Isotonicity of the Tensorial Rational Bernstein Form. In: Tucker, W., Wolff von Guddenberg, J. (eds.) Proceedings of the 16th GAMM-IMACS International Symposium on Scientific Computing, Computer Arithmetic and Validated Numerics. LNCS. Springer, Heidelberg (to appear)
6. Jeronimo, G., Perrucci, D.: On the Minimum of a Positive Polynomial over the Standard Simplex, J. Symbolic Comput 45, 434–442 (2010)
7. de Klerk, E.: The complexity of Optimizing over a Simplex, Hypercube or Sphere: a Short Survey. CEJOR Cent. Eur. J. Oper. Res. 16, 111–125 (2008)
8. de Klerk, E., den Hertog, D., Elabwabi, G.: On the Complexity of Optimization over the Standard Simplex. European J. Oper. Res. 191, 773–785 (2008)
9. de Klerk, E., Laurent, M., Parrilo, P.A.: A PTAS for the Minimization of Polynomials of Fixed Degree over the Simplex. Theor. Comput. Sci. 361, 210–225 (2006)
10. Leroy, R.: Certificats de positivité et minimisation polynomiale dans la base de Bernstein multivariée, Ph D Thesis, Université de Rennes 1 (2008)
11. Leroy, R.: Convergence under Subdivision and Complexity of Polynomial Minimization in the Simplicial Bernstein Basis. Reliab. Comput. 17, 11–21 (2012)
12. Leroy, R.: Certificates of Positivity in the Simplicial Bernstein Basis (May 3, 2011), http://hal.archives-ouvertes.fr/hal-00589945v_1/ (preprint)
13. Narkawicz, A., Garloff, J., Smith, A.P., Muñoz, C.A.: Bounding the Range of a Rational Function over a Box. Reliab. Comput. 17, 34–39 (2012)

Inferring Mean Road Axis from Big Data: Sorted Points Cloud Belonging to Traces

F. J. Ariza-López[1], D. Barrera[2], J. F. Reinoso[3], and R. Romero-Zaliz[4]

[1] Department of Cartographic, Geodesic and Photogrammetry Engineering,
University of Jaén, Spain
jfariza@ujaen.es

[2] Department of Applied Mathematics, University of Granada, Spain
dbarrera@ugr.es

[3] Department of Architectonic and Engineering Graphic Expression,
University of Granada, Spain
jreinoso@ugr.es

[4] Department of Computer Science and Artificial Intelligence,
University of Granada, Spain
rocio@decsai.ugr.es

Abstract. Roads are, probably the most important features appearing in cartography, both digital and analog one. The necessary tasks, to get accurate roads representation, were traditionally really expensive: photogrammetry and in situ differential GPS, for example. Nevertheless nowadays, the web allows people to register waypoints in their navigation device, with low accuracy and offer them to the rest of community. This way a lot of traces could be available to infer a mean road axis which, probably to be much more precise than the individual ones. In this paper we present three approaches in order to compute the representative axis above mentioned: a) Fréchet distance concept, b) B-spline least square fit and c) genetic algorithm spline-based. This paper shows that all our approaches are suitable to be deployed in a web-based application in order to support collaborative digital cartography. The dataset we used in our study is composed of 149 traces captured by a low accuracy user consumer GPS.

Keywords: geographic information, roads determination, Fréchet distance, homologous points, B-spline fitting, genetic algorithms.

1 Introduction

Linear features are the objects the more numerous one, around 80% in cartography [21], and belonging to that category roads are the most appreciate by user consulting cartography [17]. For this reason both cartographic agencies and companies producer consider the roads are strategic objects where focus their attention and where it is worth to invest money. Intelligent Transportation Systems require high geometric and accurate attributes in their roads digital database [9,11]. This required accuracy has been reached using different methodologies,

© Springer International Publishing Switzerland 2015
H.A. Le Thi et al. (eds.), *Model. Comput. & Optim. in Inf. Syst. & Manage. Sci.*,
Advances in Intelligent Systems and Computing 359, DOI: 10.1007/978-3-319-18161-5_38

e.g. photogrammetry and LiDAR boarding airplanes, like the MTN25 maps in Spain [12], or capturing directly the waypoints by precise differential GPS techniques [8]. The approach to infer a representative axis from repeated large traces set has been adopted by several authors [1,2,15,18]. The computation of a representative axis from a large dataset of traces lies on the research area of data mining [14]. Multiple GPS traces is the preferred dataset to be processed in the axis road determination [2,13,19]. The approaches for deriving the representative axis use cluster methods [8], lines fusion [19] and centroid-based computation for every set of points included in a traverse plane to the axis [22].

Nowadays exists a new possibility to produce accurate cartographical roads using the collaborative tools that are taken place on the world wide web. For example, some collaborative projects like OpenStreetMap [10,16] allows to upload geographical information in both, vector and attribute categories, and on the same time edit and improve the existing one on the web site. So, it would be easy to implement a similar system to OpenStreetMap but where the users just upload their waypoints and an application, which processes those data and computes the representative axis for the corresponding road or street. In this paper we present three algorithms to get the mentioned target: estimating the representative axis.

2   Material and Methodology

Our dataset is composed of 149 traces from a piece of the JV-2227 road located in Jaén (Spain). The piece road length is around 1460 m and its geometry is composed by a series of bends (see Figure 1). We have selected this kind of geometry because if we get a suitable fit in such a complex geometry, using our methods, it would be expected that suitable results occur in any kind of roads.

The GPS model we used to capture the traces was a Columbus V-990 which is a low-accuracy user consumer GPS. Although the Columbus had the ability to perform in differential mode (until 5 m/CEP accuracy in the 95% of time) we din't activate it, in order to reproduce the worst scenery, which is the case when a user takes waypoints with differential mode off (this is the more usual way to proceed by user consumer). The worst positioning in the traces captured was observed in the height component as we can see in Figure 1.

The three methodological processes we used for estimating the representative axis from the dataset of traces were the following:

- Identification of homologous points based on the Fréchet distance concept using the approach in [6].
- Least square B-spline fitting to a points cloud dataset.
- Genetic algorithmic in order to optimize nodes distribution in defining an approximating B-spline curve.

Fig. 1. Perspective view from Google Earth. The heights of all traces have been increased in order to be observed over the ground.

2.1   Identification of Homologous Points Based on the Fréchet Distance Concept

This method is based on identifying homologous points (Devogele criteria based on the Fréchet distance concept) between two traces, and computing the mean point for each couple of matched points, so that the set of new mean points built the new representative mean axis for the two older traces. The workflow is as follows:

1. Eliminating outliers (the traces that are far away from the central tendency), after that we obtain a set of traces called TS (see Figure 2). We considered three outliers based on the perception that they were ill-placed, i.e. we used a qualitative criterion. Note that just three outliers were considered with this criterion.
2. Matching homologous points between traces pairwise based on the discrete Fréchet distance, and computing the mean point for each couple of matched points. For each couple of traces compute their mean discrete Fréchet distance (mdFd).
3. Replacing in the TS the pair with the maximum mdFd by the mean Fréchet trace (mFt). So, an updated TS is generated, which is diminished in an unit.
4. Repet from step 2 until only one trace remains in TS.
5. Smoothing the final result by an approximating B-spline curve.

The Fréchet distance is traditionally illustrated by a man and his dog linked by a leash and everyone walking for a different and predefined way (trace). Both may vary their speed, but backtracking is not allowed. The Fréchet distance between both traces is the leash minimal length necessary to walk each one its trace keeping linked by the leash.

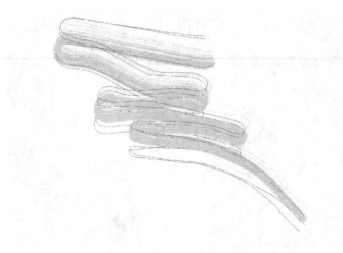

Fig. 2. Outlier (red color) to be deleted from the dataset

Considering two sets $t_1 = \langle t_{1,1}, \ldots, t_{1,n} \rangle$ and $t_2 = \langle t_{2,1}, \ldots, t_{2,m} \rangle$ of ordered vertices, Devogele proposes in [6] the following recursive computation based on the Euclidean distance d_E:

$$d_{Fd}(t_1, t_2) = \max \left(\begin{array}{l} d_E(t_{1,n}, t_{2,m}) \\ \min \left(\begin{array}{l} d_{Fd}\left(\langle t_{1,1}, t_{1,n-1} \rangle, \langle t_{2,1}, t_{2,m} \rangle \right) \forall n \neq 1 \\ d_{Fd}\left(\langle t_{1,1}, t_{1,n} \rangle, \langle t_{2,1}, t_{2,m-1} \rangle \right) \forall m \neq 1 \\ d_{Fd}\left(\langle t_{1,1}, t_{1,n-1} \rangle, \langle t_{2,1}, t_{2,m-1} \rangle \right) \forall n \neq 1, \forall m \neq 1 \end{array} \right) \end{array} \right)$$

This process continues recursively until both traces are reduced to both points $(\langle t_{1,1}, t_{2,2} \rangle)$ and $d_{Fd}\left(\langle t_{1,1}, t_{2,2} \rangle\right) = d_E(t_{1,1}, t_{2,2})$.

Devogele proposes matching the homologous points from one of the d_{Fd} ways. The d_{Fd} selected is the minimal path, defined as the case in which the average distances between its pair $(t_{1,i}, t_{2,j})$ is minimal. A graphical example coming from [18] is shown in Figure 3.

In the step 3, we combine all possible traces pairwise (t_i and t_k, $1 \leq i, k \leq$ n) and then we compute the mean Fréchet trace (t_{ik}). The t_{ik} vertices are the weighted mean points between the corresponding homologous points (in Figure 4, t_{ik} is the red line). The weight of each trace is equal to the number of traces that participated in its actual shape. Supposing a trace t_{ik} reaches its actual status after 11 matches, then its weight is 12. The value for the point between two matched points will be

$$p_{ik}^l = \frac{p_i^l w_i + p_k^l w_k}{w_i + w_k},$$

where l is the l-th matched point in traces i and k; w_i and w_k are the weights for traces i-th and k-th, respectively. In Figure 4, the weight for the green trace is 3 and for the blue trace is 1.

The resulting mean axis is shown on Figure 5 where the inferred axis has red color.

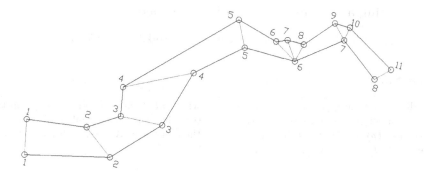

Fig. 3. Example of homologous points matched according to [18] (criteria based on Fréchet distance)

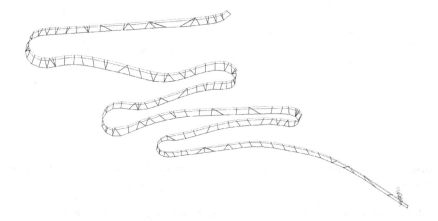

Fig. 4. Weighted mean axis from homologous points based on the Fréchet distance

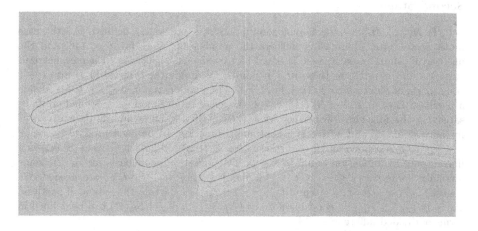

Fig. 5. Representative axis from the Fréchet distance approach

2.2   Fitting a B-spline Curve to Points Cloud Dataset

In this approach, after eliminating the outliers the workflow is as follows:

First Stage

We divide each original trace t_i ($1 \leq i \leq nt$, where nt is the number of traces) in the same number of segments (m), so that each trace has the same number ($n = m + 1$) of vertices (p), where p_{ij} is the j-th vertex of the i-th trace. We define the homologous points set $H_j = (p_{1,j}, p_{2,j}, \ldots, p_{nt,j})$.

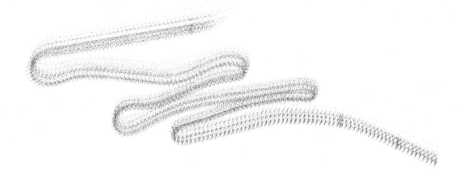

Fig. 6. Traces subdivided in equal number of segments in order to facilitate uniform partition.

The corresponding points cloud derived from this step is shown in Figure 6.

Second Stage

Let $H_1, H_2, \ldots, H_{nt}$ be the homologous points sets previously defined. For the case under study, after eliminating outliers the number of traces is $nt = 146$, and the number of points in each trace is equal to $n = m + 1 = 300$. To fit a parametrized curve $c : [0, 1] \longrightarrow \mathbb{R}^3$ to these $nt \times n$ points, we divide the interval $[0, 1]$ into $\nu + 1$ equal parts of length $h := 1/(\nu + 1)$, $\nu \geq 1$. The knots $t_i := ih$, $0 \leq i \leq \nu + 1$, define a partition Δ of $[0, 1]$. Let $S_k(\Delta)$, $k \geq 2$, the space of all functions s defined on $[0, 1]$, satisfying the following properties: a) $s \in C^{k-2}([0, 1])$, i.e. s admits derivative $s^{(r)}$ for all $r \leq k - 1$, and $s^{(k-2)}$ is a continuous function; b) on each interval (t_i, t_{i+1}), $0 \leq i \leq \nu$, s is a polynomial of degree $\leq k - 1$. It is well-known [7] that $S_k(\Delta)$ is a linear space of dimension equal to $k + \nu$. To compute a good basis for $S_k(\Delta)$ some auxiliary knots $t_{-k+1} \leq \cdots \leq t_{-1} \leq 0$ and $1 \leq t_{\nu+2} \leq \cdots \leq t_{\nu+k}$ are needed. We choose $t_{-k+1} = \cdots = t_{-1} = 0$ and $t_{\nu+2} = \cdots = t_{\nu+k} = 1$. From the extended partition $\Delta_* := (t_i)_{-k+1 \leq i \leq \nu+k}$, a good basis $\mathcal{B}_k := (N_{j,k})_{-k+1 \leq j \leq \nu}$ of $S_k(\Delta)$ can be defined in terms of divided differences:

$$N_{j,k}(t) := (t_{j+k} - t_j)[t_j, t_{j+1}, \ldots, t_{j+k}](\cdot - t)_+^{k-1},$$

where $(\cdot)_+^{k-1}$ stands for the truncated power of degree $k-1$. We recall that

$$(z)_+^{k-1} := \begin{cases} z^{k-1}, & \text{if } z \geq 0, \\ 0, & \text{otherwise.} \end{cases}$$

The B-splines $N_{j,k} \in S_k(\Delta)$ are nonnegative on $[0,1]$ and positive on (t_j, t_{j+k}), and form a partition of unity. It is well-known that the B-splines can be computed by the recurrence relation

$$N_{i,r}(t) = \frac{t - t_i}{t_{i+r} - t_i} N_{i,r-1}(t) + \frac{t_{i+r+1} - t}{t_{i+r+1} - t_{i+1}} N_{i+1,r-1}(t), \ 2 \leq r \leq k,$$

starting with

$$N_{i,1}(t) = \begin{cases} 1, t_i \leq t < t_{i+1}, \\ 0, \text{otherwise.} \end{cases}$$

However, Δ is a uniform partition and explicit expressions can be given for all B-splines with support inside the interval $[0,1]$, i.e. $N_{0,k}, \ldots, N_{\nu+1-k,k}$. More explicitly [20],

$$N_{i,k}(t) = B\left(\frac{t}{h} - i\right), \ 0 \leq i \leq \nu + 1 - k,$$

where

$$B(t) := \frac{1}{(k-1)!} \sum_{\ell=0}^{k} \binom{k}{\ell} (t - \ell)_+^{k-1}$$

is the B-spline of order k and supported on $[0,k]$ associated with the partition of the real line induced by the integer numbers. The remaining B-splines can be computed by the given recurrence relation or from the definition in terms of divided differences taking into account the following properties:

1. If $x_0 \neq x_k$, then

$$[x_0, x_1, \ldots, x_k] f = \frac{[x_1, \ldots, x_k] f - [x_0, \ldots, x_{k-1}] f}{x_k - x_0};$$

2. For a enough regular function f, it holds

$$[x_0, \overset{(k)}{\ldots}, x_0] f = \frac{1}{k!} f^{(k)}(x_0).$$

The parametrized curve c can be constructed in the space $S_k(\Delta)$ from the basis $\mathcal{B}_k$ and then

$$c(t) = \sum_{i=-k+1}^{\nu} Q_i N_{i,k}(t)$$

for some control points $Q_i := (q_{i,1}, q_{i,2}, q_{i,3}) \in \mathbb{R}^3$. It can be computed by least-squares fitting, i.e. by minimizing the objective function

$$F(Q_{-k+1}, \ldots, Q_n) := \sum_{j=0}^{m} \sum_{i=1}^{nt} \|c(\tau_j) - p_{i,j}\|^2,$$

where the evaluation points τ_j are obtained by dividing the interval $[0,1]$ into m equal parts of length $\lambda := \frac{1}{m}$, i.e. $\tau_j = j\lambda$, $0 \leq j \leq m$. This minimization problem results in the solution of the corresponding normal equations.

This method provides the solution shown in the Figure 7. The fitted cubic B-spline curve uses $\nu + 1 = 53$ knots.

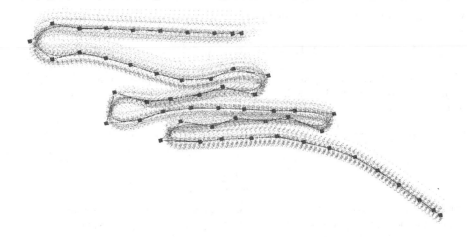

Fig. 7. B-spline (blue color) fitted on the points cloud uniformly distributed

2.3   Optimization of Nodes Distribution in B-spline Using a Multi-objective Genetic Algorithm

Previous section showed how to obtain the mean road axis by fitting a B-spline curve over a fixed set of uniformly distributed knots. In this section, we will optimize the selection of the knots used in the calculation of the B-spline curve (using the same configuration as in Section 2.2) applying a multi-objective genetic algorithm (MOGA).

As the reader may infer, increasing the number of knots usually give a more precise B-spline curve with the drawback of a slow computation, and vice versa. We argue that if we carefully select knots in the proper positions we may be able to reduce the total number of knots and provide an accurate solution simultaneously.

In this kind of problems where several objectives have to be simultaneously optimized, there is usually not a single best solution solving the problem (i.e., being better than the remainder with respect to every objective) as in single-objective optimization. Instead, there is a set of solutions that are superior to the remainder when all the objectives are considered, these solutions are known as *non-dominated solutions* [3,4]. A solution x is said to *dominate* another solution y using objective function vector $f = (f_1, \ldots, f_k))$ if and only if

$$\forall i \in \{1, \ldots, k\}, f_i(x) \leq f_i(y) \land \exists i \in \{1, \ldots, k\} : f_i(x) < f_i(y),$$

(supposing we are minimizing all objectives). Since none of the non-dominated solutions is absolutely better than the other non-dominated solutions, all of them are equally acceptable as regards to the satisfaction of all the objectives.

The multi-objective minimization process will be carried out by a MOGA as genetic algorithms are well suited for this kind of problems [5]. Our MOGA is based on the well known NSGA-II developed in [5]. In order to customize it for our particular problem we used a variable-length chromosome composed of a set of real values that represent the knots positions (e.g., chromosome [0, 0.5, 1] correspond to three knots, the first and last knots are located at the beginning and end position in 3D of the road, while the second knot is located in the middle of the road trajectory). We selected a simple one-point crossover and several mutation operators: add a knot (increases the number

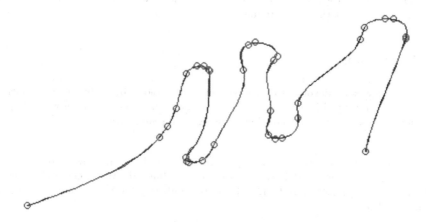

Fig. 8. A non-dominated solution with 32 knots (blue line). B-spline obtained with a fixed set of 53 uniform distributed knots is color-coded in red.

Fig. 9. A non-dominated solution with 35 knots (blue line). B-spline obtained with a fixed set of 53 uniform distributed knots is color-coded in red.

of knots by 1 placing a new knot in a random position), remove a knot (decreases the number of knots by 1 removing a knot randomly) and move a knot (changes the position of a knot to the left of right randomly from the original position).

In our road problem, we will minimize $k = 2$ objectives: number of knots (f_1) and precision (f_2). The latter objective is calculated as:

$$precision = \sqrt{\frac{\sum \left((\hat{x} - x)^2 + (\hat{y} - y)^2 + (\hat{z} - z)^2\right)}{n}}$$

where n is the number of vertices dividing the original traces, $\hat{x}$, $\hat{y}$ and $\hat{z}$ are the values of the B-spline evaluation (over the selected knots in the chromosome) on the $[0, 1]$ interval dividing in $n - 1$ segments, for each dimension; and x, y and z are the values in the database for each dimension.

Preliminary results using a population of 50 chromosomes and run for 500 generations show that solutions with an acceptable accuracy can be achieved with only 30 knots approximately (see Figure 8). In Figures 8 and 9 we can see a solution with $f_1 = 32$ and $f_2 = 0.057$ and with $f_1 = 35$ and $f_2 = 0.057$, respectively. Notice that both f_2 values are rounded to the millimeter, thus showing the same precision, but they are actually not the same original value.

3   Discussion

Future work will be devoted to improve our methodology by including information about the complexity of each area of the road. Also, we expect to improve the optimization process for the knot positioning using local search strategies.

Acknowledgements. Work partially carried out within the framework of Research Projects FQM 191 and FQM 1861 supported by the Junta de Andalucía, and Projects E3DLIN (BIA 2011-23271) and TIN2012-38805 funded by the Science and Innovation Ministry of Spain.

References

1. Agamennoni, G., Nieto, J., Nebot, E.: Robust Inference of Principal Road Paths for Intelligent Transportation Systems. IEEE Transactions on Intelligent Transportation Systems 12(1), 298–308 (2011)
2. Biagioni, J., Eriksson, J.: Inferring road maps from GPS traces: Survey and comparative evaluation. In: Transportation Research Board, 91st Annual 2291, pp. 61–71 (2012)
3. Coello-Coello, C., Lamont, G.B., van Veldhuizen, D.A.: Evolutionary Algorithms for Solving Multi-Objective Problems. Springer, New York (2002)
4. Deb, K.: Multi-Objective Optimization using Evolutionary Algorithms. Wiley-Interscience Series in Systems and Optimization. John Wiley & Sons, Chichester (2001)
5. Deb, K., Pratap, A., Agarwal, S., Meyarivan, T.: A fast and elitist multiobjective genetic algorithm: NSGA-II. IEEE Transactions on Evolutionary Computation 6, 182–197 (2002)
6. Devogele, T.: A new merging process for data integration based on the discrete Fréchet distance. In: Richardson, D., van Oosterom, P. (eds.) Advances in Spatial Data Handling, pp. 167–181. Springer, Berlin (2002)
7. DeVore, R.A., Lorentz, G.G.: Constructive approximation. Springer, Berlin (1993)
8. Edelkamp, S., Schrödl, S.: Route Planning and Map Inference with Global Positioning Traces. In: Klein, R., Six, H.-W., Wegner, L. (eds.) Computer Science in Perspective. LNCS, vol. 2598, pp. 128–151. Springer, Heidelberg (2003)
9. EDMap Team: IVI Light Vehicle Enabling Research Program."Enhanced digital mappingproject final report" (2004)
10. Haklay, M., Weber, P.: OpenStreetMap: User-Generated Street Maps. IEEE Pervasive Computing 7(4), 12–18 (2008)
11. Haskitt, P.: Map based safety applications: from research to reality (a review of the IVI-EDMap project). In: Proceedings of the 12th World Congress on Intelligent Transport Systems, San Francisco, USA, pp. 6–10 (2005)
12. IGN: Instituto Geográfico Nacional, Hoja 1011-I. Guadix. Mapa Topográfico Nacional de España 1:25.000, Ministerio de Fomento, Madrid (2000)

13. Karagiorgou, S., Pfoser, D.: On vehicle tracking data based road network generation. In: Proceedings of the 20th International Conference on Advances in Geographic Information Systems (SIGSPATIAL 2012), Redondo Beach, USA, pp. 89–98 (2012)
14. Lima, F., Ferreira, M.: Mining spatial data from GPS traces for automatic road network extraction. In: 6th International Symposium on Mobile Mapping Technology, Presidente Prudente, Sao Paulo, Brazil (2009)
15. Liu, X., Biagioni, J., Eriksson, J., Wang, Y., Forman, G., Zhu, Y.: Mining large scale, sparse GPS traces for map inference: comparison of approaches. In: Proceedings of the 18th ACM SIGKDD International Conference on Knowledge Discovery and Data Mining (KDD 2012), Beijing, China, pp. 669–677 (2012)
16. http://www.openstreetmap.org (last accessed October 16, 2014)
17. Reinoso, J.F., Moncayo, M., Pasadas, M., Ariza, F.J., García, J.L.: The Frenet frame beyond classical differential geometry: Application to cartographic generalization of roads. Mathematics and Computers in Simulation 79(12), 3556–3566 (2009)
18. Reinoso, F.J., Moncayo, M., Ariza, F.J.: A new iterative algorithm for creating a mean 3D axis of a road from a set of GNSS traces. Mathematics and Computers in Simulation (2015), doi:10.1016/j.matcom.2014.12.003
19. Schuessler, N., Axhausen, K.: Map-Matching of GPS traces on high resolution navigation networks using the multiple hypothesis technique (MHT). Working Paper 568, Institute for Transport Planning and System (IVT), ETH Zurich, Zurich (2009)
20. Schoenberg, I.J.: Cardinal spline interpolation (second printing). SIAM, Philadelphia (1993)
21. Thapa, T.: Optimization approach for generalization and data abstraction. International Journal of Geographical Information Science 19(8-9), 871–987 (1990)
22. Zhang, L., Thiemann, F., Sester, M.: Integration of GPS traces with road map. In: Proceedings of the Second International Workshop on Computational Transportation Science (IWCTS 2010), San José, USA, pp. 17–22 (2010)

Interval-Krawczyk Approach for Solving Nonlinear Equations Systems in B-spline Form

Dominique Michel and Ahmed Zidna

Laboratory of Theoretical and Applied Computer Science
UFR MIM, University of Lorraine, Ile du Saulcy, 57045 Metz, France
domic62@hotmail.com, ahmed.zidna@univ-lorraine.fr

Abstract. In this paper we propose a new approach based on Interval-Newton and Interval-Krawczyk operators for solving non linear systems of equations given in B-spline form. The proposed algorithm is making great benefit of geometric properties of B-spline functions to avoid unnecessary computations. Since B-spline functions can provide an accurate approximation for a wide range of functions, the algorithm can be made available for those functions by prior conversion/approximation to B-spline basis. It has successfully been used for solving various multivariate nonlinear equations systems.

Keywords: Nonlinear System equations, Newton-Interval, Krawczyk, Range computation, B-spline functions.

1 Introduction

Solving nonlinear systems of equations is critical in many research fields. Computing all solutions of a system of nonlinear polynomial equations within some finite domain is a fundamental problem in computer-aided design, manufacturing, engineering, and optimization. For example, in computer graphics, it is common to compute and display the intersection of two surfaces. Also the fundamental ray tracing algorithm needs to compute intersections between rays and primitive solids or surfaces [14]. This generally results in computing the roots of an equation or a system of equations, which in general case, is nonlinear. In global optimization, for equality constraints problem $minf(x)$ s.t. $g(x) = 0$, where g is nonlinear and x belongs to a finite domain, we may seek a feasible initial point to start a numerical algorithm. This requires to compute a solution of $\|g(x)\| = 0$. In this paper we focus on finding all zeroes of a function $f : x \to f(x)$, where x belongs to a n-dimensional box U, $f(x)$ belongs to $\mathbb{R}^n$ and f is C^1 over the inside of U. In [10,11],[19], the authors investigated the application of binary subdivision and the variation diminishing property of polynomials in the Bernstein basis to eliciting the real roots and extrema of a polynomial within an interval. In [3,4], the authors extended this idea to general non uniform subdivision of B-splines. In [7], the authors use homotopy perturbation method for solving systems of nonlinear algebraic equations. The approach of homotopy methods consists in gradually transforming a starting system of equations whose solution

H.A. Le Thi et al. (eds.), *Model. Comput. & Optim. in Inf. Syst. & Manage. Sci.*,
Advances in Intelligent Systems and Computing 359, DOI: 10.1007/978-3-319-18161-5_39

is known to the original system. The idea is that the solutions of the successive systems converge to a solution of the original system. At each step, the current system is solved by using a Newton-type method to find a starting solution for the next stage system. Unfortunately, this approach may not converge to a finite solution. In [12], a hybrid approach for solving system of nonlinear equations that combine a chaos optimization algorithm and quasi-Newton method has been proposed. The essence of the proposed method is to search an initial guess which should entry the convergent regions of quasi-Newton method. In the approach presented by the authors in [9], a system of nonlinear equations was converted to the minimization problem. Also they suggested to solve the minimization problem using a new particle swarm optimization algorithm. In [18], the authors use recursive subdivision to solve nonlinear systems of polynomial equations in Bernstein basis. They exploit the convex hull property of this basis to improve the bisection step. In the same context, in [2], the authors propose to use blending operators to rule out the no-root containing domains resulting from recursive subdivision. In [6], the authors use recursive subdivision to solve nonlinear systems of polynomial equations expressed in Bernstein basis. They use a heuristic sweep direction selection rule based on the greatest magnitude of partial derivatives in order to minimize the number of sub-boxes to be searched for solutions. Again, in [16], the authors combine interval Newton operator and subdivision for solving nonlinear system of polynomial equations in Bernstein basis.

B-splines functions are commonly used as a low-computation cost extension of polynomials [5]. For this reason, in [8], the author proposes a method to find all zeroes of a univariate spline function using the interval Newton's method. Unlike the method proposed in [11], interval division is used to speed up bisection. In previous work [20], we proposed a generalization of the method described in [8]. It results in a hybrid algorithm that combines recursive bisection and interval Newton method. In this paper, we improve the latter by replacing the Newton operator by the well known Krawczyk operator.

This paper is organized as follows : In section 2, we briefly recall the spline function model. In section 3, the main properties of arithmetic interval are presented. In section 4, the new hybrid algorithm that combines divide-and-conquer strategy and Krawzcyk operator is described. In section 5, some numerical comparisons are presented along with performance results. Finally, in the last section, we conclude.

2   B-spline Functions

Let m and k be two integer numbers such that $1 \leq k \leq m + 1$. Let $U = \{u_0, u_1, u_2, \ldots, u_{m+k}\}$ be a finite sequence of real numbers such that :

1. $u_i \leq u_{i+1}$ $\forall i = 0 \ldots m + k - 1$,
2. $u_i < u_{i+k}$ $\forall i = 0 \ldots m$.

Under these hypotheses, the u_i are called knots and U a knot vector. Then, for any integer numbers r and i such that $1 \leq r \leq k$ and $0 \leq i \leq m + k - r$, a

B-spline function $N_i^r : \mathbb{R} \to \mathbb{R}$ is defined as follows:

$$N_i^1(u) = \begin{cases} 0, \text{ for } \quad u_i \leq u < u_{i+1}, \\ 1, \text{ otherwise}, \end{cases} \tag{1}$$

and for $2 \leq r \leq k$, we have

$$N_i^r(u) = \frac{u - u_i}{u_{i+r-1} - u_i} N_i^{r-1}(u) + \frac{u_{i+r} - u}{u_{i+r} - u_{i+1}} N_{i+1}^{r-1}(u). \tag{2}$$

The functions $u \mapsto N_i^k(u)$ for $i = 0 \dots m$ satisfy the following properties :

1. Each B-spline N_i^k is finitely supported on $[u_i, u_{i+k}[$.
2. Each B-spline is positive in the interior of its support.
3. The $\{N_i^k\}_{i=0}^m$ form a partition of unity, i.e, $\sum_{i=0}^m N_i^k(u) = 1$ when u draws $[u_{k-1}, u_{m+1}[$.

Given a second knot vector $V = \{v_0, v_1, v_2, \dots, v_{n+l}\}$, a bivariate B-spline function $f : \mathbb{R}^2 \to \mathbb{R}^2$ can be defined by tensor product as follows :

$$f(u, v) = \sum_{i=0}^m \sum_{j=0}^n c_{ij} N_i^k(u) N_j^l(v), \tag{3}$$

where the coefficients c_{ij} are given in the two-dimensional space. $u \mapsto f(u, v)$, resp. $v \mapsto f(u, v)$, is a piecewise polynomial of degree $k - 1$ in u, resp $l - 1$ in v. This above definition can straightforwardly be generalized to higher dimensional cases. Let us recall that f can be evaluated or subdivided by using Cox-De Boor algorithm [3],[5] and that the partial derivatives of f are still expressed as B-spline functions. From properties 2 and 3, it is obvious that, for any (u, v) in $[u_{k-1}, u_{m+1}[\times[v_{l-1}, v_{n+1}[$, $f(u, v)$ is a convex linear combination of the coefficients c_{ij}, thus the graph of f entirely lies in the convex hull of its coefficients c_{ij}. Consequently, the ranges of f and of its derivatives can be calculated at low computational cost [13]. This last property makes the B-spline functions very attractive whenever interval arithmetics is involved.

3   Interval Arithmetic

The interval arithmetic was introduced by Moore[15]. Here we give a brief summary of the necessary definition of interval arithmetic. An interval number is represented by a lower and upper bound, $[a, b]$ and corresponds to a range of real values [1][15]. Let $\mathbb{I}$ be a set of real intervals given by $\mathbb{I} = \{[a, b]/a \leq b, a, b \in \mathbb{R}\}$. The basic operations can be defined on $\mathbb{I}$ instead of floating point numbers by :

$$K = I \vartriangle J = \{a \vartriangle b/a \in I \text{ and } b \in J\}$$

where $\vartriangle$ represents any operations $+, -, *, /$. The width of the interval $I = [a, b]$ is defined by $w(I) = b - a$. The midpoint of the interval I is the real number $mid(A) = (a + b)/2$.

An ordinary real number can be represented by a degenerate interval $[a, a]$. The basic operations $+, -, *, /$ are used on intervals instead of floating point numbers. Let us remind these 4 operations : let a, b, c, d be 4 real numbers (finite or infinite) such as $a \leq b$ and $c \leq d$.

1. $[a, b] + [c, d] = [a + b, c + d]$
2. $[a, b] - [c, d] = [a - d, b - c]$
3. $[a, b] * [c, d] = [min(a * c, a * d, b * c, b * d), max(a * c, a * d, b * c, b * d)]$
4. $\dfrac{[a, b]}{[c, d]} = \begin{cases}] - \infty, min(\dfrac{a}{c}, \dfrac{b}{d})] \cup [max(\dfrac{a}{d}, \dfrac{b}{c}), +\infty[, & \text{if } 0 \in \,]c, d[\\ [a, b] * [\dfrac{1}{d}, \dfrac{1}{c}], & \text{otherwise.} \end{cases}$

We define the absolute value of a real interval $[a, b]$ as the mignitude function, that is : $|[a, b]| = \begin{cases} a & \text{if} & 0 \leq a, \\ -b & \text{if} & b \leq 0, \\ 0 & \text{otherwise.} \end{cases}$

Let us point out that $*$ and $/$ dramatically increase the interval span. When used repeatedly, they yield useless results. Associativity is not true with interval operators $+$ and $*$. Also the distributive multiplication law of real arithmetic is not valid in general. Only the following so-called sub-distributive law holds in $\mathbb{I}$,

$$\forall I, J, K \in \mathbb{I}, \quad I * (J + K) \subseteq I * J + I * K.$$

An important and useful property for constructing the inclusion functions in global optimization method is the so-called inclusion isotonicity property

$$\forall I, J, K, L \in \mathbb{I}, \text{ if } I \subseteq K \text{ and } J \subseteq L, \text{ then } I \vartriangle J \subseteq K \vartriangle L.$$

Let $I = (I_1, I_2, \ldots, I_n)^T$ an n-dimensional interval I in $\mathbb{I}^n$, the set of n-dimensional interval column vectors. The midpoint of I is given by $(mid(I_1), \ldots, mid(I_n))$ and the width of I is given by $\max_{i=1}^{n} w(I_i)$.

Definition 1. *A function $F : \mathbb{I}^n \to \mathbb{I}$ is called an inclusion function of $f : \mathbb{R}^n \to \mathbb{R}$, provided $\{f(x)/x \in X\} \subseteq F(X)$ for all box $X \in \mathbb{I}^n$ within the domain of f.*

Definition 2. *Let $f : X \to \mathbb{R}^n$ where $X \in \mathbb{I}^n$, and let $x \in X$, $J_f(x)$ denote the Jacobian matrix of f at x, then $J(X)$ is a natural interval extension of the Jacobian matrix $J_f(x)$ to X, that is, each x is replaced by X in $J_f(x)$.*

Different algorithms for inverting an $n \times n$ interval matrix M might not find the same matrix M^{-1}.

4   Root-Finding Algorithm in B-spline Form

In [8], the author proposed an algorithm to find all roots of a spline function f over an interval I in the one dimensional case. Starting from the interval I, the algorithm builds a binary tree of searching intervals.

4.1   Multivariate Interval Newton Method

In the n-dimensional case, the general form of interval Newton method is given by :

$$I_1 = I_0 \cap N(I_0); \quad \text{with} \quad N(I_0) = m_{I_0} - [f'(I_0)]^{-1} f(m_{I_0}), \tag{4}$$

where I_0 denotes the current searching hyper-interval, I_1 denotes the next searching hyper-interval, and $[f'(I_0)]^{-1}$ is $n \times n$ interval matrix resulting from calculating the inverse of the interval jacobian matrix of f over I_0. Meaning $[f'(I_0)]$ is such as if x lies in I_0, all the elements of the jacobian matrix $[f'(x)]$ belong to the matching interval-elements of the matrix $[f'(I_0)]$. Accordingly, for all x in the interval I_0, all the elements of the inverse of the jacobian matrix belong to the matching elements of the $n \times n$ interval matrix $[f'(I_0)]^{-1}$. The critical step of the equation (4) is the computation of the inverse of the jacobian matrix. Obviously, the algorithm to be used to compute the inverse of $[f'(I_0)]$ is the Gauss-Jordan algorithm, keeping in mind that every operation is done in interval arithmetic. The jacobian matrix might not be invertible. In the one dimensional version, this situation was used to split the searching interval. However for $n \geq 2$, if the Jacobian is not invertible, the method fails.

4.2   Multivariate Interval Krawczyk Method

Let assume that, during the elimination process of the Gaussian algorithm applied to a nonsingular $n \times n$ interval matrix $f'(I_0)$ and an n interval vector $f(m_{I_0})$, no division by an interval containing zero occurs. The interval Newton method may not converge and its feasibility is not guaranteed. Indeed, even if the interval arithmetic evaluation $f'(I_0)$ of the Jacobian contains no singular matrix, $(f'(I_0)^{-1} * f(m_{I_0}))$ can in general only be computed if the width of the interval I_0 is sufficiently small. It is why we introduce the Krawczyk operator instead of the Newton operator:

$$K(I_0) = m_{I_0} - Cf(m_{I_0}) + (Id - C * [f'(I_0)])(I_0 - m_{I_0}), \tag{5}$$

where C a nonsingular real matrix close to the inverse of the real Jacobian matrix computed at some point in I_0 and where m_{I_0} is the center of I_0. Generally C is chosen as the inverse of the center of the Jacobian matrix over I_0. Therefore equation (4) can be replaced by the following :

$$I_1 = I_0 \cap K(I_0). \tag{6}$$

4.3   Range Function and Jacobian Matrix Computing in B-spline Form

A simple way to verify that an interval can be eliminated as not containing any root is to use interval extensions for function range testing. Consider a search for solutions of $f(x) = 0$ in I_0. If an interval extension $F(x)$ of $f(x)$ over I_0 does not contain zero, that is, $0 \notin F(X)$, then the range of $f(x)$ over I_0 does not contain zero, and it is not possible for I_0 to contain a solution of $f(x) = 0$. Thus, I_0 can be eliminated from the search space.

Let us point out that the convex hull property of B-spline function makes the filtering test straightforward. Moreover, In B-spline form, the computation of the interval extension of $f(x)$ (i.e $F(I_0)$) is straightforward. Likewise, the computation of the Jacobian matrix $F'(I_0)$ can easily be performed since the derivatives are still expressed in B-spline form.

4.4   Proposed Hybrid Algorithm

In multidimensional case, when the jacobian matrix is not invertible, the Gauss-Jordan algorithm produces a matrix of which the interval elements have infinite bounds which makes the results useless. Therefore in multidimensional case, we use recursive bisection to separate the roots. The Gauss-Jordan method is only used when the jacobian matrix is invertible.

Let $\{I_0^1, I_0^2, \ldots, I_0^\alpha\}$, be a partition of I_0, we propose the following hybrid iterations instead of the previous equations (4) and (6) :

$$\bigsqcup_{j=1}^{\alpha} I_1^j = \bigsqcup_{j=1}^{\alpha} \left(I_0^j \cap N(I_0) \right), \quad \text{or} \quad \bigsqcup_{j=1}^{\alpha} I_1^j = \bigsqcup_{j=1}^{\alpha} \left(I_0^j \cap K(I_0) \right). \tag{7}$$

The above formula is only used when the jacobian matrix is invertible, otherwise it is replaced by a simple bisection :

$$\bigsqcup_{j=1}^{\alpha} I_1^j = \bigsqcup_{j=1}^{\alpha} I_0^j. \tag{8}$$

In the following, either $\alpha = 2$ or $\alpha = 2^n$. In the former option, I_0 is split along the widest component. In the latter, I_0 is equally split along every direction. The general algorithm for solving nonlinear systems in B-spline form with the interval Krawczyk method is described by the method **Solve** in Algorithm 1 while the specific hybrid iteration task is performed by the method **Succ** in Algorithm 2. For inverting the $n \times n$ interval jacobian matrix, we propose to use the Gauss-Jordan which has arithmetic complexity of $\mathcal{O}(n^3)$. The numerical precision of the elements of the resulting matrix is improved by using total pivot. This strategy implies, at each step, to seek the input matrix element of greatest mignitude [15].

5   Numerical Results

The proposed algorithm has been implemented in $C++$, using finite precision floating point numbers. On the next few examples, we compare the Newton form and the Krawczyk form of the proposed algorithm, for $\alpha = 2$ and for $\alpha = 2^n$. In this purpose, several various functions have been selected in the cases

---

Algorithm 1. Recursive root finding algorithm

Function **Solve**(I_0, f)
Input: I_0 : an n-hyperinterval, $f : I_0 \to \mathbb{R}^n$
Output: A list of hyper-intervals, each of them contains one solution
begin
   | **If** $I_0 = \emptyset$ or $(0, ..., 0) \notin f(I_0)$ **Return** $\emptyset$
   | **ElseIf** $| I_0 | \leq \varepsilon$ **Return** I_0
   | **Else begin**
   | | (* Let L be a list of hyper-intervals of dimension n *)
   | | $L \leftarrow Succ(I_0, f)$
   | | **Return** $\cup_{l \in L} solve(l, f)$
   | **end**
end

Algorithm 2. Krawczyk iteration algorithm

Function **Succ**(**Input** I_0 : an n-hyperinterval,$f : I_0 \to \mathbb{R}^n$): **Output**   A list of hyper-intervals
begin
   | **try**
   | **begin**
   | | (* Let J be the jacobian matrix *)
   | | $J \leftarrow f'(I_0)$
   | | (* Let $\mathcal{L}$ be the list of α hyper-intervals resulting from bisection of I_0 *)
   | | $\mathcal{L} \leftarrow dichotomie(I_0)$
   | | $R \leftarrow K(I_0)$
   | | **return** $\sqcup_{L \in \mathcal{L}}(L \cap R)$
   | **end**
   | **catch**(Exception)
   | **begin**
   | | (* Capture exception produced when computing the C matrix fails *)
   | | **return** $\mathcal{L}$
   | **end**
end

$n = 2, 4$ and 5. For performance comparisons, the following criteria are taken in consideration : the number of iterations, the CPU-time and the number of hyper-intervals explored.

Example 1. In this example we calculate the intersection points between a hyperbola and an ellipse given by the following equations :
$$\begin{cases} -2.077u^2 - 2.077v^2 + 5.692uv - 5.384u - 5.384v + 17.846 = 0 & \text{(Hyperbola)} \\ 23.692u^2 + 23.692v^2 - 41.23uv - 21.538u - 21.538v + 39.385 = 0 & \text{(Ellipse)} \end{cases}$$
The two curves intersect at four points : $(1.5, 2.5)$, $(2.5, 1.5)$, $(4.5, 5.5)$, $(5.5, 4.5)$. The equations have been translated in B-spline form through quasi-interpolation techniques [13],[17]. The resulting function is a bi-cubic B-spline with 10×10

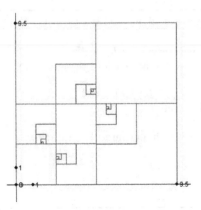

Fig. 1. Sequences of sub-domains explored

control points. Fig. 1 shows the four sequences of sub-domains converging to-wards the found roots.

Example 2. In this example, the algorithm has been used to find the extrema of a functional $f(u, v)$. Here f is bivariate B-spline function of degree 2×2 defined by a grid of 5×5 control points (real numbers) over uniform and clamped knot vectors. The grid is $[(0, -1, 0, 1, 0)(-1, -2, 0, 2, 1)(0, 0, 0, 0, 0)(1, 2, 0, -2, -1)$ $(0, 1, 0, -1, 0)]$. The system of equations to be solved is $\frac{\partial_f}{\partial_u} = 0$ and $\frac{\partial_f}{\partial_v} = 0$. The functional has 4 extrema located at $(2.466, 2.466)$, $(0.534, 0.534)$, $(2.466, 0.534)$, $(0.534, 2.466)$, and a saddle point at $(1.5, 1.5)$.

Example 3. In this example, we compute the intersection of a pyramidal surface and a line (see Fig. 2). The surface is a biquadratic patch defined by 25×25 control points.

Example 4. In this example, eigenvalues and eigenvectors of a 3×3 matrix are computed. Given a real $d \times d$ square matrix A, let λ be an eigenvalue of it and $\mathbf{v}$ a unit eigenvector associated to λ.

$(\lambda, \mathbf{v})$ is then a solution of the $(d + 1) \times (d + 1)$ non linear system :

$$\begin{cases} A\mathbf{v} - \lambda\mathbf{v} = \mathbf{0} \\ \|\mathbf{v}\|^2 - 1 = 0 \end{cases} \tag{9}$$

which yields a B-Spline function of degree $1 \times 2 \times 2 \times 2$ over a grid de-fined by $2 \times 3 \times 3 \times 3$ control points. The starting searching hyper-interval is $[-\|A\|_\infty, \|A\|_\infty] \times [-1, 1] \times \cdots \times [-1, 1]$. A is chosen as $A = P \times D \times P^{-1}$ with

$$D = \begin{pmatrix} 4 & 0 & 0 \\ 0 & -2 & 0 \\ 0 & 0 & 7 \end{pmatrix} \text{ and } P = \begin{pmatrix} -1 & 11 & 2 \\ 4 & 0 & -5 \\ -13 & 9 & -3 \end{pmatrix}$$

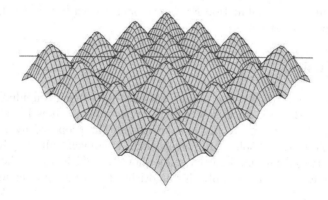

Fig. 2. Computation of the intersection of a pyramidal B-Spline function and a line. The solution are located at : $(21.5, 1.5)$, $(19.5, 3.5)$, $(15.5, 7.5)$, $(13.5, 9.5)$, $(9.5, 13.5)$, $(7.5, 15.5)$, $(3.5, 19.5)$, $(1.5, 21.5)$

Table 1. Performances comparison betwen the proposed Krawczyk algorithm and the Newton algorithm

Problem and dimension n	N_{sub}	Interval Newton method			Interval Krawczyk method		
		N_{iter}	time	$N_{explored}$	N_{iter}	time	$N_{explored}$
Hyperbola/Ellipse Intersection	2	16	1	235	16	0	203
$n = 2$	2^n	11	0	110	11	0	110
Functional extrema	2	8.8	0	63	9.6	0	79
$n = 2$	2^n	6.6	0	44	7.4	0	54
Line/surface Intersection	2	16	1	311	16	1	315
$n = 2$	2^n	11	1	147	11	0	135
Eigenvalues computation	2	19,33	30	3223	15,33	8	937
$n = 4$	2^n	9,5	11	415	9	10	392
Eigenvalues computation	2	41	46	601619	20,5	1	7349
$n = 5$	2^n	13,75	7	40297	10,44	2	17601

Example 5. Likewise, in this example, the eigen values and the eigen vectors of a 4×4 matrix are computed. Though, this time, the resulting function is directly expressed as a quadratic polynomial.

According to the numerical results summarized in table1, we can make the following observations : For $n \leq 2$, the performances of Newton operator and Krawczyk are similar. Beyond $n \geq 4$, the performances of Newton operator deteriorate much faster than Krawczyk operator performances. In the same way, we observe that it is more efficient to associate $\alpha = 2$ with Krawczyk whereas it is better to associate $\alpha = 2^n$ with Newton method. This is explained by the fact that the jacobian interval matrix inverse is generally only available for tiny hyper-intervals. The combination Krawczyk operator with $\alpha = 2$ is the overall best solution amongst all.

6   Conclusion

In this paper we have proposed a new approach for solving a nonlinear system of equations in B-spline form over a finite hyper-interval. It is based on Interval Krawczyk Method and recursive subdivision. The proposed hybrid method strongly exploits the B-spline properties such as convex hull and diminishing variation features. The numerical results show the efficiency of the proposed method. Further work is to apply this algorithm to surface-surface intersection problems.

References

1. Alefeld, G., Mayer, G.: Interval Analysis: Theory and Applications. Journal of Computational and Applied Mathematics 121, 421–464 (2000)
2. Bartoň, M.: Solving Polynomial Systems Using No-root Eliminations Blending Schemes. Computer Aided Design 43, 1870–1878 (2001)
3. Boehm, W.: Inserting new Knots into B-spline Curves. Computer Aided Design 12, 199–201 (1980)
4. Cohen, E., Lyche, T., Riesenfeld, R.F.: Discrete B-splines and Subdivision Techniques in Computer-Aided Geometric Design and Computer Graphics. Comput. Graphic Image Processing 14, 87–111 (1980)
5. De Boor, C.: A Practical Guide to Splines. Springer, New York (1978)
6. Garloff, J., Smith, A.P.: Solution of Systems of Polynomial Equations by Using Bernstein Expansion. In: Alefeld, G., Rump, S., Rohn, J., Yamamoto, T. (eds.) Symbolic Algebraic Methods and Verification Methods, pp. 87–97. Springer (2001)
7. Golbabai, A., Javidi, M.: A New Family of Iterative Methods for Solving System of Nonlinear Algebric Equations. Applied Mathematics and Computation 190, 1717–1722 (2007)
8. Grandine, T.A.: Computing Zeroes of Spline Functions. Computer Aided Geometric 6, 129–136 (1989)
9. Jaberipour, M., Khorram, E., Karimi, B.: Particle Swarm Algorithm for Solving Systems of Nonlinear Equations. Computers and Mathematics with Applications 62, 566–576 (2011)
10. Kiciak, P., Zidna, A.: Recursive de Casteljau Bisection and Rounding Errors. Computer Aided Geometric Design 21, 683–695 (2004)
11. Lane, J.M., Riesenfeld, R.F.: Bounds on a Polynomial. BIT: Nordisk Tidskrift for Haformations-Behandling 21, 112–117 (1981)
12. Luo, Y.Z., Tang, G.Z., Zhou, L.N.: Hybrid Approach for Solving Systems of Nonlinear Equations Using Chaos Optimization and Quasi-Newton Method. Applied Soft Computing 8, 1068–1073 (2008)
13. Michel, D., Mraoui, H., Sbibih, D., Zidna, A.: Computing the Range of Values of Real Functions Using B-spline Form. Applied Mathematics and Computation 233, 85–102 (2014)
14. Mitchell, D.P.: Robust Ray Intersection with Interval Arithmetic. Proceedings on Graphics Interface 1990, 68–74 (1990)

15. Moore, R.E., Kearfott, R.B., Cloud, M.J.: Introduction to Interval Analysis. SIAM, Philadelphia (2009)
16. Nataraj, P.S.V., Arounassalame, M.: An Interval Newton Method Based on the Bernstein Form for Bounding the Zeros of Polynomial Systems. Reliable Computing 15, 109–119 (2011)
17. Sablonnière, P.: Univariate Spline Quasi-Interpolants and Applications to Numerical Analysis. Rend. Sem. Mat. Univ. Pol. Torino 63(3) (2010)
18. Sherbrooke, E.C., Patrikalakis, N.M.: Computation of the Solutions of Nonlinear Polynomial Systems. Computer Aided Geometric Design 10, 379–405 (1993)
19. Zidna, A., Michel, D.: A Two-Steps Algorithm for Approximating Real Roots of a Polynomial in Bernstein Basis. Mathematics and Computers in Simulation 77, 313–323 (2008)
20. Zidna, A., Michel, D.: A New Approach Based on Interval Analysis and B-splines Properties for Solving Bivariate Nonlinear Equations Systems. In: van Do, T., Le Thi, H.A., Nguyen, N.T. (eds.) Advanced Computational Methods for Knowledge Engineering. AISC, vol. 282, pp. 341–351. Springer, Heidelberg (2014)

Solving Nonconvex Optimization Problems in Systems and Control: A Polynomial B-spline Approach

Deepak Gawali[1,3,*], Ahmed Zidna[2], and Paluri S.V. Nataraj[3]

[1] Vidyavardhini's College of Engineering and Technology, Maharashtra, India
ddgawali2002@iitb.ac.in
[2] Theoretical and Applied Computer Science Laboratory,
University of Lorraine, France
ahmed.zidna@univ-lorraine.fr
[3] Systems and control Engineering, Indian Institute of Technology Bombay,
Mumbai, India
nataraj@sc.iitb.ac.in

Abstract. Many problems in systems and control engineering can be formulated as constrained optimization problems with multivariate polynomial objective functions. We propose algorithms based on polynomial B-spline form for constrained global optimization of multivariate polynomial functions. The proposed algorithms are based on a branch-and-bound framework. We tested the proposed basic constrained global optimization algorithms by considering three test problems from systems and control. The obtained results agree with those reported in literature.

Keywords: Polynomial B-spline, Global optimization, Polynomial optimization, Constrained optimization.

1 Introduction

Global optimization of nonlinear programming problems (NLP) is the study of how to find the best (optimum) solution to a problem. The constrained global optimization of NLPs is stated as follows:

$$\min_{x \in \mathbf{x}} f(x) \tag{1}$$

$$\text{s.t.} \quad g_i(x) \leq 0, i = 1, 2, ..., p \tag{2}$$

$$h_j(x) = 0, j = 1, 2, ..., q \tag{3}$$

Branch-and-bound framework is commonly used for solving constrained global optimization problems [13]. For instance, several interval methods [14–17] use this framework to find the global minimum of a given NLP. In this work, we propose B-spline based algorithms for solving nonconvex nonlinear multivariate polynomial programming problems in systems and control, where the objective function f and constraints (g_i & h_j) are limited to being *polynomial* functions.

---

* Corresponding author.

© Springer International Publishing Switzerland 2015   467
H.A. Le Thi et al. (eds.), *Model. Comput. & Optim. in Inf. Syst. & Manage. Sci.*,
Advances in Intelligent Systems and Computing 359, DOI: 10.1007/978-3-319-18161-5_40

The polynomial objective function and constraints in power form are transformed into the polynomial B-spline form [2–4]. Then, the B-spline coefficients provide a bound on the range of the objective function and constraints.

In this paper, we investigate three applications of the basic constrained global optimization algorithm: the robust stability analysis problem, the minimum distance problem, and the domain of attraction problem. These problems are reduced to strict inequalities (or equations) involving multivariate polynomials and solved using the proposed algorithm for constrained global optimization.

The merits of the proposed approach are: (*i*) it avoids evaluation of the objective function and constraints; (*ii*) an initial guess to start optimization is not required; only an initial search box bounding the region of interest; (*iii*) it guarantees that the global minimum is found to a user-specified accuracy, and (*iv*) prior knowledge of stationary points is not required.

2 Multivariate B-spline Form

Let $s \in \mathbb{N}$ be the number of variables and $x = (x_1, x_2, ..., x_s) \in \mathbb{R}^s$. A multi-index I is defined as $I = (i_1, i_2, ..., i_s) \in (\mathbb{N} \cup \{0\})^s$ and multi-power x^I is defined as $x^I = (x_1^{i_1}, x_2^{i_2}, ..., x_s^{i_s})$. Given a multi-index $N = (n_1, n_2, ..., n_s)$ and an index r, we define $N_{r,-l} = (n_1, ..., n_{r-1}, n_r - l, n_{r+1}, ..., n_s)$, where $0 \leq n_r - l \leq n_r$. Inequalities $I \leq N$ for multi-indices are meant componentwise, i.e. $i_l \leq n_l$, $l = 1, 2, ..., s$. With $I = (i_1, ..., i_{r-1}, i_r, i_{r+1}, ..., i_s)$ we associate the index $I_{r,l}$ given by $I_{r,l} = (i_1, ..., i_{r-1}, i_r + l, i_{r+1}, ..., i_s)$, where $0 \leq i_r + l \leq n_r$. A real bounded and closed interval $\mathbf{x}_r$ is defined as $\mathbf{x}_r \equiv [\underline{\mathbf{x}_r}, \overline{\mathbf{x}_r}] := [\inf \mathbf{x}_r = \min \mathbf{x}_r, \sup \mathbf{x}_r = \max \mathbf{x}_r] \in \mathbb{IR}$, where $\mathbb{IR}$ denotes the set of *compact intervals*. Let $\operatorname{wid} \mathbf{x}_r$ denotes the *width* of $\mathbf{x}_r$, that is $\operatorname{wid} \mathbf{x}_r := \overline{\mathbf{x}_r} - \underline{\mathbf{x}_r}$.

We will follow the procedure by Lin and Rokne [2, 3] to obtain the B-spline representation of a multivariate polynomial

$$p(x) = \sum_{I \leq N} a_I x^I, x \in \mathbb{R}^s, \tag{4}$$

of degree N, in order to derive bounds for its range over an s-dimensional box $x = (x_1, x_2, ..., x_s)$. Firstly, we consider a univariate polynomial

$$p(x) := \sum_{t=0}^{n} a_t x^t, \ x \in [a, b], \tag{5}$$

to be expressed in terms of the B-spline basis of the space of polynomial splines of degree $m \geq n$ (i.e. order $m + 1$). In the following, we give some preliminary results about the construction of B-spline bases. First of all, we consider the following uniform grid partition

$$\mathbf{u} = \{x_0 < x_1 < ... < x_{k-1} < x_k\}$$

of the interval $I = [a, b]$, where $x_i = a + ih, 0 \leq i \leq k$, and $h = (b - a)/k$. Let $\mathbb{P}_m$ be the space of polynomials of degree at most m. Then the space of splines of degree m and class C^{m-1} on $[a, b]$ associated with $\mathbf{u}$ is defined by

$$S_m(I, \mathbf{u}) = \{S \in C^{m-1}(I) : S|_{[x_i, x_{i+1}]} \in \mathbb{P}_m, i = 0, ..., k - 1\}.$$

It is well known that $S_m(I, \mathbf{u})$ is a linear space of dimension equal to $k + m$ [25]. In order to construct a basis of locally supported splines for $S_m(I, \mathbf{u})$, some auxiliary knots $x_{-m} \leq \cdots \leq x_{-1} \leq a$ and $b \leq x_{k+1} \leq \cdots \leq x_{k+m}$ are needed. Taking into account that $\mathbf{u}$ is a uniform partition, we choose $x_i := a + ih$ for $i \in \{-m, \ldots, -1\} \cup \{k + 1, \ldots, m + k\}$.

$$x_{-m} \leq \cdots \leq x_{-1} \leq a = x_0 < x_1 < \cdots < x_{k-1} < x_k = b \leq x_{k+1} \leq \cdots \leq x_{k+m}.$$

From the extended partition, a basis $\left(N_j^m\right)_{-m \leq j \leq k-1}$ of $S_m(I, \mathbf{u})$ can be defined in terms of divided differences:

$$N_j^m(x) := (x_{j+m} - x_j)[x_j, x_{j+1}, \ldots, x_{j+m+1}](\cdot - x)_+^m,$$

where $(\cdot)_+^m$ stands for the truncated power of degree m. It is easy to prove that

$$N_j^m(x) = \Omega_m\left(\frac{x - a}{h} - j\right), \quad -m \leq j \leq k - 1,$$

where

$$\Omega_m(x) := \frac{1}{m!}\sum_{\ell=0}^{m+1}(-1)^\ell\binom{m+1}{\ell}(x - \ell)_+^m$$

is the B-spline of degree m associated with the partition of the real line induced by the integer numbers and supported on the interval $[0, m + 1]$. The B-splines can be computed by the recurrence formula

$$N_i^m(x) = \gamma_{i,m}(x)N_i^{m-1}(x) + (1 - \gamma_{i+1,m}(x))N_{i+1}^{m-1}(x), m \geq 1,$$

where

$$\gamma_{i,m}(x) = \begin{cases} \dfrac{x - x_i}{x_{i+m} - x_i}, & \text{if } x_i \leq x_{i+m}, \\ 0, & \text{otherwise}, \end{cases}$$

and

$$N_i^0(x) := \begin{cases} 1, & \text{if } x \in [x_i, x_{i+1}), \\ 0, & \text{otherwise}. \end{cases}$$

It is well known that the set $\{N_i^m\}_{i=-m}^{k-1}$ is a basis for $S_m(I, \mathbf{u})$ that satisfies interesting properties; for example, each N_i^m is positive on its support and $\{N_i^m\}_{i=-m}^{k-1}$ form a partition of unity.

On the other hand, as $\mathbb{P}_m \subset S_m(I, \mathbf{u})$, the power basis functions $\{x^r\}_{r=0}^m$ can be expressed in terms of B-splines through the relations

$$x^t = \sum_{j=-m}^{k-1}\pi_j^t N_j^m(x), t = 0, \ldots, m, \tag{6}$$

where π_j^t are the symmetric polynomials given by

$$\pi_j^t = \left(\sum_{j+1 \leq j_1 < \cdots < j_t \leq j+m} x_{j_1} x_{j_2} \cdots x_{j_t} \right) / \binom{m}{t}, \quad \text{for } t = 0, 1, \ldots, m.$$

By substituting (6) into (5) we get

$$p(x) = \sum_{t=0}^{n} a_t \sum_{j=-m}^{k-1} \pi_j^{(t)} N_j^m(x) = \sum_{j=-m}^{k-1} \left[\sum_{t=0}^{n} a_t \pi_j^{(t)} \right] N_j^m(x) = \sum_{j=-m}^{k-1} d_j N_j^m(x),$$

$$(7)$$

where

$$d_j \triangleq \sum_{t=0}^{n} a_t \pi_j^{(t)}. \tag{8}$$

Now, we derive the B-spline representation of a given multivariate polynomial

$$p(x_1, x_2, \ldots, x_s) = \sum_{i_1=0}^{n_1} \cdots \sum_{i_s=0}^{n_s} a_{i_1 \ldots i_s} x_1^{i_1} \ldots x_s^{i_s} = \sum_{I \leq N} a_I x^I, \tag{9}$$

where $I := (i_1, i_2, \ldots, i_s)$, and $N := (n_1, n_2, \ldots, n_s)$. By substituting (6) for each x^t, (9) can be written as

$$p(x_1, x_2, \ldots, x_s) = \sum_{i_1=0}^{n_1} \cdots \sum_{i_s=0}^{n_s} a_{i_1 \ldots i_s} \sum_{j_1=-m_1}^{k_1-1} \pi_{j_1}^{(i_1)} N_{j_1}^{m_1}(x_1) \ldots \sum_{j_s=-m_s}^{k_s-1} \pi_{j_s}^{(i_s)} N_{j_s}^{m_s}(x_s)$$

$$= \sum_{j_1=-m_1}^{k_1-1} \cdots \sum_{j_s=-m_s}^{k_s-1} \left(\sum_{i_1=0}^{n_1} \cdots \sum_{i_s=0}^{n_s} a_{i_1 \ldots i_s} \pi_{j_1}^{(i_1)} \ldots \pi_{j_s}^{(i_s)} \right) N_{j_1}^{m_1}(x_1) \ldots N_{j_s}^{m_s}(x_s)$$

$$= \sum_{j_1=-m_1}^{k_1-1} \cdots \sum_{j_s=-m_s}^{k_s-1} d_{j_1 \ldots j_s} N_{j_1}^{m_1}(x_1) \ldots N_{j_s}^{m_s}(x_s),$$

$$(10)$$

we have expressed p as

$$p(x) = \sum_{I \leq N} d_I(x) N_I^N(x), \tag{11}$$

with the coefficients $d_I(x)$ given by

$$d_{j_1, \ldots, j_s} = \sum_{i_1=0}^{n_1} \cdots \sum_{i_s=0}^{n_s} a_{i_1 \ldots i_s} \pi_{j_1}^{(i_1)} \ldots \pi_{j_s}^{(i_s)}. \tag{12}$$

The B-spline coefficients are collected in an array $D(\mathbf{x}) = (d_I(\mathbf{x}))_{I \in S}$, where $S = \{I : I \leq N\}$. This array is called a *patch*. We denote S_0 as a special subset of the index set S comprising indices of the vertices of this array, that is

$$S_0 := \{0, n_1 + k_1 - 1\} \times \{0, n_2 + k_2 - 1\} \times \ldots \times \{0, n_s + k_s - 1\}.$$

The following lemma describes the range enclosure property of the B-spline coefficients.

Lemma 1. *[2, 3],[18, 19] Let p be a polynomial of degree N and let $\bar{p}(\mathbf{x})$ denote the range of p on the given domain $\mathbf{x}$. Then, for a patch $D(\mathbf{x})$ of B-spline coefficients it holds*

$$\bar{p}(\mathbf{x}) \subseteq [\min D(\mathbf{x}), \max D(\mathbf{x})]$$

Obtaining the B-spline coefficients of multivariate polynomials by transforming the polynomial from power form to B-spline form, provides an enclosure of the range of the multivariate polynomial p on $\mathbf{x}$. Then by Lemma 1, the minimum and the maximum values of B-spline coefficient provide lower and upper bounds for the range of polynomial p. This range enclosure will be sharp if and only if $\min(d_I(\mathbf{x}))_{I \in S}$ (respectively $\max(d_I(\mathbf{x}))_{I \in S}$) is attained at the indices of the vertices of the array $D(\mathbf{x})$, as described in following lemma,

Lemma 2. *[20] Let p be a polynomial of degree N and let $\bar{p}(\mathbf{x}) = [a, b]$. Then*

$$a = \min_{0 \leq I \leq N} d_I(\mathbf{x}) \qquad iff \qquad \min_{0 \leq I \leq N} d_I(\mathbf{x}) = \min_{I \in S_0} d_I(\mathbf{x})$$

and

$$b = \max_{0 \leq I \leq N} d_I(\mathbf{x}) \qquad iff \qquad \max_{0 \leq I \leq N} d_I(\mathbf{x}) = \max_{I \in S_0} d_I(\mathbf{x}).$$

Based on Bernstein coefficients range enclosure proofs [20], the proofs of Lemma 1 and Lemma 2 are given in [5].

3 Basic B-spline Constrained Global Optimization Algorithm Summary

The basic B-spline algorithm for constrained global optimization of multivariate nonlinear polynomials [6], is similar to the one described in [21, 22]. The algorithm can be summarized as follows.

Step 1: The basic algorithm uses the polynomial coefficients array of the objective function, A_o, the inequality constraints, A_{g_i} and the equality constraints, A_{h_j}. These coefficient arrays are stored in a cell structure A_c.

Step 2: A cell structure N_c, contains degree vectors N, N_{g_i} and N_{h_j}, $i = 0, \ldots, p$, $j = 0, \ldots, q$, where these degree vector represents the degree of each variable occurring in objective function, the inequality constraints and the equality constraints respectively.

Step 3: The vector degree is used to compute the B-spline segment number, as the B-spline is constructed with the number of segments equal to order of the B-spline plus one. The vectors K_o, K_{g_i}, and K_{h_j} are computed using degree vectors N, N_{g_i} and N_{h_j} as $K = N+2$, and stored in a cell structure K_c.

Step 4: Then we compute the B-spline coefficients of the objective, inequality and equality constraint polynomials on the initial search box $\mathbf{x}$. We store them in arrays $D_o(\mathbf{x})$, $D_{g_i}(\mathbf{x})$ and $D_{h_j}(\mathbf{x})$, respectively. The algorithm in [6] is suggested for the computation of B-spline coefficients.

Step 5: We initialize the current minimum estimate $\tilde{p}$ to the maximum B-spline coefficient of the objective function on $\mathbf{x}$, i.e. $\tilde{p} = \max D_o(\mathbf{x})$.

Step 6: Next, we initialize a flag vector F with each component to zero as $F := (F_1, \ldots, F_p, F_{p+1}, \ldots, F_{p+q}) = (0, \ldots, 0)$. The flag vector F is used to make the algorithm more efficient. Consider, i^{th} inequality constraint is satisfied for x in a the box $\mathbf{b}$, i.e. $g_i(x) \leq 0$ for $x \in \mathbf{b}$. Then there is no need to check again $g_i(x) \leq 0$ for any subbox $\mathbf{b}_0 \subseteq \mathbf{b}$. The same holds true for $h_j(x)$. To handle this information, we use flag vector $F = (F_1, \ldots, F_p, F_{p+1}, \ldots, F_{p+q})$, where the components F_f, takes the value 0 or 1, as follows

a) $F_f = 1$ if the fth inequality or equality constraint is satisfied for the box.

b) $F_f = 0$ if the fth inequality or equality constraint has not yet been verified for the box.

Step 7: Initialize a working list $\mathcal{L}$ with the item $\mathcal{L} \leftarrow \{\mathbf{x}, D_o(\mathbf{x}), D_{g_i}(\mathbf{x}), D_{h_j}(\mathbf{x}), F\}$, and a solution list $\mathcal{L}^{sol}$ to the empty list.

Step 8: Sort the list $\mathcal{L}$ in descending order of $(\min D_o(\mathbf{x}))$.

Step 9: Start iteration. If $\mathcal{L}$ is empty go to Step 14. Otherwise pick the last item from $\mathcal{L}$, denote it as $\{\mathbf{b}, D_o(\mathbf{b}), D_{g_i}(\mathbf{b}), D_{h_j}(\mathbf{b}), F\}$, and delete this item entry from $\mathcal{L}$.

Step 10: Perform the cut-off test. As mentioned in Lemma 2, the minimum and maximum B-spline coefficients provide the range enclosure of the function. Let $\tilde{p}$ be the current minimum estimate, and $\{\mathbf{b}, D(\mathbf{b})\}$ be the current item for processing, for which $\tilde{p} \leq \min D(\mathbf{b})$. Then, this item surely cannot contain the global minimizer and can be discarded. Discard the item $\{\mathbf{y}, D_o(\mathbf{y}), D_{g_i}(\mathbf{y}), D_{h_j}(\mathbf{y}), F\}$ if $\min D_o(\mathbf{y}) > \tilde{p}$ and return to Step 9.

Step 11: Subdivision decision. If

$$(\text{wid } \mathbf{b}) \ \& \ (\max D_o(\mathbf{b}) - \min D_o(\mathbf{b})) < \epsilon$$

then add the item $\{\mathbf{b}, \min D_0(\mathbf{b})\}$ to $\mathcal{L}^{sol}$ and go to step 9. Else go to Step 12. Here ϵ is a tolerance number.

Step 12: Generate two sub boxes. Choose the subdivision direction along the longest direction of $\mathbf{b}$ and the subdivision point as the midpoint. Subdivide $\mathbf{b}$ into two subboxes $\mathbf{b}_1$ and $\mathbf{b}_2$ such that $\mathbf{b} = \mathbf{b}_1 \cup \mathbf{b}_2$.

Step 13: For $r = 1, 2$
   1. Set $F^r := (F_1^r, \ldots, F_p^r, F_{p+1}^r, \ldots, F_{p+q}^r) = F$
   2. Compute the B-spline coefficient arrays of objective and constraints polynomial on the box $\mathbf{b}_r$ and compute corresponding B-spline range enclosure $\mathbb{D}_o(\mathbf{b}_r), \mathbb{D}_{g_i}(\mathbf{b}_r)$ and $\mathbb{D}_{h_j}(\mathbf{b}_r)$ for objective and constraints polynomial.

3. Set $\tilde{p}_{local} = \min(\mathbb{D}_o(\mathbf{b}_r))$.
4. If $\tilde{p}_{local} > \tilde{p}$ go to sub Step 9.
5. for $i = 1, \ldots, p$ if $F_i = 0$ then
   (a) If $\mathbb{D}_{g_i}(\mathbf{b}_r) > 0$ then go to sub Step 6.
   (b) If $\mathbb{D}_{g_i}(\mathbf{b}_r) \leq 0$ then set $F_i^r = 1$.
6. for $j = 1, \ldots, q$ if $F_{p+j} = 0$ then
   (a) If $0 \notin \mathbb{D}_{h_j}(\mathbf{b}_r)$ then go to sub Step 9.
   (b) If $\mathbb{D}_{h_j}(\mathbf{b}_r) \subseteq [-\epsilon_{zero}, \epsilon_{zero}]$ then set $F_{p+j}^r = 1$.
7. If $F^r = (1, \ldots, 1)$ then set $\tilde{p} := \min(\tilde{p}, \max(\mathbb{D}_o(\mathbf{b}_r)))$.
8. Enter $\{\mathbf{b}_r, D_o(\mathbf{b}_r), D_{g_i}(\mathbf{b}_r), D_{h_j}(\mathbf{b}_r), F^r\}$ into the list $\mathcal{L}$.
9. End (of r-loop)

Step 14: Set the global minimum to the current minimum estimate, $\hat{p} = \tilde{p}$.

Step 15: Find all those items in $\mathcal{L}^{sol}$ for which $\min D_o(\mathbf{b}) = \hat{p}$. The first entries of these items are the global minimizer(s) $\mathbf{z}^{(i)}$.

Step 16: Return the global minimum $\hat{p}$ and all the global minimizers $\mathbf{z}^{(i)}$ found above.

4   Numerical Results

In this section, we give three applications of the basic constrained global optimization algorithm: the robust stability analysis problem, the minimum distance problem, and the domain of attraction problem. These problems are reduced to strict inequalities (or equations) involving multivariate polynomials and solved using the basic algorithms for constrained global optimization. The computations are done on a PC Intel i3-370M 2.40 GHz processor, 6 GB RAM, while the algorithms are implemented in MATLAB [24]. An accuracy $\epsilon = 10^{-6}$ is prescribed for computing the global minimum and minimizer(s). The time in second required to solve the problems is reported.

1. **Robust Stability Analysis**
   It is well known that the roots of the closed loop characteristic equation determine the stability of the closed loop system. The characteristic equation with parameter uncertainties can be written as a polynomial equation, and the uncertainty bounds can be considered as constraints for the system. In linear system, robust stability analysis means finding the region of parameter uncertainties for which controller stabilize any disturbance in the system [7]. Consider $G_P(s)$ and $G_C(s)$ are the transfer functions of the plant and controller. The characteristic equation of the feedback system is

$$\det(I - G_P(s)G_C(s)) = 0.$$

   Now consider that there is parametric uncertainty, with $\mathbf{q}$ as the vector of uncertain parameters. Then, the uncertain transfer functions for the plant and controller are $G_P(s, \mathbf{q})$ and $G_C(s, \mathbf{q})$ respectively. The characteristic equation with this uncertainties is given by

$$\det(I - G_P(s, \mathbf{q})G_C(s, \mathbf{q})) = 0.$$

This determinant can be expanded as a polynomial

$$F(s, \mathbf{q}) = a_n(\mathbf{q})s^n + a_{n-1}(\mathbf{q})s^{n-1} + \ldots + a_1(\mathbf{q})s + a_0(\mathbf{q}),$$

where the coefficients $a_i(\mathbf{q}), i = 0, \ldots, n$ are typically multivariate polynomial functions. A stability margin k_m can be defined as

$$k_m(j\omega) = \inf\{k : F(j\omega, \mathbf{q}(k)) = 0, \forall\, \mathbf{q} \in Q\}.$$

Robust stability margin is then guaranteed if and only if $k_m \geq 1$. The problem of finding robust stability of a linear system with characteristic equation $F(j\omega, \mathbf{q})$, becomes the following constrained optimization problem

$$\min_{\mathbf{q}_i, k \geq 0, \omega \geq 0} k$$

$$s.t. \qquad \Re[F(j\omega, \mathbf{q}] = 0,$$
$$\Im[F(j\omega, \mathbf{q}] = 0,$$
$$q_i^N - \triangle q_i^- k \leq q_i \leq q_i^N + \triangle q_i^+ k,\ i = 1, \ldots, n,$$

where q^N is a stable nominal point for the uncertain parameters and $\triangle q_i^+$, $\triangle q_i^-$ are the estimated bounds [7].

The above is a constrained optimization problem involving multivariate polynomial functions. In this problem, it is necessary to find the global minimum, otherwise the stability margin might be overestimated. An overestimate can lead to wrong conclusion that the given system is stable, when actually it is not [7]. Hence, it is necessary to use a proven global optimization technique to ensure that the global minimum of k is indeed found. The algorithm in [6] can be used to correctly assess the robust stability of the system, due to its ability to find global minima. We illustrate this ability via the following example.

Example 1. We examine the l_∞ stability margin for a closed-loop system [7]. The global optimization problem is given by

$$\min k$$

$$s.t. \qquad q_1^4 q_2^4 - q_1^4 - q_2^4 q_3 = 0,$$
$$1.4 - 0.25k \leq q_1 \leq 1.4 + 0.25k,$$
$$1.5 - 0.20k \leq q_2 \leq 1.5 + 0.20k,$$
$$0.8 - 0.20k \leq q_3 \leq 0.8 + 0.20k.$$

The problem has 4 continuous variables q_1, q_2, q_3, and k. There are one equality constraint and six inequality constraints. The basic algorithm for an accuracy of 10^{-6}, finds the global minimum as $k = 1.0899$ and the global minimizer as

$$q_1 = 1.1275, q_2 = 1.282, q_3 = 1.018.$$

These results agree with those reported in [7]. The time required to solve this problem is 58.85 seconds.

2. Minimum Distance Problem

Another key problem in system analysis is to determine the minimum distance of a point to the surface defined by a polynomial constraint $p(x) = 0$. We can pose it as the constrained optimization problem

$$\rho^* = \min_{x \in R^n} \|x\|_2^2$$

$$s.t. \ p(x) = 0.$$

Most methods in literature for solving the minimum distance problem are based on LMI relaxation techniques [1],[8]. These methods are based on a suitable representation of the polynomials in homogeneous forms. We shall next investigate the ability of the basic algorithm to solve a minimum distance problem.

Example 2. This problem is from [1],[9]. Consider the state-space system

$$\dot{z} = A(x)z(t),$$

where $z \in \mathbb{R}^n$ is the state vector and $x = (x_1, x_2, ..., x_n)' \in \mathbb{R}^n$ is the vector of uncertain parameters. Assuming $A(0)$ to be a *Hurwitz matrix*, the l_2 parametric stability margin is given by

$$\rho_2 = \sqrt{\rho^*} = \sqrt{min\{\rho_R, \rho_I\}}.$$

Where ρ_R is the solution of the equality constrained optimization problem

$$\rho_R = \min_{x \in \mathbb{R}^n} x_1^2 + x_2^2$$

$$s.t. \ \det[A(x)] = 0,$$

and ρ_I is the solution of another equality constrained optimization problem

$$\rho_I = \min_{x \in \mathbb{R}^n} x_1^2 + x_2^2$$

$$s.t. \ H_{n-1}[A(x)] = 0.$$

If $A(x)$ is a polynomial in x, then this minimum distance problem becomes a quadratic optimization problem. For the particular example reported in [9], we have

$$\det[A(x)] = -3x_1^3 - 7x_1^2x_2 - 2x_1x_2^2 - 2x_2^3 - 4x_1^2 + x_2^2 + 2x_1 + 2x_2 - 1,$$
$$H_{n-1}[A(x)] = -8x_1^3 - 4x_1o^2x_2 - 2x_1x_2^2 - 28x_1^2 + x_1x_2 - 3x_2 - 22x_1 - 7x_2 + 8,$$
$$\mathbf{x}_1 = [0, 0.5], \ \mathbf{x}_2 = [0, 0.5].$$

The proposed algorithm finds the global minimum to the first constrained optimization problem as

$$\rho_R = 0.2083,$$

while it finds the global minimum to the second constrained optimization problem as

$$\rho_I = 0.0664.$$

The global minimum of the stability margin is therefore

$$\rho^* = \min\{\rho_R, \rho_I\} = 0.0664,$$

giving the l_2 parametric stability margin as

$$\rho_2 = \sqrt{\rho^*} = 0.2576.$$

These results agree with those reported in [1],[9]. The first problem is solved in 83.33 seconds and second in 81.48 seconds.

3. Domain of Attraction

Consider the differential equation with equilibrium point x_0 at t_0 given by

$$\dot{x} = f(x, t), x(t_0) = x_0,$$

where $x \in \mathbb{R}^n$ and $t \geq 0$. The domain of attraction of x_0 at t_0 is the set of all initial conditions x at time t_0, denote $g(t, t_0, x)$ satisfying

$$\lim_{t \to \infty} g(t, t_0, x) = x_0.$$

The domain of attraction of the equilibrium point x_0 is a set that is independent of the initial time, since the flow only depends on the time difference $t - t_0$ [11].

The domain of attraction can be determined by formulating the problem as an optimization problem. To do this, we need to find a positive invariant subset on which the time derivative of *Lyapunov function* is negative [12]. In turn, this leads to the following constrained optimization problem

$$\gamma^2 = \min V(x), \tag{13}$$
$$\text{s.t. } \dot{V}(x) = 0, \ (x \neq 0). \tag{14}$$

We applied the basic algorithm to solve the above optimization problem and obtain lower bounds on γ which immediately provide the domain of attraction. The example below demonstrates this application.

Example 3. Consider the following example [12]

$$\dot{x}_1 = -x_1 + x_2,$$
$$\dot{x}_2 = 0.1x_1 - 2x_2 - x_1^2 - 0.1x_1^3.$$

Let us choose the Lyapunov function as $V = x_1^2 + x_2^2$. The first derivative of Lyapunov function is then

$$\dot{V} = -2x_1^2 - 4x_2^2 + 2.2x_1x_2 - 2x_1^3x_2 - 0.2x_1^3x_2.$$

We apply the basic algorithm to solve this problem over the domain $[-3, -0.001]^2$ to the specified accuracy of 10^{-6}. The algorithm finds the global minimum value and the global minimizer as

$$\gamma = 2.6664, \ (x_1, x_2) = (-2.2610, -1.4140).$$

These results agree with those reported in [12]. The time required to solve this problem is 353.65 seconds.

5   Conclusion

We solved successfully three well known problems in systems and control engineering with the basic algorithm for constrained global optimization of multivariate polynomial function. The algorithm does not need any linearization or relaxation techniques and solves the problem to specified accuracy.

References

1. Henrion, D., Lasserre, J.-B.: Solving Nonconvex Optimization Problems. IEEE Control Systems 24, 72–83 (2004)
2. Lin, Q., Rokne, J.G.: Methods for Bounding the Range of a Polynomial. J. Computational and Applied Mathematics 58, 193–199 (1995)
3. Lin, Q., Rokne, J.G.: Interval Approximation of Higher Order to the Ranges of Functions. Computers & Mathematics with Applications 31, 101–109 (1996)
4. Michel, D., Mraoui, H., Sbibih, D., Zidna, A.: Computing the Range of Values of Real Functions using B-spline Form. Applied Mathematics and Computation 233, 85–102 (2014)
5. http://www.sc.iitb.ac.in/~ddgawali/Proof_of_Lemma.pdf (2014)
6. http://www.sc.iitb.ac.in/~ddgawali/Algorithm_pdf_file.pdf (2014)
7. Floudas, C.A.: Handbook of test problems in local and global optimization (1999)
8. Chesi, G., Garulli, A., Tesi, A.: Solving Quadratic Distance Problems: An LMI Based Approach. IEEE Trans. Automatic Control 48, 200–212 (2003)
9. Chesi, G., Tesi, A., Vicinio, A., Genesio, R.: An LMI Approach to Constrained Optimization with Homogenous Forms. Systems and Control Letters 42, 11–19 (2001)
10. Chesi, G.: Estimating the Domain of Attraction via Union of Continuous Families of Lyapunov Estimates. Systems and Control Letters 56, 326–333 (2007)
11. Sastry, S.: Nonlinear Systems Analysis, Stability and Control. Springer, Berlin (1999)
12. Parrilo, P.A.: Structured Semidefinite Programs and Semialgebric Geometric Methods in Robustness and Optimization. PhD thesis, California Insitiute of Technology, Pasadena, California (2000)
13. Horst, R., Pardalos, P.M.: Handbook of Global Optimization (1995)

14. Hansen, E., Walster, G.W.: Global Optimization using Interval Analysis: Revised and Expanded, vol. 264. CRC Press (2003)
15. Jaulin, L.: Applied Interval Analysis: with Examples in Parameter and State Estimation, Robust Control and Robotics, vol. 1. Springer (2001)
16. Kearfott, R.B.: Rigorous Global Search Continuous Problems. Springer (1996)
17. Vaidyanathan, R., El-Halwagi, M.: Global Optimization of Nonconvex Nonlinear Programs via Interval Analysis. Computers & Chemical Engineering 18, 889–897 (1994)
18. Lyche, T., Morken, K.: Spline Methods Draft. Department of Informatics, Centre of Mathematics for Applications, University of Oslo (2008)
19. Park, S.: Approximate Branch-and-Bound Global Optimization using B-spline Hypervolumes. Advances in Engineering Software 45, 11–20 (2012)
20. Ratschek, H., Rokne, J.: Computer Methods for the Range of Functions. Ellis Horwood Limited Publishers, Chichester (1984)
21. Patil, B.V., Nataraj, P.S.V., Bhartiya, S.: Global Optimization of Mixed-Integer Nonlinear (Polynomial) Programming Problems: the Bernstein Polynomial Approach. Computing 94, 325–343 (2012)
22. Ratschek, H., Rokne, J.: New Computer Methods for Global Optimization. Ellis Horwood series in Mathematics and its Applications. Horwood (1988)
23. Rogers, D.F.: An Introduction to NURBS: With Historical Perspective. Morgan Kaufmann (2001)
24. Mathworks Inc. MATLAB version 8.0.0.783 (R 2012 b) (2012)
25. DeVore, R.A., Lorentz, G.G.: Constructive Approximation, vol. 303. Springer Science & Business Media (1993)

Part VIII
Variational Principles and Applications

Approximation of Weak Efficient Solutions in Vector Optimization

Lidia Huerga[1], César Gutiérrez[2], Bienvenido Jiménez[1], and Vicente Novo[1]

[1] Departamento de Matemática Aplicada, E.T.S.I. Industriales, UNED,
c/ Juan del Rosal 12, Ciudad Universitaria, 28040 Madrid, Spain
lhuerga@bec.uned.es, bjimenez@ind.uned.es, vnovo@ind.uned.es
[2] Departamento de Matemática Aplicada, E.T.S. de Ingenieros de Telecomunicación,
Universidad de Valladolid, Paseo de Belén 15,
Campus Miguel Delibes, 47011 Valladolid, Spain
cesargv@mat.uva.es

Abstract. In this paper, a well-known concept of ε-efficient solution due to Kutateladze is studied, in order to approximate the weak efficient solutions of vector optimization problems. In particular, it is proved that the limit, in the Painlevé-Kuratowski sense, of the ε-efficient sets when the precision ε tends to zero is the set of weak efficient solutions of the problem. Moreover, several nonlinear scalarization results are derived to characterize the ε-efficient solutions in terms of approximate solutions of scalar optimization problems. Finally, the obtained results are applied not only to propose a kind of penalization scheme for Kutateladze's approximate solutions of a cone constrained convex vector optimization problem but also to characterize ε-efficient solutions of convex multiobjective problems with inequality constraints via multiplier rules.

Keywords: Vector optimization, weak efficient solution, ε-efficient solution, nonlinear scalarization, Kuhn-Tucker optimality conditions, ε-subgradients.

1 Introduction

During the last two decades, there has been a growing interest on approximate solutions of vector optimization problems since, from a theoretical point of view, they play an important role in many concepts and results of vector optimization, such as the Ekeland Variational Principle, the well-posedness properties and the ε-subdifferential (see, for instance, [1,4,7] and the references therein).

In this work, we focus on the concept of ε-efficient solution defined by Kutateladze [9], which is the most popular ε-efficiency notion of the literature. To be exact, in Section 2 we show that these ε-efficient solutions provide suitable approximations to the set of weak efficient solutions by following the approach introduced in [6]. In Section 3, inspired by Weirbicky's approach based on order representation and monotonicity properties (see [10]), we derive nonlinear scalarization results through generic functionals whose sublevel sets at

© Springer International Publishing Switzerland 2015
H.A. Le Thi et al. (eds.), *Model. Comput. & Optim. in Inf. Syst. & Manage. Sci.*,
Advances in Intelligent Systems and Computing 359, DOI: 10.1007/978-3-319-18161-5_41

zero approximate the ordering cone of the problem. Then, as a consequence of these results and a new kind of penalization technique, we characterize the ε-efficient solutions of cone constrained vector optimization problems via the so-called Tammer-Weidner nonlinear separation functional. Finally, in Section 4, as an application of the obtained results, we state a type of penalization scheme for Kutateladze's approximate solutions of a cone constrained convex vector optimization problem and also we derive Kuhn-Tucker optimality conditions for ε-efficient solutions of nondifferentiable convex Pareto multiobjective problems with inequality constraints.

2 εv-Efficient Solutions

Let X, Y be real locally convex Hausdorff topological linear spaces. The closure and the topological interior of a set $M \subset Y$ are denoted by $\operatorname{cl} M$ and $\operatorname{int} M$, respectively, and the nonnegative orthant of $\mathbb{R}^p$ by $\mathbb{R}^p_+$. Moreover, $\mathbb{R}_+ := \mathbb{R}^1_+$. In the sequel, we consider the vector optimization problem

$$\operatorname{Min}_D\{f(x) : x \in S\}, \tag{1}$$

where $f : X \to Y$, $\emptyset \neq S \subset X$ and $D \subset Y$ is the ordering cone, which is assumed to be convex, proper (i.e., $D \neq Y$) and solid (i.e., it has a nonempty topological interior). In order to deal with weak solutions of problem (1), we consider the weak (Pareto) generalized ordering relation:

$$y_1, y_2 \in Y, \quad y_1 <_D y_2 \iff y_2 - y_1 \in \operatorname{int} D. \tag{2}$$

Definition 1. *A point $x_0 \in S$ is said to be a weak efficient solution of problem (1), denoted by $x_0 \in \operatorname{WE}(f, S, D)$, if there does not exist a point $x \in S$ such that $f(x) <_D f(x_0)$.*

Next, we recall the approximate efficiency notion due to Kutateladze [9].

Definition 2. *Let $v \in Y \backslash \{0\}$ and $\varepsilon \geq 0$. It is said that a point $x_0 \in S$ is an εv-efficient solution of problem (1), denoted by $x_0 \in \operatorname{WE}(f, S, \varepsilon v, D)$, if there does not exist a point $x \in S$ such that $f(x) <_D f(x_0) - \varepsilon v$.*

It is clear that for each v, $\operatorname{WE}(f, S, \varepsilon v, D)$ reduces to the set of weak efficient solutions of problem (1) whenever $\varepsilon = 0$. Moreover, it follows that

$$\operatorname{WE}(f, S, \varepsilon v, D) = \{x_0 \in S : (f(S) - f(x_0)) \cap (-\varepsilon v - \operatorname{int} D) = \emptyset\}. \tag{3}$$

The next theorem collects the main properties that relate the sets of εv-efficient solutions for different precisions ε. Let us underline that these properties work whenever the direction v belongs to $\operatorname{int} D$. By $\operatorname{Limsup}_{\varepsilon \to 0} \operatorname{WE}(f, S, \varepsilon v, D)$ and $\operatorname{Lim}_{\varepsilon \to 0} \operatorname{WE}(f, S, \varepsilon v, D)$ we denote, respectively, the upper limit and the limit of the set-valued mapping $\mathbb{R}_+ \ni \varepsilon \to WE(f, S, \varepsilon v, D)$ in the Painlevé-Kuratowski sense (see [3]). To be precise,

$$\underset{\varepsilon \to 0}{\text{Limsup}}\, WE(f, S, \varepsilon v, D) := \{x_0 \in X : \text{there exist nets } (x_i) \subset X, (\varepsilon_i) \subset \mathbb{R}_+\backslash\{0\},$$

$$\varepsilon_i \to 0 \text{ such that } x_i \in WE(f, S, \varepsilon_i v, D), x_i \to x_0\}\,,$$

$$\underset{\varepsilon \to 0}{\text{Liminf}}\, WE(f, S, \varepsilon v, D) := \{x_0 \in X : \text{for each net } (\varepsilon_i) \subset \mathbb{R}_+\backslash\{0\}, \varepsilon_i \to 0,$$

$$\text{there exist } (\varepsilon_{\varphi(j)}), (x_j) \in WE(f, S, \varepsilon_{\varphi(j)} v, D),$$

$$x_j \to x_0\}$$

$((\varepsilon_{\varphi(j)})$ denotes a subnet of $(\varepsilon_i))$ and

$$\underset{\varepsilon \to 0}{\text{Lim}}\, WE(f, S, \varepsilon v, D) := \underset{\varepsilon \to 0}{\text{Limsup}}\, WE(f, S, \varepsilon v, D) = \underset{\varepsilon \to 0}{\text{Liminf}}\, WE(f, S, \varepsilon v, D)$$

whenever

$$\underset{\varepsilon \to 0}{\text{Limsup}}\, WE(f, S, \varepsilon v, D) = \underset{\varepsilon \to 0}{\text{Liminf}}\, WE(f, S, \varepsilon v, D)\,.$$

Theorem 1. *Consider* $q \in \text{int}\, D$. *The following holds:*

1. $WE(f, S, \varepsilon_1 q, D) \subset WE(f, S, \varepsilon_2 q, D)$, *for all* $\varepsilon_1, \varepsilon_2 \geq 0$, $\varepsilon_1 < \varepsilon_2$.
2. $WE(f, S, \bar{\varepsilon} q, D) = \bigcap_{\varepsilon > \bar{\varepsilon}} WE(f, S, \varepsilon q, D)$.
3. *Let* $x_0 \in S$, $\bar{\varepsilon} \geq 0$ *and nets* $(x_i) \subset S$ *and* $(\varepsilon_i) \subset \mathbb{R}_+$ *such that* $x_i \in WE(f, S, \varepsilon_i q, D)$, $\varepsilon_i \to \bar{\varepsilon}$ *and* $f(x_i) \to f(x_0)$. *Then* $x_0 \in WE(f, S, \bar{\varepsilon} q, D)$.
4. *Suppose that* S *is closed,* f *is continuous and* $\bar{\varepsilon} \geq 0$. *Then* $WE(f, S, \bar{\varepsilon} q, D)$ *is closed and*

$$\underset{\varepsilon \to \bar{\varepsilon}}{\text{Limsup}}\, WE(f, S, \varepsilon q, D) = WE(f, S, \bar{\varepsilon} q, D)\,. \tag{4}$$

Proof. Parts 1 and 2 follow from the definition of $WE(f, S, \varepsilon q, D)$.

3. Suppose that $x_0 \notin WE(f, S, \bar{\varepsilon} q, D)$. As

$$\bar{\varepsilon} q + \text{int}\, D = \bigcup_{\varepsilon > \bar{\varepsilon}} (\varepsilon q + \text{int}\, D)\,, \tag{5}$$

there exist $x \in S$ and $\varepsilon_0 > \bar{\varepsilon}$ such that

$$f(x) - f(x_0) \in -\varepsilon_0 q - \text{int}\, D\,. \tag{6}$$

Since $f(x_i) \to f(x_0)$ and $-\varepsilon_0 q - \text{int}\, D$ is open, there exists i_1 such that

$$f(x) - f(x_i) \in -\varepsilon_0 q - \text{int}\, D, \; \forall i \succeq i_1\,. \tag{7}$$

On the other hand, as $\varepsilon_i \to \bar{\varepsilon}$ and $\varepsilon_0 > \bar{\varepsilon}$ there exists i_2 such that $\varepsilon_i < \varepsilon_0$, for all $i \succeq i_2$. Let i_0 be such that $i_0 \succeq i_1$, $i_0 \succeq i_2$. It follows that

$$f(x) - f(x_{i_0}) \in -\varepsilon_0 q - \text{int}\, D \subset -\varepsilon_{i_0} q - \text{int}\, D\,, \tag{8}$$

which is a contradiction, since $x_{i_0} \in \text{WE}(f, S, \varepsilon_{i_0}q, D)$. Thus, $x_0 \in \text{WE}(f, S, \bar\varepsilon q, D)$ and part *3* is proved.

4. Let us define

$$H_x := \{x_0 \in X : f(x) - f(x_0) \notin -\bar\varepsilon q - \text{int } D\}, \quad \forall x \in S . \tag{9}$$

For every $x \in S$, the set H_x is closed, since $H_x = f^{-1}(Y \backslash (f(x) + \bar\varepsilon q + \text{int } D))$. On the other hand, it is clear that

$$\text{WE}(f, S, \bar\varepsilon q, D) = \left(\bigcap_{x \in S} H_x \right) \cap S , \tag{10}$$

and so $\text{WE}(f, S, \bar\varepsilon q, D)$ is closed.

Moreover, if $\varepsilon_0 > \bar\varepsilon$, then

$$\underset{\varepsilon \to \bar\varepsilon}{\text{Limsup}}\, \text{WE}(f, S, \varepsilon q, D) \subset \text{WE}(f, S, \varepsilon_0 q, D) . \tag{11}$$

Indeed, since $\varepsilon_0 > \bar\varepsilon$, if $x_i \to x$, with $x_i \in \text{WE}(f, S, \varepsilon_i q, D)$ and $\varepsilon_i \to \bar\varepsilon$, then there exists i_0 such that $x_i \in \text{WE}(f, S, \varepsilon_0 q, D)$ for all $i \succeq i_0$, and as $\text{WE}(f, S, \varepsilon_0 q, D)$ is closed it follows that $x \in \text{WE}(f, S, \varepsilon_0 q, D)$.

Finally, taking into account part *2* we obtain that

$$\text{WE}(f, S, \bar\varepsilon q, D) \subset \underset{\varepsilon \to \bar\varepsilon}{\text{Limsup}}\, \text{WE}(f, S, \varepsilon q, D)$$
$$\subset \bigcap_{\varepsilon > \bar\varepsilon} \text{WE}(f, S, \varepsilon q, D) = \text{WE}(f, S, \bar\varepsilon q, D) , \tag{12}$$

and the proof is complete. $\square$

By applying these properties we see that the sets of εq-efficient solutions are a suitable approximation for the set of weak efficient solutions, as it is established in the following corollary.

Corollary 1. *Assume that S is closed, f is continuous and consider $q \in \text{int } D$. Then*

$$\underset{\varepsilon \to \bar\varepsilon}{\text{Lim}}\, \text{WE}(f, S, \varepsilon q, D) = \bigcap_{\varepsilon > 0} \text{WE}(f, S, \varepsilon q, D) = \text{WE}(f, S, D) . \tag{13}$$

3   Scalarization

Consider an scalarization mapping $\varphi : Y \to \mathbb{R}$ and $\delta \geq 0$. We denote

$$\delta - \text{argmin}_S(\varphi \circ f) = \{x_0 \in S : (\varphi \circ f)(x_0) - \delta \leq (\varphi \circ f)(x), \forall x \in S\} . \tag{14}$$

Let us observe that (14) is the set of suboptimal solutions with precision δ of the scalar optimization problem defined by the objective function $\varphi \circ f$ and the feasible set S.

The next two propositions relate the sets $\mathrm{WE}(f, S, \varepsilon v, D)$ and $\delta-\mathrm{argmin}_S(\varphi \circ f)$ when φ satisfies simple properties. The first one states a sufficient condition for εv-efficient solutions of problem (1) by a scalarization and the second one a necessary condition. For each $x_0 \in S$, let $f_{x_0} : X \to Y$, $f_{x_0}(x) = f(x) - f(x_0)$, for all $x \in X$.

Proposition 1. *Consider $v \in Y \backslash \{0\}$ and $\varepsilon \geq 0$, and suppose that*

$$\varphi(0) \geq 0, \quad -\varepsilon v - \mathrm{int}\, D \subset \{y \in Y : \varphi(y) < 0\}. \tag{15}$$

If $x_0 \in \varphi(0)-\mathrm{argmin}_S(\varphi \circ f_{x_0})$, then $x_0 \in \mathrm{WE}(f, S, \varepsilon v, D)$.

Proposition 2. *Consider $v \in Y \backslash (-\mathrm{int}\, D)$ and $\varepsilon \geq 0$, and suppose that*

$$\{y \in Y : \varphi(y) < 0\} \subset -\varepsilon v - \mathrm{int}\, D. \tag{16}$$

If $x_0 \in \mathrm{WE}(f, S, \varepsilon v, D)$, then $x_0 \in \varphi(0)-\mathrm{argmin}_S(\varphi \circ f_{x_0})$.

Let us observe that statement (16) implies that $\varphi(0) \geq 0$.

In order to apply the previous scalarization results we need mappings $\varphi : Y \to \mathbb{R}$ such that

$$\{y \in Y : \varphi(y) < 0\} = -\varepsilon v - \mathrm{int}\, D, \tag{17}$$

where $v \notin -\mathrm{int}\, D$. By the so-called Tammer-Weidner nonlinear separation functional [2] we can define a mapping satisfying this property (in [5] the reader can find other functionals satisfying (17)). Indeed, denote by $\varphi_e : Y \to \mathbb{R}$ the Tammer-Weidner functional defined by $e \in \mathrm{int}\, D$, i.e.,

$$\varphi_e(y) = \inf\{t \in \mathbb{R} : y \in te - \mathrm{cl}\, D\}, \quad \forall y \in Y. \tag{18}$$

It is not hard to check that

$$\{y \in Y : \varphi_e(y) < 0\} = -\mathrm{int}\, D \tag{19}$$

and so we have the following result (see [3] for other properties of φ_e).

Theorem 2. *Consider $v \in Y \backslash (-\mathrm{int}\, D)$, $\varepsilon \geq 0$ and the mapping $\varphi : Y \to \mathbb{R}$, $\varphi(y) = \varphi_e(y + \varepsilon v)$, for all $y \in Y$. It follows that*

$$x_0 \in \mathrm{WE}(f, S, \varepsilon v, D) \iff x_0 \in \varepsilon \varphi_e(v)-\mathrm{argmin}_S(\varphi \circ f_{x_0}). \tag{20}$$

Proof. By (19) it is clear that

$$\{y \in Y : \varphi(y) < 0\} = -\varepsilon v - \mathrm{int}\, D, \tag{21}$$

and by Propositions 1 and 2 we deduce that

$$x_0 \in \mathrm{WE}(f, S, \varepsilon v, D) \iff x_0 \in \varphi(0)-\mathrm{argmin}_S(\varphi \circ f_{x_0}). \tag{22}$$

Moreover, $\varphi(0) = \varphi_e(\varepsilon v) = \varepsilon \varphi_e(v)$, and the proof finishes. $\qquad \square$

4   An Application

In this last section, we apply Theorem 2 in order to state a kind of penalization scheme for Kutateladze's approximate solutions of a cone constrained convex vector optimization problem. To be exact, let Z be a real locally convex Hausdorff topological linear space, $K \subset Z$ be a proper convex cone with nonempty topological interior and consider a mapping $g : X \to Z$ and the set $S_K = \{x \in X : g(x) \in -K\}$. It said that f is D-convex if

$$\alpha f(x_1) + (1-\alpha)f(x_2) \in f(\alpha x_1 + (1-\alpha)x_2) + D, \quad \forall x_1, x_2 \in X, \forall \alpha \in (0,1) . \quad (23)$$

Then we will characterize the εq-efficient solutions of the vector optimization problem

$$\mathrm{Min}_D\{f(x) : g(x) \in -K\} , \quad (24)$$

where $q \in \mathrm{int}\, D$, f, g are assumed to be D-convex and K-convex, respectively, and the so-called Slater constraint qualification is satisfied, i.e., there exists $\bar{x} \in S_K$ such that $g(\bar{x}) \in -\mathrm{int}\, K$. The following lemma is needed.

Lemma 1. *Let $\varepsilon > 0$.*

1. We have that

$$x_0 \in \mathrm{WE}(f, S_K, \varepsilon v, D) \Rightarrow x_0 \in \mathrm{WE}((f,g), X, \varepsilon(v, (1/\varepsilon)g(x_0)), D \times K) \cap S_K . \quad (25)$$

2. Consider $q \in \mathrm{int}\, D$ and suppose that f is D-convex and g is K-convex, and the Slater constraint qualification is satisfied. Then

$$x_0 \in \mathrm{WE}(f, S_K, \varepsilon q, D) \iff x_0 \in \mathrm{WE}((f,g), X, \varepsilon(q, (1/\varepsilon)g(x_0)), D \times K) \cap S_K . \quad (26)$$

Proof. Part 1 follows directly from the definitions.

2. Let us prove the sufficient condition, since the necessary condition is given by part *1*. Consider

$$x_0 \in \mathrm{WE}((f,g), X, \varepsilon(q, (1/\varepsilon)\, g(x_0)), D \times K) \cap S_K \quad (27)$$

and suppose that $x_0 \notin \mathrm{WE}(f, S_K, \varepsilon q, D)$. By part *2* of Theorem 1 we deduce that there exists $\varepsilon_0 > \varepsilon$ such that $x_0 \notin \mathrm{WE}(f, S_K, \varepsilon_0 q, D)$ and so there exists $x \in S_K$ such that

$$f(x) - f(x_0) \in -\varepsilon_0 q - \mathrm{int}\, D . \quad (28)$$

Let $\bar{x} \in X$ be such that $g(\bar{x}) \in -\mathrm{int}\, K$ and consider $x_t := (1-t)x + t\bar{x}, t \in (0,1)$. As g is K-convex we have that

$$g(x_t) \in (1-t)g(x) + tg(\bar{x}) - K \subset -\mathrm{int}\, K . \quad (29)$$

Moreover,

$$
\begin{aligned}
f(x_t) - f(x_0) &\in (1-t)f(x) + tf(\bar{x}) - f(x_0) - D \\
&= t(f(\bar{x}) - f(x)) + f(x) - f(x_0) - D \\
&\subset t(f(\bar{x}) - f(x)) - \varepsilon_0 q - \operatorname{int} D \\
&\subset -\varepsilon q - \operatorname{int} D
\end{aligned} \tag{30}
$$

for t small enough. Then

$$
(f,g)(x_t) - (f,g)(x_0) \in -\varepsilon\,(q, (1/\varepsilon)g(x_0)) - \operatorname{int}(D \times K)\,, \tag{31}
$$

which is a contradiction. Thus, $x_0 \in \mathrm{WE}(f, S, \varepsilon q, D)$. □

Let $e \in \operatorname{int} D$ and $k \in \operatorname{int} K$. Next we consider the mapping $\varphi_{e,k} : Y \times Z \to \mathbb{R}$ defined as follows:

$$
\varphi_{e,k}(y,z) = \inf\{t \in \mathbb{R} : y \in te - \operatorname{cl} D, z \in tk - \operatorname{cl} K\}, \quad \forall y \in Y, z \in Z\,. \tag{32}
$$

Theorem 3. *Let $q \in \operatorname{int} D$, $\varepsilon > 0$, $x_0 \in S_K$ and suppose that f is D-convex and g is K-convex, and the Slater constraint qualification is satisfied. Consider the mapping $\varphi : Y \times Z \to \mathbb{R}$, $\varphi(y,z) = \varphi_{e,k}((y,z)) + \varepsilon(q, (1/\varepsilon)g(x_0))$, for all $y \in Y$, $z \in Z$. Then*

$$
x_0 \in \mathrm{WE}(f, S_K, \varepsilon q, D) \iff x_0 \in \varepsilon\varphi_{e,k}(q, (1/\varepsilon)g(x_0)) - \operatorname{argmin}_X(\varphi \circ (f,g)_{x_0})\,. \tag{33}
$$

In order to illustrate Theorem 3, let us obtain a Kuhn-Tucker multiplier rule for εq-efficient solutions of nondifferentiable convex Pareto multiobjective problems with inequality constraints. Some preliminaries are needed.

Definition 3. *Let $h : \mathbb{R}^n \to \mathbb{R}$ be convex, $x_0 \in \mathbb{R}^n$ and $\varepsilon \geq 0$. It is said that $x^* \in \mathbb{R}^n$ is an ε-subgradient of h at x_0, denoted by $x^* \in \partial_\varepsilon h(x_0)$, if*

$$
h(x) \geq h(x_0) - \varepsilon + \langle x^*, x - x_0\rangle, \quad \forall x \in \mathbb{R}^n\,. \tag{34}
$$

For a complete description of this concept, the reader can see [8]. In particular, it follows that:

$$
\partial_\varepsilon(h(\cdot) + c)(x_0) = \partial_\varepsilon h(x_0), \quad \forall c \in \mathbb{R}\,, \tag{35}
$$

$$
x_0 \in \varepsilon\text{-}\operatorname{argmin}_X h \iff 0 \in \partial_\varepsilon h(x_0)\,. \tag{36}
$$

Theorem 4. *Consider $x_0 \in \mathbb{R}^n$, $\varepsilon \geq 0$ and $\Psi = \max_{1 \leq l \leq r}\{h_l\}$, where $h_l : \mathbb{R}^n \to \mathbb{R}$ is convex for all $l = 1, 2, \ldots, r$. It follows that*

$$
\partial_\varepsilon \Psi(x_0) = \left\{ \sum_{l=1}^r \partial_{\varepsilon_l}(\alpha_l h_l)(x_0) : \alpha_l \geq 0, \sum_{l=1}^r \alpha_l = 1, \varepsilon_l \geq 0, \right.
$$
$$
\left. \sum_{l=1}^r \varepsilon_l + \Psi(x_0) - \sum_{l=1}^r \alpha_l h_l(x_0) \leq \varepsilon \right\}. \tag{37}
$$

Assume that (24) is a Pareto multiobjective problem with inequality constraints, i.e., $X = \mathbb{R}^n$, $Y = \mathbb{R}^p$, $Z = \mathbb{R}^m$, $D = \mathbb{R}^p_+$ and $K = \mathbb{R}^m_+$. Then, we have the following Kuhn-Tucker multiplier rule for Kutateladze's approximate solutions in the convex case. We denote $f = (f_1, f_2, \ldots, f_p)$, $g = (g_1, g_2, \ldots, g_m)$ and $q = (q_1, q_2, \ldots, q_p)$.

Theorem 5. *Suppose that f_i, g_j are convex for all i, j, $q_i > 0$ for all i, and the Slater constraint qualification is satisfied. Let $\varepsilon > 0$ and $x_0 \in \mathbb{R}^n$ such that $g_j(x_0) \le 0$ for all j. Then $x_0 \in \mathrm{WE}(f, S_{\mathbb{R}^m_+}, \varepsilon q, \mathbb{R}^p_+)$ if and only if there exist $\varepsilon_1, \varepsilon_2, \ldots, \varepsilon_p \ge 0$, $\gamma_1, \gamma_2, \ldots, \gamma_m \ge 0$, $\lambda_1, \lambda_2, \ldots, \lambda_p \ge 0$ not all zero and $\mu_1, \mu_2, \ldots, \mu_m \ge 0$ such that $\sum_{i=1}^p \lambda_i + \sum_{j=1}^m \mu_j = 1$, and*

$$0 \in \sum_{i=1}^p \partial_{\varepsilon_i}(\lambda_i f_i)(x_0) + \sum_{j=1}^m \partial_{\gamma_j}(\mu_j g_j)(x_0) , \tag{38}$$

$$\sum_{i=1}^p \varepsilon_i + \sum_{j=1}^m \gamma_j - \varepsilon \sum_{i=1}^p \lambda_i q_i \le \sum_{j=1}^m \mu_j g_j(x_0) . \tag{39}$$

Acknowledgments. The authors are very grateful to Prof. Q.T. Bao, Chr. Tammer and A. Soubeyran for their kind invitation to contribute in the special session "Variational Principles and Applications", and also to the anonymous referees for their helpful comments and suggestions.

This work was partially supported by Ministerio de Economía y Competitividad (Spain) under project MTM2012-30942.

References

1. Bao, T.Q., Mordukhovich, B.S.: Relative Pareto Minimizers for Multiobjective Problems: Existence and Optimality Conditions. Math. Program. 122, 301–347 (2010)
2. Gerth, C., Weidner, P.: Nonconvex Separation Theorems and Some Applications in Vector Optimization. J. Optim. Theory Appl. 67, 297–320 (1990)
3. Göpfert, A., Riahi, H., Tammer, C., Zălinescu, C.: Variational Methods in Partially Ordered Spaces. Springer, Berlin (2003)
4. Gutiérrez, C., Huerga, L., Jiménez, B., Novo, V.: Proper Approximate Solutions and ε-Subdifferentials in Vector Optimization: Basic Properties and Limit Behaviour. Nonlinear Anal. 79, 52–67 (2013)
5. Gutiérrez, C., Jiménez, B., Novo, V.: Optimality Conditions via Scalarization for a New ε-Efficiency Concept in Vector Optimization Problems. European J. Oper. Res. 201, 11–22 (2010)
6. Gutiérrez, C., Jiménez, B., Novo, V.: A Generic Approach to Approximate Efficiency and Applications to Vector Optimization with Set-Valued Maps. J. Glob. Optim. 49, 313–342 (2011)

7. Gutiérrez, C., Miglierina, E., Molho, E., Novo, V.: Pointwise Well-Posedness in Set Optimization with Cone Proper Sets. Nonlinear Anal. 75, 1822–1833 (2012)
8. Hiriart-Urruty, J.-B., Lemaréchal, C.: Convex Analysis and Minimization Algorithms II. Springer, Berlin (1993)
9. Kutateladze, S.S.: Convex ϵ-Programming. Soviet. Math. Dokl. 20, 391–393 (1979)
10. Wierzbicki, A.P.: On the Completeness and Constructiveness of Parametric Characterizations to Vector Optimization Problems. OR Spektrum 8, 73–87 (1986)

Characterization of Set Relations by Means of a Nonlinear Scalarization Functional

Elisabeth Köbis[1] and Christiane Tammer[2]

[1] Department of Mathematics, Friedrich-Alexander-University Erlangen-Nuremberg,
Cauerstr. 11, 91058 Erlangen, Germany
elisabeth.koebis@fau.de
[2] Institute of Mathematics, Martin-Luther-University of Halle-Wittenberg,
Theodor-Lieser-Str. 5, 06120 Halle-Saale, Germany
christiane.tammer@mathematik.uni-halle.de

Abstract. In this paper we show how a nonlinear scalarization functional can be used in order to characterize set order relations. We will show that this functional plays a key role in set optimization. As set order relations, we consider the upper set less order relation and the lower set less order relation introduced by Kuroiwa [10,9] and the set less order relation which was introduced independently by Young [13] and Nishnianidze [11]. Our approaches do not rely on any convexity assumptions on the considered sets.

Keywords: Nonlinear scalarization functionals, set optimization, set order relations.

1 Introduction

Set optimization has become a very important field in various applications, see, for instance, [8]. For some applications, it is necessary to assume a feasible element to be associated with a whole set of function values instead of just one vector. For example, certain concepts of robustness for dealing with uncertainties in vector optimization can be described using approaches from set-valued optimization (see [5]).

Set optimization deals with the process of obtaining minimal sets, where the map to be minimized is set-valued (see [6]). In order to obtain minimal solutions of a set optimization problem, one uses set order relations. We are interested in characterizing certain set relations by using a very broad and manageable functional. Such characterizations of set order relations via scalarization are important for deriving numerical methods for solving set-valued optimization problems.

In Section 2, we recall the nonlinear scalarizing functional and give some important properties that this functional satisfies under very general assumptions. In Subsection 2.2 we very shortly discuss how set order relations have been discussed in the context of scalarization functionals in the literature. Section 3 deals with the characterization of the upper set, lower set and set less order relations

© Springer International Publishing Switzerland 2015
H.A. Le Thi et al. (eds.), *Model. Comput. & Optim. in Inf. Syst. & Manage. Sci.*,
Advances in Intelligent Systems and Computing 359, DOI: 10.1007/978-3-319-18161-5_42

by means of the nonlinear functional introduced in Section 2. Finally, in Section 4 we describe minimal solutions of set optimization problems by means of our findings.

2　Nonlinear Scalarization Functional

2.1　Preliminaries

Let Y be a linear topological space, $k \in Y \setminus \{0\}$ and let $\mathcal{F}, C$ be proper subsets of Y. We assume that C is closed and

$$C + [0, +\infty) \cdot k \subset C. \tag{1}$$

Now we introduce the functional $z^{C,k} : Y \to \mathbb{R} \cup \{+\infty\} \cup \{-\infty\} =: \bar{\mathbb{R}}$

$$z^{C,k}(y) := \inf\{t \in \mathbb{R} | y \in tk - C\}. \tag{2}$$

Then we formulate the problem of minimizing the functional $z^{C,k}$ as

$$z^{C,k}(y) \to \inf_{y \in \mathcal{F}}. \tag{$P_{k,C,\mathcal{F}}$}$$

Figure 1 visualizes the functional $z^{C,k}$ for $C = \mathbb{R}^2_+$ and a given vector $k \in \operatorname{int} C$. We can see that the set $-C$ is moved along the ray $t \cdot k$ up until y belongs to $tk - C$. The functional $z^{C,k}$ is assigned the smallest value t such that the property $y \in tk - C$ is fulfilled. By a variation of the set C and the vector $k \in Y \setminus \{0\}$ all Pareto minimal elements of a vector optimization problem without any convexity assumptions can be found.

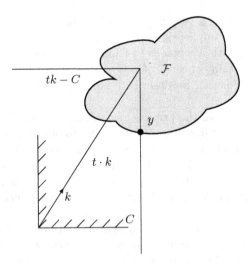

Fig. 1. Illustration of the functional $z^{C,k}(y) := \inf\{t \in \mathbb{R} | y \in tk - C\}$

The scalarizing functional $z^{C,k}$ was used in [1] to prove separation theorems for nonconvex sets. Applications of $z^{C,k}$ include coherent risk measures in financial mathematics (see, for instance, [4]). Many properties of $z^{C,k}$ were studied in [1], [2] and [12]. First let us recall the C-monotonicity of a functional.

Definition 1. Let Y be a linear topological space, $C \subset Y$, $C \neq \emptyset$. A functional $z : Y \to \bar{\mathbb{R}}$ is C-**monotone**, if for

$$y_1, \ y_2 \in Y : \ y_1 \in y_2 - C \Rightarrow z(y_1) \leq z(y_2).$$

Moreover, z is said to be **strictly** C-**monotone**, if for

$$y_1, y_2 \in Y : \ y_1 \in y_2 - C \setminus \{0\} \Rightarrow z(y_1) < z(y_2).$$

Below we provide some properties of the functional $z^{C,k}$ introduced in (2). As usual, we denote the topological boundary of C with bd C and the topological interior by int C.

Theorem 1 ([1], [2]). *Let Y be a linear topological space, $C \subset Y$ a closed proper set and $D \subset Y$. Furthermore, let $k \in Y \setminus \{0\}$ be such that (1) is satisfied. Then the following properties hold for $z = z^{C,k}$:*

(a) z is lower semi-continuous.
(b) z is convex $\iff C$ is convex,
    $[\forall \ y \in Y, \ \forall \ r > 0 : \ z(ry) = rz(y)] \iff C$ is a cone.
(c) z is proper $\iff C$ does not contain lines parallel to k, i.e., $\forall \ y \in Y \ \exists \ r \in \mathbb{R} : \ y + rk \notin C$.
(d) z is D-monotone $\iff C + D \subset C$.
(e) z is subadditive $\iff C + C \subset C$.
(f) $\forall \ y \in Y, \ \forall \ r \in \mathbb{R} : \ z(y) \leq r \iff y \in rk - C$.
(g) $\forall \ y \in Y, \ \forall \ r \in \mathbb{R} : \ z(y + rk) = z(y) + r$.
(h) z is finite-valued $\iff C$ does not contain lines parallel to k and $\mathbb{R}k - C = Y$.

Let furthermore $C + (0, +\infty) \cdot k \subset$ int C. Then

(i) z is continuous.
(j) $\forall \ y \in Y, \ \forall \ r \in \mathbb{R} : \ z(y) = r \iff y \in rk -$ bd C,
    $\forall \ y \in Y, \ \forall \ r \in \mathbb{R} : \ z(y) < r \iff y \in rk -$ int C.
(k) Assume furthermore that z is proper. Then z is D-monotone $\iff C + D \subset C \iff$ bd $C + D \subset C$.
(l) If z is finite-valued, then z is strictly D-monotone $\iff C + (D \setminus \{0\}) \subset$ int $C \iff$ bd $C + (D \setminus \{0\}) \subset$ int C.
(m) Suppose z is proper. Then z is subadditive $\iff C + C \subset C \iff$ bd $C +$ bd $C \subset C$.

For the proof, see [2, Theorem 2.3.1].

The following corollary unifies the features given in Theorem 1 in case C is a proper closed convex cone.

Corollary 1 ([2, Corollary 2.3.5.]) *Let C be a proper closed convex cone and $k \in \operatorname{int} C$. Then $z = z^{C,k}$, defined by (2), is a finite-valued continuous sublinear and strictly $(\operatorname{int} C)$-monotone functional such that*

$$\forall\, y \in Y,\ \forall\, r \in \mathbb{R}: \ z(y) \leq r \Longleftrightarrow y \in rk - C, \tag{3}$$
$$\forall\, y \in Y,\ \forall\, r \in \mathbb{R}: \ z(y) < r \Longleftrightarrow y \in rk - \operatorname{int} C.$$

Throughout this paper, let the following Assumption 1 be satisfied.

Assumption 1. Let Y be a linear topological space, $C \subset Y$ a proper closed pointed convex cone and $k \in \operatorname{int} C$. Furthermore, let two sets $A, B \in \mathcal{P}(Y) := \{A \subseteq Y \mid A \text{ is nonempty}\}$ be given. When we consider the terms

$$
\begin{aligned}
&\sup_{a \in A} \inf_{b \in B} z^{C,k}(a - b), \\
&\sup_{b \in B} \inf_{a \in A} z^{C,k}(a - b), \\
&\sup_{a \in A} \sup_{b \in B} z^{C,k}(a - b), \\
&\inf_{a \in A} \inf_{b \in B} z^{C,k}(a - b)
\end{aligned}
\tag{4}
$$

for two sets $A, B \in \mathcal{P}(Y)$, we assume that the respective suprema and infima are attained for all $k \in \operatorname{int} C$.

The partial ordering in Y will be denoted by $\leq_C$, where for $y_1, y_2 \in Y$

$$y_1 \leq_C y_2 \ :\Longleftrightarrow\ y_1 \in y_2 - C.$$

Notice that the suprema and infima in the terms (4) are attained if $z^{C,k}$ is a lower semi-continuous functional on compact sets A and B (compare [14, Theorem 38.B]).

In this paper we aim at using the functional $z^{C,k}$ in order to characterize known set relations. In the following subsection we briefly discuss the literature on this subject.

2.2   Notes on the Literature

Hernández and Rodríguez-Marín [3] introduced an extension of the functional $z^{C,k}$ (see equation (2)) in order to characterize the set order relation $B \subseteq A + C$. They consider a function

$$Z^{C,k}(A, B) := \sup_{b \in B} \inf \{ t \in \mathbb{R} \mid b \in tk + A + C \}$$

$$\left(= \sup_{b \in B} z^{-(C+A),k}(b) \ \text{ with the notations from Subsection 2.1} \right)$$

and they show that

$$B \subseteq A + C \ \Longleftrightarrow\ \text{for some } k \in \operatorname{int} C: \ Z^{C,k}(A, B) \leq 0,$$

compare [3, Thm. 3.10]. We will show in the proceeding section that it is not necessary to introduce such a functional $Z^{C,k}$ to characterize the set order relation
$B \subseteq A + C$. Instead, it is possible to use the given functional $z^{C,k}$ (see (2)) in
its traditional form to characterize the relation $B \subseteq A + C$. This approach even
enables us to present a full characterization of other set order relations known
from the literature, as the upper set less order relation and the set less order
relation. Of course, the functional $z^{C,k}$ in (2) has an easier structure than the
functional $Z^{C,k}$. Furthermore, the functional (2) has nice geometrical interpretations and useful continuity properties that are important for deriving optimality
condition (see [8, Chapter 5]).

Jahn [7] showed that

$$A \subseteq B - C \iff \forall\, l \in C^* \setminus \{0\} : \sup_{a \in A} l(a) \le \sup_{b \in B} l(b) \tag{5}$$

and

$$A \subseteq B + C \iff \forall\, l \in C^* \setminus \{0\} : \inf_{b \in B} l(b) \le \inf_{a \in A} l(a)$$

if the sets $B - C$ and $B + C$ are closed and convex, where C^* is the dual cone
of C.

We will show in this paper that is is possible to use the functional $z^{C,k}$ to
characterize the relations $A \subseteq B - C$ and $A \subseteq B + C$ without any convexity
assumptions on the considered sets.

3   Characterizations of Set Order Relations by Means of the Nonlinear Scalarization Functional $z^{C,k}$

A well known set order relation is the upper set less order relation introduced
by Kuroiwa [10,9].

Definition 2 (Upper set less order relation, [10,9]). Let Y be a linear
topological space and let $C \subset Y$ be a proper closed pointed convex cone. The
upper set less order relation $\preceq_C^u$ is defined for two sets $A, B \in \mathcal{P}(Y)$ by

$$A \preceq_C^u B :\iff A \subseteq B - C,$$

which is equivalent to

$$\forall\, a \in A\, \exists\, b \in B : a \le_C b.$$

The following theorem shows a first connection between the upper set less
order relation and the nonlinear scalarizing functional $z^{C,k}$.

Theorem 2. *Let Assumption 1 be fulfilled. Then we have the implication*

$$A \subseteq B - C \implies \forall\, k \in \operatorname{int} C : \sup_{a \in A} z^{C,k}(a) \le \sup_{b \in B} z^{C,k}(b).$$

(a)   (b)

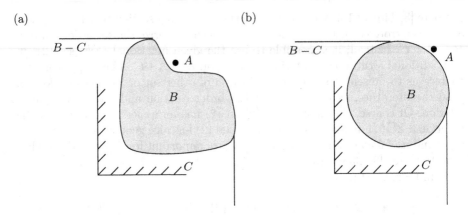

Fig. 2. Illustration of the set $B - C$ and the single-valued set $A = \{a\}$

Proof. Let $A \subseteq B - C$. This corresponds to

$$\forall\, a \in A\ \exists\, b \in B:\ a \in b - C.$$

Now choose a vector $k \in \operatorname{int} C$ arbitrarily, but fixed. Because $z^{C,k}$ is C-monotone (see (d) in Theorem 1), we have

$$\forall\, a \in A\ \exists\, b \in B:\ z^{C,k}(a) \le z^{C,k}(b),$$

resulting in

$$\sup_{a \in A} z^{C,k}(a) \le \sup_{b \in B} z^{C,k}(b).$$

$\square$

The inverse direction in Theorem 2 is generally not fulfilled, as the following example illustrates.

Example 1. Consider the special case $C = \mathbb{R}_+^2$, a single-valued set $A = \{a\}$ and the set B in Figure 2 (a). Apparently, we have $z^{C,k}(a) \le \sup_{b \in B} z^{C,k}(b)$ for every $k \in \operatorname{int} C$, but obviously $A \nsubseteq B - C$. The same holds if the set $B - C$ is convex, as the illustration in Figure 2 (b) shows.

We can, however, express the inclusion $A \subseteq B - C$ for two arbitrary sets $A, B \in \mathcal{P}(Y)$ by using the nonlinear scalarization functional $z^{C,k}$, as the following theorem verifies.

Theorem 3. *Let Assumption 1 be fulfilled. Then*

$$A \subseteq B - C \quad \Longleftrightarrow \quad \forall\, k \in \operatorname{int} C:\ \sup_{a \in A}\ \inf_{b \in B} z^{C,k}(a - b) \le 0 \qquad (6)$$

$$\Longleftrightarrow \quad \sup_{k \in \operatorname{int} C}\ \sup_{a \in A}\ \inf_{b \in B} z^{C,k}(a - b) \le 0. \qquad (7)$$

Proof. Choose an arbitrary $k \in \operatorname{int} C$ and let $A \subseteq B - C$. This corresponds to

$$\forall\, a \in A \ \exists\, b \in B : \ a \in b - C,$$

which is equivalent to for all $a \in A$ there exists some $b \in B$ such that $a - b \in -C$. Because of Theorem 1 (f) with $r = 0$ and $y = a - b$, we have

$$\forall\, a \in A \ \exists\, b \in B : \ a - b \in -C$$
$$\Longleftrightarrow \forall\, a \in A \ \exists\, b \in B : \ z^{C,k}(a - b) \le 0,$$

and this implies

$$\sup_{a \in A} \ \inf_{b \in B} \ z^{C,k}(a - b) \le 0.$$

Conversely, let for all $k \in \operatorname{int} C$ $\sup_{a \in A} \inf_{b \in B} z^{C,k}(a - b) \le 0$, and suppose that $A \not\subseteq B - C$. Thus, there exists some $a \in A$ such that $\{a\} \cap (B - C) = \emptyset$. This is equivalent to

$$\exists\, a \in A \ \forall\, b \in B : \ \{a - b\} \cap (-C) = \emptyset.$$

Now consider the functional $z^{C,k}$ with an arbitrary vector $k \in \operatorname{int} C$. Due to Theorem 1 (f) with $r = 0$ and $y = a - b$, we obtain

$$\exists\, a \in A \ \forall\, b \in B : \ z^{C,k}(a - b) > 0,$$

resulting in

$$\sup_{a \in A} \ \inf_{b \in B} \ z^{C,k}(a - b) > 0,$$

in contradiction to the assumption. The second equivalence (7) is obvious.   □

From Theorem 3 we directly deduce that

$$\forall\, k \in \operatorname{int} C : \quad \sup_{a \in A, b \in B} z^{C,k}(a - b) \le 0$$

implies that $A \subseteq B - C$.

Example 2. Consider again Example 1 with $C = \mathbb{R}^2_+$. We can see in Figure 3 that for all $b \in B$, $z^{C,k}(a - b) > 0$, and thus $\inf_{b \in B} z^{C,k}(a - b) > 0$. Due to Theorem 3, this is equivalent to $\{a\} \not\subseteq B - C$, as it was assumed here.

Example 3. Consider Example 2 with $C = \mathbb{R}^2_+$, but now we assume that $A = \{a\} \subseteq B - C$. We can see in Figure 4 that for all $b \in B$, $z^{C,k}(a - b) \le 0$, and thus $\inf_{b \in B} z^{C,k}(a - b) \le 0$.

From Theorem 3 we get directly the following characterization of the upper set less order relation.

Corollary 1. *Let Assumption 1 be satisfied. Then*

$$A \preceq^u_C B \quad \Longleftrightarrow \quad \forall\, k \in \operatorname{int} C : \ \sup_{a \in A} \ \inf_{b \in B} z^{C,k}(a - b) \le 0$$

$$\Longleftrightarrow \quad \sup_{k \in \operatorname{int} C} \ \sup_{a \in A} \ \inf_{b \in B} z^{C,k}(a - b) \le 0.$$

Furthermore, we study the lower set less order relation (see Kuroiwa [10,9]).

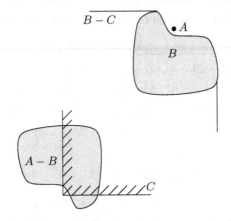

Fig. 3. Illustration of the set $A - B$, where $A = \{a\}$. It is obvious that for all $k \in \operatorname{int} C$, $\inf_{b \in B} \inf\{t \in \mathbb{R} \mid a - b \in tk - C\} > 0$.

Definition 3 (Lower set less order relation, [10,9]). Let Y be a linear topological space and let $C \subset Y$ be a proper closed pointed convex cone. The **lower set less order relation** $\preceq^l_C$ is defined for two sets $A, B \in \mathcal{P}(Y)$ by

$$A \preceq^l_C B :\Longleftrightarrow B \subseteq A + C,$$

which is equivalent to

$$\forall\, b \in B\, \exists\, a \in A : a \leq_C b.$$

Below we show how the set relation $B \subseteq A + C$ corresponds to the functional $z^{C,k}$. We refrain from giving the proofs of the following Theorems 4 and 5, since they can be deduced in a similar way as the proofs of Theorems 2 and 3.

Theorem 4. *Let Assumption 1 be fulfilled. Then*

$$B \subseteq A + C \implies \forall\, k \in \operatorname{int} C : \inf_{a \in A} z^{C,k}(a) \leq \inf_{b \in B} z^{C,k}(b).$$

Theorem 5. *Let Assumption 1 be fulfilled. Then*

$$B \subseteq A + C \quad \Longleftrightarrow \quad \forall\, k \in \operatorname{int} C : \sup_{b \in B}\, \inf_{a \in A} z^{C,k}(a - b) \leq 0$$

$$\Longleftrightarrow \quad \sup_{k \in \operatorname{int} C}\, \sup_{b \in B}\, \inf_{a \in A} z^{C,k}(a - b) \leq 0.$$

Theorem 5 directly yields the following characterization for the lower set less order relation.

Corollary 2. *Let Assumption 1 be satisfied. Then*

$$A \preceq^l_C B \quad \Longleftrightarrow \quad \forall\, k \in \operatorname{int} C : \sup_{b \in B}\, \inf_{a \in A} z^{C,k}(a - b) \leq 0$$

$$\Longleftrightarrow \quad \sup_{k \in \operatorname{int} C}\, \sup_{b \in B}\, \inf_{a \in A} z^{C,k}(a - b) \leq 0.$$

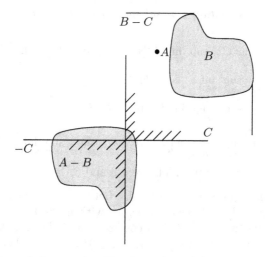

Fig. 4. Illustration of the set $A - B$, where $A = \{a\}$. It can be seen that for all $k \in \text{int } C$, $\inf_{b \in B} \inf\{t \in \mathbb{R} | a - b \in tk - C\} \leq 0$.

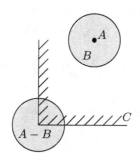

Fig. 5. Illustration of the set $A - B = B - A$, where $A = \{a\} \subseteq B$

Example 4. Here we set $C = \mathbb{R}^2_+$. Consider the sets $\{a\} = A \subset B \in \mathcal{P}(Y)$ in Figure 5, and we deduce $A \subseteq B - C$ and $A \subseteq B + C$. Then $A - B = B - A$ and we conclude with

$$\forall\, k \in \text{int } C : \ \inf_{b \in B} z^{C,k}(a - b) \leq 0 \text{ and } \inf_{b \in B} z^{C,k}(b - a) \leq 0,$$

in correspondence with Theorems 3 and 5.

Definition 4 (Set less order relation, [13,11]). Let Y be a linear topological space and let $C \subset Y$ be a proper closed pointed convex cone. The **set less order relation** $\preceq^s_C$ is defined for two sets $A, B \in \mathcal{P}(Y)$ by

$$A \preceq^s_C B :\Longleftrightarrow A \subseteq B - C \text{ and } A + C \supseteq B,$$

being equivalent to

$$(\forall\, a \in A \,\exists\, b \in B : a \leq_C b) \text{ and } (\forall\, b \in B \,\exists\, a \in A : a \leq_C b).$$

Corollary 3. *Let Assumption 1 be satisfied. Then we have*

$$A \preceq^s_C B \iff \forall\, k \in \operatorname{int} C : \sup_{a \in A} \inf_{b \in B} z^{C,k}(a-b) \leq 0 \text{ and } \sup_{b \in B} \inf_{a \in A} z^{C,k}(a-b) \leq 0$$

$$\iff \sup_{k \in \operatorname{int} C} \sup_{a \in A} \inf_{b \in B} z^{C,k}(a-b) \leq 0$$

$$\text{and } \sup_{k \in \operatorname{int} C} \sup_{b \in B} \inf_{a \in A} z^{C,k}(a-b) \leq 0$$

$$\iff \max\left(\sup_{k \in \operatorname{int} C} \sup_{a \in A} \inf_{b \in B} z^{C,k}(a-b),\ \sup_{k \in \operatorname{int} C} \sup_{b \in B} \inf_{a \in A} z^{C,k}(a-b) \right) \leq 0.$$

Proof. The assertions follow from Theorem 3 and 5. □

Example 5 (Weighted Sum Scalarization). Let $Y := \mathbb{R}^m$, let A, B be closed bounded subsets in $\mathbb{R}^m$, $w := (w_1, \ldots, w_m)^T$ with $w_i > 0$, $i = 1, \ldots, m$, $C := \{y \in \mathbb{R}^m \mid w^T y \geq 0\}$ (note that C is a convex cone; indeed, a closed half space and thus C is not pointed) and $k := (k_1, \ldots, k_m)^T \in \operatorname{int} C$. Then we have

$$
\begin{aligned}
z^{C,k}(a-b) &= \inf\{t \in \mathbb{R} \mid (a-b) \in tk - C\} \\
&= \inf\{t \in \mathbb{R} \mid w^T(a-b) \leq w^T(tk)\} \\
&= \inf\{t \in \mathbb{R} \mid w^T(a-b) \leq t \cdot (w^T k)\} \\
&\overset{k \in \operatorname{int} C}{=} \inf\{t \in \mathbb{R} \mid \frac{1}{w^T k} \cdot \sum_{i=1}^m w_i(a_i - b_i) \leq t\} \\
&= \frac{1}{w^T k} \cdot \sum_{i=1}^m w_i(a_i - b_i).
\end{aligned}
$$

This leads to

$$
\begin{aligned}
\sup_{a \in A} \inf_{b \in B} z^{C,k}(a-b) &= \sup_{a \in A} \inf_{b \in B} \frac{1}{w^T k} \cdot \sum_{i=1}^m w_i(a_i - b_i) \\
&= \sup_{a \in A} \frac{1}{w^T k} \cdot \sum_{i=1}^m w_i a_i - \sup_{b \in B} \frac{1}{w^T k} \cdot \sum_{i=1}^m w_i b_i \\
&= \frac{1}{w^T k} \cdot \left(\sup_{a \in A} \sum_{i=1}^m w_i a_i - \sup_{b \in B} \sum_{i=1}^m w_i b_i \right).
\end{aligned}
$$

Hence, with the above definitions of C and k and different weights $w_i > 0$, $i = 1, \ldots, m$, condition (6) is equivalent to

$$A \subseteq B - C \iff \forall\, k \in \operatorname{int} C : \frac{1}{w^T k} \sup_{a \in A} \sum_{i=1}^m w_i a_i \leq \frac{1}{w^T k} \sup_{b \in B} \sum_{i=1}^m w_i b_i$$

$$\iff \sup_{a \in A} \sum_{i=1}^m w_i a_i \leq \sup_{b \in B} \sum_{i=1}^m w_i b_i$$

This example shows that the condition (6) coincides with the equivalence (5) in the special case $C = \{y \in \mathbb{R}^m \mid w^T y \geq 0\}$, because the sets $A - C$ and $B - C$ are closed and convex (compare the assertion in [7, Remark 1]).

Example 6 (Natural Ordering). Let Assumption 1 be satisfied. Specifically, let $Y := \mathbb{R}^m$, $C := \mathbb{R}_+^m$ and $k := (k_1, \ldots, k_m)^T \in \operatorname{int} C$. Then we have

$$
\begin{aligned}
z^{C,k}(a - b) &= \inf\{t \in \mathbb{R} \mid (a - b) \in tk - C\} \\
&= \inf\{t \in \mathbb{R} \mid (a - b) - tk \in -C\} \\
&= \inf\{t \in \mathbb{R} \mid \forall\, i = 1, \ldots, m : (a - b)_i - tk_i \leq 0\} \\
&= \inf\{t \in \mathbb{R} \mid \forall\, i = 1, \ldots, m : \frac{(a - b)_i}{k_i} \leq t\} \\
&= \sup_{i=1,\ldots,m} \frac{(a - b)_i}{k_i}.
\end{aligned}
$$

Hence, with the above definitions of C and k and different weights $w_i > 0$, $i = 1, \ldots, m$, condition (6) is equivalent to

$$
A \subseteq B - C \iff \forall\, k \in \operatorname{int} C : \sup_{a \in A} \inf_{b \in B} \sup_{i=1,\ldots,m} \frac{(a - b)_i}{k_i} \leq 0.
$$

4   Characterization of Minimal Elements of Set Optimization Problems by Means of the Nonlinear Scalarizing Functional $z^{C,k}$

Set optimization deals with finding feasible solution sets that are not dominated by another feasible set w.r.t. a certain set order relation. Here we use the pre-orders $\preceq_C^u$, $\preceq_C^l$, and $\preceq_C^s$ introduced in Definitions 2, 3 and 4.

Thus, here it is our goal to obtain minimal solutions of a system of nonempty subsets of Y. The following definition introduces minimal solutions (see [8, Chapter 2] and references therein).

Definition 5 (Minimal Solutions). Let $\mathcal{A}$ be a system of nonempty subsets of Y and let $\preceq$ be a preorder. $\overline{A}$ is called a **minimal element** of $\mathcal{A}$ w.r.t. $\preceq$ if

$$
A \preceq \overline{A}, \ A \in \mathcal{A} \implies \overline{A} \preceq A.
$$

Corollary 4. *Let $\mathcal{A}$ be a system of nonempty subsets of Y. $\overline{A}$ is a minimal element of $\mathcal{A}$ w.r.t. $\preceq_C^u$ if there does not exist any $A \in \mathcal{A}$ such that*

$$
\forall\, k \in \operatorname{int} C : \sup_{a \in A} \inf_{\overline{a} \in \overline{A}} z^{C,k}(a - \overline{a}) \leq 0 \text{ and}
$$

$$
\exists\, \widehat{k} \in \operatorname{int} C : \sup_{\overline{a} \in \overline{A}} \inf_{a \in A} z^{C,\widehat{k}}(\overline{a} - a) > 0.
$$

Proof

$\overline{A}$ is minimal element of $\mathcal{A}$ w.r.t. $\preceq_C^u$

$\Longleftrightarrow A \preceq_C^u \overline{A}, \ A \in \mathcal{A} \implies \overline{A} \preceq_C^u A$

$\Longleftrightarrow \left(A \in \mathcal{A} \text{ and } \forall k \in \text{int } C : \ \sup_{a \in A} \ \inf_{\overline{a} \in \overline{A}} z^{C,k}(a - \overline{a}) \leq 0 \right.$

$\implies \forall k \in \text{int } C : \ \sup_{\overline{a} \in \overline{A}} \ \inf_{a \in A} z^{C,k}(\overline{a} - a) \leq 0 \Big)$

$\Longleftrightarrow \nexists A \in \mathcal{A} : \left(\forall k \in \text{int } C : \ \sup_{a \in A} \ \inf_{\overline{a} \in \overline{A}} z^{C,k}(a - \overline{a}) \leq 0 \right.$

$\text{and } \exists \, \widehat{k} \in \text{int } C : \ \sup_{\overline{a} \in \overline{A}} \ \inf_{a \in A} z^{C,\widehat{k}}(\overline{a} - a) > 0 \Big).$

$\square$

The following corollary can be established in a similar manner as Corollary 4.

Corollary 5. *Let $\mathcal{A}$ be a system of nonempty subsets of Y. $\overline{A}$ is a minimal element of $\mathcal{A}$ w.r.t. $\preceq_C^l$ if there does not exist any $A \in \mathcal{A}$ such that*

$$\forall k \in \text{int } C : \ \sup_{\overline{a} \in \overline{A}} \ \inf_{a \in A} z^{C,k}(a - \overline{a}) \leq 0 \ \text{and}$$

$$\exists \, \widehat{k} \in \text{int } C : \ \sup_{a \in A} \ \inf_{\overline{a} \in \overline{A}} z^{C,\widehat{k}}(\overline{a} - a) > 0.$$

Finally, we are able to present a characterization of minimal elements of a system of sets w.r.t. the set less order relation. This result is deduced from Corollaries 4 and 5.

Corollary 6. *Let $\mathcal{A}$ be a system of nonempty subsets of Y. $\overline{A}$ is a minimal element of $\mathcal{A}$ w.r.t. $\preceq_C^s$ if there does not exist any $A \in \mathcal{A}$ such that*

$$\forall k \in \text{int } C : \ \sup_{a \in A} \ \inf_{\overline{a} \in \overline{A}} z^{C,k}(a - \overline{a}) \leq 0 \ \text{and} \ \sup_{\overline{a} \in \overline{A}} \ \inf_{a \in A} z^{C,k}(a - \overline{a}) \leq 0 \ \text{and}$$

$$\exists \, \widehat{k} \in \text{int } C : \ \sup_{\overline{a} \in \overline{A}} \ \inf_{a \in A} z^{C,\widehat{k}}(\overline{a} - a) > 0 \ \text{or} \ \sup_{a \in A} \ \inf_{\overline{a} \in \overline{A}} z^{C,\widehat{k}}(\overline{a} - a) > 0.$$

5   Conclusion

In this paper we propose a unifying approach for representing set order relations via a nonlinear scalarizing functional. It would be interesting to investigate whether other set order relations that are known in the literature can be comprised by this concept as well. The characterization of set order relations via scalarization can be used for deriving optimality conditions and corresponding numerical methods for solving set-valued optimization problems.

Acknowledgments. The authors would like to thank two anonymous referees for their thorough advice and valuable comments which contributed to improving the manuscript significantly.

References

1. Gerth (Tammer), C., Weidner, P.: Nonconvex separation theorems and some applications in vector optimization. J. Optim. Theory Appl. 67, 297–320 (1990)
2. Göpfert, A., Riahi, H., Tammer, C., Zălinescu, C.: Variational Methods in Partially Ordered Spaces. Springer, New York (2003)
3. Hernández, E., Rodríguez-Marín, L.: Nonconvex scalarization in set optimization with set-valued maps. J. Math. Anal. Appl. 325(1), 1–18 (2007)
4. Heyde, F.: Coherent risk measures and vector optimization. In: Küfer, K.-H., et al. (eds.) Multicriteria Decision Making and Fuzzy Systems. Theory, Methods and Applications, pp. 3–12. Shaker Verlag, Aachen (2006)
5. Ide, J., Köbis, E., Kuroiwa, D., Schöbel, A., Tammer, C.: The relationship between multi-objective robustness concepts and set-valued optimization. Fixed Point Theory Appl. 83 (2014)
6. Jahn, J.: Vector Optimization - Introduction, Theory, and Extensions, 2nd edn. Springer, Heidelberg (2011)
7. Jahn, J.: Vectorization in set optimization. J. Optim. Theory Appl., 1–13 (2013)
8. Khan, A.A., Tammer, C., Zălinescu, C.: Set-Valued Optimization – An Introduction with Applications. Springer, Berlin (2015)
9. Kuroiwa, D.: Some duality theorems of set-valued optimization with natural criteria. In: Proceedings of the International Conference on Nonlinear Analysis and Convex Analysis, pp. 221–228. World Scientific (1999)
10. Kuroiwa, D.: The natural criteria in set-valued optimization. Sūrikaisekikenkyūsho Kōkyūroku (1031), 85–90 (1997), Research on nonlinear analysis and convex analysis, Kyoto
11. Nishnianidze, Z.G.: Fixed points of monotone multivalued operators. Soobshch. Akad. Nauk Gruzin. SSR 114(3), 489–491 (1984)
12. Weidner, P.: Ein Trennungskonzept und seine Anwendung auf Vektoroptimierungsverfahren. Martin-Luther-Universität Halle-Wittenberg (Dissertation B) (1990)
13. Young, R.C.: The algebra of many-valued quantities. Math. Ann. 104(1), 260–290 (1931)
14. Zeidler, E.: Nonlinear Functional Analysis and its Applications III. Springer, New York (1985)

Scalarization of Set-Valued Optimization Problems in Normed Spaces

César Gutiérrez[1], Bienvenido Jiménez[2], Enrico Miglierina[3], and Elena Molho[4]

[1] Departamento de Matemática Aplicada, E.T.S. de Ingenieros de Telecomunicación,
Universidad de Valladolid, Paseo de Belén 15,
Campus Miguel Delibes, 47011 Valladolid, Spain
cesargv@mat.uva.es
[2] Departamento de Matemática Aplicada, E.T.S.I. Industriales, UNED,
c/ Juan del Rosal 12, Ciudad Universitaria, 28040 Madrid, Spain
bjimenez@ind.uned.es
[3] Dipartimento di Discipline Matematiche, Finanza Matematica ed Econometria,
Università Cattolica del Sacro Cuore, via Necchi 9, 20123 Milano, Italy
enrico.miglierina@unicatt.it
[4] Dipartimento di Scienze Economiche e Aziendali,
Università degli Studi di Pavia, via S. Felice 5, 27100 Pavia, Italy
molhoe@eco.unipv.it

Abstract. This work focuses on scalarization processes for nonconvex set-valued optimization problems whose solutions are defined by the so-called l-type less order relation, the final space is normed and the ordering cone is not necessarily solid. A scalarization mapping is introduced, which generalizes the well-known oriented distance, and its main properties are stated. In particular, by choosing a suitable norm it is shown that it coincides with the generalization of the so-called Tammer-Weidner nonlinear separation mapping to this kind of optimization problems. After that, two concepts of solution are characterized in terms of solutions of associated scalar optimization problems defined through the new scalarization mapping.

Keywords: Set-valued optimization, l-type less order relation, minimal solution, strict minimal solution, scalarization, oriented distance, optimality conditions.

1 Introduction

Roughly speaking, an optimization problem is said to be "set-valued" if its objective mapping is set-valued. They are a natural generalization of the vector optimization problems. In [15] and the references therein, the reader can find a complete description on this type of optimization problems.

There exist in the literature two approaches to solve a set-valued optimization problem: the vector criterion and the set criterion. The first one considers the minimal boundary of the whole image set defined by the feasible set and the objective mapping (see [3,14,15,17]). The second one is based on minimality

© Springer International Publishing Switzerland 2015   505
H.A. Le Thi et al. (eds.), *Model. Comput. & Optim. in Inf. Syst. & Manage. Sci.*,
Advances in Intelligent Systems and Computing 359, DOI: 10.1007/978-3-319-18161-5_43

notions defined by quasi orders on the values (sets) of the objective mapping. In this paper we focuses on the second approach.

During the last decade, important results of nonconvex vector optimization have been extended to set-valued optimization with the set criterion. In particular, the scalarization processes have been extensively studied (see [1,9,10,11,12,15,16,19,21]). However, the attained results cannot be applied in problems whose ordering cone is not solid, i.e., its topological interior is empty. As it is well-known, the natural ordering of some important spaces (for instance, the spaces ℓ^p and L^p, $1 \leq p < +\infty$) have this drawback, and so it is very important for applications to overcome this handicap.

In vector optimization problems whose final space is normed, one can overcome the problem by considering the so-called oriented distance (see [13,15,20]). The main objective of this work is to generalize this scalarization scheme to set-valued optimization.

The paper is structured as follows. Section 2 collects the main notations, concepts and mathematical tools used in the sequel. In particular, the set-valued optimization problem and two optimality concepts are fixed, and some well-known scalarization mappings of vector optimization are recalled, in order to illustrate their extension to set-valued optimization. In Section 3, a scalarization mapping for set-valued optimization problems is introduced, which works for not necessarily solid ordering cones. Its main properties are derived, and its relation with the generalization of the so-called Tammer-Weidner nonlinear separation mapping to set-valued optimization is stated. Finally, in Section 4, some optimality conditions in set-valued optimization are obtained via the new scalarization scheme.

2 Preliminaries

Let (Y,p) be a normed space. The topological interior and the closure of a set $M \subset Y$ are denoted by int M and cl M, respectively. Let $K \subset Y$ be the ordering cone of Y, which is assumed to be convex and proper ($\{0\} \neq \text{cl } K \neq Y$). In the sequel we deal with the following set-valued optimization problem:

$$\text{Min}\{F(x) : x \in S\}, \tag{1}$$

where $F : X \to 2^Y$, X is an arbitrary decision space and $\emptyset \neq S \subset X$. We denote by dom$F$ and ImF the domain and the image of F, i.e.,

$$\text{dom}F = \{x \in X : F(x) \neq \emptyset\}, \quad \text{Im}F = \{F(x) : x \in \text{dom}F\}. \tag{2}$$

We assume that F is proper in S (i.e., dom$F \cap S \neq \emptyset$), in order to deal with a nontrivial problem.

To study (1), we consider optimality concepts based on the next quasi order on 2^Y, which is called l-type less order relation:

$$A_1, A_2 \in 2^Y, \quad A_1 \lesssim A_2 \iff A_2 \subset A_1 + K. \tag{3}$$

If $f : X \to Y$ and $F = f$ (i.e., F is single-valued), then (1) is a usual vector optimization problem, and relation (3) applied to $\mathrm{Im}F$ reduces to the well-known ordering in linear spaces:

$$y_1, y_2 \in Y, \quad y_1 \leq_K y_2 \iff y_2 - y_1 \in K . \tag{4}$$

Definition 1. *It is said that $x_0 \in S$ is a minimal (resp. strict minimal) solution of (1), denoted by $x_0 \in \mathrm{Min}(F, S)$ (resp. $x_0 \in \mathrm{SMin}(F, S)$), if*

$$x \in S, \ F(x) \lesssim F(x_0) \Rightarrow F(x_0) \lesssim F(x) \tag{5}$$
$$(x \in S, \ F(x) \lesssim F(x_0) \Rightarrow x = x_0) . \tag{6}$$

It is clear that $\mathrm{SMin}(F, S) \subset \mathrm{Min}(F, S)$. Observe that (5) is equivalent to the following statement:

$$x \in S, \ F(x) \lesssim F(x_0) \Rightarrow F(x_0) + K = F(x) + K . \tag{7}$$

Let us recall the so-called Tammer-Weidner nonlinear separation mapping (see [5,6,15]): $\varphi_{e,b} : Y \to \mathbb{R} \cup \{\pm\infty\}$,

$$\varphi_{e,b}(y) = \inf\{t \in \mathbb{R} : y \leq_K te + b\}, \quad \forall y \in Y, \tag{8}$$

where $b \in Y$ and $e \in K\backslash(-\mathrm{cl}K)$.

It is well-known for researchers and practitioners in vector optimization that $\varphi_{e,b}$ is useful to scalarize a nonconvex vector optimization problem whenever $e \in \mathrm{int}\, K$. In other words, Tammer-Weidner mapping works as a scalarization tool for vector optimization problems whenever the ordering cone is solid, i.e., it has nonempty topological interior.

When the ordering cone is not solid and the final space is normed, it is possible to consider the so-called oriented distance (see [13,15,20]). Let us recall its definition. Given a set $M \subset Y$ and $y \in Y$, we denote

$$d(y, M) = \begin{cases} +\infty & \text{if } M = \emptyset \\ \inf_{z \in M}\{p(z - y)\} & \text{otherwise} \end{cases} . \tag{9}$$

Then, the oriented distance is the mapping $\delta_{p,b} : Y \to \mathbb{R}$,

$$\delta_{p,b}(y) = d(y - b, -K) - d(y - b, Y\backslash -K), \quad \forall y \in Y , \tag{10}$$

where $b \in Y$ (see [4] and the references therein for computations). This mapping can be considered as an extension of Tammer-Weidner mapping $\varphi_{e,b}$, since both of them coincide whenever the ordering cone K is pointed (i.e., $K \cap (-K) = \{0\}$), closed and solid. Specifically, we have the following result (see [2,3,14]).

Theorem 1. *Assume that K is pointed, closed and solid and let $e \in \mathrm{int}\, K$.*

1. *The mapping $p_e : Y \to \mathbb{R}$,*

$$p_e(y) = \inf\{t \geq 0 : y \in (-te + D) \cap (te - D)\}, \quad \forall y \in Y , \tag{11}$$

is a norm on Y.
2. *It follows that $\varphi_{e,b}(y) = \delta_{p_e,b}(y)$, for all $y \in Y$.*

3 Scalarization

In [11] (see also [1,10,7,12,16,19,21]), the scalarization mapping $\varphi_{e,b}$ has been generalized to the framework of set-valued optimization with the set criterion by the mapping $\Psi_{e,B} : 2^Y \to \mathbb{R} \cup \{\pm\infty\}$,

$$\Psi_{e,B}(A) = \begin{cases} +\infty & \text{if } \Lambda_{e,B}(A) = \emptyset \\ \inf\{t : t \in \Lambda_{e,B}(A)\} & \text{if } \Lambda_{e,B}(A) \neq \emptyset \end{cases}, \tag{12}$$

where $B \subset Y$, $B \neq \emptyset$, $e \in K\backslash(-\mathrm{cl}K)$ and

$$\Lambda_{e,B}(A) = \{t \in \mathbb{R} : A \lesssim te + B\}, \quad \forall A \in 2^Y. \tag{13}$$

Next, we introduce an extension of the scalarization $\delta_{p,b}$ to the same framework. In [7], the reader can find the proofs for the results formulated in this section.

Given two sets $A, B \subset Y$, $B \neq \emptyset$, we denote

$$d(B, A) = \inf_{b \in B} d(b, A), \quad \xi_B(A) = \begin{cases} +\infty & \text{if } A = \emptyset \\ \sup_{b \in B} d(b, A) & \text{otherwise} \end{cases}. \tag{14}$$

Let us observe that $\xi_B(A)$ define the excess of B over A (see [18]). Consider a nonempty set $B \subset Y$. We refer by generalized oriented distance with respect to B the mapping $\Delta_{p,B} : 2^Y \to \mathbb{R} \cup \{\pm\infty\}$,

$$\Delta_{p,B}(A) = \xi_B(A + K) - d(B, Y\backslash(A + K)), \quad \forall A \in 2^Y. \tag{15}$$

It is easy to check that $\Delta_{p,\{b\}}(\{y\}) = \delta_{p,b}(y)$, for all $b, y \in Y$. Moreover, the generalized oriented distance with respect to a set B coincides with $\Psi_{e,B}$ whenever the norm p_e is considered. Next theorem formulates this result, which generalizes Theorem 1.

Theorem 2. *Suppose that K is pointed, closed and solid and let $e \in \mathrm{int}\, K$. It follows that $\Delta_{p_e,B}(A) = \Psi_{e,B}(A)$, $\forall A \in 2^Y$.*

The following propositions collect the main properties of the scalarization mapping $\Delta_{p,B}$. The first one shows that $\Delta_{p,B}(A)$ can be defined by assuming that A, B coincide with their conical extensions.

Proposition 1. *Let $A \in 2^Y$. The next equalities are true.*

1. $\Delta_{p,B}(A) = \Delta_{p,B}(A + K) = \Delta_{p,B+K}(A) = \Delta_{p,B+K}(A + K)$.
2. $\Delta_{p,B}(A) = \Delta_{p,\mathrm{cl}(B+K)}(A)$.

With respect to the monotonicity and the sublevel set at zero of mapping $\Delta_{p,B+K}$ we have the following general results.

Proposition 2. *Let $A, A_1, A_2 \subset Y$.*

1. If $A_1 \lesssim A_2$ then $\Delta_{p,B}(A_1) \leq \Delta_{p,B}(A_2)$.
2. If $\Delta_{p,B}(A) < 0$ then $A \lesssim B$.
3. If $A \lesssim B$ then $\Delta_{p,B}(A) \leq 0$.

Some previous properties are clarified under additional assumptions, as it is shown in the next proposition. We say that a set $A \subset Y$ is K-proper if $A + K \neq Y$ (see [10]) and K-closed if $A + K$ is closed.

Proposition 3. *Let $A \subset Y$. The following properties hold:*

1. *If B is K-proper then $\Delta_{p,B}(B) = 0$.*
2. *If A is K-closed then $A \lesssim B \iff \Delta_{p,B}(A) \leq 0$.*
3. *If B is K-proper then $\Delta_{p,B}(A) < 0 \Rightarrow A \lesssim B$ and $A + K \neq B + K$.*

4 Optimality Conditions

In this last section we state minimality and strict minimality conditions through mapping $\Delta_{p,B}$. They are consequence of the properties obtained in Section 3. We say that mapping F is K-closed (resp. K-proper) valued on S if $F(x)$ is K-closed (resp. K-proper), for all $x \in S$. Let us recall that $x_0 \in S$ is a strict solution of the scalar optimization problem

$$\text{Min}\{h(x) : x \in S\}, \tag{16}$$

where $h : S \subset X \to \mathbb{R} \cup \{+\infty\}$, if $h(x_0) < h(x)$, for all $x \in S \backslash \{x_0\}$. The set of all solutions (resp. strict solutions) of problem (16) is denoted by $\text{Sol}(h, S)$ (resp. $\text{Str}(h, S)$).

Problem (1) is trivial whenever F is not K-proper valued on S, as it is shown in the following proposition.

Proposition 4. *Suppose that there exists $x_0 \in S$ such that $F(x_0)$ is not K-proper. Then*

$$\text{Min}(F, S) = \{x \in S : F(x) + K = Y\}, \tag{17}$$

$$\text{SMin}(F, S) = \begin{cases} \{x_0\} & \text{if } \text{Min}(F, S) = \{x_0\} \\ \emptyset & \text{otherwise} \end{cases}. \tag{18}$$

Proof. As $F(x_0) + K = Y$ it is clear that

$$F(x_0) \lesssim F(x), \quad \forall x \in S. \tag{19}$$

Thus, if $\bar{x} \in \text{Min}(F, S)$ then $F(\bar{x}) + K = F(x_0) + K = Y$ and

$$\text{Min}(F, S) \subset \{x \in S : F(x) + K = Y\}. \tag{20}$$

Reciprocally, let $x \in S$ such that $F(x) + K = Y$ and suppose that there exists $u \in S$ such that $F(u) \lesssim F(x)$. Then $F(x) + K \subset F(u) + K$ and so $F(u) + K = Y$. Thus $F(x) + K = F(u) + K$ and $x \in \text{Min}(F, S)$.

On the other hand, assume that $\text{Min}(F, S) = \{x_0\}$ and consider $x \in S$ such that $F(x) \lesssim F(x_0)$. Then $F(x) + K = F(x_0) + K = Y$ and so $x \in \text{Min}(F, S)$, i.e., $x = x_0$. Therefore, $x_0 \in \text{SMin}(F, S)$ and $\text{SMin}(F, S) = \{x_0\}$, since $\text{SMin}(F, S) \subset \text{Min}(F, S)$.

In the other case, assume that there exists $x \in \mathrm{Min}(F, S)\backslash\{x_0\}$ and suppose that $\mathrm{SMin}(F, S) \neq \emptyset$. Let $\bar{x} \in \mathrm{SMin}(F, S)$. Then $F(x) \lesssim F(\bar{x})$ in view of (17), and $F(x_0) \lesssim F(\bar{x})$ by (19). As $\bar{x}$ is a strict minimal solution of (1) it follows that $x = \bar{x}$ and $x_0 = \bar{x}$, and so $x = x_0$, that is a contradiction. This finishes the proof.   □

In the next two theorems we state minimality and strict minimality conditions through the generalized oriented distance, when F is K-proper valued on S. Let us observe that, from a practical point of view, the concept of strict minimal solution of problem (1) is suitable, since they can be characterized by scalarization. The same fact happens with the notion of weak minimal solution of problem (1) via the scalarization mapping $\Psi_{e,B}$, whenever the ordering cone K is solid (see [8]).

Theorem 3. *Assume that F is K-closed and K-proper valued on S, and consider $x_0 \in \mathrm{dom}F$. Then $x_0 \in \mathrm{SMin}(F, S)$ if and only if $x_0 \in \mathrm{Str}(\Delta_{p,F(x_0)} \circ F, S)$.*

Proof. It is not hard to check that

$$(\Delta_{p,F(x_0)} \circ F)(x) = -\infty \iff F(x) + K = Y. \qquad (21)$$

Therefore, $(\Delta_{p,F(x_0)} \circ F)(x) > -\infty$, for all $x \in S$.

Assume that $x_0 \in \mathrm{SMin}(F, S)$ and suppose by contradiction that there exists $x \in S\backslash\{x_0\}$ such that

$$(\Delta_{p,F(x_0)} \circ F)(x) \leq (\Delta_{p,F(x_0)} \circ F)(x_0) = 0, \qquad (22)$$

since $F(x_0)$ is K-proper (see part 1 of Proposition 3). As F is K-closed on S and $x \in S$ we have that $F(x)$ is K-closed, and by part 2 of Proposition 3 we deduce that $F(x) \lesssim F(x_0)$. Thus, $x_0 \notin \mathrm{SMin}(F, S)$, which is a contradiction.

Reciprocally, assume that

$$(\Delta_{p,F(x_0)} \circ F)(x) > (\Delta_{p,F(x_0)} \circ F)(x_0), \quad \forall x \in S\backslash\{x_0\}, \qquad (23)$$

and suppose by contradiction that $x_0 \notin \mathrm{SMin}(F, S)$. Then, there exists $u \in S\backslash\{x_0\}$ such that $F(u) \lesssim F(x_0)$. By part 2 of Proposition 3 we see that

$$(\Delta_{p,F(x_0)} \circ F)(u) = \Delta_{p,F(x_0)}(F(u)) \leq 0 = (\Delta_{p,F(x_0)} \circ F)(x_0), \qquad (24)$$

which is a contradiction. Then x_0 is a strict minimal solution of problem (1), and the proof finishes.   □

For each $x_0 \in S$ we denote

$$S(x_0) = \{x \in S : F(x) + K \neq F(x_0) + K\} \cup \{x_0\}. \qquad (25)$$

Theorem 4. *Assume that F is K-closed and K-proper valued on S, and let $x_0 \in \mathrm{dom}F$. The following statements are satisfied.*

1. If $x_0 \in \mathrm{Str}(\Delta_{p,F(x_0)} \circ F, S(x_0))$, then $x_0 \in \mathrm{Min}(F, S)$.
2. If $x_0 \in \mathrm{Min}(F, S)$, then $x_0 \in \mathrm{Sol}(\Delta_{p,F(x_0)} \circ F, S)$.

Proof. Part 1. Suppose that $x_0 \notin \mathrm{Min}(F, S)$. Then there exists $x \in S(x_0)\backslash\{x_0\}$ such that $F(x) \lesssim F(x_0)$. By parts 1 and 2 of Proposition 3 we deduce that $(\Delta_{p,F(x_0)} \circ F)(x) \leq (\Delta_{p,F(x_0)} \circ F)(x_0)$, which is a contradiction.

Part 2. Suppose that $x_0 \notin \mathrm{Sol}(\Delta_{p,F(x_0)} \circ F, S)$. Then there exists $x \in S$ such that

$$(\Delta_{p,F(x_0)} \circ F)(x) < (\Delta_{p,F(x_0)} \circ F)(x_0) = 0 . \tag{26}$$

By part 3 of Proposition 3, it follows that $F(x) \lesssim F(x_0)$ and $F(x) + K \neq F(x_0) + K$. Thus, $x_0 \notin \mathrm{Min}(F, S)$, that is a contradiction, and the proof is complete. □

Acknowledgments. The authors are very grateful to Prof. Q.T. Bao, C. Tammer and A. Soubeyran for your kind invitation to contribute in the special session "Variational Principles and Applications", and also to the anonymous referees for their helpful comments and suggestions.

This work was partially supported by Ministerio de Economía y Competitividad (Spain) under project MTM2012-30942. The fourth author was also partially supported by MIUR PRIN MISURA Project, 2013-2015, Italy.

References

1. Araya, Y.: Four Types of Nonlinear Scalarizations and Some Applications in Set Optimization. Nonlinear Anal. 75, 3821–3835 (2012)
2. Crespi, G.P., Ginchev, I., Rocca, M.: Points of Efficiency in Vector Optimization with Increasing-Along-Rays Property and Minty Variational Inequalities. Lecture Notes in Econom. and Math. Systems, vol. 583, pp. 209–226. Springer, Berlin (2007)
3. Chen, G.Y., Huang, X., Yang, X.: Vector Optimization. Set-Valued and Variational Analysis. Lecture Notes in Econom. and Math. Systems, vol. 541. Springer, Berlin (2005)
4. Dapogny, C., Frey, P.: Computation of the Signed Distance Function to a Discrete Contour on Adapted Triangulation. Calcolo 49, 193–219 (2012)
5. Gerth, C., Weidner, P.: Nonconvex Separation Theorems and Some Applications in Vector Optimization. J. Optim. Theory Appl. 67, 297–320 (1990)
6. Göpfert, A., Riahi, H., Tammer, C., Zălinescu, C.: Variational Methods in Partially Ordered Spaces. Springer, New York (2003)
7. Gutiérrez, C., Jiménez, B., Miglierina, E., Molho, E.: Scalarization in Set Optimization with Solid and Nonsolid Ordering Cones. J. Global Optim. 61, 525–552 (2015)
8. Gutiérrez, C., Jiménez, B., Novo, V.: Nonlinear Scalarizations of Set Optimization Problems with Set Orderings. In: Hamel, A., Heyde, F., Löhne, A., Rudloff, B., Schrage, C. (eds.) Set Optimization and Applications in Finance. The State of the Art. PROMS series. Springer, Berlin (forthcoming, 2015)
9. Gutiérrez, C., Jiménez, B., Novo, V., Thibault, L.: Strict Approximate Solutions in Set-Valued Optimization with Applications to the Approximate Ekeland Variational Principle. Nonlinear Anal. 73, 3842–3855 (2010)

10. Gutiérrez, C., Miglierina, E., Molho, E., Novo, V.: Pointwise Well-Posedness in Set Optimization with Cone Proper Sets. Nonlinear Anal. 75, 1822–1833 (2012)
11. Hamel, A., Löhne, A.: Minimal Element Theorems and Ekeland's Principle with Set Relations. J. Nonlinear Convex Anal. 7, 19–37 (2006)
12. Hernández, E., Rodríguez-Marín, L.: Nonconvex Scalarization in Set Optimization with Set-Valued Maps. J. Math. Anal. Appl. 325, 1–18 (2007)
13. Hiriart-Urruty, J.-B.: Tangent Cones, Generalized Gradients and Mathematical Programming in Banach Spaces. Math. Oper. Res. 4, 79–97 (1979)
14. Jahn, J.: Vector Optimization. Theory, Applications and Extensions. Springer, Berlin (2011)
15. Khan, A.A., Tammer, C., Zălinescu, C.: Set-Valued Optimization. An Introduction with Applications. Springer, Berlin (2015)
16. Kuwano, I., Tanaka, T., Yamada, S.: Unified Scalarization for Sets and Set-Valued Ky Fan Minimax Inequality. J. Nonlinear Convex Anal. 11, 513–525 (2010)
17. Luc, D.T.: Theory of Vector Optimization. Lecture Notes in Econom. and Math. Systems, vol. 319. Springer, Berlin (1989)
18. Lucchetti, R.: Convexity and Well-Posed Problems. Springer, New York (2006)
19. Maeda, T.: On Optimization Problems with Set-Valued Objective Maps: Existence and Optimality. J. Optim. Theory Appl. 153, 263–279 (2012)
20. Zaffaroni, A.: Degrees of Efficiency and Degrees of Minimality. SIAM J. Control Optim. 42, 1071–1086 (2003)
21. Zhang, W.Y., Li, S.J., Teo, K.L.: Well-Posedness for Set Optimization Problems. Nonlinear Anal. 71, 3769–3778 (2009)

Vectorial Ekeland Variational Principles: A Hybrid Approach

Q. Bao Truong

Department of Mathematics & Computer Science, Northern Michigan University,
Marquette, Michigan 49855, USA
btruong@nmu.edu

Abstract. This paper proposes a hybrid approach in establishing vectorial versions of Ekeland's variational principle. It bases on both the nonlinear scalarization functional in Tammer (Gerth) and Weidner's nonconvex separation theorem [14] from a scalarization approach and Bao and Mordukhovich's iterative scheme in [5] from a variational approach. Examples are provided to illustrate improvements of new results.

Keywords: Ekeland's variational principle, nonlinear separation theorem, metric, quasimetric, lower semicontinuity, decreasing closedness.

1 Introduction

Ekeland's variational principle [12] (which provides a characterization of complete metric spaces) illustrates a method for getting existence results in analysis without compactness. Let φ be a lower semi-continuous function defined on a complete metric space (X, d), with values in the extended line $\mathbb{R} \cup \{+\infty\}$, and bounded from below. Ekeland's basic principle asserts that there exists a slight perturbation of φ which attains its minimum on X. More precisely, there exists a point $\bar{x}$ such that $\varphi(x) + d(\bar{x}, x) > \varphi(\bar{x})$ for all $x \neq \bar{x}$; this says that the function has a strict minimum on X at $\bar{x}$.

This variational principle has several equivalent geometric formulations. It has been extended to vector-valued functions and set-valued mappings. To the best of our knowledge, there are two main approaches: (primal) scalarized and (dual) variational. The first approach bases on some scalarization technique to convert a given vector-valued/set-valued mapping into a scalar one. Doing so allows us to use the original Ekeland's variational principle. However, we might need some "more restrictive" assumptions imposed on ordering cones and vector-valued functions. Among others are: (1) the ordering cone has a nonempty interior and (2) the vector-valued funtion is bounded from below. These drawbacks were motivations for Bao and Mordukhovich to propose the second approach in [4,5]. It drops the nonempty interiority condition (1) while weakens the boundedness condition (2) to quasiboundedness from below.

This paper has a two-fold focus. First, we show that the nonempty interiority condition is not essential in the scalarization approach in the proof of Theorem 5.

© Springer International Publishing Switzerland 2015
H.A. Le Thi et al. (eds.), *Model. Comput. & Optim. in Inf. Syst. & Manage. Sci.*,
Advances in Intelligent Systems and Computing 359, DOI: 10.1007/978-3-319-18161-5_44

Then, we present a hybrid approach which uses both scalarization and variational techniques in order to further weaken the boundedness condition in the known vector-valued versions of Ekeland's variational principle.

The rest of the paper is organized as follows. Section 2 contains some basic definitions and preliminaries from nonlinear separation theorems and two vectorial versions of Ekeland's variational principles in scalarization and variational approaches. In the main Section 3, we establish new versions of Ekeland's variational principles which are better than the known results. Examples are provided to illustrate improvements.

2   Preliminaries

Let Z be a real linear topological space and C be a nonempty set in Z. The notations $int(C)$, $cl(C)$, and $bd(C)$ stand for the topological interior, the topological closure, and the topological boundary of the set C, respectively. The set C is said to be *solid* iff $int(C) \neq \emptyset$, *proper* iff $C \neq \emptyset$ and $C \neq Z$, *pointed* iff $C \cap (-C) \subset \{0\}$, and *a cone* iff $tc \in C$ for all $c \in C$ and $t \geq 0$. See [16,15,18,19] for basic definitions and concepts of vector optimization.

Let $\Theta \subset Z$ be a nonempty subset of a real topological linear space Z. We can define a binary relation $\leq_\Theta$ on Z by

$$z_1 \leq_\Theta z_2 \text{ if and only if } z_2 \in z_1 - \Theta.$$

It is well known that if $0 \in \Theta$, then $\leq_\Theta$ is *reflexive* [$\forall z \in Z : z \leq_\Theta z$], that if Θ is a convex cone, then $\leq_\Theta$ is *transitive* [$\forall z_1, z_2, z_3 \in Z : z_1 \leq_\Theta z_2 \wedge z_2 \leq_\Theta z_3 \implies z_1 \leq_\Theta z_3$], and that if Θ is pointed, then $\leq_\Theta$ is *antisymmetric* [$\forall z_1, z_2 \in Z : z_1 \leq_\Theta z_2 \wedge z_2 \leq_\Theta z_1 \implies z_1 = z_2$]. When $\Theta = \mathbb{R}_+^n$, the nonnegative orthant of $\mathbb{R}^n$, $\leq_\Theta$ is the Pareto ordering relation. When Θ is a convex cone, $\leq_\Theta$ is known as a generalized-Pareto order. Next, we recall the concept of efficiency with respect to $\leq_\Theta$.

Definition 1. *Let $\Xi, \Theta \subset Z$ be nonempty subsets of Z. An element $\bar{x} \in \Xi$ is said to be an* EFFICIENT *point of Ξ with respect to Θ if there exists no element $x \in \Xi$ such that $x \in \bar{x} - \Theta \setminus \{0\}$.*

*Assume that there is some element $k \in \Theta \setminus \{0\}$ such that $\Theta + [0, \infty) \cdot k \subset \Theta$. Given $\varepsilon > 0$ and set $\Theta_{\varepsilon k} := \varepsilon k + \Theta \setminus \{0\}$. An element $\bar{x} \in \Xi$ is said to be an εk-*EFFICIENT *point of Ξ with respect to Θ if there exists no element $x \in \Xi$ such that $x \in \bar{x} - \Theta_{\varepsilon k}$.*

We will denote the efficient point set of Ξ with respect to Θ by $\mathrm{Eff}\,(\Xi; \Theta)$ and the εk-efficient point set by $\mathrm{Eff}\,(\Xi; \Theta_{\varepsilon k})$. These efficiency concepts reduce to the known Pareto optimality/efficiency ones in vector optimization when Θ is a convex ordering cone of Z.

Next, let us recall a powerful nonlinear scalarization tool from [14]; cf. [15].

Let A be a nonempty subset of Z and $k \neq 0$ be a nonzero element of Z. The functional $s_{A,k} : Z \to \mathbb{R} \cup \{\pm\infty\}$ defined by

$$s_{A,k}(z) := \inf\{t \in \mathbb{R} \mid z \in tk - A\} \tag{1}$$

is called a *nonlinear (separating) scalarization* function (with respect to the set A and the direction k). The next lemma provides important properties of $s_{A,k}$.

Lemma 1. ([15, Theorem 2.3.1]) *Let $\emptyset \neq A \subset Z$ and $k \in Z \setminus \{0\}$ satisfy $A + [0, +\infty) \cdot k \subset A$. Then the following hold:*

(a) *The functional $s_{A,k}$ is l.s.c. over its domain $\operatorname{dom}\varphi_{A,k} = \mathbb{R}k - A$ iff A is a closed set. Moreover, its t-level set is given by $\{z \in Z \mid s_{A,k}(z) \leq t\} = tk - A, \ \forall t \in \mathbb{R}$ and the transformation of $\varphi_{A,k}$ along the direction k is calculated by $\varphi_{A,k}(z + tk) = \varphi_{A,k}(z) + t, \ \forall z \in Z, \ \forall t \in \mathbb{R}$.*

(b) *$s_{A,k}$ is convex if and only if the set A is convex, and $s_{A,k}$ is positively homogeneous, i.e. $s_{A,k}(tz) = ts_{A,k}(z)$ for all $t \geq 0$ and $z \in Z$, if and only if A is a cone.*

(c) *$s_{A,k}$ is proper if and only if A does not contain lines parallel to k, i.e. $\forall z \in Z, \ \exists t \in \mathbb{R} : z + tk \notin A$.*

(d) *$s_{A,k}$ is finite-valued, i.e. $\operatorname{dom} s_{A,k} = Z$, if and only if $\mathbb{R}k - A = Z$ and A does not contain lines parallel to k.*

(e) *Given $B \subset Z$. $s_{A,k}$ is B-monotone, i.e. $\big[a \in b - B \implies s_{A,k}(a) \leq s_{A,k}(b) \big]$ if and only if $A + B \subset A$.*

(f) *$s_{A,k}$ is subadditive if and only if $A + A \subset A$.*

Using the scalarization functional $s_{A,k}$, Tammer sucessfully established the next result.

Theorem 1. ([22, Theorem 4.1]) *Let $B, D \subset Z$ be proper subsets of Z and k be a nonzero element of D satisfying*

(T1) *B is an open convex subset of Z with $0 \in \operatorname{cl} B \setminus B$ and $Z = \operatorname{cl} B + \mathbb{R} \cdot k$,*
(T2) *$(0, \infty) \cdot k \subset D \setminus \{0\}$ and $0 \in \operatorname{cl} D \setminus D$,*
(T3) *$\operatorname{cl} B + (D \setminus \{0\}) \subset B$,*
(T4) *$\operatorname{bd} B + \operatorname{bd} B \subset \operatorname{cl} B$.*

Assume that $f : X \to Z$ is a (k, B)-lower semicontinuous and bounded from below on a Banach space X. Then, for every $\varepsilon > 0$ and for every $x_0 \in X$ satisfying $f(x_0) \in \operatorname{Eff}(f(X), B_{\varepsilon k})$, there exists $\bar{x} \in X$ such that

(i) *$f(\bar{x}) \in \operatorname{Eff}(f(X), D_{\varepsilon k})$,*
(ii) *$\|x_0 - \bar{x}\| \leq \sqrt{\varepsilon}$,*
(iii) *$f_{\varepsilon k}(\bar{x}) \in \operatorname{Eff}(f_{\varepsilon k}(X), B)$ with $f_{\varepsilon k}(x) := f(x) + \sqrt{\varepsilon}\|\bar{x} - x\|k$.*

It is important to emphasize that although the result was formulated for a Banach space X, its proof holds for any complete metric space (X, d) where d plays the role of the norm.

Remark 1. **(on the conical property of the sets B and D).** Any set B satisfying conditions (T1) and (T4) is a cone. Arguing by contradiction. Assume that B is not a cone. Then, there exist $b \in B$ and $t > 0$ such that $tb \notin B$. By (T1), i.e., $0 \in \operatorname{cl} B$ and B is a convex set, $\tau b \in B$ for all $\tau \in (0, 1]$. Set $\bar{t} := \inf\{\tau > 1 \mid \tau b \notin B\}$. We have $\bar{t}b \in \operatorname{bd} B$, and thus $2\bar{t}b \in \operatorname{bd} B$ due to (T4). Taking into account the convexity of the set B in (T1), the midpoint of $2\bar{t}b \in \operatorname{bd} B$ and $b \in B$ belongs to B, i.e., $(1/2 + \bar{t})b \in B$ which contradics the choice of $\bar{t}$ and thus verifies the conical property of the set B.

In [4,5] Bao and Mordukhovich proposed a new approach, which is now known as the variational approach in establishing versions of Ekeland's variational principle for vector-valued functions (indeed, set-valued mappings) without performing any scalarization technique. One significant feature of their approach is that the ordering cone does not necessarily have a nonempty interior. Below is a vectorial version obtained in this direction; cf. [2,3,4,5,6,7,8].

Theorem 2. *Let $f : (X,d) \to Z$ be a vector-valued function acting from a complete metric space (X,d) to a real linear spaces Z which is ordered by a proper, closed, and convex cone $\Theta \subset Z$ with $\Theta \setminus (-\Theta) \neq \emptyset$, i.e., Θ is not a linear subspace of Z. Assume furthermore that f is quasibounded from below, i.e., there is a bounded set M such that $f(x) \in M + \Theta$ for all $x \in X$, and is decreasingly closed in the sense that for any convergent sequence $\{x_n\}$ with the limit $\bar{x}$, if $f(x_{n+1}) \leq_\Theta f(x_n)$ for all $n \in \mathbb{N}$, then $f(\bar{x}) \leq_\Theta f(x_n)$ for all $n \in \mathbb{N}$. Then for any $\lambda > 0$, $k \in \Theta \setminus (-\Theta)$, and $x_0 \in X$, there is $\bar{x} \in X$ satisfying*

(i) $f(\bar{x}) + \lambda d(x_0, \bar{x})k \leq_\Theta f(x_0)$.
(ii) $f(x) + \lambda d(\bar{x}, x)k \not\leq_\Theta f(\bar{x})$ *for all* $x \neq \bar{x}$.

If furthermore x_0 is an εk-efficient solution of f, i.e., $x_0 \in \mathrm{Eff}(f(X); \Theta_{\varepsilon,k})$, then $\bar{x}$ can be chosen such that in addition to (i) and (ii) with ε/λ instead of λ we have **(iii)** $d(x_0, \bar{x}) \leq \lambda$.

Note that Theorem 2 is valid for a broader class of quasimetric spaces with notions of convergence, closedness, limit, completeness, and separation in Definitions 2–7; cf. [3,6,7,8].

Definition 2. *A functional $q : X \times X \to \mathbb{R}$ is said to be a* QUASIMETRIC *iff it satisfies* (q1) $q(x,y) \geq 0$ *(nonnegativity);* (q2) *if $x = y$, then $q(x,y) = 0$ (equality implies indistancy);* (q3) $q(x,z) \leq q(x,y) + q(y,z)$ *(triangularity).*
If a quasimetric q enjoys also (q4) $q(x,y) = q(y,x)$ *(symmetry), it is a* METRIC.

A quasimetric space X with a quasimetric q will be denoted by (X, q) and a metric space X with a metric d will be denoted by (X, d). For example, a quasimetric on the real numbers can be defined by $q(x,y) = x - y$ if $x \geq y$ and 1 otherwise.

Definition 3. *For a quasimetric space (X, q), a sequence $\{x_n\}$ in X is said to be* LEFT-SEQUENTIALLY CONVERGENT *to a point $x_* \in X$, denoted by $x_n \to x_*$, iff the quasidistances $q(x_n, x_*)$ tend to zero as $n \to \infty$, i.e.* $\lim_{n\to\infty} q(x_n, x_*) = 0$.

Definition 4. *For a quasimetric space (X, q), a subset $\Omega \subset X$ is said to be* LEFT-SEQUENTIALLY CLOSED *iff for any sequence $\{x_n\} \subset X$ converging to $x_* \in X$, the limit x_* belongs to Ω.*

Definition 5. *For a quasimetric space (X, q), a sequence $\{x_n\} \subset X$ is said to be* LEFT-SEQUENTIAL CAUCHY *iff for each $k \in \mathbb{N}$ there exists $N_k \in \mathbb{N}$ such that*

$$q(x_n, x_m) < 1/k \quad \text{for all} \ \ m \geq n \geq N_k.$$

Definition 6. *A quasimetric space is said to be* LEFT-SEQUENTIALLY COM-PLETE *iff each left-sequential Cauchy sequence is left-sequential convergent.*

Definition 7. *X is said to be* HAUSDORFF *iff every left-sequentially convergent sequence has the unique limit, i.e.*

$$\text{if } \lim_{n\to\infty} q(x_n, x_*) = 0 \wedge \lim_{n\to\infty} q(x_n, y_*) = 0, \text{ then } x_* = y_*.$$

For simplicity, 'left-sequential' in these terminologies will not be mentioned .

3 Main Results

In this section, we first establish a new version of Ekeland's variational principle for extended-real-valued functionals with weaker assumptions by using both the nonlinear scalarization functional (1) and Bao and Mordukhovich's iterative procedure in [4,5]. To the best of our knowledge, [7, Corollary 3.3] says that the original Ekeland's variational principle still holds for a class of decreasingly closed functionals defined on quasimetric spaces which is broader than the class of lower semicontinuous functionals defined on metric spaces.

Definition 8. *Let $\varphi : X \to \mathbb{R} \cup \{+\infty\}$ be a function. It is said to be* DECREAS-INGLY CLOSED *iff for any convergent sequence $\{x_k\} \subset X$ with the limit $\bar{x}$, if the sequence of numbers $\{\varphi(x_k)\}$ is decreasing and bounded, $\varphi(\bar{x}) \leq \varphi(x_k)$, $\forall k \in \mathbb{N}$.*

Obviously, if φ is lower semicontinuous (known also as level-closed) on dom $\varphi \backslash$ Max(φ), then it is decreasingly closed, where Max(φ) is the collection of all the local maxima of φ.

The first result in this paper can be viewed as a far-going extension of Dancs et al.'s fixed point theorem [11] which is widely used to prove vectorial versions of Ekeland's variational principles; the reader is referred to [9] for discussions and relations for previous developments in [7,8,17,20,21].

Theorem 3. *Let (X, q) be a quasimetric space, and $S : X \rightrightarrows X$ be a set-valued map with $x \in S(x)$ for all $x \in X$. Given a sequence $\{x_n\}_{n=0}^{\infty} \subset X$. Assume that*

(C1) *there is $\bar{x} \in X$ such that $S(\bar{x}) \subset S(x_n)$ for all $n \in \mathbb{N} \cup \{0\}$.*
(C2) $\lim_{n\to\infty} \sup_{x \in S(x_n)} q(x_n, x) = 0.$
(C3) $\lim_{n\to\infty} q(x_n, x_*) = \lim_{n\to\infty} q(x_n, x^*) = 0 \implies x_* = x^*.$

Then, $\bar{x} \in S(x_0)$ and $S(\bar{x}) = \{\bar{x}\}$.

Proof. The proof of the theorem is complete provided that

$$\bigcap_{n\in\mathbb{N}\cup\{0\}} S(x_n) \subset \{\bar{x}\} \tag{2}$$

due to assumption (C1) and the nonempty images of S. To justify (2), assume, in addition to $\bar{x}$, that an element $\bar{y}$ lies in the intersection of (2). Then, (C2) ensures that $\lim_{n\to\infty} q(x_n, \bar{x}) = \lim_{n\to\infty} q(x_n, \bar{y}) = 0$ which yields $\bar{x} = \bar{y}$ due to (C3), i.e., (2) holds.

Note that the Cantor theorem can not be applied in the proof of Theorem 3 even in the complete metric setting since the images of the mapping are not imposed to be closed.

Example 1. **(on assumptions (C1)–(C2)).** Let $X = \mathbb{R}$, $q(x, y) = d(x, y) = |x - y|$, $x_n = 1/(n + 1)$, and $S_1, S_2 : \mathbb{R} \rightrightarrows \mathbb{R}$ be defined by

$$S_1(x) = \begin{cases} [0, 2x) & \text{if } x > 0, \\ \{0\} & \text{if } x \leq 0, \end{cases} \quad S_2(x) = \begin{cases} (0, 2x) & \text{if } x > 0, \\ \{-1\} & \text{if } x \leq 0, \end{cases}$$

$$S_3(x) = (-\infty, x], \ \forall \, x \in \mathbb{R}, \quad S_4(x) = \begin{cases} [-1, 1] & \text{if } x \neq 0, \\ \{0\} & \text{if } x = 0, \end{cases}$$

Obviously, condition (C3) holds for the usual distance of real numbers.
– S_1 satisfying conditions (C1)–(C3) in Theorem 3 has an invariant point $\bar{x} = 0$.
– S_2 does not satisfy condition (C1), but it has an invariant point at $\bar{x} = -1$.
– S_3 does not satisfy condition (C2), and it doesn't have an invariate point.
– S_4 does not satisfy condition (C2), but it has an invariate point at $\bar{x} = 0$.

Theorem 4. *Let (X, q) be a quasimetric space, and let $\varphi : X \to \mathbb{R} \cup \{+\infty\}$ be a proper extended-real-valued functional. Given $x_0 \in \mathrm{dom}\,\varphi$ and $\lambda > 0$. Consider a set-valued mapping $S_{\varphi, \lambda} : X \rightrightarrows X$ defined by*

$$S_{\varphi, \lambda}(x) := \{u \in X \mid \varphi(u) + \lambda q(x, u) \leq \varphi(x)\} \quad \forall \, x \in X. \tag{3}$$

Assume that

(D1) φ *is bounded from below on* $\mathrm{dom}\,\varphi \neq \emptyset$.
(D2) *for any Cauchy sequence $\{x_n\}_{n=0}^{\infty}$ with $x_n \in S_{\varphi, \lambda}(x_{n-1})$ for all $n \in \mathbb{N}$, there exists $\bar{x} \in \mathrm{dom}\,\varphi$ satisfying $S_{\varphi, \lambda}(\bar{x}) \subset S_{\varphi, \lambda}(x_n)$ for all $n \in \mathbb{N} \cup \{0\}$.*
(D3) q *enjoys the limit uniqueness respect to the decreasing monotonicity of φ, i.e., for any Cauchy sequence $\{x_n\}$, if $\{\varphi(x_n)\}$ is decreasing, then* $\lim_{n \to \infty} q(x_n, x_*) = \lim_{n \to \infty} q(x_n, x^*) = 0 \implies x_* = x^*$.

Then, there is $\bar{x} \in X$ such that

(i) $\varphi(\bar{x}) + \lambda q(x_0, \bar{x}) \leq \varphi(x_0)$;
(ii) $\varphi(x) + \lambda q(\bar{x}, x) > \varphi(\bar{x}), \quad \forall \, x \neq \bar{x}$.

Proof. Since (i) and (ii) are equivalent to $\bar{x} \in S_{\varphi, \lambda}(x_0)$ and $S_{\varphi, \lambda}(\bar{x}) = \{\bar{x}\}$, respectively, the theorem is proved provided that we can construct a sequence $\{x_n\}_{n=0}^{\infty}$ satisfying conditions (C1)–(C3) in Theorem 4 with $S = S_{\varphi, \lambda}$. Such a sequence can be defined by induction. Suppose that x_n is known, then x_{n+1} can be chosen such that

$$x_{n+1} \in S_{\varphi, \lambda}(x_n) \text{ and } q(x_n, x_{n+1}) \geq \sup_{x \in S_{\varphi, \lambda}(x_n)} q(x_n, x) - 2^{-n}. \tag{4}$$

By the structure of $S_{\varphi,\lambda}$, $x_{n+1} \in S_{\varphi,\lambda}(x_n) \iff \varphi(x_{n+1}) + \lambda q(x_n, x_{n+1}) \leq \varphi(x_n)$ for all $n \in \mathbb{N} \cup \{0\}$. Summing up these inequalities from $n = l$ to $m - 1$ gives

$$\lambda \sum_{n=l}^{m-1} q(x_n, x_{n+1}) \leq \varphi(x_l) - \varphi(x_m), \text{ and thus} \tag{5}$$

$$\lambda \sum_{n=0}^{\infty} q(x_n, x_{n+1}) \leq \varphi(x_0) - \inf_{x \in X} \varphi(x) < \infty \tag{6}$$

where the last estimate holds due to the boundedness from below of φ in (D1). The convergence of the series clearly implies that $\lim_{n \to \infty} q(x_n, x_{n+1}) = 0$. This together with the inequality in (4) gives

$$\lim_{n \to \infty} \sup_{x \in S_{\varphi,\lambda}(x_n)} q(x_n, x) \leq \lim_{n \to \infty} \left(q(x_n, x_{n+1}) + 2^{-n} \right) = 0$$

which implies that the sequence $\{x_n\}$ satisfies condition (C2).

The convergence of the series in (6) says that for every $\varepsilon > 0$, there is $N_\varepsilon \in \mathbb{N}$ such that for all $m > l \geq N_\varepsilon$ we have

$$q(x_l, x_m) \leq \sum_{n=l}^{m-1} q(x_n, x_{n+1}) \leq \sum_{n=l}^{\infty} q(x_n, x_{n+1}) < \varepsilon,$$

i.e., $\{x_n\}$ is a Cauchy sequence in (X, q). Thus, conditions (C1) and (C3) follow conditions (D2) and (D3) applied to the chosen sequence $\{x_n\}$ in (4), respectively.

Remark 2. (**other iterative schemes**) The iterative scheme (4) used for the first time in [4] allows us to work with vector-valued functions since it does not depend on images of the function under consideration. It is different from other known procedures; for example, [13, Ekeland and Turnbull (1983)], [1, Aubin and Frankowska (1990)], and [10, Borwein and Zhu (2005)]. These procedures heavily depend on the complete order for real numbers and thus it seems to be not possible to extend them to partially ordered vector spaces.

In [7, Corollary 3.3] the authors worked on complete and Hausdorff quasi-metric spaces so that they could imposed assumptions on the function φ under consideration which are sufficient conditions for (D1)–(D3). Obviously, the Hausdorff condition imposed on q, i.e.,

$$\forall \{x_n\}, \lim_{n \to \infty} q(x_n, x_*) = \lim_{n \to \infty} q(x_n, x^*) = 0 \implies x_* = x^*$$

is more restrictive than (D3).

Proposition 1. (**a sufficient condition for** (D2)). *Assume that the quasi-metric space (X, q) is complete and φ is bounded from below and decreasingly closed in the sense of Definition 8. Then, condition (D2) in Theorem 4 holds.*

Proof. Fix a Cauchy sequence $\{x_n\}_{n=0}^{\infty}$ with $x_{n+1} \in S_{\varphi,\lambda}(x_n)$ for all $n \in \mathbb{N}\cup\{0\}$. Then,

$$\varphi(x_{n+1}) + \lambda q(x_n, x_{n+1}) \leq \varphi(x_n) \ \forall \, n \in \mathbb{N} \cup \{0\}. \tag{7}$$

which clearly implies that $\varphi(x_{n+1}) \leq \varphi(x_n)$ for all $n \in \mathbb{N} \cup \{0\}$. The completeness of the quasimetric space (X, q) ensures the existence of $\bar{x} \in X$ such that $\lim\limits_{n\to\infty} q(x_n, \bar{x}) = 0$. Then, we get from the decreasing closedness of φ that

$$\varphi(\bar{x}) \leq \varphi(x_n) \ \forall \, n \in \mathbb{N}\cup\{0\}. \tag{8}$$

Summing up the inequalities in (7) from $n = l$ to $m - 1$ gives

$$\lambda \sum_{n=l}^{m-1} q(x_n, x_{n+1}) \leq \varphi(x_l) - \varphi(x_m). \tag{9}$$

Adding (8) and (9) while taking into account the triangle inequality gives

$$\lambda(q(x_l, \bar{x}) - q(x_m, \bar{x})) \leq \lambda q(x_l, x_m) \leq \lambda \sum_{n=l}^{m-1} q(x_n, x_{n+1}) \leq \varphi(x_l) - \varphi(x_m) \leq \varphi(x_l) - \varphi(\bar{x})$$

which reduces, by passing m to infinity, to $\lambda q(x_l, \bar{x}) \leq \varphi(x_l) - \varphi(\bar{x})$, i.e., $\bar{x} \in S(x_l)$ which implies $S_{\varphi,\lambda}(\bar{x}) \subset S_{\varphi,\lambda}(x_l)$ since for any $x \in S_{\varphi,\lambda}(\bar{x})$ and $\bar{x} \in S_{\varphi,\lambda}(x_l)$ we have

$$\varphi(x) + \lambda q(x_l, x) \leq \big(\varphi(x) + \lambda q(\bar{x}, x)\big) + \lambda q(x_l, \bar{x}) \leq \varphi(\bar{x}) + \lambda q(x_l, \bar{x}) \leq \varphi(x_l).$$

Since l was arbitrary, (D2) holds for $\bar{x}$ as the limit of the sequence $\{x_n\}$. $\quad\square$

This proposition says that Theorem 4 implies [7, Corollary 3.3] and thus the original Ekeland's variational principle [12]. Next, let us establish a new vector version of Ekeland's variational principle by using the nonlinear scalarization technique in [14]. It is important to mention that we do not require the solidness of the ordering cone Θ.

Theorem 5. *Let $f : X \to Z$ be a vector-valued function acting between a quasimetric space (X, q) and a vector space Z equipped with a binary relation $\leq_\Theta$, where Θ be a nontrivial convex cone with vertex in Z. Assume that there is a nonzero element $k \in \Theta \setminus \{0\}$ and $\mathbb{R} \cdot k - \Theta = H$, where $H := \Theta - \Theta$ is the spanning space of Θ. Consider the nonlinear scalarization functional $s := s_{\Theta,k} : Z \to \mathbb{R} \cup \{\pm\infty\}$ defined by (1) and the level-set mapping $S_{f,\lambda} : X \rightrightarrows X$ with respect to $\lambda > 0$ defined by*

$$S_{f,\lambda}(x) := \{u \in X| \ f(u) + \lambda q(x, u)k \leq_\Theta f(x)\}, \tag{10}$$

Given $x_0 \in X$. Impose the following conditions on $\varphi := s \circ (f - f(x_0))$:

(H1) φ is bounded from below on $S_{f,\lambda}(x_0)$; in particular, $s \circ f$ is bounded from below on it.

(H2) $S_{\varphi,\lambda}$ is decreasingly closed on $S_{f,\lambda}(x_0)$.

(H3) q enjoys the limit uniqueness w.r.t. the decreasing monotonicity of φ.

Then, there exists some element $\bar{x} \in X$ such that

(i) $f(\bar{x}) + \lambda q(x_0, \bar{x})k \leq_\Theta f(x_0)$,

(ii) $f(x) + \lambda q(\bar{x}, x)k \not\leq_\Theta f(\bar{x}), \ \forall\, x \in X \setminus \{\bar{x}\}$.

Proof. By Lemma 1 (d), the functional s is finite-valued over $H := \Theta - \Theta$ and thus $\varphi = s \circ (f - f(x_0))$ is finite-valued over $S_{f,\lambda}(x_0)$ due to the imposed requirement $\mathbb{R} \cdot k - \Theta = H$ and $f(S_{f,\lambda}(x_0)) \subset H$.

Under the imposed conditions (H1)–(H3), the functional φ satisfies conditions (D1)–(D3) of Theorem 4 on the quasimetric space $(S_{f,\lambda}(x_0), q)$, where the set $S_{f,\lambda}(x_0)$ is described in (10). Therefore, the theorem ensures the existence of $\bar{x} \in S_{f,\lambda}(x_0)$ satisfying

$$\big(s \circ (f - f(x_0))\big)(x) + \lambda q(\bar{x}, x) > \big(s \circ (f - f(x_0))\big)(\bar{x}), \ \forall x \neq S_{f,\lambda}(x_0) \setminus \{\bar{x}\} \quad (11)$$

Taking into account the properties of s in Lemma 1, we get from (11) that

$$\big(s \circ (f - f(x_0))\big)(x) + \lambda q(\bar{x}, x) > \big(s \circ (f - f(x_0))\big)(\bar{x})$$

$$\overset{\text{(a)}}{\Longleftrightarrow} s(f(x) - f(x_0) + \lambda q(\bar{x}, x)k) > s(f(\bar{x}) - f(x_0))$$

$$\overset{\text{(e)}}{\Longrightarrow} f(x) - f(x_0) + \lambda q(\bar{x}, x)k \not\in f(\bar{x}) - f(x_0) - \Theta$$

$$\overset{\text{def.}}{\Longleftrightarrow} f(x) + \lambda q(\bar{x}, x)k \not\leq_\Theta f(\bar{x}).$$

Obviously, (i) holds since $\bar{x} \in S_{f,\lambda}(x_0)$. Thus, it remains to prove (ii). Arguing by contradiction, assume that there is some element $x \in X \setminus \{\bar{x}\}$ such that

$$f(x) + \lambda q(\bar{x}, x)k \leq_\Theta f(\bar{x}) \overset{\text{def.}}{\Longleftrightarrow} f(x) + \lambda q(\bar{x}, x)k \in f(\bar{x}) - \Theta. \quad (12)$$

We get from (12) and $\bar{x} \in S_{f,\lambda}(x_0)$ that

$$f(x) + \lambda q(x_0, x)k$$

$$= f(x) + \lambda q(\bar{x}, x)k + \lambda q(x_0, \bar{x})k - \lambda\big(q(\bar{x}, x) + q(x_0, \bar{x}) - q(x_0, x)\big)k$$

$$\overset{\text{(q3)}}{\in} f(x) + \lambda q(\bar{x}, x)k + \lambda q(x_0, \bar{x})k - \Theta$$

$$\overset{\text{(12)}}{\subset} f(\bar{x}) - \Theta + \lambda q(x_0, \bar{x}) - \Theta \overset{\text{(10)}}{\subset} f(x_0) - \Theta$$

which implies $x \in S_{f,\lambda}(x_0)$ contradicting to (ii') due to Lemma 1(e). Details below:

$$(12) \Longleftrightarrow f(x) + \lambda q(\bar{x}, x)k \in f(\bar{x}) - \Theta$$

$$\Longleftrightarrow \big(f(x) - f(x_0)\big) + \lambda q(\bar{x}, x)k \in \big(f(\bar{x}) - f(x_0)\big) - \Theta$$

$$\overset{(e)}{\Longrightarrow} s\big(f(x) - f(x_0) + \lambda q(\bar{x}, x)k\big) \le s\big(f(\bar{x}) - f(x_0)\big)$$

$$\overset{(a)}{\Longleftrightarrow} s\big(f(x) - f(x_0)\big) + \lambda q(\bar{x}, x) \le s\big(f(\bar{x}) - f(x_0)\big).$$

The proof is complete.

Proposition 1 says that (H2) can be replaced by

(H2′) for every convergent sequence $\{x_n\} \subset X$ with the limit $\bar{x}$, if $s\big(f(x_{n+1}) - f(x_0)\big) \le s\big(f(x_n) - f(x_0)\big)$ for all $n \in \mathbb{N}$, then $s\big(f(\bar{x}) - f(x_0)\big) \le s\big(f(x_n) - f(x_0)\big)$ for all $n \in \mathbb{N}$.

In [4,5], Bao and Mordukhovich used the following closedness assumption:

(H2″) for every convergent sequence $\{x_n\} \subset X$ with the limit $\bar{x}$, if $f(x_{n+1}) \le_\Theta f(x_n)$ for all $n \in \mathbb{N}$, then $f(\bar{x}) \le_\Theta f(x_n)$ for all $n \in \mathbb{N}$.

By Lemma 1 (e), we have only the validity of the implication $a \le_\Theta b \Longrightarrow s(a) \le s(b)$, but it is not difficult to show by examples that the reverse implication does not hold. Therefore, conditions (H1′) and (H1″) are incomparable. In the next result we show that (H2″) is also an alternative of (H2).

Theorem 6. *Assume the (X, q) is a complete quasimetric space and (H2) is replaced by either (H2′) or (H2″). Then, Theorem 5 holds.*

Proof. It is sufficient to prove the case with (H2″) since the other case follows from Proposition 1. Since we use arguments in the proofs of Theorem 4 and Proposition 1, we might omit some details.

We construct a sequence $\{x_n\}_{n=0}^\infty$ satisfying conditions (C1)–(C3) in Theorem 3 by using a modified iterative procedure:

$$x_{n+1} \in S_{f,\lambda}(x_n) \text{ and } q(x_n, x_{n+1}) \ge \sup_{x \in S(x_n)} q(x_n, x) - 2^{-n}. \tag{13}$$

Since $x \in S_{f,\lambda}(x)$ for all $x \in X$, the sequence is well-defined. Let us check conditions (C1)–(C3).

By the structure of $S_{f,\lambda}$, $x_{n+1} \in S(x_n)$ implies $f(x_{n+1}) + \lambda q(x_n, x_{n+1})k \le_\Theta f(x_n)$. Summing up these inequalities from $n = l$ to $m - 1$ gives

$$\lambda \sum_{n=l}^{m} q(x_n, x_{n+1})k \le_\Theta f(x_l) - f(x_m). \tag{14}$$

By Lemma 1 (a) and (e) for the functional $s := s_{\Theta,k}$, we have

$$\lambda \sum_{n=0}^{m-1} q(x_n, x_{n+1}) \le s(f(x_0)) - s(f(x_m)) \tag{15}$$

and thus

$$\lambda \sum_{n=0}^{\infty} q(x_n, x_{n+1}) \leq s(f(x_0)) - \sup_{x \in X} s(f(x)) < \infty$$

where the last estimate holds due to the boundedness from below of $s \circ f$ in (H1). The convergence of the series clearly implies that $\lim_{n \to \infty} q(x_n, x_{n+1}) = 0$. This together with the inequality in (13) that

$$\lim_{n \to \infty} \sup_{x \in S_{f,\lambda}(x_n)} q(x_n, x) \leq \lim_{n \to \infty} \left(q(x_n, x_{n+1}) + 2^{-n} \right) = 0$$

clearly verifying that the sequence $\{x_k\}$ satisfies condition (C2).

The convergence of the series $\sum_{n=0}^{\infty} q(x_n, x_{n+1})$ implies that the sequence $\{x_n\}$ is a Cauchy sequence in (X, q). Details can be found in Theorem 4. By the structure of $S_{f,\lambda}$, $x_{n+1} \in S_{f,\lambda}(x_n) \implies f(x_{n+1}) \leq_{\Theta} f(x_n)$ for all $n \in \mathbb{N} \cup \{0\}$. By (H2'') and the completeness of the space (X, q) there exists an element $\bar{x} \in X$ such that $\lim_{n \to \infty} q(x_n, \bar{x}) = 0$ and

$$f(\bar{x}) \leq_{\Theta} f(x_n) \text{ for all } n \in \mathbb{N} \cup \{0\}. \tag{16}$$

Adding (16) and (14) while taking into account the triangle inequality we obtain

$$\lambda \big(q(x_l, \bar{x}) - q(x_m, \bar{x}) \big) k \leq_{\Theta} \lambda q(x_l, x_m) k \leq_{\Theta} \lambda \sum_{n=l}^{m-1} q(x_n, x_{n+1}) k \leq_{\Theta} f(x_l) - f(\bar{x})$$

which reduces, by passing m to infinity, to $\lambda q(x_l, \bar{x}) k \leq_{\Theta} f(x_l) - f(\bar{x})$, i.e., $\bar{x} \in S_{f,\lambda}(x_l)$. It is not dificult to check that $S_{f,\lambda}(\bar{x}) \subset S_{f,\lambda}(x_l)$. Since l was arbitrary, condition (C1) holds.

As it was justified above, the sequence $\{x_n\}$ is Cauchy and satisfies $x_{n+1} \in S_{f,\lambda}(x_n)$ for all $n \in \mathbb{N}$. Thus, (C3) follows (H3). The proof is complete.

This result improves the corresponding results in [4,5,6,7,8] in which the function f is required bounded from below by a bounded set; known also as quasi-bounded.

Proposition 2. (a sufficient condition for (H2)). *If f is bounded from below by a bounded set M and $k \in \Theta \setminus (-\text{cl}\,\Theta)$, then $s \circ (f - f(x_0))$ is bounded from below over $S_{f,\lambda}(x_0)$.*

Proof. It is quite straigtforward.

To conclude this section we present a vector-valued function which is not (quasi)-bounded from below but the scalarization is bounded from below.

Example 2. Define $f : \mathbb{R} \to \mathbb{R}^2$ by

$$f(x) := (e^x + 1, e^x + 1 + x) \ \forall \ x \in \mathbb{R}.$$

Let $\Theta = \mathbb{R}^2_+$ (the nonnegative orthant), $k = (1, 1) \in \Theta$ and $x_0 = 0$. It is easy to check that

$$s_{\Theta,k} \circ (f(x) - f(0)) = s_{\Theta,k}((0, x) + e^x k) = e^x + s_{\Theta,k}(0, x) \geq e^x \geq 0 \ \forall \ x \in \mathbb{R}.$$

Therefore, $s_{\Theta,k} \circ (f - f(0))$ is bounded from below over $\mathbb{R}$. Fix $n > 0$ and consider the output of f at $-n$. Since the second component of $f(-n)$ is negative and diverges to negative infinity as $n \to \infty$, f can not be bounded from below by any bounded set M in $\mathbb{R}^2$.

References

1. Aubin, J.-P., Frankowska, H.: Set-Valued Analysis. Systems and Control: Foundations and Applications. Birkhäuser, Boston (1990)
2. Bao, T.Q., Eichfelder, G., Soleimani, B., Tammer, C.: Ekeland's Variational Principle for Vector Optimization with Variable Ordering Structure. Preprint No. M 14/08. Technische Universität Ilmenau Institut für Mathematik (2014)
3. Bao, T.Q., Khanh, P.Q., Soubeyran, A.: Variational Principles with Generalized Distances and Applications to Behavioral Sciences (2015)
4. Bao, T.Q., Mordukhovich, B.S.: Variational Principles for Set-Valued Mappings with Applications to Multiobjective Optimization. Control Cybern. 36, 531–562 (2007)
5. Bao, T.Q., Mordukhovich, B.S.: Relative Pareto Minimizers for Multiobjective Problems: Existence and Optimality Conditions. Math. Progr. 122, 301–347 (2010)
6. Bao, T.Q., Mordukhovich, B.S., Soubeyran, A.: Variational Analysis in Psychological Modeling. J. Optim. Theory Appl. 164, 290–315 (2015)
7. Bao, T.Q., Mordukhovich, B.S., Soubeyran, A.: Fixed Points and Variational Principles with Applications to Capability Theory of Wellbeing via Variational Rationality. Set-Valued Var. Anal. (2015), doi:10.1007/s11228-014-0313-4
8. Bao, T.Q., Mordukhovich, B.S., Soubeyran, A.: Minimal Points, Variational Principles, and Variable Preferences in Set Optimization. To appear in J. Nonlinear Convex Anal. (2015)
9. Bao, T.Q., Théra, M.: On Extended Versions of Dancs-Hegedüs-Medvegyev's Fixed Point Theorem (2015)
10. Borwein, J.M., Zhu, Q.J.: Techniques of variational analysis. Springer, New York (2005)
11. Dancs, S., Hegedüs, M., Medvegyev, P.: A General Ordering and Fixed-Point Principle in Complete Metric Space. Acta Sci. Math. 46, 381–388 (1983)
12. Ekeland, I.: Nonconvex Minimization Problems. Bull. Amer. Math. Soc. 1, 443–474 (1979)
13. Ekeland, I., Turnbull, T.: Infinite-Dimensional Optimization and Convexity. University of Chicago Press, Chicago (1983)
14. Gerth (Tammer), C., Weidner, P.: Nonconvex Separation Theorems and Some Applications in Vector Optimization. J. Optim. Theory Appl. 67, 297–320 (1990)

15. Göpfert, A., Riahi, H., Tammer, C., Zălinescu, C.: Variational Methods in Partially Ordered Spaces. Springer, New York (2003)
16. Jahn, J.: Vector Optimization: Theory, Applications and Extensions. Springer, New York (2004)
17. Khanh, P.Q., Quy, D.N.: A Generalized Distance and Enhanced Ekeland's Variational Principle for Vector Functions. Nonlinear Anal. 73, 2245–2259 (2010)
18. Luc, D.T.: Theory of Vector Optimization. Springer (1989)
19. Mordukhovich, B.S.: Variational Analysis and Generalized Differentiation, I: Basic Theory, II: Applications. Springer, New York (2006)
20. Qiu, J.H.: A Preorder Principle and Set-Valued Ekeland Variational Principle. J. Math. Anal. Appl. 419, 904–937 (2014)
21. Qui, J.-H.: A Revised Preorder Principle and Set-Valued Ekeland Variational Principles. arXiv (2014)
22. Tammer, C.: A Generalization of Ekeland's Variational Principle. Optimization 5, 129–141 (1992)

Author Index

Printed in the United States
By Bookmasters